MATHEMATICAL LOGIC

by

Joseph R. Shoenfield
Department of Mathematics
Duke University

ASSOCIATION FOR SYMBOLIC LOGIC

A K Peters, Ltd. • Natick, Massachusetts

Sales and Customer Service:
A K Peters, Ltd.
63 South Avenue
Natick, Massachusetts 01760

Association for Symbolic Logic:
C. Ward Henson, Publisher
Mathematics Department
University of Illinois
1409 West Green Street
Urbana, Illinois 61801

Library of Congress Cataloging-in-Publication Data

Shoenfield, Joseph R. (Joseph Robert), 1927-
Mathematical logic / Joseph R. Shoenfield
p. cm.
Originally published: Reading, Mass. : Addison-Wesley Pub. Co, 1967, in series: Addison-Wesley series in logic.
Includes index.
ISBN 1-56881-135-7 (pbk. : acid-free paper)
1. Logic, Symbolic and mathematical. I. Title.

QA9 .S52 2001
511.3–dc21

00-065277

Publisher's Note: Joseph R. Shoenfield died on November 15, 2000, while his book was being reprinted. ASL publication of *Mathematical Logic* was approved by the Editorial Board of the Lecture Notes in Logic; the Editors are Samuel R. Buss, Managing Editor, Lance Fortnow, Shaughan Lavine, Steffen Lempp, Anand Pillay, and W. Hugh Woodin. It is published, with the permission of the author, as a photographic copy of the second printing of *Mathematical Logic*, published by Addison Wesley in 1973, for which copyright had reverted to the author. This book was printed by VictorGraphics, Inc., Baltimore, Maryland, on acid-free paper. The cover design is by Richard Hannus, Hannus Design Associates, Boston, Massachusetts.

10 09 08 07 06 05 10 9 8 7 6 5 4 3 2

PREFACE

It is rare for an author of a mathematical text not to complain that space does not permit an adequate treatment of his subject. Today it is impossible in an introductory text in analysis, algebra, or topology to treat even all the central topics in the field. It is barely possible in mathematical logic; but it requires the total omission of many interesting side topics. I have therefore attempted to collect the principal results in what seem to me to be the central topics of mathematical logic: proof theory, model theory, recursion theory, axiomatic number theory, and set theory. I do not wish to create a prejudice against many valued logics, recursive equivalence types, and other special topics; but the reader will have to study them elsewhere. An especially important topic which has been omitted is the subject of intuitionism; the availability of suitable introductory texts on the subject and the incompetence of the author in this field are the chief excuses which I offer.

Even the main topics could not be covered completely; but I have tried to present some nontrivial theorems and proofs in each topic. The number of results is considerably increased by the problems. These are not routine exercises, but significant results whose proofs often require considerable extensions of the methods of the text. I hope that the hints will always be sufficient to enable a good student to solve the problems; but he should not be discouraged if many of them seem quite difficult.

Mathematical logic has always been closely connected with the philosophy of mathematics. I have generally avoided philosophical issues except when they were closely connected with the mathematical material. I will offer no apology, however, if I have occasionally stated a philosophical opinion without observing that the contrary opinion is also widely held.

This volume is essentially an expansion of a set of notes for a course in mathematical logic offered several times at Duke University since 1958. Some material also originated in a course in recursion theory offered at Stanford University in 1964–1965. The material included is somewhat more than I usually cover in a one-year course.

Since I have intended this book to be a text for a first-year graduate course, I have assumed a reasonable amount of mathematical maturity. On the other hand, only a very limited knowledge of mathematics is presupposed. A knowledge of the simplest properties of natural numbers, real numbers, and sets and a slight acquaintance with modern algebra should be sufficient. Some of the problems require acquaintance with more advanced topics.

Only a minute fraction of the results in the text and the problems is due to the author. I have made no attempt to credit each result to its author; the names attached to the principal theorems are there simply to give the reader some idea of the people who have created the subject. I have also omitted all bibliographical references. I should, however, mention one book which has proved especially valuable to me; this is Kleene's *Introduction to Metamathematics*.

In addition to using published sources, I have made much use of conversations and correspondence with many logicians. I would particularly like to thank John Addison, Solomon Feferman, Azriel Lévy, Andrzej Mostowski, Richard Platek, Hartley Rogers, Dana Scott, Clifford Spector, W. W. Tait, and Robert Vaught. This list is not exhaustive; but I hope the many others will accept my collective thanks for their contribution.

I owe a very special debt to Georg Kreisel. It is in discussions and correspondence with him over a number of years that I have come to realize that mathematical logic is not a collection of vaguely related results, but a method of attacking some of the most interesting problems which face the mathematician. I feel that the measure of success of this volume will be the extent to which it conveys this idea to the reader.

Durham, North Carolina J. R. S.
May 1967

CONTENTS

Dedicated to
CLIFFORD SPECTOR (1930–1961)

CHAPTER 1

THE NATURE OF MATHEMATICAL LOGIC

1.1 AXIOM SYSTEMS

Logic is the study of reasoning; and mathematical logic is the study of the type of reasoning done by mathematicians. To discover the proper approach to mathematical logic, we must therefore examine the methods of the mathematician.

The conspicuous feature of mathematics, as opposed to other sciences, is the use of proofs instead of observations. A physicist may prove physical laws from other physical laws; but he usually regards agreement with observation as the ultimate test of a physical law. A mathematician may, on occasions, use observation; for example, he may measure the angles of many triangles and conclude that the sum of the angles is always 180°. However, he will accept this as a law of mathematics only when it has been proved.

Nevertheless, it is clearly impossible to prove all mathematical laws. The first laws which one accepts cannot be proved, since there are no earlier laws from which they can be proved. Hence we have certain first laws, called *axioms*, which we accept without proof; the remaining laws, called *theorems*, are proved from the axioms.

For what reason do we accept the axioms? We might try to use observation here; but this is not very practical and is hardly in the spirit of mathematics. We therefore attempt to select as axioms certain laws which we feel are evident from the nature of the concepts involved.

We thus have a reduction of a large number of laws to a small number of axioms. A rather similar reduction takes place with mathematical concepts. We find that we can define certain concepts in terms of other concepts. But again, the first concepts which we use cannot be defined, since there are no earlier concepts in terms of which they can be defined. We therefore have certain concepts, called *basic concepts*, which are left undefined; the remaining concepts, called *derived concepts*, are defined in terms of these. We have a criterion for basic concepts similar to that for axioms: they should be so simple and clear that we can understand them without a precise definition.

In any statement, we can replace the derived concepts by the basic concepts in terms of which they are defined. In particular, we may do this for axioms. Hence we may suppose that all the concepts which appear in the axioms are basic concepts.

We may now describe what a mathematician does as follows. He presents us with certain basic concepts and certain axioms about these concepts. He then explains these concepts to us until we understand them sufficiently well to see that the axioms are true. He then proceeds to define derived concepts and to prove theorems about both basic and derived concepts. The entire edifice which he constructs, consisting of basic concepts, derived concepts, axioms, and theorems, is called an *axiom system.* It may be an axiom system for all of mathematics, or for a part of mathematics, such as plane geometry or the theory of real numbers.

We have so far supposed that we have definite concepts in mind. Even so, it may be possible to discover other concepts which make the axioms true. In this case, all the theorems proved will also be true for these new concepts. This has led mathematicians to frame axiom systems in which the axioms are true for a large number of concepts. A typical example is the set of axioms for a group. We call such axiom systems *modern* axiom systems, as opposed to the *classical* axiom systems discussed above. Of course, the difference is not really in the axiom system, but in the intentions of the framer of the system.

Guided by this discussion, we shall begin the study of mathematical logic by studying axiom systems. This will eventually lead us to a variety of problems, some of them only faintly related to axiom systems.

1.2 FORMAL SYSTEMS

An axiom (or theorem) may be viewed in two ways. It may be viewed as a sentence, i.e., as the object which appears on paper when we write down the axiom, or as the meaning of a sentence, i.e., the fact which is expressed by the axiom. At first sight, the latter appears much more important. The obvious purpose of the sentence is to convey the meaning of the sentence in a clear and precise manner. This is a useful purpose, but it does not seem to have much to do with the foundations of mathematics.

Nevertheless, there are two good reasons for studying axioms and theorems as sentences. The first is that if we choose the language for expressing the axioms suitably, then the structure of the sentence will reflect to some extent the meaning of the axiom. Thus we can study the concepts of the axiom system by studying the structure of the sentences expressing the axioms. This is particularly valuable for modern axiom systems, since for them our initial understanding of the basic concepts may be very weak.

The second reason is that the concepts of mathematics are usually very abstract and therefore difficult to understand. A sentence, on the other hand, is a concrete object; so by studying axioms as sentences, we approach the abstract through the concrete.

One point is apparent: there is no value in studying concrete (rather than abstract) objects unless we approach them in a concrete or constructive manner. For example, when we wish to prove that a concrete object with a certain property exists, we should actually construct such an object, not merely show that the nonexistence of such an object would lead to a contradiction.

Proofs which deal with concrete objects in a constructive manner are said to be *finitary*. Another description, suggested by Kreisel, is this: a proof is finitary if we can *visualize* the proof. Of course, neither description is very precise; but we can apply them in many cases to decide whether or not a particular proof is finitary.

Once the fundamental difference between concrete and abstract objects is appreciated, a variety of questions are suggested which can only be answered by a study of finitary proofs. For example Hilbert, who first instituted this study, felt that only finitary mathematics was immediately justified by our intuition. Abstract mathematics is introduced in order to obtain finitary results in an easier or more elegant manner. He therefore suggested as a program to show that all (or a considerable part) of the abstract mathematics commonly accepted can be viewed in this way. The question of how far such a program can be carried out is of obvious interest, even to those who do not find Hilbert's view of abstract mathematics congenial.

The study of axioms and theorems as sentences is called the *syntactical* study of axiom systems; the study of the meaning of these sentences is called the *semantical* study of axiom systems. For the above reasons, we shall often keep the syntactical and the semantical parts of our investigations separate. When it is possible and reasonably convenient, we shall carry out our syntactical investigations in a finitary manner. We shall always consider axioms and theorems to be sentences, and hence syntactical objects; when we wish to study them semantically, we speak of the meaning of the axiom or theorem.

To guide us in our syntactical study, we introduce the notion of a *formal system*. Roughly, a formal system is the syntactical part of an axiom system. We shall give a precise definition.

The first part of a formal system is its *language*. As previously stated, this should be chosen so that, as far as possible, the structure of the sentences reflects their meaning. For this reason among others, we generally use artificial languages for our formal systems.

To specify a language, we must first of all specify its *symbols*. In the case of English, the symbols would be the letters, the digits, and the punctuation marks. Most of our artificial languages will have infinitely many symbols.

Any finite sequence of symbols of a language is called an *expression* of that language. It is understood that a symbol may appear several times in an expression; each such appearance is called an *occurrence* of that symbol in the expression. The number of occurrences of symbols in an expression is called the *length* of that expression. (Thus the English expression *boot* has length 4.) We allow the empty sequence of symbols as an expression; it is the only expression of length 0.

It is possible for one expression to appear within another expression. Each such appearance is called an *occurrence* of the first expression in the second expression. Thus the English expression *on* has two occurrences in the English expression *confront*. However, we do not count it as an occurrence when the symbols of the first expression appear in the second expression in a different order or separated by other symbols. Thus *on* has no occurrences in *not* or in *corn*.

Most expressions in English are meaningless. Among the meaningful ones are the (declarative) sentences, which may be roughly described as those expressions which state some fact. We shall require that in each language, certain expressions of the language are designated as the *formulas* of the language; it is intended that these be the expressions which assert some fact.

We consider a language to be completely specified when its symbols and formulas are specified. This makes a language a purely syntactical object. Of course, most of our languages will have a meaning (or several meanings); but the meaning is not considered to be a part of the language. We shall designate the language of a formal system F by $L(F)$.

The next part of a formal system consists of its *axioms*. Our only requirement on these is that each axiom shall be a formula of the language of the formal system.

We need a third part of a formal system which will enable us to conclude theorems from the axioms. This is provided by the *rules of inference*, which we often call simply *rules*. Each rule of inference states that under certain conditions, one formula, called the *conclusion* of the rule, can be *inferred* from certain other formulas, called the *hypotheses* of the rule.

How should we define the *theorems* of a formal system F? Obviously they should satisfy the two laws:

i) the axioms of F are theorems of F;

ii) if all of the hypotheses of a rule of F are theorems of F, then the conclusion of the rule is a theorem of F.

Moreover, we want a formula to be a theorem of F only if it follows from these laws that it is a theorem. We can therefore define a theorem of F to be a formula of F which can be seen to be a theorem on the basis of laws (i) and (ii).

We can give a somewhat more explicit description of the theorems of F. Let S_0 be the set of axioms; these are the formulas which can be seen to be theorems on the basis of (i). Let S_1 be the set of formulas which are conclusions of rules whose hypotheses are all in S_0; these are some of the formulas which can be seen to be theorems on the basis of (ii). Let S_2 be the set of formulas which are conclusions of rules whose hypotheses are all in S_0 and S_1; these are also theorems on the basis of (ii). In this way, we can construct sets $S_3, S_4, \ldots$. Let S_ω be the set of formulas which are conclusions of rules whose hypotheses are all in at least one of $S_0, S_1, \ldots$; these are again theorems by (ii). We continue in this way until no new theorems can be obtained by (ii); and we then have all of the theorems.

A definition of the type just given is called a *generalized inductive definition.* A generalized inductive definition of a collection C of objects consists of a set of laws, each of which says that, under suitable hypotheses, an object x is in C. Some of the hypotheses may say that certain objects (related to x in a certain way) are in C. When we give such a definition, we always understand that an object shall be in C only if it follows from the laws that it is in C. We can then give a more explicit description of C along the above lines.

As another example, suppose that we have defined 0 and *successor*, and wish to define *natural number*. (The natural numbers are the nonnegative integers: 0, 1, 2, ... The successor of a natural number is the next larger natural number.) We can give the following generalized inductive definition:

i) 0 is a natural number;

ii) if y is a natural number, then the successor of y is a natural number.

In order to prove that every theorem of F has a property P, it suffices to prove that the formulas having property P satisfy the laws in the definition of *theorem*. In other words, it suffices to prove:

i′) every axiom of F has property P;

ii′) if all of the hypotheses of a rule of F have property P, then the conclusion of the rule has property P.

For (i′) and (ii′) imply that each member of the sets $S_0, S_1, \ldots, S_\omega, \ldots$ constructed above has property P; so every theorem of F has property P. A proof by this method is called a proof *by induction on theorems;* the assumption in (ii′) that the hypotheses of the rule have property P is called the *induction hypothesis.*

More generally, suppose that a collection C is defined by a generalized inductive definition. Then in order to prove that every object in C has property P, it suffices to prove that the objects having property P satisfy the laws of the definition. Such a proof is called a proof *by induction on objects in C.* The hypotheses in the laws that certain objects belong to C become, in such a proof, hypotheses that certain objects have property P; these hypotheses are called *induction hypotheses.* The reader will easily see that if C is the collection of natural numbers with the generalized inductive definition given above, then *proof by induction* and *induction hypothesis* have their usual meaning.

A rule in a formal system F is *finite* if it has only finitely many hypotheses. Almost all the rules which we will consider will be finite.

Let F be a formal system in which all the rules are finite. By a *proof* in F, we mean a finite sequence of formulas, each of which either is an axiom or is the conclusion of a rule whose hypotheses precede that formula in the proof. If A is the last formula in a proof P, we say that P is a proof *of A.*

We will show that a formula A of F is a theorem iff* there is a proof of A. First of all, it follows from the rules (i) and (ii) that every formula in a proof is a theorem; so if A has a proof, then it is a theorem. We prove the converse by induction on theorems. If A is an axiom, then A by itself is a proof of A; so A has a proof. Now suppose that A can be inferred from $B_1, \ldots, B_n$ by some rule of F. By the induction hypothesis, each of the B_i has a proof. If we put these proofs one after the other, and add A to the end of this sequence, we obtain a proof of A.

---

* We use *iff* as an abbreviation for *if and only if.*

We shall write $\vdash_F \ldots$ as an abbreviation for ... *is a theorem of F*. When no confusion results, we omit the subscript *F*.

The basic concepts of an axiom system will correspond to certain symbols or expressions in the associated formal system. The derived concepts, since they are defined in terms of the basic concepts, will generally correspond to more complicated expressions. If an important derived concept corresponds to a rather complicated expression, it may be desirable to introduce a new symbol as an abbreviation for that expression. We may also wish to introduce abbreviations to make certain expressions shorter or more readable.

For these reasons, we allow ourselves to introduce in any language new symbols, called *defined symbols*. Each such symbol is to be combined in certain ways with symbols of the language and previously introduced defined symbols to form expressions called *defined formulas*. Each defined formula is to be an abbreviation of some formula of the language. (In this terminology, an abbreviation does not have to be shorter than the expression which it abbreviates.) With each defined symbol, we must give a *definition* of that symbol; this is a rule which tells how to form defined formulas with the new symbol and how to find, for each such defined formula, the formula of the given language which it abbreviates.

We emphasize that defined symbols are *not* symbols of the language, and that defined formulas are *not* formulas of the language. Moreover, when we say anything about a defined formula, we are really talking about the formula of the language which it abbreviates (provided that it makes any difference). Thus the length of a defined formula is not the number of occurrences of symbols in the defined formula, but the number of occurrences of symbols in the formula which the defined formula abbreviates.

1.3 SYNTACTICAL VARIABLES

In our study of formal systems, we shall be studying expressions, just as an analyst studies real numbers. In both cases, the investigation is carried out in English augmented by certain special symbols specially suited to the investigation. We shall examine some of the special symbols used in analysis texts, and introduce analogous special symbols to be used in the investigation of formal systems.

First of all, an analysis text uses names for certain real numbers, for example, $3, -\frac{1}{2}, \pi$. Similarly, we shall need names for expressions. We are in the fortunate position of being able to provide a name for each expression with one convention: each expression shall be used as a name for itself. This convention is not available to writers of analysis texts; for a name must be an expression, and a real number is not an expression.

There is, however, a danger to this convention. The expression may (in the language being discussed) be a name of some object; now it is also a name for itself. Thus *Boston* is the name of a city; according to our convention, it is also the name of an English word. We are saved from this danger because we only discuss artificial languages and because we discuss them in English. Thus when the expression occurs in a context written in the artificial language, it is a name of some object; when it occurs in a context written in English, it is a name of that expression.*

Variables are another important type of symbol used in analysis texts. Unlike a name, which has only one meaning, a variable has many meanings. In an analysis text, a variable may mean any real number; or, as we shall say, a variable *varies through* the real numbers. However, a variable keeps the same meaning throughout any one context. A formula containing variables also has many meanings, one for each assignment of a real number as a meaning to each variable occurring in the formula. For example, $x = x$ has $2 = 2$ and $\pi = \pi$ among its meanings; $x = y$ has these meanings, and also $2 = 5$. When a writer of an analysis text asserts a formula containing variables, he is claiming that all of its meanings are true.

We use *syntactical variables* in a similar manner, except that they vary through the expressions of the language being discussed instead of through the real numbers. Thus a syntactical variable may mean any expression of the language; but its meaning remains fixed throughout any one context. A formula containing syntactical variables has many meanings, one for each assignment of an expression as a meaning to each syntactical variable occurring in the formula. If we assert such a formula, we are claiming that all of its meanings are true.

To give an example of the use of syntactical variables and of expressions as names for themselves, suppose that x is a symbol of the formal system F. Suppose that it turns out that whenever we add the symbol x to the right of a formula of F, we obtain a new formula of F. If we have agreed to use $\mathbf{u}$ as a syntactical variable, we can express this fact as follows: if $\mathbf{u}$ is a formula, then the expression obtained by adding x to the right of $\mathbf{u}$ is a formula.

In an analysis text, some variables are restricted to vary through only certain real numbers. For example, it is common to restrict i and j to vary through integers only. We shall often use syntactical variables which vary through only certain expressions of the language being discussed. If we use $\mathbf{A}$ as a syntactical variable which varies through formulas, then the statement at the end of the previous paragraph can be abbreviated to: the expression obtained by adding x to the right of $\mathbf{A}$ is a formula.

In an analysis text, xy stands for the result of multiplying x by y. If $\mathbf{u}$ and $\mathbf{v}$ are syntactical variables, we shall use $\mathbf{uv}$ to stand for the expression obtained by juxtaposing $\mathbf{u}$ and $\mathbf{v}$, that is, by writing down $\mathbf{u}$ and then writing down $\mathbf{v}$ immediately after it. The same convention is used with other syntactical variables. It is also used to combine syntactical variables with names of expressions. As an example, we may shorten the statement at the end of the previous paragraph to: $\mathbf{A}x$ is a formula.

* To avoid any possibility of confusion, some books replace our convention by another convention: as a name for an expression, that expression enclosed in quotation marks is used.

We shall use boldface letters as syntactical variables. In particular, **u** and **v** will be syntactical variables which vary through all expressions, and **A**, **B**, **C**, and **D** will be syntactical variables which vary through formulas. Other syntactical variables will be introduced later. When we introduce a boldface letter as a syntactical variable, we shall understand that we may form new syntactical variables by adding primes or subscripts, and that these new syntactical variables vary through the same expressions as the old ones. Thus $\mathbf{A}'$ and $\mathbf{A}_1$ are syntactical variables which vary through formulas.

We add two words of caution. First, if two different syntactical variables occur in the same context, they do not necessarily represent different expressions (just as, in an analysis text, x and y do not necessarily represent different real numbers). Second, syntactical variables are *not* symbols of the language being discussed; they are symbols added to English to aid in the discussion of the language.

CHAPTER 2

FIRST-ORDER THEORIES

2.1 FUNCTIONS AND PREDICATES

As we have remarked, we want to study formal systems in which the structure of the language is related to the intended meaning of the language. We will now select a class of formal systems having this property which is sufficiently large to contain formalizations of the usual axiom systems of mathematics.

Our first task is to describe the languages to be used in these formal systems. Before doing this in a precise manner, we investigate informally the concepts which appear in mathematical axiom systems and introduce some notation for them. Some of these concepts are common to all the axiom systems. We call these *logical concepts;* they will be discussed in the next two sections. In this section, we discuss the nature of the remaining concepts, which are called *nonlogical concepts.*

Let us begin with an example. Suppose that we wish to construct a language for discussing the natural numbers. A typical formula in such a language would be $2 + 1 < 4$. What does each of the symbols in this formula represent?

Clearly 2, 1, and 4 represent particular natural numbers. We can think of $+$ as representing an object which associates with each pair (a, b) of natural numbers a third natural number, namely, the sum of a and b. We want to think of $<$ as representing some object which distinguishes between those pairs (a, b) of natural numbers for which a is less than b and those pairs (a, b) for which a is not less than b. We therefore take it to represent the collection of pairs (a, b) such that a is less than b.

We can explain this more succinctly by introducing some terminology from set theory.* A *set* or *class* is a collection of objects. If A and B are sets, a *mapping from A to B* is an assignment of an object in B to each object in A. If F designates the mapping, and F assigns the element b of B to the element a of A, we say that b is the *value* of F for the *argument* a, and write $F(a)$ for b. An *n-tuple* in A is a sequence of n (not necessarily distinct) objects in A. We write $(a_1, a_2, \ldots, a_n)$ for the n-tuple consisting of the objects $a_1, a_2, \ldots, a_n$ in that order. A mapping

* This paragraph is not intended as an introduction to elementary set theory, with which we assume the reader is already familiar. Our object here is merely to establish the terminology and notation. An axiomatic treatment of set theory is given in Chapter 9; but only very elementary results will be needed before then.

from the set of n-tuples in A to B is called an *n-ary function from A to B*. A subset of the set of n-tuples in A is called an *n-ary predicate in A*. If P represents such a predicate, then $P(a_1, \ldots, a_n)$ means that the n-tuple $(a_1, \ldots, a_n)$ is in P. We say *unary* for 1-ary and *binary* for 2-ary. Note that a unary function from A to B is a mapping from A to B, and that a unary predicate in A is a subset of A.

We agree that there is exactly one 0-tuple in A, and we designate it by $(\ )$. A 0-ary function from A to B is then completely determined by its value for the argument $(\ )$. We shall identify the function with this value. This means that a 0-ary function from A to B is simply an element of B.

Returning to our example, let N be the set of natural numbers. Then $+$ represents a binary function from N to N, and $<$ represents a binary predicate in N. We can also think of 1, 2, and 4 as representing 0-ary functions from N to N.

In a mathematical axiom system, we will have a certain set of objects which replaces the set of natural numbers in the above example. This set is called the *universe* of the axiom system, and its elements are called the *individuals* of the system. Functions from the universe to the universe are called *individual functions;* predicates in the universe are called *individual predicates*. Among the symbols needed in formalizing the axiom system are names for certain individuals, individual functions, and individual predicates. (In view of our convention on 0-ary functions, the first of these is a special case of the second.)

It might be thought that for some axiom systems we would need more than one universe. For example, in plane geometry we would have the set of points and the set of lines. However, we can take the universe to be the set of all points and lines, and then introduce symbols for the set of points and the set of lines (each considered as a unary individual predicate).

We shall restrict ourselves to axiom systems in which the universe is not empty, i.e., in which there is at least one individual. This is technically convenient; and it obviously does not exclude any interesting cases.

In any set A, we can form the binary predicate which consists of all 2-tuples whose first and second elements are the same element of A. This predicate is called the *equality predicate* in A. We shall use $=$ to designate the equality predicate in the universe.

The discussion so far assumes that we are formalizing a classical axiom system. For a modern axiom system the results are the same, except that we have several different universes in mind. Thus if we are formalizing the theory of fields, the possible universes are the various fields. We would then introduce symbols for the addition function of the field, the multiplication function of the field, and so on.

2.2 TRUTH FUNCTIONS

The symbols discussed in the last section enable us to build some simple formulas. We now introduce some further symbols which enable us to build more complicated formulas from these simple formulas. For example, we introduce the symbol & to mean *and*. Then we may build the formula $2 < 4 \,\&\, 6 = 3$ from the formulas $2 < 4$ and $6 = 3$.

A noteworthy feature of the formula **A** & **B** is that in order to know whether **A** & **B** is true or false, we only need to know whether **A** is true or false and whether **B** is true or false; we do not have to know what **A** and **B** mean. We can express this more simply by introducing some terminology. We select two objects, **T** and **F**, which we call *truth values*. It does not matter what these objects are, so long as they are distinct from each other. We then assign a truth value to each formula as follows: we assign **T** to each true formula and **F** to each false formula. Then we see that the truth value of **A** & **B** is determined by the truth values of **A** and **B**.

A *truth function* is a function from the set of truth values to the set of truth values. We can restate our remark as follows: there is a binary truth function $H_\&$ such that if a and b are the truth values of **A** and **B** respectively, then $H_\&(a, b)$ is the truth value of **A** & **B**. This truth function is described by the equations

$$H_\&(\mathbf{T}, \mathbf{T}) = \mathbf{T},$$
$$H_\&(\mathbf{T}, \mathbf{F}) = H_\&(\mathbf{F}, \mathbf{T}) = H_\&(\mathbf{F}, \mathbf{F}) = \mathbf{F}.$$

Next we introduce the symbol $\vee$ to mean *or*. Is the truth value of $\mathbf{A} \vee \mathbf{B}$ determined by the truth values of **A** and **B**? Certainly $\mathbf{A} \vee \mathbf{B}$ is false when **A** and **B** are both false, and is true when exactly one of **A** and **B** is true. If **A** and **B** are both true, we might or might not regard **A** *or* **B** as true in everyday speech; but we would certainly regard it as true in mathematics. We give $\vee$ this mathematical meaning of *or*. Thus if a and b are the truth values of **A** and **B** respectively, then the truth value of $\mathbf{A} \vee \mathbf{B}$ is $H_\vee(a, b)$, where $H_\vee$ is the binary truth function defined by

$$H_\vee(\mathbf{T}, \mathbf{T}) = H_\vee(\mathbf{T}, \mathbf{F}) = H_\vee(\mathbf{F}, \mathbf{T}) = \mathbf{T},$$
$$H_\vee(\mathbf{F}, \mathbf{F}) = \mathbf{F}.$$

Now we introduce $\rightarrow$ to mean *if . . . then*, so that $\mathbf{A} \rightarrow \mathbf{B}$ means *if* **A**, *then* **B**. In mathematics, we regard a statement *if* **A**, *then* **B** as being incorrect only when **A** is true and **B** is nevertheless false. If **A** is false, then *if* **A**, *then* **B** is a correct, although uninteresting, result. We adopt this mathematical meaning of *if . . . then* as our meaning of $\rightarrow$. Then $\rightarrow$ is associated in the above manner with the truth function $H_\rightarrow$ defined by

$$H_\rightarrow(\mathbf{T}, \mathbf{T}) = H_\rightarrow(\mathbf{F}, \mathbf{T}) = H_\rightarrow(\mathbf{F}, \mathbf{F}) = \mathbf{T},$$
$$H_\rightarrow(\mathbf{T}, \mathbf{F}) = \mathbf{F}.$$

Now we introduce $\leftrightarrow$ to mean *iff*. Clearly $\mathbf{A} \leftrightarrow \mathbf{B}$ is true if **A** and **B** are both true or both false, and is false otherwise. Hence $\leftrightarrow$ is associated with the truth function $H_\leftrightarrow$ defined by

$$H_\leftrightarrow(\mathbf{T}, \mathbf{T}) = H_\leftrightarrow(\mathbf{F}, \mathbf{F}) = \mathbf{T},$$
$$H_\leftrightarrow(\mathbf{T}, \mathbf{F}) = H_\leftrightarrow(\mathbf{F}, \mathbf{T}) = \mathbf{F}.$$

Next we introduce $\neg$ to mean *not;* so $\neg$**A** means *not* **A**. If **A** has the truth value a, then $\neg$**A** has the truth value $H_\neg(a)$, where $H_\neg$ is the unary truth function defined by

$$H_\neg(\mathbf{T}) = \mathbf{F}, \qquad H_\neg(\mathbf{F}) = \mathbf{T}.$$

A little thought shows that some of these symbols can be defined in terms of the others. For example, **A** $\rightarrow$ **B** means that either **A** is false or **B** is true; so it means the same as $\neg$**A** $\vee$ **B**. This may be seen more formally as follows. If a and b are the truth values of **A** and **B** respectively, then the truth value of **A** $\rightarrow$ **B** is $H_\rightarrow(a, b)$ and the truth value of $\neg$**A** $\vee$ **B** is $H_\vee(H_\neg(a), b)$. But for every a and b,

$$H_\rightarrow(a, b) = H_\vee(\mathrm{H}_\neg(a), b),$$

as we see by checking all the possibilities. In the same way, we see that **A** & **B** means the same as $\neg$(**A** $\rightarrow$ $\neg$**B**) and that **A** $\leftrightarrow$ **B** means the same as

$$(\mathbf{A} \rightarrow \mathbf{B}) \ \&\ (\mathbf{B} \rightarrow \mathbf{A}).$$

Hence all our symbols can be defined in terms of $\neg$ and $\vee$. (One can actually show that every symbol having a truth function associated with it in the above manner can be defined in terms of $\neg$ and $\vee$; see Problem 1.)

2.3 VARIABLES AND QUANTIFIERS

Using the notation introduced so far, we can express quite complicated facts about particular natural numbers. We cannot, however, express even so simple a *general* law as: *every natural number is equal to itself.*

For this purpose, we introduce *individual variables.* These are like the variables in analysis texts which we discussed earlier, except that they vary through the individuals instead of through the real numbers. Thus an individual variable can mean any individual, but its meaning remains fixed throughout any one context. A formula containing an individual variable has many meanings, one for each assignment of an individual as meaning to each individual variable in the formula. If we assert such a formula, we are asserting that all of its meanings are correct.

Since individual variables are the only variables which will occur in our languages, we shall call them simply *variables.* As variables, we use the symbols x, y, z, and w, adding primes to form new variables when they are needed. Thus if the universe is the set of natural numbers, we can assert that every natural number is equal to itself by asserting $x = x$.

While we can now *assert* that every individual has a certain property, we have no formula which *means* that every individual has the property. To see the disadvantage of this, suppose that we assert $x = 0$. We would then be asserting, incorrectly, that every natural number is equal to 0. We might hope to make this into a correct assertion by placing $\neg$ in front. But to assert $\neg(x = 0)$ is to assert that every natural number is unequal to 0; and this is also incorrect.

To overcome this difficulty, we introduce the symbol $\forall$, which means *for all individuals*. Thus $\forall x(x = 0)$ means *for all natural numbers* x, $x = 0$, that is, *every natural number is equal to* 0. To make the correct assertion that not every natural number is equal to 0, we assert $\neg\forall x(x = 0)$.

Most of our previous remarks about variables are false when applied to the x in $\forall x(x = 0)$. This formula has only one meaning, while $x = 0$ has many meanings. We get a particular meaning of $x = 0$ by substituting 2 for x; but if we substitute 2 for x in $\forall x(x = 0)$, we get the meaningless expression $\forall 2(2 = 0)$. On the other hand, if we substitute y for x in $\forall x(x = 0)$, we get a formula $\forall y(y = 0)$ which has the same meaning as $\forall x(x = 0)$ (namely, that every natural number is equal to 0). By contrast, $x = 0$ and $y = 0$ do not necessarily have the same meaning; it depends upon the meanings chosen for x and y.

To distinguish between these cases, we call the occurrence of x in $x = 0$ a *free* occurrence, and call the occurrences of x in $\forall x(x = 0)$ *bound* occurrences. If we use the differences described above to distinguish between free and bound occurrences of variables, we see that there are also bound occurrences of variables in analysis texts. Thus consider the x in $\int_0^1 \sin x\, dx$. This expression has a single meaning; it stands for a particular real number. If we substitute 2 for x, we get the meaningless expression $\int_0^1 \sin 2\, d2$. If we substitute y for x, we get $\int_0^1 \sin y\, dy$, which means exactly the same thing as $\int_0^1 \sin x\, dx$ (although it is sometimes hard to convince freshmen calculus students of this fact). Another example of a bound occurrence of a variable is the n in $\sum_{n=1}^{k} x_n$.

We now introduce the symbol $\exists$ to mean *for some individual*. Thus $\exists x(x = 0)$ means *for some natural number* x, $x = 0$, that is, *some natural number is equal to* 0. We may also read $\exists x$ as *there exists an individual* x *such that;* for example, $\exists x(x = 0)$ means *there exists a natural number* x *such that* $x = 0$. If we apply the above criteria, we see that the occurrences of x in $\exists x(x = 0)$ are bound.

An occurrence of $\forall x$ or $\exists x$ governs only the free occurrences of x in the formula immediately following. Thus in

$$\forall x(x = 5) \vee \exists x(x < 2) \vee x = 7,$$

the first two occurrences of x have no connection with the next two occurrences, and the last occurrence has no connection with the first four occurrences. Again, in

$$\forall x(x = 5 \rightarrow \exists x(x < 7)),$$

the first two occurrences of x have no connection with the last two. We can always avoid such formulas by a change of variable. Thus we could replace the first formula by

$$\forall y(y = 5) \vee \exists z(z < 2) \vee x = 7$$

and the second formula by

$$\forall x(x = 5 \rightarrow \exists y(y < 7)).$$

However, it would be inconvenient to exclude such formulas altogether. If $\forall x$ or $\exists x$ is placed before a formula which has no free occurrences of x, then the meaning of the formula is unchanged.

It is not necessary to take both $\forall$ and $\exists$ as undefined symbols; we may define $\forall$ in terms of $\exists$. First note that $\exists x \neg(x = 0)$ means *some natural number is unequal to* 0. Hence $\neg\exists x \neg(x = 0)$ means *no natural number is unequal to* 0, i.e., *every natural number is equal to* 0. This is just what $\forall x(x = 0)$ means. The same argument shows that $\forall x\mathbf{A}$ always has the same meaning as $\neg\exists x\neg\mathbf{A}$; so we may define $\forall x\mathbf{A}$ to mean $\neg\exists x\neg\mathbf{A}$. (Similarly, we could define $\exists x\mathbf{A}$ to mean $\neg\forall x\neg\mathbf{A}$.)

2.4 FIRST-ORDER LANGUAGES

We now have all the necessary concepts to proceed to a precise definition of the type of language which we wish to consider.

A *first-order language* has as symbols the following:

a) the *variables*

$$x, y, z, w, x', y', z', w', x'', \ldots ;$$

b) for each n, the *n-ary function symbols* and the *n-ary predicate symbols;*
c) the symbols $\neg$, $\vee$, and $\exists$.

For each n, the number of n-ary function symbols may be zero or nonzero, finite or infinite. The same holds for predicate symbols, except that among the binary predicate symbols must be the *equality symbol* $=$.

A 0-ary function symbol is called a *constant.* A function symbol or a predicate symbol other than $=$ is called a *nonlogical* symbol; other symbols are called *logical* symbols.

It will sometimes be convenient to have a fixed ordering of the variables to refer to. We call the order in which they are listed above *alphabetical order.*

Note that we have not included parentheses and commas, which have been used in earlier sections to indicate grouping. It turns out that the grouping is uniquely determined without them, provided that we make one change in our notation by writing $\vee\mathbf{AB}$ instead of $\mathbf{A} \vee \mathbf{B}$. We can then, of course, introduce the notation $\mathbf{A} \vee \mathbf{B}$ by means of a definition.

We use $\mathbf{x}$, $\mathbf{y}$, $\mathbf{z}$ and $\mathbf{w}$, as syntactical variables which vary through variables; $\mathbf{f}$ and $\mathbf{g}$ as syntactical variables which vary through function symbols; $\mathbf{p}$ and $\mathbf{q}$ as syntactical variables which vary through predicate symbols; and $\mathbf{e}$ as a syntactical variable which varies through constants.

Assume that we have a collection of symbols as described above. We define the *terms* by the generalized inductive definition:

i) a variable is a term;

ii) if $\mathbf{u}_1, \ldots, \mathbf{u}_n$ are terms and $\mathbf{f}$ is n-ary, then $\mathbf{f}\mathbf{u}_1 \ldots \mathbf{u}_n$ is a term.

(As part of rule (ii), a constant is a term.) It is clear that the terms are just the expressions which designate individuals. We use **a**, **b**, **c**, and **d** as syntactical variables which vary through terms.

An *atomic formula* is an expression of the form $\mathbf{pa}_1 \ldots \mathbf{a}_n$ where **p** is n-ary. We define the *formulas* by the generalized inductive definition:

i) an atomic formula is a formula;

ii) if **u** is a formula, then $\neg\mathbf{u}$ is a formula;

iii) if **u** and **v** are formulas, then $\vee\mathbf{uv}$ is a formula;

iv) if **u** is a formula, then $\exists\mathbf{xu}$ is a formula.

Corresponding to these two generalized inductive definitions we have forms of proof by induction. However, it is usually simpler to use induction on the length of the term or formula. Sometimes we use induction on the height of a formula, where the *height* is defined to be the number of occurrences of $\neg$, $\vee$, and $\exists$ in the formula.

A *first-order language* is now defined to be a language in which the symbols and formulas are as described above. A first-order language is thus completely determined by its nonlogical symbols. These symbols may be any symbols which are not already assigned to another purpose. However, we agree that if a symbol is used as an n-ary function symbol in one first-order language, then it will not be used in any other first-order language except as an n-ary function symbol; and similarly for predicate symbols. This ensures that if two first-order languages have the same nonlogical symbols, then they are identical.

A *designator* is an expression which is either a term or a formula. As one sees from the definition of *term* and *formula*, every designator has the form $\mathbf{uv}_1 \ldots \mathbf{v}_n$, where **u** is a symbol, $\mathbf{v}_1, \ldots, \mathbf{v}_n$ are designators, and n is a natural number determined by **u**. For example, if **u** is a variable, then $n = 0$; if **u** is a k-ary function symbol, then $n = k$; if **u** is $\exists$, then $n = 2$. We call n the *index* of **u**.

We say that two expressions are *compatible* if one of them can be obtained by adding some expression (possibly the empty expression) to the right end of the other. If **uv** and **u′v′** are compatible, then **u** and **u′** are compatible; if **uv** and **uv′** are compatible, then **v** and **v′** are compatible.

Lemma 1. If $\mathbf{u}_1, \ldots, \mathbf{u}_n, \mathbf{u}'_1, \ldots, \mathbf{u}'_n$ are designators and $\mathbf{u}_1 \ldots \mathbf{u}_n$ and $\mathbf{u}'_1 \ldots \mathbf{u}'_n$ are compatible, then $\mathbf{u}_i$ is $\mathbf{u}'_i$ for $i = 1, \ldots, n$.

Proof. We use induction on the length of $\mathbf{u}_1 \ldots \mathbf{u}_n$. Write $\mathbf{u}_1$ as $\mathbf{vv}_1 \ldots \mathbf{v}_k$, where **v** is a symbol of index k and $\mathbf{v}_1, \ldots, \mathbf{v}_k$ are designators. Since $\mathbf{u}'_1$ begins with **v**, it has the form $\mathbf{vv}'_1 \ldots \mathbf{v}'_k$ where $\mathbf{v}'_1, \ldots, \mathbf{v}'_k$ are designators. Now $\mathbf{u}_1$ is compatible with $\mathbf{u}'_1$; so $\mathbf{v}_1 \ldots \mathbf{v}_k$ is compatible with $\mathbf{v}'_1 \ldots \mathbf{v}'_k$. Hence by induction hypothesis, $\mathbf{v}_i$ is $\mathbf{v}'_i$ for $i = 1, \ldots, k$; so $\mathbf{u}_1$ is $\mathbf{u}'_1$. From this, it follows that $\mathbf{u}_2 \ldots \mathbf{u}_n$ is compatible with $\mathbf{u}'_2 \ldots \mathbf{u}'_n$; so by induction hypothesis, $\mathbf{u}_i$ is $\mathbf{u}'_i$ for $i = 2, \ldots, n$.

The next theorem is a precise version of our statement that in a first-order language, commas and parentheses are not necessary to determine grouping.

Formation Theorem. Every designator can be written in the form $\mathbf{uv}_1 \ldots \mathbf{v}_n$, where $\mathbf{u}$ is a symbol of index n and $\mathbf{v}_1, \ldots, \mathbf{v}_n$ are designators, in one and only one way.

Proof. We need only prove that it can be done in only one way. Now $\mathbf{u}$ must be the first symbol of the designator; so $\mathbf{u}$ and n are uniquely determined. Thus it remains to show that if $\mathbf{uv}_1 \ldots \mathbf{v}_n$ is the same as $\mathbf{uv}'_1 \ldots \mathbf{v}'_n$, where $\mathbf{v}_1, \ldots, \mathbf{v}_n$, $\mathbf{v}'_1, \ldots, \mathbf{v}'_n$ are designators, then $\mathbf{v}_i$ is $\mathbf{v}'_i$ for $i = 1, \ldots, n$. This follows from Lemma 1.

Lemma 2. Every occurrence of a symbol in a designator $\mathbf{u}$ begins an occurrence of a designator in $\mathbf{u}$.

Proof. We use induction on the length of $\mathbf{u}$. Write $\mathbf{u}$ as $\mathbf{vv}_1 \ldots \mathbf{v}_k$, where $\mathbf{v}$ is a symbol of index k and $\mathbf{v}_1, \ldots, \mathbf{v}_k$ are designators. If the occurrence of a symbol in question is the initial $\mathbf{v}$, then it begins $\mathbf{u}$. Otherwise, the occurrence is in some $\mathbf{v}_i$, and hence, by induction hypothesis, begins an occurrence of a designator in $\mathbf{v}_i$. Hence it begins an occurrence of a designator in $\mathbf{u}$.

Occurrence Theorem. Let $\mathbf{u}$ be a symbol of index n, and let $\mathbf{v}_1, \ldots, \mathbf{v}_n$ be designators. Then any occurrence of a designator $\mathbf{v}$ in $\mathbf{uv}_1 \ldots \mathbf{v}_n$ is either all of $\mathbf{uv}_1 \ldots \mathbf{v}_n$ or a part of one of the $\mathbf{v}_i$.

Proof. Suppose that the initial symbol of the occurrence of $\mathbf{v}$ is the initial $\mathbf{u}$ of $\mathbf{uv}_1 \ldots \mathbf{v}_n$. Then $\mathbf{v}$ is $\mathbf{uv}'_1 \ldots \mathbf{v}'_n$, where $\mathbf{v}'_1, \ldots, \mathbf{v}'_n$ are designators. Since $\mathbf{v}$ is compatible with $\mathbf{uv}_1 \ldots \mathbf{v}_n$, $\mathbf{v}'_1 \ldots \mathbf{v}'_n$ is compatible with $\mathbf{v}_1 \ldots \mathbf{v}_n$. By Lemma 1, $\mathbf{v}_i$ is $\mathbf{v}'_i$ for $i = 1, \ldots, n$; so $\mathbf{v}$ is all of $\mathbf{uv}_1 \ldots \mathbf{v}_n$.

Now suppose that the initial symbol of the occurrence of $\mathbf{v}$ is within $\mathbf{v}_i$. This symbol begins an occurrence of a designator $\mathbf{v}'$ in $\mathbf{v}_i$ by Lemma 2. Clearly $\mathbf{v}$ and $\mathbf{v}'$ are compatible; so by Lemma 1, $\mathbf{v}$ is $\mathbf{v}'$. Hence $\mathbf{v}$ is a part of $\mathbf{v}_i$.

We now give precise definitions of free and bound occurrences. An occurrence of $\mathbf{x}$ in $\mathbf{A}$ is *bound in* $\mathbf{A}$ if it occurs in a part of $\mathbf{A}$ of the form $\exists\mathbf{xB}$; otherwise, it is *free in* $\mathbf{A}$. We say that $\mathbf{x}$ is *free* (*bound*) *in* $\mathbf{A}$ if some occurrence of $\mathbf{x}$ is free (bound) in $\mathbf{A}$. (Note that $\mathbf{x}$ may be both free and bound in $\mathbf{A}$.) It follows from the occurrence theorem that if $\mathbf{y}$ is distinct from $\mathbf{x}$, then the free occurrences of $\mathbf{x}$ in $\neg\mathbf{A}$, $\vee\mathbf{AB}$, and $\exists\mathbf{yA}$ are just the free occurrences of $\mathbf{x}$ in $\mathbf{A}$ and $\mathbf{B}$. Of course, $\mathbf{x}$ has no free occurrences in $\exists\mathbf{xA}$.

We use $\mathbf{b_x[a]}$ to designate the expression obtained from $\mathbf{b}$ by replacing each occurrence of $\mathbf{x}$ by $\mathbf{a}$; and we use $\mathbf{A_x[a]}$ to designate the expression obtained from $\mathbf{A}$ by replacing each *free* occurrence of $\mathbf{x}$ by $\mathbf{a}$. Using induction on the length of $\mathbf{b}$ and $\mathbf{A}$, we easily prove that $\mathbf{b_x[a]}$ is a term and that $\mathbf{A_x[a]}$ is a formula.

In general, $\mathbf{A_x[a]}$ says the same thing about the individual designated by $\mathbf{a}$ that $\mathbf{A}$ says about the individual designated by $\mathbf{x}$; but this is not always the case. Thus suppose that $\mathbf{A}$ is $\exists y(x = 2 \cdot y)$, $\mathbf{x}$ is x, and $\mathbf{a}$ is $y + 1$. Then $\mathbf{A}$ says that x is even; but $\mathbf{A_x[a]}$, which is $\exists y(y + 1 = 2 \cdot y)$, does not say that $y + 1$ is even.

The difficulty is, of course, that the y in the substituted $y + 1$ has become bound. We wish to exclude such possibilities.

We say that **a** is *substitutible for* **x** *in* **A** if for each variable **y** occurring in **a**, no part of **A** of the form $\exists\mathbf{yB}$ contains an occurrence of **x** which is free in **A**. We now agree that whenever $\mathbf{A_x[a]}$ appears, **A**, **x**, and **a** are restricted to represent expressions such that **a** is substitutible for **x** in **A**. Note that this is certainly the case if **a** contains no variables or if **A** contains no bound occurrences of variables.

We now extend this notation to several variables. We let $\mathbf{b}_{\mathbf{x}_1,\ldots,\mathbf{x}_n}[\mathbf{a}_1, \ldots, \mathbf{a}_n]$ designate the term obtained from **b** by replacing all occurrences of $\mathbf{x}_1, \ldots, \mathbf{x}_n$ by $\mathbf{a}_1, \ldots, \mathbf{a}_n$ respectively; and we let $\mathbf{A}_{\mathbf{x}_1,\ldots,\mathbf{x}_n}[\mathbf{a}_1, \ldots, \mathbf{a}_n]$ designate the formula obtained from **A** by replacing all free occurrences of $\mathbf{x}_1, \ldots, \mathbf{x}_n$ by $\mathbf{a}_1, \ldots, \mathbf{a}_n$ respectively. Whenever either of these is used, $\mathbf{x}_1, \ldots, \mathbf{x}_n$ are restricted to represent distinct variables. Whenever $\mathbf{A}_{\mathbf{x}_1,\ldots,\mathbf{x}_n}[\mathbf{a}_1, \ldots, \mathbf{a}_n]$ appears, **A**, $\mathbf{x}_1, \ldots, \mathbf{x}_n$, $\mathbf{a}_1, \ldots, \mathbf{a}_n$ are restricted to represent expressions such that $\mathbf{a}_i$ is substitutible for $\mathbf{x}_i$ in **A** for $i = 1, \ldots, n$. We shall omit the subscripts $\mathbf{x}_1, \ldots, \mathbf{x}_n$ when they are immaterial or clear from the context.

We now introduce some defined symbols. In all theories, $(\mathbf{A} \vee \mathbf{B})$ is an abbreviation of $\vee\mathbf{AB}$; $(\mathbf{A} \rightarrow \mathbf{B})$ is an abbreviation of $(\neg\mathbf{A} \vee \mathbf{B})$; $(\mathbf{A} \,\&\, \mathbf{B})$ is an abbreviation of $\neg(\mathbf{A} \rightarrow \neg\mathbf{B})$; $(\mathbf{A} \leftrightarrow \mathbf{B})$ is an abbreviation of $((\mathbf{A} \rightarrow \mathbf{B}) \,\&\, (\mathbf{B} \rightarrow \mathbf{A}))$; and $\forall\mathbf{xA}$ is an abbreviation of $\neg\exists\mathbf{x}\neg\mathbf{A}$. If **u** is a binary predicate or function symbol, then **(aub)** is an abbreviation of **uab**; if **u** is a binary predicate symbol, then $(\mathbf{a}\not\mathbf{u}\mathbf{b})$ is an abbreviation of $\neg(\mathbf{aub})$. In each case, the defined formulas are those expressions which give formulas when the defined symbols are eliminated according to the rules just given.

We could show that the parentheses introduced in the above definitions are sufficient to determine the grouping. However, as we do not intend to study defined formulas, such a general result is unnecessary. All that we require is that each defined formula which we write abbreviate a unique formula of the language. For this reason, we shall omit parentheses when they are not necessary to determine the grouping; e.g., we write $\mathbf{x} = \mathbf{y} \rightarrow \mathbf{y} = \mathbf{x}$ instead of

$$((\mathbf{x} = \mathbf{y}) \rightarrow (\mathbf{y} = \mathbf{x})).$$

On the other hand, we may add superfluous parentheses and commas to increase readability. Thus we shall often write $\mathbf{u}(\mathbf{a}_1, \ldots, \mathbf{a}_n)$ instead of $\mathbf{ua}_1 \ldots \mathbf{a}_n$ when **u** is a function or predicate symbol. To enable us to omit more parentheses, we agree that a formula shall be of the form $\mathbf{A} \rightarrow \mathbf{B}$ or $\mathbf{A} \leftrightarrow \mathbf{B}$ rather than $\mathbf{A} \vee \mathbf{B}$ or $\mathbf{A} \,\&\, \mathbf{B}$ whenever there is a choice. Thus $\mathbf{A} \rightarrow \mathbf{B} \vee \mathbf{C}$ is to be read as

$$\mathbf{A} \rightarrow (\mathbf{B} \vee \mathbf{C}),$$

and $\mathbf{A} \,\&\, \mathbf{B} \leftrightarrow \mathbf{C}$ is to be read as $(\mathbf{A} \,\&\, \mathbf{B}) \leftrightarrow \mathbf{C}$.

We also adopt the convention of *association to the right* for omitting parentheses. This means that $\mathbf{A} \vee \mathbf{B} \vee \mathbf{C}$ is to be read as $\mathbf{A} \vee (\mathbf{B} \vee \mathbf{C})$; $\mathbf{A} \vee \mathbf{B} \vee \mathbf{C} \vee \mathbf{D}$ is to be read as $\mathbf{A} \vee (\mathbf{B} \vee (\mathbf{C} \vee \mathbf{D}))$; and so on. The same convention is used for a sequence of formulas connected by &, or for a sequence of formulas connected by $\rightarrow$. Note that $\mathbf{A}_1 \vee \cdots \vee \mathbf{A}_n$ means that at least one of $\mathbf{A}_1, \ldots, \mathbf{A}_n$ is true; $\mathbf{A}_1 \mathbin{\&} \cdots \mathbin{\&} \mathbf{A}_n$ means that all of $\mathbf{A}_1, \ldots, \mathbf{A}_n$ are true; and $\mathbf{A}_1 \rightarrow \cdots \rightarrow \mathbf{A}_n \rightarrow \mathbf{B}$ means that if all of $\mathbf{A}_1, \ldots, \mathbf{A}_n$ are true, then **B** is true. In each of these we allow $n = 1$. If we write $\mathbf{A}_1 \rightarrow \cdots \rightarrow \mathbf{A}_n \rightarrow \mathbf{B}$, we even allow n to be 0; in this case, the formula is just **B**.

We call $\neg\mathbf{A}$ the *negation* of **A**; $\mathbf{A} \vee \mathbf{B}$ the *disjunction* of **A** and **B**; $\mathbf{A} \mathbin{\&} \mathbf{B}$ the *conjunction* of **A** and **B**; $\mathbf{A} \rightarrow \mathbf{B}$ the *implication* of **B** by **A**; $\mathbf{A} \leftrightarrow \mathbf{B}$ the *equivalence* of **A** and **B**; $\exists\mathbf{x}\mathbf{A}$ the *instantiation* of **A** by **x**; and $\forall\mathbf{x}\mathbf{A}$ the *generalization* of **A** by **x**. We also call $\mathbf{A}_1 \vee \cdots \vee \mathbf{A}_n$ the *disjunction* of $\mathbf{A}_1, \ldots, \mathbf{A}_n$ and $\mathbf{A}_1 \mathbin{\&} \cdots \mathbin{\&} \mathbf{A}_n$ the *conjunction* of $\mathbf{A}_1, \ldots, \mathbf{A}_n$. However, when we say that a formula is a disjunction or a conjunction without specifying of what, we mean that it is a disjunction or conjunction of two formulas. The expressions $\exists\mathbf{x}$ and $\forall\mathbf{x}$ are called *quantifiers on* **x**; the former is an *existential* quantifier and the latter is a *universal* quantifier.

2.5 STRUCTURES

We now turn to a precise description of the semantics of first-order languages. As already indicated, a meaning for a first-order language consists of a universe and a meaning of the appropriate sort for each nonlogical symbol. Writing this out in detail, we arrive at the following definition.

Let L be a first-order language. A *structure* $\mathcal{A}$ for L consists of the following things:

i) A nonempty set $|\mathcal{A}|$, called the *universe* of $\mathcal{A}$. The elements of $|\mathcal{A}|$ are called the *individuals* of $\mathcal{A}$.

ii) For each n-ary function symbol **f** of L, an n-ary function $\mathbf{f}_{\mathcal{A}}$ from $|\mathcal{A}|$ to $|\mathcal{A}|$. (In particular, for each constant **e** of L, $\mathbf{e}_{\mathcal{A}}$ is an individual of $\mathcal{A}$.)

iii) For each n-ary predicate symbol **p** of L other than $=$, an n-ary predicate $\mathbf{p}_{\mathcal{A}}$ in $|\mathcal{A}|$.

We want to define a formula **A** to be valid in $\mathcal{A}$ if all the meanings of **A** are true in $\mathcal{A}$. It would therefore be convenient if for each meaning of **A**, we had a formula which expressed exactly that meaning. Since a meaning of **A** is obtained by assigning an individual as meaning to each variable free in **A**, it is clear that what we require are names for the individuals. This leads us to the following definitions.

Let $\mathcal{A}$ be a structure for L. For each individual a of $\mathcal{A}$, we choose a new constant, called the *name* of a. It is understood that different names are chosen for different individuals. The first-order language obtained from L by adding all the names of individuals of $\mathcal{A}$ is designated by $L(\mathcal{A})$. We use **i** and **j** as syntactical variables which vary through names.

An expression is *variable-free* if it contains no variables. We shall now define an individual $\mathcal{A}(\mathbf{a})$ of $\mathcal{A}$ for each variable-free term **a** of $L(\mathcal{A})$. The definition is by induction on the length of **a**. If **a** is a name, $\mathfrak{A}(\mathbf{a})$ is the individual of which **a** is the name. If **a** is not a name, then (since it is variable-free) it must be $\mathbf{fa}_1 \ldots \mathbf{a}_n$ with **f** a function symbol of L. We then let $\mathfrak{A}(\mathbf{a})$ be $\mathbf{f}_{\mathfrak{A}}(\mathfrak{A}(\mathbf{a}_1), \ldots, \mathfrak{A}(\mathbf{a}_n))$.

Remark. In the above definition, we have used the formation theorem tacitly to see that **a** could be written in the form $\mathbf{fa}_1 \ldots \mathbf{a}_n$ in only one way. We will frequently make such tacit use of the formation theorem in definitions by induction on the length of a term or formula.

A formula **A** is *closed* if no variable is free in **A**. (This means that **A** has only one meaning.) We shall now define a truth value $\mathfrak{A}(\mathbf{A})$ for each closed formula **A** in $L(\mathfrak{A})$. The definition is by induction on the length of **A**. If **A** is $\mathbf{a} = \mathbf{b}$, then **a** and **b** must be variable-free (since **A** is closed). We let

$$\mathfrak{A}(\mathbf{A}) = \mathsf{T} \qquad \text{iff} \qquad \mathfrak{A}(\mathbf{a}) = \mathfrak{A}(\mathbf{b})$$

(i.e., iff $\mathfrak{A}(\mathbf{a})$ and $\mathfrak{A}(\mathbf{b})$ are the same). If **A** is $\mathbf{pa}_1 \ldots \mathbf{a}_n$, where **p** is not $=$, we let

$$\mathfrak{A}(\mathbf{A}) = \mathsf{T} \qquad \text{iff} \qquad \mathbf{p}_{\mathfrak{A}}(\mathfrak{A}(\mathbf{a}_1), \ldots, \mathfrak{A}(\mathbf{a}_n))$$

(i.e., iff the n-tuple $(\mathfrak{A}(\mathbf{a}_1), \ldots, \mathfrak{A}(\mathbf{a}_n))$ belongs to the predicate $\mathbf{p}_{\mathfrak{A}}$). If **A** is $\neg\mathbf{B}$, then $\mathfrak{A}(\mathbf{A})$ is $H_\neg(\mathfrak{A}(\mathbf{B}))$. If **A** is $\mathbf{B} \vee \mathbf{C}$, then $\mathfrak{A}(\mathbf{A})$ is $H_\vee(\mathfrak{A}(\mathbf{B}), \mathfrak{A}(\mathbf{C}))$. If **A** is $\exists\mathbf{xB}$, then $\mathfrak{A}(\mathbf{A}) = \mathsf{T}$ iff $\mathfrak{A}(\mathbf{B_x[i]}) = \mathsf{T}$ for some **i** in $L(\mathfrak{A})$.

It is clear that $\mathfrak{A}(\mathbf{A} \to \mathbf{B}) = H_\to(\mathfrak{A}(\mathbf{A}), \mathfrak{A}(\mathbf{B}))$, and similarly for & and $\leftrightarrow$. We have

$$\mathfrak{A}(\mathbf{A}_1 \vee \cdots \vee \mathbf{A}_n) = \mathsf{T} \qquad \text{iff} \qquad \mathfrak{A}(\mathbf{A}_i) = \mathsf{T} \text{ for at least one } i$$

and

$$\mathfrak{A}(\mathbf{A}_1 \mathbin{\&} \cdots \mathbin{\&} \mathbf{A}_n) = \mathsf{T} \qquad \text{iff} \qquad \mathfrak{A}(\mathbf{A}_i) = \mathsf{T} \text{ for all } i.$$

Also,

$$\mathfrak{A}(\forall\mathbf{xA}) = \mathsf{T} \qquad \text{iff} \qquad \mathfrak{A}(\mathbf{A_x[i]}) = \mathsf{T} \text{ for every } \mathbf{i} \text{ in } L(\mathfrak{A}).$$

If **A** is a formula of L, an $\mathfrak{A}$-*instance* of **A** is a closed formula of the form $\mathbf{A}[\mathbf{i}_1, \ldots, \mathbf{i}_n]$ in $L(\mathfrak{A})$. A formula **A** of L is *valid* in $\mathfrak{A}$ if $\mathfrak{A}(\mathbf{A}') = \mathsf{T}$ for every $\mathfrak{A}$-instance $\mathbf{A}'$ of **A**. In particular, a closed formula **A** of L is valid in $\mathfrak{A}$ iff $\mathfrak{A}(\mathbf{A}) = \mathsf{T}$.

We will now prove a lemma which shows that $\mathbf{b_x[a]}$ and $\mathbf{A_x[a]}$ have the proper meaning.

Lemma. Let $\mathfrak{A}$ be a structure for L; **a** a variable-free term in $L(\mathfrak{A})$; **i** the name of $\mathfrak{A}(\mathbf{a})$. If **b** is a term of $L(\mathfrak{A})$ in which no variable except **x** occurs, then $\mathfrak{A}(\mathbf{b_x[a]}) = \mathfrak{A}(\mathbf{b_x[i]})$. If **A** is a formula of $L(\mathfrak{A})$ in which no variable except **x** is free, then $\mathfrak{A}(\mathbf{A_x[a]}) = \mathfrak{A}(\mathbf{A_x[i]})$.

Proof. We prove the first conclusion by induction on the length of **b**. If **b** is a name, then $\mathbf{b_x[a]}$ and $\mathbf{b_x[i]}$ are both **b**; so the conclusion is evident. If **b** is a variable, it must be **x**; so $\mathbf{b_x[a]}$ is **a** and $\mathbf{b_x[i]}$ is **i**. But $\mathfrak{A}(\mathbf{a}) = \mathfrak{A}(\mathbf{i})$ by the choice of **i**. If **b** is $\mathbf{fb}_1 \ldots \mathbf{b}_n$ with $\mathbf{f}$ a function symbol of L, then, using the induction hypothesis, we have

$$\begin{aligned}\mathcal{A}(\mathbf{b}[\mathbf{a}]) &= \mathcal{A}(\mathbf{fb}_1[\mathbf{a}] \ldots \mathbf{b}_n[\mathbf{a}])\\ &= \mathbf{f}_{\mathcal{A}}(\mathcal{A}(\mathbf{b}_1[\mathbf{a}]), \ldots, \mathcal{A}(\mathbf{b}_n[\mathbf{a}]))\\ &= \mathbf{f}_{\mathcal{A}}(\mathcal{A}(\mathbf{b}_1[\mathbf{i}]), \ldots, \mathcal{A}(\mathbf{b}_n[\mathbf{i}]))\\ &= \mathcal{A}(\mathbf{fb}_1[\mathbf{i}] \ldots \mathbf{b}_n[\mathbf{i}])\\ &= \mathcal{A}(\mathbf{b}[\mathbf{i}]).\end{aligned}$$

We now prove the second conclusion by induction on the length of $\mathbf{A}$. If $\mathbf{A}$ is $\mathbf{b} = \mathbf{c}$, then, using the first conclusion,

$$\begin{aligned}\mathcal{A}(\mathbf{A}[\mathbf{a}]) = \mathbf{T} &\leftrightarrow \mathcal{A}(\mathbf{b}[\mathbf{a}]) = \mathcal{A}(\mathbf{c}[\mathbf{a}])\\ &\leftrightarrow \mathcal{A}(\mathbf{b}[\mathbf{i}]) = \mathcal{A}(\mathbf{c}[\mathbf{i}])\\ &\leftrightarrow \mathcal{A}(\mathbf{A}[\mathbf{i}]) = \mathbf{T}.\end{aligned}$$

If $\mathbf{A}$ is $\mathbf{pb}_1 \ldots \mathbf{b}_n$ where $\mathbf{p}$ is not $=$, the proof is quite similar. If $\mathbf{A}$ is $\neg\mathbf{B}$, then

$$\begin{aligned}\mathcal{A}(\mathbf{A}[\mathbf{a}]) &= H_{\neg}(\mathcal{A}(\mathbf{B}[\mathbf{a}]))\\ &= H_{\neg}(\mathcal{A}(\mathbf{B}[\mathbf{i}]))\\ &= \mathcal{A}(\mathbf{A}[\mathbf{i}]).\end{aligned}$$

If $\mathbf{A}$ is $\mathbf{B} \vee \mathbf{C}$, the proof is similar. Now suppose that $\mathbf{A}$ is $\exists\mathbf{yB}$. We may suppose that $\mathbf{y}$ is not $\mathbf{x}$, since otherwise $\mathbf{A}_{\mathbf{x}}[\mathbf{a}]$ and $\mathbf{A}_{\mathbf{x}}[\mathbf{i}]$ are both $\mathbf{A}$. Then

$$\begin{aligned}\mathcal{A}(\mathbf{A}_{\mathbf{x}}[\mathbf{a}]) = \mathbf{T} &\leftrightarrow \mathcal{A}(\exists\mathbf{yB}_{\mathbf{x}}[\mathbf{a}]) = \mathbf{T}\\ &\leftrightarrow \mathcal{A}(\mathbf{B}_{\mathbf{x},\mathbf{y}}[\mathbf{a}, \mathbf{j}]) = \mathbf{T} \quad \text{for some } \mathbf{j}\\ &\leftrightarrow \mathcal{A}(\mathbf{B}_{\mathbf{x},\mathbf{y}}[\mathbf{i}, \mathbf{j}]) = \mathbf{T} \quad \text{for some } \mathbf{j}\\ &\leftrightarrow \mathcal{A}(\exists\mathbf{yB}_{\mathbf{x}}[\mathbf{i}]) = \mathbf{T}\\ &\leftrightarrow \mathcal{A}(\mathbf{A}_{\mathbf{x}}[\mathbf{i}]) = \mathbf{T}.\end{aligned}$$

2.6 LOGICAL AXIOMS AND RULES

If we are formulating a classical axiom system in a first-order language L, we have in mind a particular meaning for L, i.e., a particular structure $\mathcal{A}$ for L. We then want all of the theorems of our formal system to be valid in $\mathcal{A}$. In order to ensure this, we require that the axioms be valid and that the rules be such that the validity of the conclusion follows from the validity of the hypotheses, or, as we shall say, such that the conclusion is a *consequence* of the hypotheses.

Certain formulas of L are valid simply because of the meaning of the logical symbols; i.e., they are valid in *every* structure for L. For example, $x = x$ has this property. Such formulas are said to be *logically valid*. Certain of our axioms, called the *logical axioms*, will be logically valid. The others, called the *nonlogical axioms*, will be valid because of particular properties of the structure $\mathcal{A}$.

We say that $\mathbf{A}$ is a *logical consequence* of a set Γ of formulas if it is a consequence of Γ because of the meaning of the logical symbols, i.e., if $\mathbf{A}$ is valid in every structure for L in which all of the formulas in Γ are valid. We might expect our rules also to divide into two classes: *logical rules*, in which the conclusion is a logical consequence of the hypotheses, and *nonlogical rules*, in which the conclusion is a consequence of the hypotheses only because of special properties of $\mathfrak{A}$. However, we can dispense with nonlogical rules altogether. For suppose that we want to be able to infer **B** from $\mathbf{A}_1, \ldots, \mathbf{A}_n$. Then **B** will be a consequence of $\mathbf{A}_1, \ldots, \mathbf{A}_n$; so $\mathbf{A}_1 \rightarrow \cdots \rightarrow \mathbf{A}_n \rightarrow \mathbf{B}$ is valid and may be adopted as a nonlogical axiom. But **B** is a logical consequence of $\mathbf{A}_1, \ldots, \mathbf{A}_n$, and $\mathbf{A}_1 \rightarrow \cdots \rightarrow \mathbf{A}_n \rightarrow \mathbf{B}$; so if we have sufficiently many logical rules, we can infer **B** from $\mathbf{A}_1, \ldots, \mathbf{A}_n$ and the axiom $\mathbf{A}_1 \rightarrow \cdots \rightarrow \mathbf{A}_n \rightarrow \mathbf{B}$.

We shall now describe the logical axioms and rules. Let L be a first-order language. A *propositional axiom* is a formula of the form $\neg\mathbf{A} \vee \mathbf{A}$. A *substitution axiom* is a formula of the form $\mathbf{A_x[a]} \rightarrow \exists\mathbf{xA}$. An *identity axiom* is a formula of the form $\mathbf{x} = \mathbf{x}$. An *equality axiom* is a formula of the form

$$\mathbf{x}_1 = \mathbf{y}_1 \rightarrow \cdots \rightarrow \mathbf{x}_n = \mathbf{y}_n \rightarrow \mathbf{fx}_1 \ldots \mathbf{x}_n = \mathbf{fy}_1 \ldots \mathbf{y}_n$$

or of the form

$$\mathbf{x}_1 = \mathbf{y}_1 \rightarrow \cdots \rightarrow \mathbf{x}_n = \mathbf{y}_n \rightarrow \mathbf{px}_1 \ldots \mathbf{x}_n \rightarrow \mathbf{py}_1 \ldots \mathbf{y}_n.$$

A *logical axiom* is a formula which is a propositional axiom, a substitution axiom, an identity axiom, or an equality axiom.

We now show that the logical axioms are valid. Let $\mathfrak{A}$ be a structure for L. An $\mathfrak{A}$-instance of a propositional axiom has the form $\neg\mathbf{A} \vee \mathbf{A}$; and

$$\mathfrak{A}(\neg\mathbf{A} \vee \mathbf{A}) = H_{\vee}(H_{\neg}(\mathfrak{A}(\mathbf{A})), \mathfrak{A}(\mathbf{A})) = \mathsf{T}.$$

An $\mathfrak{A}$-instance of a substitution axiom has the form $\mathbf{A_x[a]} \rightarrow \exists\mathbf{xA}$. Suppose that $\mathfrak{A}(\mathbf{A_x[a]} \rightarrow \exists\mathbf{xA}) = \mathsf{F}$. Then $\mathfrak{A}(\mathbf{A_x[a]}) = \mathsf{T}$ and $\mathfrak{A}(\exists\mathbf{xA}) = \mathsf{F}$. If **i** is the name of $\mathfrak{A}(\mathbf{a})$, the latter implies that $\mathfrak{A}(\mathbf{A_x[i]}) = \mathsf{F}$ while the former with the lemma of §2.5 implies that $\mathfrak{A}(\mathbf{A_x[i]}) = \mathsf{T}$. This is a contradiction. An $\mathfrak{A}$-instance of an identity axiom has the form $\mathbf{i} = \mathbf{i}$; since $\mathfrak{A}(\mathbf{i}) = \mathfrak{A}(\mathbf{i})$, $\mathfrak{A}(\mathbf{i} = \mathbf{i}) = \mathsf{T}$. We leave the equality axioms to the reader.

We now introduce five rules of inference. (Note that these rules are finite.)

Expansion Rule. *Infer* $\mathbf{B} \vee \mathbf{A}$ *from* **A**.

Contraction Rule. *Infer* **A** *from* $\mathbf{A} \vee \mathbf{A}$.

Associative Rule. *Infer* $(\mathbf{A} \vee \mathbf{B}) \vee \mathbf{C}$ *from* $\mathbf{A} \vee (\mathbf{B} \vee \mathbf{C})$.

Cut Rule. *Infer* $\mathbf{B} \vee \mathbf{C}$ *from* $\mathbf{A} \vee \mathbf{B}$ *and* $\neg\mathbf{A} \vee \mathbf{C}$.

∃-Introduction Rule. *If* **x** *is not free in* **B**, *infer* $\exists\mathbf{xA} \rightarrow \mathbf{B}$ *from* $\mathbf{A} \rightarrow \mathbf{B}$.

We now wish to see that the conclusion of each rule is a logical consequence of the hypotheses of the rule. We check this for the last two rules, leaving the first three to the reader.

Suppose that $\mathbf{A} \vee \mathbf{B}$ and $\neg\mathbf{A} \vee \mathbf{C}$ are valid in $\mathfrak{A}$. Let $\mathbf{B}' \vee \mathbf{C}'$ be an $\mathfrak{A}$-instance of $\mathbf{B} \vee \mathbf{C}$. We can clearly choose an $\mathfrak{A}$-instance $\mathbf{A}'$ of **A** so that $\mathbf{A}' \vee \mathbf{B}'$

is an $\mathcal{A}$-instance of $\mathbf{A} \vee \mathbf{B}$ and $\neg\mathbf{A}' \vee \mathbf{C}'$ is an $\mathcal{A}$-instance of $\neg\mathbf{A} \vee \mathbf{C}$. Then

$$\mathcal{A}(\mathbf{A}' \vee \mathbf{B}') = \mathcal{A}(\neg\mathbf{A}' \vee \mathbf{C}') = \mathsf{T}.$$

Hence $\mathcal{A}(\mathbf{A}') = \mathsf{T}$ or $\mathcal{A}(\mathbf{B}') = \mathsf{T}$, and $\mathcal{A}(\mathbf{A}') = \mathsf{F}$ or $\mathcal{A}(\mathbf{C}') = \mathsf{T}$. From this, it follows that either $\mathcal{A}(\mathbf{B}') = \mathsf{T}$ or $\mathcal{A}(\mathbf{C}') = \mathsf{T}$ (according as $\mathcal{A}(\mathbf{A}') = \mathsf{F}$ or $\mathcal{A}(\mathbf{A}') = \mathsf{T}$). Hence $\mathcal{A}(\mathbf{B}' \vee \mathbf{C}') = \mathsf{T}$.

Suppose that $\mathbf{A} \rightarrow \mathbf{B}$ is valid in $\mathcal{A}$, and that $\mathbf{x}$ is not free in $\mathbf{B}$. An $\mathcal{A}$-instance of $\exists\mathbf{x}\mathbf{A} \rightarrow \mathbf{B}$ has the form $\exists\mathbf{x}\mathbf{A}' \rightarrow \mathbf{B}'$. Suppose that $\mathcal{A}(\exists\mathbf{x}\mathbf{A}' \rightarrow \mathbf{B}') = \mathsf{F}$. Then $\mathcal{A}(\exists\mathbf{x}\mathbf{A}') = \mathsf{T}$ and $\mathcal{A}(\mathbf{B}') = \mathsf{F}$. From the former, $\mathcal{A}(\mathbf{A}'_\mathbf{x}[\mathbf{i}]) = \mathsf{T}$ for some $\mathbf{i}$; so $\mathcal{A}(\mathbf{A}'_\mathbf{x}[\mathbf{i}] \rightarrow \mathbf{B}') = \mathsf{F}$. This is impossible, since $\mathbf{A}'_\mathbf{x}[\mathbf{i}] \rightarrow \mathbf{B}'$ is an $\mathcal{A}$-instance of $\mathbf{A} \rightarrow \mathbf{B}$.

We can now define the class of formal systems which we are going to study. A *first-order theory*, or simply a *theory*, is a formal system T such that

i) the language of T is a first-order language;

ii) the axioms of T are the logical axioms of $L(T)$ and certain further axioms, called the *nonlogical axioms;*

iii) the rules of T are the expansion rule, the contraction rule, the associative rule, the cut rule, and the $\exists$-introduction rule.

In order to specify a theory, we have only to specify its nonlogical symbols and its nonlogical axioms; everything else is given by the definition of a theory. Note also that the logical axioms and the rules are determined as soon as the language is chosen; they are independent of the nonlogical axioms.

We give two examples of theories. The first, which we designate by N, formalizes a classical axiom system for the natural numbers. The nonlogical symbols of N are the constant 0, the unary function symbol S (which designates the successor function), the binary function symbols $+$ and $\cdot$, and the binary predicate symbol $<$. The nonlogical axioms of N are:

N1. $Sx \neq 0$.

N2. $Sx = Sy \rightarrow x = y$.

N3. $x + 0 = x$.

N4. $x + Sy = S(x + y)$.

N5. $x \cdot 0 = 0$.

N6. $x \cdot Sy = (x \cdot y) + x$.

N7. $\neg(x < 0)$.

N8. $x < Sy \leftrightarrow x < y \vee x = y$.

N9. $x < y \vee x = y \vee y < x$.

Our second example formalizes the modern axiom system for groups. It is called the *elementary theory of groups*, and is designated by G. The only nonlogical symbol of G is the binary function symbol $\cdot$. The nonlogical axioms of G are:

G1. $(x \cdot y) \cdot z = x \cdot (y \cdot z)$.

G2. $\exists x\big(\forall y(x \cdot y = y)\ \&\ \forall y\, \exists z(z \cdot y = x)\big)$.

By a *model* of a theory T, we mean a structure for $L(T)$ in which all the nonlogical axioms of T are valid. A formula is *valid in T* if it is valid in every model of T; equivalently, if it is a logical consequence of the nonlogical axioms of T.

Examples

1. We construct a model of N by taking the universe to be the set of natural numbers and assigning the obvious individuals, functions, and predicates to the nonlogical symbols of N. This model is called the *standard model* of N, and is designated by $\mathfrak{N}$.

2. A structure for $L(G)$ may be described as a nonempty set (the universe) together with a binary operation (the function assigned to $\cdot$). Such a structure will be a model of G iff it is a group.

Validity Theorem. If T is a theory, then every theorem of T is valid in T.

Proof. We use induction on theorems. If **A** is a nonlogical axiom, the result is true by the definition of a model. If **A** is a logical axiom, the result was proved above. If **A** is the conclusion of a rule, the result follows from the induction hypothesis and the facts proved above.

If we have a structure or several structures in mind when we formulate the nonlogical axioms, then we shall certainly choose these axioms to be valid in these structures. The structures are then models of the theory; so the validity theorem shows that the theorems of the theory are all valid in these structures. Thus we see that our logical axioms and rules are correct.

The question now arises: can we profitably add more logical axioms or rules? One way to establish that we cannot do so is to prove the converse of the validity theorem: every formula which is valid in T is a theorem of T. For then if a new logical axiom or rule gave a new theorem, this theorem would not be a logical consequence of the nonlogical axioms, and hence the new axiom or rule would be incorrect. We shall establish this converse of the validity theorem later; but we will find it necessary to first investigate some of the consequences of the logical axioms and rules.

PROBLEMS

1. An n-ary truth function H is *definable in terms of* the truth functions $H_1, \ldots, H_k$ if H has a definition

$$H(a_1, \ldots, a_n) = \ldots,$$

where the right-hand side is built up from $H_1, \ldots, H_k, a_1, \ldots, a_n$, and commas and parentheses.

a) Let $H_{d,n}$ be the truth function defined by setting $H_{d,n}(a_1, \ldots, a_n) = \mathbf{T}$ iff $a_i = \mathbf{T}$ for at least one i, and let $H_{c,n}$ be the truth function defined by setting

$$H_{c,n}(a_1, \ldots, a_n) = \mathbf{T} \quad \text{iff} \quad a_i = \mathbf{T} \text{ for all } i.$$

Show that every truth function is definable in terms of $H_\neg$ and certain of the $H_{d,n}$ and $H_{c,n}$.

b) Show that every truth function is definable in terms of $H_{\neg}$ and $H_{\vee}$. [Use (a).]

c) Show that every truth function is definable in terms of $H_{\neg}$ and $H_{\rightarrow}$. [Use (b).]

d) Show that every truth function is definable in terms of $H_{\neg}$ and $H_{\&}$. [Use (b).]

e) Show that $H_{\neg}$ is not definable in terms of $H_{\vee}$, $H_{\rightarrow}$, $H_{\&}$, and $H_{\leftrightarrow}$.

2. a) Let H_d be the truth function defined by

$$H_d(a, b) = \mathsf{T} \quad \text{iff} \quad a = b = \mathsf{F}.$$

Show that every truth function is definable in terms of H_d. [Use 1(b).]

b) Let H_s be the truth function defined by

$$H_s(a, b) = \mathsf{F} \quad \text{iff} \quad a = b = \mathsf{T}.$$

Show that every truth function is definable in terms of H_s. [Use 1(b).]

c) A truth function H is *singulary* if there is a truth function H' and an i such that $H(a_1, \ldots, a_n) = H'(a_i)$ for all $a_1, \ldots, a_n$. Show that if H is singulary, then every truth function definable in terms of H is singulary.

d) Show that if H is a binary truth function such that every truth function is definable in terms of H, then H is H_d or H_s. [Show that $H(\mathsf{T}, \mathsf{T}) = \mathsf{F}$ and $H(\mathsf{F}, \mathsf{F}) = \mathsf{T}$, and use (c).]

3. Show that if **uv** and **vv′** are designators, then either **v** or **v′** is the empty expression.

4. Show that the result of replacing **a** by **x** in a term is a term, and that the result of replacing **a** by **x** in a formula is a formula.

5. Let T be the theory with no nonlogical symbols and no nonlogical axioms.

a) Show that $\neg\neg(x = x) \vee \neg(x = x)$ is a theorem of T not provable without propositional axioms. [Let f be a mapping from the set of formulas to the set of truth values such that $f(\mathbf{A}) = \mathsf{T}$ for **A** atomic; $f(\neg\mathbf{A}) = \mathsf{F}$; $f(\mathbf{A} \vee \mathbf{B}) = f(\mathbf{B})$; $f(\exists\mathbf{x}\mathbf{A}) = \mathsf{T}$. Show that if **A** is provable without propositional axioms, then $f(\mathbf{A}) = \mathsf{T}$.]

b) Show that $x = x \rightarrow \exists x(x = x)$ is a theorem of T not provable without substitution axioms. [Proceed as in (a), letting $f(\mathbf{A}) = \mathsf{T}$ for **A** atomic; $f(\neg\mathbf{A}) = H_{\neg}(f(\mathbf{A}))$; $f(\mathbf{A} \vee \mathbf{B}) = H_{\vee}(f(\mathbf{A}), f(\mathbf{B}))$; $f(\exists\mathbf{x}\mathbf{A}) = \mathsf{F}$.]

c) Show that $x = x$ is a theorem of T not provable without identity axioms. [Let $f(\mathbf{A}) = \mathsf{F}$ for **A** atomic; $f(\neg\mathbf{A}) = H_{\neg}(f(\mathbf{A}))$; $f(\mathbf{A} \vee \mathbf{B}) = H_{\vee}(f(\mathbf{A}), f(\mathbf{B}))$; $f(\exists\mathbf{x}\mathbf{A}) = f(\mathbf{A})$. To treat substitution axioms, prove that $f(\mathbf{A}_\mathbf{x}[\mathbf{a}]) = f(\mathbf{A})$.]

d) Show that $x = y \rightarrow x = z \rightarrow x = x \rightarrow y = z$ is a theorem of T not provable without equality axioms. [Obtain L' from $L(T)$ by adding constants $\mathbf{e}_1$, $\mathbf{e}_2$, $\mathbf{e}_3$. For **A** a closed formula of L', define $f(\mathbf{A})$ so that

$$f(\mathbf{e}_i = \mathbf{e}_j) = \mathsf{T} \quad \text{iff} \quad i \leqslant j;$$
$$f(\neg\mathbf{A}) = H_{\neg}(f(\mathbf{A}));$$
$$f(\mathbf{A} \vee \mathbf{B}) = H_{\vee}(f(\mathbf{A}), f(\mathbf{B}));$$
$$f(\exists\mathbf{x}\mathbf{A}) = \mathsf{T} \quad \text{iff} \quad f(\mathbf{A}_\mathbf{x}[\mathbf{e}_i]) = \mathsf{T} \text{ for some } i.$$

Show that if **A** is provable in T without equality axioms, then $f(\mathbf{A}') = \mathsf{T}$ for every formula obtained from **A** by replacing each variable by some $\mathbf{e}_i$ at all its free occurrences.]

e) Show that $x = x \vee (x \neq x \vee x = x)$ is a theorem of T not provable without the expansion rule. [Let $f(\mathbf{A}) = \mathbf{T}$ for $\mathbf{A}$ atomic;

$$f(\neg\mathbf{A}) = H_\neg(f(\mathbf{A}));$$
$$f(\mathbf{A} \vee \mathbf{B}) = H_\leftrightarrow(f(\mathbf{A}), H_\neg(f(\mathbf{B})));$$
$$f(\exists\mathbf{x}\mathbf{A}) = f(\mathbf{A}).]$$

f) Show that $\neg\neg(x = x)$ is a theorem of T not provable without the contraction rule. [Let $f(\mathbf{A}) = \mathbf{T}$ for $\mathbf{A}$ atomic; $f(\neg\mathbf{A}) = f(\exists\mathbf{x}\mathbf{A}) = \mathbf{F}$; $f(\mathbf{A} \vee \mathbf{B}) = \mathbf{T}$. To prove $\neg\neg(x = x)$ in T, obtain $x = x \vee \neg\neg(x = x)$ as a conclusion of the cut rule, and then use the cut rule to prove $\neg\neg(x = x) \vee \neg\neg(x = x)$.]

g) Show that $\neg(x \neq x \vee x \neq x)$ is a theorem of T not provable without the associative rule. [Let f be a mapping from the set of formulas to the set of integers such that

$$f(\mathbf{A}) = 0 \qquad \text{for } \mathbf{A} \text{ atomic};$$
$$f(\neg\mathbf{A}) = 1 - f(\mathbf{A});$$
$$f(\mathbf{A} \vee \mathbf{B}) = f(\mathbf{A}) \cdot f(\mathbf{B}) \cdot (1 - f(\mathbf{A}) - f(\mathbf{B}));$$
$$f(\exists\mathbf{x}\mathbf{A}) = f(\mathbf{A}).$$

Show that if $\mathbf{A}$ is provable without the associative rule, then $f(\mathbf{A}) = 0$.]

h) Show that $\neg\neg(x = x)$ is a theorem of T not provable without the cut rule. [Let $f(\mathbf{A}) = \mathbf{T}$ for $\mathbf{A}$ atomic; $f(\neg\mathbf{A}) = \mathbf{T}$ if $f(\mathbf{A}) = \mathbf{F}$ or $\mathbf{A}$ is atomic, $f(\neg\mathbf{A}) = \mathbf{F}$ otherwise; $f(\mathbf{A} \vee \mathbf{B}) = H_\vee(f(\mathbf{A}), f(\mathbf{B}))$; $f(\exists\mathbf{x}\mathbf{A}) = f(\mathbf{A})$.]

i) Show that $\exists y(x \neq x) \rightarrow x \neq x$ is a theorem of T not provable without the $\exists$-introduction rule. [Let $f(\mathbf{A}) = \mathbf{T}$ for $\mathbf{A}$ atomic; $f(\neg\mathbf{A}) = H_\neg(f(\mathbf{A}))$; $f(\mathbf{A} \vee \mathbf{B}) = H_\vee(f(\mathbf{A}), f(\mathbf{B}))$; $f(\exists\mathbf{x}\mathbf{B}) = \mathbf{T}$.]

CHAPTER 3

THEOREMS IN FIRST-ORDER THEORIES

3.1 THE TAUTOLOGY THEOREM

We shall suppose throughout this chapter that a theory T is fixed, and we shall examine some of the theorems which can be proved in T.

We shall show in the next chapter that if $\mathbf{B}$ is a logical consequence of $\mathbf{A}_1, \ldots, \mathbf{A}_n$, and if $\mathbf{A}_1, \ldots, \mathbf{A}_n$ are theorems, then $\mathbf{B}$ is a theorem. In this section, we prove a particular case of this result. Roughly, this is the case in which $\mathbf{B}$ can be seen to be a consequence of $\mathbf{A}_1, \ldots, \mathbf{A}_n$ by utilizing only the rules for computing the truth values of $\neg\mathbf{C}$ and $\mathbf{C} \vee \mathbf{D}$ from the truth values of $\mathbf{C}$ and $\mathbf{D}$.

A formula is *elementary* if it is either an atomic formula or an instantiation. A *truth valuation* for T is a mapping from the set of elementary formulas in T to the set of truth values.

Let V be a truth valuation for T. We shall define a truth value $V(\mathbf{A})$ for every formula $\mathbf{A}$ of T by induction on the length of $\mathbf{A}$. If $\mathbf{A}$ is elementary, then $V(\mathbf{A})$ is already defined. If $\mathbf{A}$ is $\neg\mathbf{B}$, then $V(\mathbf{A}) = H_\neg(V(\mathbf{B}))$. If $\mathbf{A}$ is $\mathbf{B} \vee \mathbf{C}$, then $V(\mathbf{A}) = H_\vee(V(\mathbf{B}), V(\mathbf{C}))$. From this definition and the definitions of $\rightarrow$, &, and $\leftrightarrow$, we see that $V(\mathbf{B} \rightarrow \mathbf{C}) = H_\rightarrow(V(\mathbf{B}), V(\mathbf{C}))$ and similarly for & and $\leftrightarrow$. Moreover,

$$V(\mathbf{A}_1 \vee \cdots \vee \mathbf{A}_n) = \mathbf{T} \quad \text{iff} \quad V(\mathbf{A}_i) = \mathbf{T} \text{ for at least one } i,$$

and

$$V(\mathbf{A}_1 \,\&\, \cdots \,\&\, \mathbf{A}_n) = \mathbf{T} \quad \text{iff} \quad V(\mathbf{A}_i) = \mathbf{T} \text{ for all } i.$$

We say that $\mathbf{B}$ is a *tautological consequence* of $\mathbf{A}_1, \ldots, \mathbf{A}_n$ if $V(\mathbf{B}) = \mathbf{T}$ for every truth valuation V such that $V(\mathbf{A}_1) = \cdots = V(\mathbf{A}_n) = \mathbf{T}$. A formula $\mathbf{A}$ is a *tautology* if it is a tautological consequence of the empty sequence of formulas, i.e., if $V(\mathbf{A}) = \mathbf{T}$ for every truth valuation V. It is easily seen that $\mathbf{B}$ is a tautological consequence of $\mathbf{A}_1, \ldots, \mathbf{A}_n$ iff $\mathbf{A}_1 \rightarrow \cdots \rightarrow \mathbf{A}_n \rightarrow \mathbf{B}$ is a tautology.

We shall now show that, given $\mathbf{A}_1, \ldots, \mathbf{A}_n, \mathbf{B}$, we can determine in a finite number of steps whether or not $\mathbf{B}$ is a tautological consequence of $\mathbf{A}_1, \ldots, \mathbf{A}_n$. By the last statement of the previous paragraph, it will suffice to show that, given $\mathbf{A}$, we can determine whether or not $\mathbf{A}$ is a tautology. We shall show how to determine whether $\mathbf{A}_1 \vee \cdots \vee \mathbf{A}_n$ is a tautology, using induction on the sum of the lengths of the $\mathbf{A}_i$; the above result is then obtained by taking $n = 1$.

First suppose that each $\mathbf{A}_i$ is either elementary or the negation of an elementary formula. We claim that $\mathbf{A}_1 \vee \cdots \vee \mathbf{A}_n$ is a tautology iff some $\mathbf{A}_j$ is the negation of some $\mathbf{A}_i$. If this condition holds, then for every truth valuation V, either $V(\mathbf{A}_i) = \mathsf{T}$ or $V(\mathbf{A}_j) = \mathsf{T}$; so $V(\mathbf{A}_1 \vee \cdots \vee \mathbf{A}_n) = \mathsf{T}$. Suppose that the condition does not hold. Define a truth valuation V by letting $V(\mathbf{A}) = \mathsf{T}$ iff $\neg\mathbf{A}$ is an $\mathbf{A}_i$. It is easy to see that $V(\mathbf{A}_i) = \mathsf{F}$ for all i; so $V(\mathbf{A}_1 \vee \cdots \vee \mathbf{A}_n) = \mathsf{F}$.

Suppose that some $\mathbf{A}_i$ is neither elementary nor the negation of an elementary formula. Since

$$V(\mathbf{A}_1 \vee \cdots \vee \mathbf{A}_n) = V(\mathbf{A}_i \vee \cdots \vee \mathbf{A}_n \vee \mathbf{A}_1 \vee \cdots \vee \mathbf{A}_{i-1})$$

for all truth valuations V, $\mathbf{A}_1 \vee \cdots \vee \mathbf{A}_n$ is a tautology iff

$$\mathbf{A}_i \vee \cdots \vee \mathbf{A}_n \vee \mathbf{A}_1 \vee \cdots \vee \mathbf{A}_{i-1}$$

is a tautology. Hence we may as well suppose that $\mathbf{A}_1$ is neither elementary nor the negation of an elementary formula. Then $\mathbf{A}_1$ is either a disjunction or a negation; and, in the latter case, $\mathbf{A}_1$ is either the negation of a negation or the negation of a disjunction.

Suppose that $\mathbf{A}_1$ is $\mathbf{B} \vee \mathbf{C}$. Since

$$V(\mathbf{A}_1 \vee \cdots \vee \mathbf{A}_n) = V(\mathbf{B} \vee \mathbf{C} \vee \mathbf{A}_2 \vee \cdots \vee \mathbf{A}_n)$$

for every truth valuation V, $\mathbf{A}_1 \vee \cdots \vee \mathbf{A}_n$ is a tautology iff

$$\mathbf{B} \vee \mathbf{C} \vee \mathbf{A}_2 \vee \cdots \vee \mathbf{A}_n$$

is a tautology. Hence it suffices to determine whether $\mathbf{B} \vee \mathbf{C} \vee \mathbf{A}_2 \vee \cdots \vee \mathbf{A}_n$ is a tautology; and this can be done by the induction hypothesis.

Suppose that $\mathbf{A}_1$ is $\neg\neg\mathbf{B}$. Then for every truth valuation V, $V(\mathbf{A}_1) = V(\mathbf{B})$ and hence $V(\mathbf{A}_1 \vee \cdots \vee \mathbf{A}_n) = V(\mathbf{B} \vee \mathbf{A}_2 \vee \cdots \vee \mathbf{A}_n)$. It follows that it suffices to determine whether $\mathbf{B} \vee \mathbf{A}_2 \vee \cdots \vee \mathbf{A}_n$ is a tautology; and this can be done by the induction hypothesis.

Suppose that $\mathbf{A}_1$ is $\neg(\mathbf{B} \vee \mathbf{C})$. Then for every truth valuation V,

$$V(\mathbf{A}_1) = \mathsf{T} \quad \text{iff} \quad V(\neg\mathbf{B}) = V(\neg\mathbf{C}) = \mathsf{T};$$

so

$$V(\mathbf{A}_1 \vee \cdots \vee \mathbf{A}_n) = \mathsf{T}$$

iff

$$V(\neg\mathbf{B} \vee \mathbf{A}_2 \vee \cdots \vee \mathbf{A}_n) = V(\neg\mathbf{C} \vee \mathbf{A}_2 \vee \cdots \vee \mathbf{A}_n) = \mathsf{T}.$$

Hence it suffices to determine whether

$$\neg\mathbf{B} \vee \mathbf{A}_2 \vee \cdots \vee \mathbf{A}_n \quad \text{and} \quad \neg\mathbf{C} \vee \mathbf{A}_2 \vee \cdots \vee \mathbf{A}_n$$

are tautologies; and this can be done by the induction hypothesis.

We now turn to the main result of this section.

Tautology Theorem (*Post*). If $\mathbf{B}$ is a tautological consequence of $\mathbf{A}_1, \ldots, \mathbf{A}_n$, and $\vdash\mathbf{A}_1, \ldots, \vdash\mathbf{A}_n$, then $\vdash\mathbf{B}$.

Corollary. Every tautology is a theorem.

Our first step is to reduce the theorem to the corollary.

Lemma 1. If $\vdash \mathbf{A} \lor \mathbf{B}$, then $\vdash \mathbf{B} \lor \mathbf{A}$.

Proof. Since $\neg\mathbf{A} \lor \mathbf{A}$ is a propositional axiom, $\vdash \mathbf{A} \lor \mathbf{B}$ and $\vdash \neg\mathbf{A} \lor \mathbf{A}$. Hence $\vdash \mathbf{B} \lor \mathbf{A}$ by the cut rule.

Detachment Rule. If $\vdash \mathbf{A}$ and $\vdash \mathbf{A} \rightarrow \mathbf{B}$, then $\vdash \mathbf{B}$.

Proof. From $\vdash \mathbf{A}$, we get $\vdash \mathbf{B} \lor \mathbf{A}$ by the expansion rule and hence $\vdash \mathbf{A} \lor \mathbf{B}$ by Lemma 1. From $\vdash \mathbf{A} \lor \mathbf{B}$ and $\vdash \mathbf{A} \rightarrow \mathbf{B}$, we get $\vdash \mathbf{B} \lor \mathbf{B}$ by the cut rule (and the definition of $\rightarrow$); so $\vdash \mathbf{B}$ by the contraction rule.

Corollary. If $\vdash \mathbf{A}_1, \ldots, \vdash \mathbf{A}_n$, and $\vdash \mathbf{A}_1 \rightarrow \cdots \rightarrow \mathbf{A}_n \rightarrow \mathbf{B}$, then $\vdash \mathbf{B}$.

Proof. By induction on n.

It is clear that the tautology theorem follows from the corollary to the tautology theorem and the corollary to the detachment rule. It thus remains to prove that every tautology is a theorem. Now $V(\mathbf{A} \lor \mathbf{A}) = V(\mathbf{A})$ for every truth valuation V; so if $\mathbf{A}$ is a tautology, then $\mathbf{A} \lor \mathbf{A}$ is a tautology. On the other hand, $\vdash \mathbf{A} \lor \mathbf{A}$ implies $\vdash \mathbf{A}$ by the contraction rule. Hence we need only show that if $\mathbf{A} \lor \mathbf{A}$ is a tautology, then $\mathbf{A} \lor \mathbf{A}$ is a theorem. This is a special case of the following result.

Lemma 2. If $n \geqslant 2$, and $\mathbf{A}_1 \lor \cdots \lor \mathbf{A}_n$ is a tautology, then $\vdash \mathbf{A}_1 \lor \cdots \lor \mathbf{A}_n$.

Our proof of Lemma 2 is by induction on the sum of the lengths of the $\mathbf{A}_i$, and parallels the method described above. As we proceed, we shall note certain results which are needed, and prove these later.

First suppose that each $\mathbf{A}_i$ is either elementary or the negation of an elementary formula. By the method described above, some $\mathbf{A}_j$ is the negation of some $\mathbf{A}_i$. Then $\vdash \mathbf{A}_j \lor \mathbf{A}_i$ by the propositional axioms. We then obtain $\vdash \mathbf{A}_1 \lor \cdots \lor \mathbf{A}_n$ by

A) If $n \geqslant 1$, $m \geqslant 1$, and $i_1, \ldots, i_m$ are among $1, \ldots, n$, and

$$\vdash \mathbf{A}_{i_1} \lor \cdots \lor \mathbf{A}_{i_m},$$

then

$$\vdash \mathbf{A}_1 \lor \cdots \lor \mathbf{A}_n.$$

Now suppose that some $\mathbf{A}_i$ is neither elementary nor the negation of an elementary formula. By (A),

$$\vdash \mathbf{A}_1 \lor \cdots \lor \mathbf{A}_n \quad \text{iff} \quad \vdash \mathbf{A}_i \lor \cdots \lor \mathbf{A}_n \lor \mathbf{A}_1 \lor \cdots \lor \mathbf{A}_{i-1}.$$

Hence, as in the method described above, we may suppose that $\mathbf{A}_1$ is neither elementary nor the negation of an elementary formula. As before, this splits into three cases.

Suppose that $\mathbf{A}_1$ is $\mathbf{B} \lor \mathbf{C}$. Then $\mathbf{B} \lor \mathbf{C} \lor \mathbf{A}_2 \lor \cdots \lor \mathbf{A}_n$ is a tautology and hence, by the induction hypothesis, a theorem. Hence $\vdash \mathbf{A}_1 \lor \cdots \lor \mathbf{A}_n$ by the associative rule.

Suppose that $\mathbf{A}_1$ is $\neg\neg\mathbf{B}$. Then $\mathbf{B} \vee \mathbf{A}_2 \vee \cdots \vee \mathbf{A}_n$ is a tautology and hence, by the induction hypothesis, a theorem. We then obtain $\vdash \mathbf{A}_1 \vee \cdots \vee \mathbf{A}_n$ by

B) If $\vdash \mathbf{A} \vee \mathbf{B}$, then $\vdash \neg\neg\mathbf{A} \vee \mathbf{B}$.

Finally, suppose that $\mathbf{A}_1$ is $\neg(\mathbf{B} \vee \mathbf{C})$. Then

$$\neg\mathbf{B} \vee \mathbf{A}_2 \vee \cdots \vee \mathbf{A}_n \qquad \text{and} \qquad \neg\mathbf{C} \vee \mathbf{A}_2 \vee \cdots \vee \mathbf{A}_n$$

are tautologies and hence, by the induction hypothesis, theorems. We then obtain $\vdash \mathbf{A}_1 \vee \cdots \vee \mathbf{A}_n$ by

C) If $\vdash \neg\mathbf{A} \vee \mathbf{C}$ and $\vdash \neg\mathbf{B} \vee \mathbf{C}$, then $\vdash \neg(\mathbf{A} \vee \mathbf{B}) \vee \mathbf{C}$.

We now prove (A) by induction on m. We first suppose that $m \geqslant 3$. Let $\mathbf{A}$ be $\mathbf{A}_1 \vee \cdots \vee \mathbf{A}_n$. From the hypothesis of (A) and the associative law,

$$\vdash (\mathbf{A}_{i_1} \vee \mathbf{A}_{i_2}) \vee \mathbf{A}_{i_3} \vee \cdots \vee \mathbf{A}_{i_m}.$$

Hence $\vdash (\mathbf{A}_{i_1} \vee \mathbf{A}_{i_2}) \vee \mathbf{A}$ by the induction hypothesis. Applying Lemma 1 and the associative law, we have $\vdash (\mathbf{A} \vee \mathbf{A}_{i_1}) \vee \mathbf{A}_{i_2}$; so $\vdash (\mathbf{A} \vee \mathbf{A}_{i_1}) \vee \mathbf{A}$ by the induction hypothesis. Again applying Lemma 1 and the associative law, we have $\vdash (\mathbf{A} \vee \mathbf{A}) \vee \mathbf{A}_{i_1}$; so by the induction hypothesis, $\vdash (\mathbf{A} \vee \mathbf{A}) \vee (\mathbf{A} \vee \mathbf{A})$. Two applications of the contraction rule now give $\vdash \mathbf{A}$.

Now suppose that $m = 1$, and write i for i_1. By the expansion rule,

$$\vdash (\mathbf{A}_{i+1} \vee \cdots \vee \mathbf{A}_n) \vee \mathbf{A}_i;$$

so by Lemma 1, $\vdash \mathbf{A}_i \vee \cdots \vee \mathbf{A}_n$. By $i - 1$ uses of the expansion rule, we obtain $\vdash \mathbf{A}_1 \vee \cdots \vee \mathbf{A}_n$.

The case $m = 2$ remains. If $i_1 = i_2$, then $\vdash \mathbf{A}_{i_1} \vee \mathbf{A}_{i_2}$ gives $\vdash \mathbf{A}_{i_1}$ by the contraction rule, and we are back in the case $m = 1$. The case $i_1 > i_2$ reduces to the case $i_1 < i_2$ by Lemma 1. Hence the result remaining to be proved is

A′) If $1 \leqslant i < j \leqslant n$ and $\vdash \mathbf{A}_i \vee \mathbf{A}_j$, then $\vdash \mathbf{A}_1 \vee \cdots \vee \mathbf{A}_n$.

We prove (A′) by induction on n. If $n = 2$, there is nothing to prove; so we suppose that $n \geqslant 3$. Letting $\mathbf{B}$ be $\mathbf{A}_3 \vee \cdots \vee \mathbf{A}_n$, the result to be proved is $\vdash \mathbf{A}_1 \vee \mathbf{A}_2 \vee \mathbf{B}$.

If $i \geqslant 2$, then $\vdash \mathbf{A}_2 \vee \mathbf{B}$ by the induction hypothesis; so $\vdash \mathbf{A}_1 \vee \mathbf{A}_2 \vee \mathbf{B}$ by the expansion rule.

If $i = 1$ and $j \geqslant 3$, then $\vdash \mathbf{A}_1 \vee \mathbf{B}$ by the induction hypothesis. By Lemma 1 and the expansion rule, $\vdash \mathbf{A}_2 \vee \mathbf{B} \vee \mathbf{A}_1$. By the associative rule and Lemma 1, $\vdash \mathbf{A}_1 \vee \mathbf{A}_2 \vee \mathbf{B}$.

If $i = 1$ and $j = 2$, then $\vdash \mathbf{A}_1 \vee \mathbf{A}_2$ by hypothesis. Hence $\vdash \mathbf{B} \vee \mathbf{A}_1 \vee \mathbf{A}_2$ by the expansion rule. Applying the associative rule and Lemma 1, we have $\vdash \mathbf{A}_2 \vee \mathbf{B} \vee \mathbf{A}_1$. Applying the associative rule and Lemma 1 again yields $\vdash \mathbf{A}_1 \vee \mathbf{A}_2 \vee \mathbf{B}$.

Now we prove (B). Since $\neg\neg\mathbf{A} \vee \neg\mathbf{A}$ is a propositional axiom, we have $\vdash\neg\mathbf{A} \vee \neg\neg\mathbf{A}$ by Lemma 1. From $\vdash\mathbf{A} \vee \mathbf{B}$ and $\vdash\neg\mathbf{A} \vee \neg\neg\mathbf{A}$ by the cut rule, we get $\vdash\mathbf{B} \vee \neg\neg\mathbf{A}$. Hence $\vdash\neg\neg\mathbf{A} \vee \mathbf{B}$ by Lemma 1.

Now we prove (C). Since $\neg(\mathbf{A} \vee \mathbf{B}) \vee \mathbf{A} \vee \mathbf{B}$ is a propositional axiom, $\vdash\mathbf{A} \vee \mathbf{B} \vee \neg(\mathbf{A} \vee \mathbf{B})$ by (A). From this and $\vdash\neg\mathbf{A} \vee \mathbf{C}$ by the cut rule, $\vdash(\mathbf{B} \vee \neg(\mathbf{A} \vee \mathbf{B})) \vee \mathbf{C}$. From this by Lemma 1, $\vdash\mathbf{C} \vee \mathbf{B} \vee \neg(\mathbf{A} \vee \mathbf{B})$; so $\vdash\mathbf{B} \vee \mathbf{C} \vee \neg(\mathbf{A} \vee \mathbf{B})$ by (A). From this and $\vdash\neg\mathbf{B} \vee \mathbf{C}$ by the cut rule, $\vdash(\mathbf{C} \vee \neg(\mathbf{A} \vee \mathbf{B})) \vee \mathbf{C}$. Applying Lemma 1, $\vdash\mathbf{C} \vee \mathbf{C} \vee \neg(\mathbf{A} \vee \mathbf{B})$; so $\vdash\neg(\mathbf{A} \vee \mathbf{B}) \vee \mathbf{C}$ by (A). This completes the proof of the tautology theorem.

When we state that a formula is a tautological consequence of other formulas or that a formula is a tautology, we generally leave the proof to the reader. As a rule, he will find an indirect proof quickest. Thus to show that $\mathbf{A} \,\&\, \mathbf{B} \rightarrow \mathbf{A}$ is a tautology, assume that there is a truth valuation V such that $V(\mathbf{A} \,\&\, \mathbf{B} \rightarrow \mathbf{A}) = \mathsf{F}$. Then $V(\mathbf{A} \,\&\, \mathbf{B}) = \mathsf{T}$ and $V(\mathbf{A}) = \mathsf{F}$. From the former, $V(\mathbf{A}) = V(\mathbf{B}) = \mathsf{T}$. Since we have two different values for $V(\mathbf{A})$, we have a contradiction. After some experience, the reader will become convinced that whenever he can see that $\mathbf{B}$ is true (or a consequence of $\mathbf{A}_1, \ldots, \mathbf{A}_n$) by using only the meaning of $\neg$, $\vee$, $\rightarrow$, $\&$, and $\leftrightarrow$, then $\mathbf{B}$ is a tautology (or a tautological consequence of $\mathbf{A}_1, \ldots, \mathbf{A}_n$).

Remark. Suppose that to each formula $\mathbf{A}$ we have associated a formula $\mathbf{A}^*$ so that $(\neg\mathbf{A})^*$ is $\neg\mathbf{A}^*$ and $(\mathbf{A} \vee \mathbf{B})^*$ is $\mathbf{A}^* \vee \mathbf{B}^*$. If $\mathbf{B}$ is a tautological consequence of $\mathbf{A}_1, \ldots, \mathbf{A}_n$, then $\mathbf{B}^*$ is a tautological consequence of $\mathbf{A}_1^*, \ldots, \mathbf{A}_n^*$. To see this, suppose that V is a truth valuation. Define a new truth valuation V' by $V'(\mathbf{A}) = V(\mathbf{A}^*)$ for $\mathbf{A}$ elementary. We then readily verify that $V'(\mathbf{A}) = V(\mathbf{A}^*)$ for all $\mathbf{A}$. Hence if

$$V(\mathbf{A}_1^*) = \cdots = V(\mathbf{A}_n^*) = \mathsf{T},$$

then

$$V'(\mathbf{A}_1) = \cdots = V'(\mathbf{A}_n) = \mathsf{T};$$

so $V'(\mathbf{B}) = \mathsf{T}$; so $V(\mathbf{B}^*) = \mathsf{T}$.

We list some frequently used cases of the tautology theorem (other than the detachment rule):

i) If $\vdash\mathbf{A} \leftrightarrow \mathbf{B}$, then $\vdash\mathbf{A}$ iff $\vdash\mathbf{B}$.

ii) If $\vdash\mathbf{A} \rightarrow \mathbf{B}$ and $\vdash\mathbf{B} \rightarrow \mathbf{C}$, then $\vdash\mathbf{A} \rightarrow \mathbf{C}$.

iii) If $\vdash\mathbf{A} \leftrightarrow \mathbf{B}$ and $\vdash\mathbf{B} \leftrightarrow \mathbf{C}$, then $\vdash\mathbf{A} \leftrightarrow \mathbf{C}$.

iv) $\vdash\mathbf{A} \,\&\, \mathbf{B}$ iff $\vdash\mathbf{A}$ and $\vdash\mathbf{B}$.

v) $\vdash\mathbf{A} \leftrightarrow \mathbf{B}$ iff $\vdash\mathbf{A} \rightarrow \mathbf{B}$ and $\vdash\mathbf{B} \rightarrow \mathbf{A}$.

vi) $\vdash\mathbf{A} \rightarrow \mathbf{B}$ iff $\vdash\neg\mathbf{B} \rightarrow \neg\mathbf{A}$.

In addition, the tautology theorem can be used to replace applications of the propositional axioms and the expansion, contraction, associative, and cut rules. Putting it another way, we can define the theorems of T by the generalized inductive definition:

i) every substitution axiom, identity axiom, equality axiom, and nonlogical axiom is a theorem;

ii) if $\mathbf{A}_1, \ldots, \mathbf{A}_n$ ($n \geqslant 0$) are theorems, and $\mathbf{B}$ is a tautological consequence of $\mathbf{A}_1, \ldots, \mathbf{A}_n$, then $\mathbf{B}$ is a theorem;

iii) if $\mathbf{A}$ is a theorem and $\mathbf{B}$ can be inferred from $\mathbf{A}$ by the $\exists$-introduction rule, then $\mathbf{B}$ is a theorem.

This gives rise to a method of proof by induction, which we also call *induction on theorems*. In this method, we prove that every theorem has a property P by proving:

i) every substitution axiom, identity axiom, equality axiom, and nonlogical axiom has property P;

ii) if $\mathbf{A}_1, \ldots, \mathbf{A}_n$ have property P, and $\mathbf{B}$ is a tautological consequence of $\mathbf{A}_1, \ldots, \mathbf{A}_n$, then $\mathbf{B}$ has property P;

iii) if $\mathbf{A}$ has property P and $\mathbf{B}$ can be inferred from $\mathbf{A}$ by the $\exists$-introduction rule, then $\mathbf{B}$ has property P.

3.2 RESULTS ON QUANTIFIERS

We now derive some rules for operating with quantifiers.

$\forall$-Introduction Rule. If $\vdash \mathbf{A} \rightarrow \mathbf{B}$ and $\mathbf{x}$ is not free in $\mathbf{A}$, then $\vdash \mathbf{A} \rightarrow \forall \mathbf{x}\mathbf{B}$.

Proof. From $\vdash \mathbf{A} \rightarrow \mathbf{B}$ by the tautology theorem, $\vdash \neg\mathbf{B} \rightarrow \neg\mathbf{A}$. Then $\vdash \exists\mathbf{x}\neg\mathbf{B} \rightarrow \neg\mathbf{A}$ by the $\exists$-introduction rule; so $\vdash \mathbf{A} \rightarrow \neg\exists\mathbf{x}\neg\mathbf{B}$ by the tautology theorem. By the definition of $\forall$, this is $\vdash \mathbf{A} \rightarrow \forall\mathbf{x}\mathbf{B}$.

Generalization Rule. If $\vdash \mathbf{A}$, then $\vdash \forall\mathbf{x}\mathbf{A}$.

Proof. From $\vdash \mathbf{A}$ by the tautology theorem, $\vdash \neg\forall\mathbf{x}\mathbf{A} \rightarrow \mathbf{A}$. Then by the $\forall$-introduction rule, $\vdash \neg\forall\mathbf{x}\mathbf{A} \rightarrow \forall\mathbf{x}\mathbf{A}$; so $\vdash \forall\mathbf{x}\mathbf{A}$ by the tautology theorem.

We say that $\mathbf{A}'$ is an *instance* of $\mathbf{A}$ if $\mathbf{A}'$ is of the form $\mathbf{A}_{\mathbf{x}_1,\ldots,\mathbf{x}_n}[\mathbf{a}_1, \ldots, \mathbf{a}_n]$.

Substitution Rule. If $\vdash \mathbf{A}$ and $\mathbf{A}'$ is an instance of $\mathbf{A}$, then $\vdash \mathbf{A}'$.

Proof. First suppose that $\mathbf{A}'$ is $\mathbf{A}_\mathbf{x}[\mathbf{a}]$. From $\vdash \mathbf{A}$ by the generalization rule, $\vdash \forall\mathbf{x}\mathbf{A}$, that is, $\vdash \neg\exists\mathbf{x}\neg\mathbf{A}$. By the substitution axioms, $\vdash \neg\mathbf{A}_\mathbf{x}[\mathbf{a}] \rightarrow \exists\mathbf{x}\neg\mathbf{A}$. Hence $\vdash \mathbf{A}_\mathbf{x}[\mathbf{a}]$ by the tautology theorem.

Now suppose that $\mathbf{A}'$ is $\mathbf{A}_{\mathbf{x}_1,\ldots,\mathbf{x}_n}[\mathbf{a}_1, \ldots, \mathbf{a}_n]$. Let $\mathbf{y}_1, \ldots, \mathbf{y}_n$ be n new variables (i.e., variables which do not appear in $\mathbf{A}$ or $\mathbf{A}'$). Using the first part of the proof, we find successively that

$$\vdash \mathbf{A}_{\mathbf{x}_1}[\mathbf{y}_1],\ \vdash \mathbf{A}_{\mathbf{x}_1,\mathbf{x}_2}[\mathbf{y}_1, \mathbf{y}_2],\ \ldots,\ \vdash \mathbf{A}_{\mathbf{x}_1,\ldots,\mathbf{x}_n}[\mathbf{y}_1, \ldots, \mathbf{y}_n].$$

Starting with the last of these and again using the first part of the proof, we find successively

$$\vdash \mathbf{A}_{\mathbf{x}_1,\ldots,\mathbf{x}_n}[\mathbf{a}_1, \mathbf{y}_2, \ldots, \mathbf{y}_n],$$
$$\vdash \mathbf{A}_{\mathbf{x}_1,\ldots,\mathbf{x}_n}[\mathbf{a}_1, \mathbf{a}_2, \ldots, \mathbf{y}_n],$$
$$\vdots$$
$$\vdash \mathbf{A}_{\mathbf{x}_1,\ldots,\mathbf{x}_n}[\mathbf{a}_1, \mathbf{a}_2, \ldots, \mathbf{a}_n].$$

(The reader will note that the use of the $\mathbf{y}_i$ in the above proof is really necessary. Starting from **A**, we could obtain $\mathbf{A}_{\mathbf{x}_1}[\mathbf{a}_1]$; but we could not then obtain $\mathbf{A}_{\mathbf{x}_1,\mathbf{x}_2}[\mathbf{a}_1, \mathbf{a}_2]$ if $\mathbf{x}_2$ happened to occur in $\mathbf{a}_1$.)

Substitution Theorem.

a) $\vdash \mathbf{A}_{\mathbf{x}_1,\ldots,\mathbf{x}_n}[\mathbf{a}_1, \ldots, \mathbf{a}_n] \rightarrow \exists \mathbf{x}_1 \ldots \exists \mathbf{x}_n \mathbf{A}$.

b) $\vdash \forall \mathbf{x}_1 \ldots \forall \mathbf{x}_n \mathbf{A} \rightarrow \mathbf{A}_{\mathbf{x}_1,\ldots,\mathbf{x}_n}[\mathbf{a}_1, \ldots, \mathbf{a}_n]$.

Proof. By the substitution axioms,

$$\vdash \mathbf{C} \rightarrow \exists \mathbf{x} \mathbf{C} \tag{1}$$

and $\vdash \neg \mathbf{C} \rightarrow \exists \mathbf{x} \neg \mathbf{C}$. From the latter by the tautology theorem and the definition of $\forall$,

$$\vdash \forall \mathbf{x} \mathbf{C} \rightarrow \mathbf{C}. \tag{2}$$

From (1),

$$\vdash \exists \mathbf{x}_{i+1} \ldots \exists \mathbf{x}_n \mathbf{A} \rightarrow \exists \mathbf{x}_i \exists \mathbf{x}_{i+1} \ldots \exists \mathbf{x}_n \mathbf{A}$$

for $i = 1, \ldots, n$. From these by the tautology theorem, we get $\vdash \mathbf{A} \rightarrow \exists \mathbf{x}_1 \ldots \exists \mathbf{x}_n \mathbf{A}$; and from this, we get (a) by the substitution rule. Similarly, we can use (2) to obtain $\vdash \forall \mathbf{x}_1 \ldots \forall \mathbf{x}_n \mathbf{A} \rightarrow \mathbf{A}$, and then get (b) by the substitution rule.

Distribution Rule. If $\vdash \mathbf{A} \rightarrow \mathbf{B}$, then $\vdash \exists \mathbf{x} \mathbf{A} \rightarrow \exists \mathbf{x} \mathbf{B}$ and $\vdash \forall \mathbf{x} \mathbf{A} \rightarrow \forall \mathbf{x} \mathbf{B}$.

Proof. By the substitution theorem, $\vdash \mathbf{B} \rightarrow \exists \mathbf{x} \mathbf{B}$. From this and $\vdash \mathbf{A} \rightarrow \mathbf{B}$, $\vdash \mathbf{A} \rightarrow \exists \mathbf{x} \mathbf{B}$ by the tautology theorem; so $\vdash \exists \mathbf{x} \mathbf{A} \rightarrow \exists \mathbf{x} \mathbf{B}$ by the $\exists$-introduction rule.

By the substitution theorem, $\vdash \forall \mathbf{x} \mathbf{A} \rightarrow \mathbf{A}$. From this and $\vdash \mathbf{A} \rightarrow \mathbf{B}$, $\vdash \forall \mathbf{x} \mathbf{A} \rightarrow \mathbf{B}$ by the tautology theorem; so $\vdash \forall \mathbf{x} \mathbf{A} \rightarrow \forall \mathbf{x} \mathbf{B}$ by the $\forall$-introduction rule.

Let **A** be a formula, and let $\mathbf{x}_1, \ldots, \mathbf{x}_n$ be the variables which are free in **A** arranged in alphabetical order. The formula $\forall \mathbf{x}_1 \ldots \forall \mathbf{x}_n \mathbf{A}$ is called the *closure* of **A**. It is clear that the closure of **A** is closed, and that if **A** is closed, then the closure of **A** is **A**.

Closure Theorem. If $\mathbf{A}'$ is the closure of **A**, then $\vdash \mathbf{A}'$ iff $\vdash \mathbf{A}$.

Proof. If $\vdash \mathbf{A}$, then $\vdash \mathbf{A}'$ by the generalization rule. By the substitution theorem, $\vdash \mathbf{A}' \rightarrow \mathbf{A}$; so if $\vdash \mathbf{A}'$, then $\vdash \mathbf{A}$ by the detachment rule.

Corollary. If $\mathbf{A}'$ is the closure of **A**, then **A** is valid in a structure $\mathcal{A}$ iff $\mathbf{A}'$ is valid in $\mathcal{A}$.

Proof. Suppose that **A** is valid in $\mathcal{A}$. If T has **A** as its only nonlogical axiom, then $\mathcal{A}$ is a model of T. By the closure theorem, $\vdash_T \mathbf{A}'$; so $\mathbf{A}'$ is valid in $\mathcal{A}$ by the validity theorem. The converse is proved similarly.

3.3 THE DEDUCTION THEOREM

If a mathematician wishes to prove a statement *if P, then Q*, he will generally assume P and then prove Q. We will show that there is a similar method of proving theorems in theories.

We designate the theory obtained from T by adding $\mathbf{A}_1, \ldots, \mathbf{A}_n$ as new nonlogical axioms by $T[\mathbf{A}_1, \ldots, \mathbf{A}_n]$. The analogue of the above method for theories is then the following: if we wish to prove $\mathbf{A} \to \mathbf{B}$ in T, we try to prove **B** in $T[\mathbf{A}]$. We shall show that if we succeed, and if **A** is closed, then $\mathbf{A} \to \mathbf{B}$ is indeed a theorem of T.

Deduction Theorem. Let **A** be a closed formula in T. For every formula **B** of T, $\vdash_T \mathbf{A} \to \mathbf{B}$ iff **B** is a theorem of $T[\mathbf{A}]$.

Proof. If $\vdash_T \mathbf{A} \to \mathbf{B}$, then **A** and $\mathbf{A} \to \mathbf{B}$ are theorems of $T[\mathbf{A}]$; so **B** is a theorem of $T[\mathbf{A}]$ by the detachment rule. We now prove that $\vdash_T \mathbf{A} \to \mathbf{B}$ for every theorem **B** of $T[\mathbf{A}]$, using induction on theorems (in the form described in §3.1).

Suppose that **B** is an axiom of $T[\mathbf{A}]$. If **B** is **A**, then $\mathbf{A} \to \mathbf{B}$ is a tautology and hence a theorem of T. Otherwise, **B** is an axiom of T; so $\vdash_T \mathbf{B}$; so $\vdash_T \mathbf{A} \to \mathbf{B}$ by the tautology theorem.

Suppose that **B** is a tautological consequence of $\mathbf{C}_1, \ldots, \mathbf{C}_n$. Then $\mathbf{A} \to \mathbf{B}$ is a tautological consequence of $\mathbf{A} \to \mathbf{C}_1, \ldots, \mathbf{A} \to \mathbf{C}_n$. By induction hypothesis, $\vdash_T \mathbf{A} \to \mathbf{C}_1, \ldots, \vdash_T \mathbf{A} \to \mathbf{C}_n$; so $\vdash_T \mathbf{A} \to \mathbf{B}$ by the tautology theorem.

Suppose that **B** is inferred by the $\exists$-introduction rule; say that **B** is $\exists\mathbf{x}\mathbf{C} \to \mathbf{D}$ and is inferred from $\mathbf{C} \to \mathbf{D}$, where **x** is not free in **D**. By induction hypothesis, $\vdash_T \mathbf{A} \to \mathbf{C} \to \mathbf{D}$; so by the tautology theorem, $\vdash_T \mathbf{C} \to \mathbf{A} \to \mathbf{D}$. Since **x** is not free in **A** or **D**, $\vdash_T \exists\mathbf{x}\mathbf{C} \to \mathbf{A} \to \mathbf{D}$ by the $\exists$-introduction rule; so $\vdash_T \mathbf{A} \to \mathbf{B}$ by the tautology theorem.

Corollary. Let $\mathbf{A}_1, \ldots, \mathbf{A}_n$ be closed formulas in T. For every formula **B** in T, $\vdash_T \mathbf{A}_1 \to \cdots \to \mathbf{A}_n \to \mathbf{B}$ iff **B** is a theorem of $T[\mathbf{A}_1, \ldots, \mathbf{A}_n]$.

Proof. By induction on n.

The deduction theorem fails if **A** is not required to be closed. Thus if we add $x = 0$ as an axiom to N, we can prove $y = 0$ by the substitution rule. But $x = 0 \to y = 0$ is not a theorem of N, since it is not valid in the model $\mathfrak{N}$. To circumvent this difficulty, we prove a theorem which enables us to replace $\mathbf{A} \to \mathbf{B}$ by an implication with a closed left member.

Theorem on Constants. Let T' be obtained from T by adding new constants (but no new nonlogical axioms). For every formula **A** of T and every sequence $\mathbf{e}_1 \ldots, \mathbf{e}_n$ of distinct new constants, $\vdash_T \mathbf{A}$ iff $\vdash_{T'} \mathbf{A}[\mathbf{e}_1, \ldots, \mathbf{e}_n]$.

Proof. If $\vdash_T \mathbf{A}$, then $\vdash_{T'} \mathbf{A}$; so $\vdash_{T'} \mathbf{A}[\mathbf{e}_1, \ldots, \mathbf{e}_n]$ by the substitution rule. Now suppose that we have a proof in T' of $\mathbf{A}[\mathbf{e}_1, \ldots, \mathbf{e}_n]$. We choose n variables $\mathbf{y}_1, \ldots, \mathbf{y}_n$ not occurring in the proof or in $\mathbf{A}$, and replace $\mathbf{e}_1, \ldots, \mathbf{e}_n$ throughout the proof by $\mathbf{y}_1, \ldots, \mathbf{y}_n$. This does not affect the nonlogical axioms (which do not contain any of the new constants). It is easy to see that every other axiom becomes an axiom of the same type, and that each application of a rule becomes a new application of the same rule. It follows that we obtain a proof in T of $\mathbf{A}[\mathbf{y}_1, \ldots, \mathbf{y}_n]$. Thus

$$\vdash_T \mathbf{A}[\mathbf{y}_1, \ldots, \mathbf{y}_n];$$

so $\vdash_T \mathbf{A}$ by the substitution rule.

Now let us return to the problem of proving $\mathbf{A} \to \mathbf{B}$ in T. Let $\mathbf{x}_1, \ldots, \mathbf{x}_n$ be the variables free in $\mathbf{A}$. Form T' from T by adding n new constants $\mathbf{e}_1, \ldots, \mathbf{e}_n$. By the theorem on constants,

$$\vdash_T \mathbf{A} \to \mathbf{B} \quad \text{iff} \quad \vdash_{T'} \mathbf{A}[\mathbf{e}_1, \ldots, \mathbf{e}_n] \to \mathbf{B}[\mathbf{e}_1, \ldots, \mathbf{e}_n].$$

Since $\mathbf{A}[\mathbf{e}_1, \ldots, \mathbf{e}_n]$ is closed, the deduction theorem shows that it suffices to prove $\mathbf{B}[\mathbf{e}_1, \ldots, \mathbf{e}_n]$ in $T'[\mathbf{A}[\mathbf{e}_1, \ldots, \mathbf{e}_n]]$.

3.4 THE EQUIVALENCE AND EQUALITY THEOREMS

The results of this section may be roughly stated as follows: equivalent formulas and equal terms may be substituted for one another.

Equivalence Theorem. Let $\mathbf{A}'$ be obtained from $\mathbf{A}$ by replacing some occurrences of $\mathbf{B}_1, \ldots, \mathbf{B}_n$ by $\mathbf{B}'_1, \ldots, \mathbf{B}'_n$ respectively. If

$$\vdash \mathbf{B}_1 \leftrightarrow \mathbf{B}'_1, \ldots, \vdash \mathbf{B}_n \leftrightarrow \mathbf{B}'_n,$$

then

$$\vdash \mathbf{A} \leftrightarrow \mathbf{A}'.$$

Proof. Consider first a special case: there is only one such occurrence, and it is all of $\mathbf{A}$. Then for some i, $\mathbf{A}$ is $\mathbf{B}_i$ and $\mathbf{A}'$ is $\mathbf{B}'_i$; so $\vdash \mathbf{A} \leftrightarrow \mathbf{A}'$ by hypothesis.

We now prove the theorem by induction on the length of $\mathbf{A}$. If $\mathbf{A}$ is atomic, then any occurrence of a formula in $\mathbf{A}$ is all of $\mathbf{A}$ by the occurrence theorem. Hence if we are not in the special case, then no occurrences are replaced, and $\mathbf{A}'$ is $\mathbf{A}$. Then $\vdash \mathbf{A} \leftrightarrow \mathbf{A}'$ by the tautology theorem.

Suppose that $\mathbf{A}$ is $\neg \mathbf{C}$. By the occurrence theorem, any occurrence of a formula in $\mathbf{A}$ is either all of $\mathbf{A}$ or is entirely contained in $\mathbf{C}$. It follows that, if we are not in the special case, then $\mathbf{A}'$ is $\neg \mathbf{C}'$, where $\mathbf{C}'$ results from $\mathbf{C}$ by replacements of the type described in the theorem. By induction hypothesis, $\vdash \mathbf{C} \leftrightarrow \mathbf{C}'$; so $\vdash \mathbf{A} \leftrightarrow \mathbf{A}'$ by the tautology theorem.

If $\mathbf{A}$ is $\mathbf{C} \vee \mathbf{D}$, the proof is similar. Now suppose that $\mathbf{A}$ is $\exists \mathbf{x} \mathbf{C}$. If we are not in the special case, then $\mathbf{A}'$ is $\exists \mathbf{x} \mathbf{C}'$, where $\vdash \mathbf{C} \leftrightarrow \mathbf{C}'$ by the induction hypothesis. Then $\vdash \mathbf{C} \to \mathbf{C}'$ and $\vdash \mathbf{C}' \to \mathbf{C}$ by the tautology theorem. By the distribution rule, $\vdash \mathbf{A} \to \mathbf{A}'$ and $\vdash \mathbf{A}' \to \mathbf{A}$; so $\vdash \mathbf{A} \leftrightarrow \mathbf{A}'$ by the tautology theorem.

We have remarked that a formula does not change its meaning if a bound variable is changed to another variable. As an application of the equivalence theorem, we now prove a syntactical version of this remark.

We say that **A′** is a *variant* of **A** if **A′** can be obtained from **A** by a sequence of replacements of the following type: replace a part $\exists\mathbf{x}\mathbf{B}$ by $\exists\mathbf{y}\mathbf{B}_\mathbf{x}[\mathbf{y}]$, where **y** is a variable not free in **B**.

Variant Theorem. If **A′** is a variant of **A**, then $\vdash\mathbf{A} \leftrightarrow \mathbf{A}'$.

Proof. In view of the equivalence theorem and the tautology theorem, we need only show that in the notation of the preceding definition, $\vdash\exists\mathbf{x}\mathbf{B} \leftrightarrow \exists\mathbf{y}\mathbf{B}_\mathbf{x}[\mathbf{y}]$. Let **B′** be $\mathbf{B}_\mathbf{x}[\mathbf{y}]$. By the substitution theorem, $\vdash\mathbf{B}' \rightarrow \exists\mathbf{x}\mathbf{B}$; so by the ∃-introduction rule,

$$\vdash\exists\mathbf{y}\mathbf{B}' \rightarrow \exists\mathbf{x}\mathbf{B}. \tag{1}$$

Using the fact that **y** is not free in **B**, we see that $\mathbf{B}'_\mathbf{y}[\mathbf{x}]$ is **B**. Hence by the substitution theorem, $\vdash\mathbf{B} \rightarrow \exists\mathbf{y}\mathbf{B}'$; so by the ∃-introduction rule, $\vdash\exists\mathbf{x}\mathbf{B} \rightarrow \exists\mathbf{y}\mathbf{B}'$. From this and (1), we get $\vdash\exists\mathbf{x}\mathbf{B} \leftrightarrow \exists\mathbf{y}\mathbf{B}'$ by the tautology theorem.

Variants are useful when we have difficulty because a term **a** is not substitutible for **x** in **A**. We can then find a variant **A′** of **A** in which none of the variables of **a** is bound. Then **a** will be substitutible for **x** in **A′**, and we can replace **A** by **A′**.

Symmetry Theorem. $\vdash\mathbf{a} = \mathbf{b} \leftrightarrow \mathbf{b} = \mathbf{a}$.

Proof. Let **x** and **y** be distinct. By the equality axioms,

$$\vdash\mathbf{x} = \mathbf{y} \rightarrow \mathbf{x} = \mathbf{x} \rightarrow \mathbf{x} = \mathbf{x} \rightarrow \mathbf{y} = \mathbf{x}.$$

Hence by the identity axioms and the tautology theorem, $\vdash\mathbf{x} = \mathbf{y} \rightarrow \mathbf{y} = \mathbf{x}$. By the substitution rule, $\vdash\mathbf{a} = \mathbf{b} \rightarrow \mathbf{b} = \mathbf{a}$ and $\vdash\mathbf{b} = \mathbf{a} \rightarrow \mathbf{a} = \mathbf{b}$; so $\vdash\mathbf{a} = \mathbf{b} \leftrightarrow \mathbf{b} = \mathbf{a}$ by the tautology theorem.

Equality Theorem. Let **b′** be obtained from **b** by replacing some occurrences of $\mathbf{a}_1, \ldots, \mathbf{a}_n$ not immediately following ∃ or ∀ by $\mathbf{a}'_1, \ldots, \mathbf{a}'_n$ respectively, and let **A′** be obtained from **A** by the same type of replacements. If $\vdash\mathbf{a}_1 = \mathbf{a}'_1, \ldots, \vdash\mathbf{a}_n = \mathbf{a}'_n$, then $\vdash\mathbf{b} = \mathbf{b}'$ and $\vdash\mathbf{A} \leftrightarrow \mathbf{A}'$.

Proof. We first prove $\vdash\mathbf{b} = \mathbf{b}'$. If the only occurrence replaced is all of **b**, then for some i, **b** is $\mathbf{a}_i$ and **b′** is $\mathbf{a}'_i$; so $\vdash\mathbf{b} = \mathbf{b}'$ by hypothesis. We now exclude this special case and proceed by induction on the length of **b**. If **b** is a variable, then no occurrences can be replaced (since the special case is excluded). Hence **b′** is **b**, and $\vdash\mathbf{b} = \mathbf{b}'$ by the identity axioms and the substitution rule. Now suppose that **b** is $\mathbf{f}\mathbf{c}_1 \ldots \mathbf{c}_k$. Then **b′** is $\mathbf{f}\mathbf{c}'_1 \ldots \mathbf{c}'_k$, where $\vdash\mathbf{c}_i = \mathbf{c}'_i$ for $i = 1, \ldots, k$ by induction hypothesis. By the equality axioms and the substitution rule,

$$\vdash\mathbf{c}_1 = \mathbf{c}'_1 \rightarrow \cdots \rightarrow \mathbf{c}_k = \mathbf{c}'_k \rightarrow \mathbf{b} = \mathbf{b}'.$$

Hence $\vdash\mathbf{b} = \mathbf{b}'$ by the detachment rule.

We now prove that $\vdash \mathbf{A} \leftrightarrow \mathbf{A}'$ by induction on the length of $\mathbf{A}$. If $\mathbf{A}$ is an atomic formula $\mathbf{pc}_1 \ldots \mathbf{c}_k$, then $\mathbf{A}'$ is $\mathbf{pc}'_1 \ldots \mathbf{c}'_k$, where $\vdash \mathbf{c}_i = \mathbf{c}'_i$ for $i = 1, \ldots, k$ by the first part of the proof. By the symmetry theorem, $\vdash \mathbf{c}'_i = \mathbf{c}_i$ for $i = 1, \ldots, k$. By the equality axioms and the substitution rule,

$$\vdash \mathbf{c}_1 = \mathbf{c}'_1 \rightarrow \cdots \rightarrow \mathbf{c}_k = \mathbf{c}'_k \rightarrow \mathbf{A} \rightarrow \mathbf{A}'$$

and

$$\vdash \mathbf{c}'_1 = \mathbf{c}_1 \rightarrow \cdots \rightarrow \mathbf{c}'_k = \mathbf{c}_k \rightarrow \mathbf{A}' \rightarrow \mathbf{A}.$$

Hence $\vdash \mathbf{A} \leftrightarrow \mathbf{A}'$ by the tautology theorem. The remaining cases are treated as in the proof of the equivalence theorem.

Corollary 1. $\vdash \mathbf{a}_1 = \mathbf{a}'_1 \rightarrow \cdots \rightarrow \mathbf{a}_n = \mathbf{a}'_n \rightarrow \mathbf{b}[\mathbf{a}_1, \ldots, \mathbf{a}_n] = \mathbf{b}[\mathbf{a}'_1, \ldots, \mathbf{a}'_n]$.

Proof. Replace each variable occurring in an $\mathbf{a}_i$ or an $\mathbf{a}'_i$ by a new constant. Suppose that $\mathbf{a}_i$, $\mathbf{a}'_i$, and $\mathbf{b}$ become $\mathbf{c}_i$, $\mathbf{c}'_i$, and $\mathbf{d}$. The result to be proved becomes

$$\mathbf{c}_1 = \mathbf{c}'_1 \rightarrow \cdots \rightarrow \mathbf{c}_n = \mathbf{c}'_n \rightarrow \mathbf{d}[\mathbf{c}_1, \ldots, \mathbf{c}_n] = \mathbf{d}[\mathbf{c}'_1, \ldots, \mathbf{c}'_n].$$

By the results of the last section, it suffices to add the $\mathbf{c}_i = \mathbf{c}'_i$ as axioms and prove $\mathbf{d}[\mathbf{c}_1, \ldots, \mathbf{c}_n] = \mathbf{d}[\mathbf{c}'_1, \ldots, \mathbf{c}'_n]$. This can be done by the theorem.

Corollary 2. $\vdash \mathbf{a}_1 = \mathbf{a}'_1 \rightarrow \cdots \rightarrow \mathbf{a}_n = \mathbf{a}'_n \rightarrow (\mathbf{A}[\mathbf{a}_1, \ldots, \mathbf{a}_n] \leftrightarrow \mathbf{A}[\mathbf{a}'_1, \ldots, \mathbf{a}'_n])$.

Proof. Like that of Corollary 1.

In applications, both the theorem and the two corollaries will be referred to simply as the equality theorem.

Corollary 3. If $\mathbf{x}$ does not occur in $\mathbf{a}$, then

$$\vdash \mathbf{A}_\mathbf{x}[\mathbf{a}] \leftrightarrow \exists \mathbf{x}(\mathbf{x} = \mathbf{a} \;\&\; \mathbf{A}).$$

Proof. By the equality theorem $\vdash \mathbf{x} = \mathbf{a} \rightarrow (\mathbf{A} \leftrightarrow \mathbf{A}_\mathbf{x}[\mathbf{a}])$; so by the tautology theorem and the $\exists$-introduction rule,

$$\vdash \exists \mathbf{x}(\mathbf{x} = \mathbf{a} \;\&\; \mathbf{A}) \rightarrow \mathbf{A}_\mathbf{x}[\mathbf{a}]. \tag{2}$$

By the substitution axioms,

$$\vdash (\mathbf{a} = \mathbf{a} \;\&\; \mathbf{A}_\mathbf{x}[\mathbf{a}]) \rightarrow \exists \mathbf{x}(\mathbf{x} = \mathbf{a} \;\&\; \mathbf{A}). \tag{3}$$

By the identity axioms and the substitution rule,

$$\vdash \mathbf{a} = \mathbf{a}. \tag{4}$$

The corollary follows from (2), (3), and (4) by the tautology theorem.

3.5 PRENEX FORM

We shall now show that every formula is equivalent to a formula in a certain special form.

A formula is *open* if it contains no quantifiers. A formula $\mathbf{A}$ is in *prenex form* if it has the form $Q\mathbf{x}_1 \ldots Q\mathbf{x}_n\mathbf{B}$, where each $Q\mathbf{x}_i$ is either $\exists \mathbf{x}_i$ or $\forall \mathbf{x}_i$; $\mathbf{x}_1, \ldots, \mathbf{x}_n$

are distinct; and **B** is open. We then call $Q\mathbf{x}_1 \ldots Q\mathbf{x}_n$ the *prefix* and **B** the *matrix* of **A**. We allow the prefix to be empty; that is, an open formula is in prenex form.

Our definition of prenex form involves the defined symbol ∀. As in all such cases, the definition really refers to the formulas obtained by eliminating the defined symbols. However, in dealing with prenex form, it is generally better not to imagine the defined symbol ∀ eliminated.

We now introduce some operations, called *prenex operations*. These are operations which may be performed on a formula **A**, possibly containing the defined symbol ∀; the result of the operation is another such formula. The prenex operations are:

a) Replace **A** by a variant.

b) Replace a part $\neg Q\mathbf{x}\mathbf{B}$ of **A** by $Q'\mathbf{x}\neg\mathbf{B}$, where $Q'\mathbf{x}$ is $\forall\mathbf{x}$ if $Q\mathbf{x}$ is $\exists\mathbf{x}$, and $Q'\mathbf{x}$ is $\exists\mathbf{x}$ if $Q\mathbf{x}$ is $\forall\mathbf{x}$.

c) Replace a part $Q\mathbf{x}\mathbf{B} \vee \mathbf{C}$ of **A** by $Q\mathbf{x}(\mathbf{B} \vee \mathbf{C})$, provided that **x** is not free in **C**.

d) Replace a part $\mathbf{B} \vee Q\mathbf{x}\mathbf{C}$ of **A** by $Q\mathbf{x}(\mathbf{B} \vee \mathbf{C})$, provided that **x** is not free in **B**.

We shall first show that if **A′** results from **A** by a prenex operation, then $\vdash \mathbf{A} \leftrightarrow \mathbf{A}'$. For (a), this follows from the variant theorem. For (b), it suffices, in view of the equivalence theorem, to show that $\vdash \neg Q\mathbf{x}\mathbf{B} \leftrightarrow Q'\mathbf{x}\neg\mathbf{B}$. Thus we must show that

$$\neg\exists\mathbf{x}\mathbf{B} \leftrightarrow \forall\mathbf{x}\neg\mathbf{B} \qquad \text{and} \qquad \neg\forall\mathbf{x}\mathbf{B} \leftrightarrow \exists\mathbf{x}\neg\mathbf{B}$$

are theorems. If we eliminate ∀, these become

$$\neg\exists\mathbf{x}\mathbf{B} \leftrightarrow \neg\exists\mathbf{x}\neg\neg\mathbf{B} \qquad \text{and} \qquad \neg\neg\exists\mathbf{x}\neg\mathbf{B} \leftrightarrow \exists\mathbf{x}\neg\mathbf{B}.$$

Both follow from the equivalence theorem and the tautology theorem.

To treat (c), it suffices to prove that $\vdash Q\mathbf{x}\mathbf{B} \vee \mathbf{C} \leftrightarrow Q\mathbf{x}(\mathbf{B} \vee \mathbf{C})$. This will follow from the tautology theorem if we prove

$$\vdash Q\mathbf{x}\mathbf{B} \rightarrow Q\mathbf{x}(\mathbf{B} \vee \mathbf{C}), \tag{1}$$

$$\vdash \mathbf{C} \rightarrow Q\mathbf{x}(\mathbf{B} \vee \mathbf{C}), \tag{2}$$

$$\vdash Q\mathbf{x}(\mathbf{B} \vee \mathbf{C}) \rightarrow Q\mathbf{x}\mathbf{B} \vee \mathbf{C}. \tag{3}$$

We obtain (1) from the tautology $\mathbf{B} \rightarrow \mathbf{B} \vee \mathbf{C}$ by the distribution rule. If $Q\mathbf{x}$ is $\exists\mathbf{x}$, we get (2) from the substitution axiom $\mathbf{B} \vee \mathbf{C} \rightarrow \exists\mathbf{x}(\mathbf{B} \vee \mathbf{C})$ by the tautology theorem. If $Q\mathbf{x}$ is $\forall\mathbf{x}$, we get (2) from the tautology $\mathbf{C} \rightarrow \mathbf{B} \vee \mathbf{C}$ by the ∀-introduction rule. Now from the substitution axiom $\mathbf{B} \rightarrow \exists\mathbf{x}\mathbf{B}$, we get $\vdash \mathbf{B} \vee \mathbf{C} \rightarrow \exists\mathbf{x}\mathbf{B} \vee \mathbf{C}$ by the tautology theorem; so $\vdash \exists\mathbf{x}(\mathbf{B} \vee \mathbf{C}) \rightarrow \exists\mathbf{x}\mathbf{B} \vee \mathbf{C}$ by the ∃-introduction rule. This is (3) when $Q\mathbf{x}$ is $\exists\mathbf{x}$. By the substitution theorem, $\vdash \forall\mathbf{x}(\mathbf{B} \vee \mathbf{C}) \rightarrow \mathbf{B} \vee \mathbf{C}$; so by the tautology theorem and the ∀-introduction rule,

$$\vdash \forall\mathbf{x}(\mathbf{B} \vee \mathbf{C}) \,\&\, \neg\mathbf{C} \rightarrow \forall\mathbf{x}\mathbf{B}.$$

From this by the tautology theorem, $\vdash \forall\mathbf{x}(\mathbf{B} \vee \mathbf{C}) \rightarrow \forall\mathbf{x}\mathbf{B} \vee \mathbf{C}$, which is (3) when $Q\mathbf{x}$ is $\forall\mathbf{x}$.

To treat (d), it suffices to show that $\vdash \mathbf{B} \vee Q\mathbf{x}\mathbf{C} \leftrightarrow Q\mathbf{x}(\mathbf{B} \vee \mathbf{C})$ if $\mathbf{x}$ is not free in $\mathbf{B}$. By the above, $\vdash Q\mathbf{x}\mathbf{C} \vee \mathbf{B} \leftrightarrow Q\mathbf{x}(\mathbf{C} \vee \mathbf{B})$; and the desired result follows by the equivalence theorem and the tautology theorem.

We now show that every formula can be converted into a formula in prenex form by applying prenex operations. The proof is by induction on the length of $\mathbf{A}$. If $\mathbf{A}$ is atomic, it is already in prenex form. Suppose that $\mathbf{A}$ is $\neg\mathbf{B}$. By induction hypothesis, we can convert $\mathbf{B}$ into a formula $\mathbf{B}'$ in prenex form by means of prenex operations. The same operations convert $\mathbf{A}$ into $\neg\mathbf{B}'$. But clearly $\neg\mathbf{B}'$ can be converted into a formula in prenex form by successive uses of operation (b).

Now suppose that $\mathbf{A}$ is $\mathbf{B} \vee \mathbf{C}$. By induction hypothesis, we can convert $\mathbf{B}$ and $\mathbf{C}$ into formulas $\mathbf{B}'$ and $\mathbf{C}'$ in prenex form. In view of operation (a), we may further suppose that the variables in the prefix of $\mathbf{B}'$ are distinct from the variables in the prefix of $\mathbf{C}'$, and that the variables in both prefixes are distinct from the variables free in $\mathbf{B}'$ and $\mathbf{C}'$. We can then convert $\mathbf{A}$ into $\mathbf{B}' \vee \mathbf{C}'$; and by means of operations (c) and (d), we can convert $\mathbf{B}' \vee \mathbf{C}'$ into a formula in prenex form.

Finally, suppose that $\mathbf{A}$ is $Q\mathbf{x}\mathbf{B}$. We can convert $\mathbf{B}$ into a formula $\mathbf{B}'$ in prenex form; and we may suppose that the variables in the prefix of $\mathbf{B}'$ are distinct from $\mathbf{x}$. Then $\mathbf{A}$ may be converted to $Q\mathbf{x}\mathbf{B}'$, which is in prenex form.

By a *prenex form of* $\mathbf{A}$, we mean a formula in prenex form to which $\mathbf{A}$ may be converted by means of prenex operations. We have just seen that every formula has a prenex form; and that if $\mathbf{A}'$ is a prenex form of $\mathbf{A}$, then $\vdash \mathbf{A} \leftrightarrow \mathbf{A}'$. One should note that the prenex operations are independent of the theory in which we are operating.

Our method for obtaining a prenex form of $\mathbf{A}$ requires us to eliminate defined symbols other than $\forall$. We can avoid eliminating $\rightarrow$ and & by introducing the following additional prenex operations:

e) Replace a part $Q\mathbf{x}\mathbf{B} \rightarrow \mathbf{C}$ of $\mathbf{A}$ by $Q'\mathbf{x}(\mathbf{B} \rightarrow \mathbf{C})$, where $Q'\mathbf{x}$ is as in (b), provided that $\mathbf{x}$ is not free in $\mathbf{C}$.

f) Replace a part $\mathbf{B} \rightarrow Q\mathbf{x}\mathbf{C}$ of $\mathbf{A}$ by $Q\mathbf{x}(\mathbf{B} \rightarrow \mathbf{C})$, provided that $\mathbf{x}$ is not free in $\mathbf{B}$.

g) Replace a part $Q\mathbf{x}\mathbf{B}$ & $\mathbf{C}$ of $\mathbf{A}$ by $Q\mathbf{x}(\mathbf{B}$ & $\mathbf{C})$, provided that $\mathbf{x}$ is not free in $\mathbf{C}$.

h) Replace a part $\mathbf{B}$ & $Q\mathbf{x}\mathbf{C}$ of $\mathbf{A}$ by $Q\mathbf{x}(\mathbf{B}$ & $\mathbf{C})$, provided that $\mathbf{x}$ is not free in $\mathbf{B}$.

One sees as above that these operations are sufficient to convert any formula containing $\rightarrow$ and & to prenex form. To see that the formula obtained as the result of these operations is equivalent to the formula operated upon, we note that each operation, upon elimination of the defined symbol, becomes a succession of operations (b), (c), and (d). For example, suppose that we use (e) to replace $\exists\mathbf{x}\mathbf{B} \rightarrow \mathbf{C}$ by $\forall\mathbf{x}(\mathbf{B} \rightarrow \mathbf{C})$. Upon eliminating $\rightarrow$, we see that we have replaced $\neg\exists\mathbf{x}\mathbf{B} \vee \mathbf{C}$ by $\forall\mathbf{x}(\neg\mathbf{B} \vee \mathbf{C})$. This can be done by first applying (b) and then applying (c). We cannot obtain any similar rules for $\leftrightarrow$; if we eliminate $\leftrightarrow$ and apply (a) through (h), we find that we cannot restore the $\leftrightarrow$.

We conclude with an example of the steps in converting a formula of N to prenex form:

$$\exists x(x = y) \rightarrow \exists x(x = 0 \vee \neg\exists y(y < 0)),$$
$$\exists x(x = y) \rightarrow \exists z(z = 0 \vee \neg\exists w(w < 0)),$$
$$\exists x(x = y) \rightarrow \exists z(z = 0 \vee \forall w\neg(w < 0)),$$
$$\exists x(x = y) \rightarrow \exists z\,\forall w(z = 0 \vee \neg(w < 0)),$$
$$\forall x\,\exists z\,\forall w(x = y \rightarrow z = 0 \vee \neg(w < 0)).$$

PROBLEMS

1. Show that if **A** is a formula which is provable without use of substitution axioms, nonlogical axioms, identity axioms, equality axioms, or the $\exists$-introduction rule, then **A** is a tautology.

2. Let T be a theory with no nonlogical axioms. For every formula **A** of T, let **A*** be the formula obtained from **A** by omitting all quantifiers and replacing all terms by a new constant **e**. Show that if $\vdash_T$ **A**, then **A*** is a tautological consequence of formulas of the form **a** = **a**. Conclude that there is no formula **A** such that $\vdash_T$ **A** and $\vdash_T \neg$**A**.

3. a) $\vdash \forall\mathbf{x}(\mathbf{A} \rightarrow \mathbf{B}) \rightarrow \exists\mathbf{x}\mathbf{A} \rightarrow \exists\mathbf{x}\mathbf{B}$.

b) $\vdash \forall\mathbf{x}(\mathbf{A} \rightarrow \mathbf{B}) \rightarrow \forall\mathbf{x}\mathbf{A} \rightarrow \forall\mathbf{x}\mathbf{B}$.

4. a) $\vdash \exists\mathbf{x}(\mathbf{A} \vee \mathbf{B}) \leftrightarrow \exists\mathbf{x}\mathbf{A} \vee \exists\mathbf{x}\mathbf{B}$.

b) $\vdash \forall\mathbf{x}(\mathbf{A} \,\&\, \mathbf{B}) \leftrightarrow \forall\mathbf{x}\mathbf{A} \,\&\, \forall\mathbf{x}\mathbf{B}$.

c) $\vdash \exists\mathbf{x}(\mathbf{A} \,\&\, \mathbf{B}) \rightarrow \exists\mathbf{x}\mathbf{A} \,\&\, \exists\mathbf{x}\mathbf{B}$.

d) $\vdash \forall\mathbf{x}\mathbf{A} \vee \forall\mathbf{x}\mathbf{B} \rightarrow \forall\mathbf{x}(\mathbf{A} \vee \mathbf{B})$.

e) Give examples of formulas of N of the form

$$\forall\mathbf{x}(\mathbf{A} \vee \mathbf{B}) \rightarrow \forall\mathbf{x}\mathbf{A} \vee \forall\mathbf{x}\mathbf{B} \quad \text{and} \quad \exists\mathbf{x}\mathbf{A} \,\&\, \exists\mathbf{x}\mathbf{B} \rightarrow \exists\mathbf{x}(\mathbf{A} \,\&\, \mathbf{B})$$

which are not valid in $\mathfrak{N}$.

5. If **x** is not free in **A**, show that $\vdash \exists\mathbf{x}\mathbf{A} \leftrightarrow \mathbf{A}$ and $\vdash \forall\mathbf{x}\mathbf{A} \leftrightarrow \mathbf{A}$.

6. a) $\vdash \exists\mathbf{x}\exists\mathbf{y}\mathbf{A} \leftrightarrow \exists\mathbf{y}\exists\mathbf{x}\mathbf{A}$.

b) $\vdash \forall\mathbf{x}\forall\mathbf{y}\mathbf{A} \leftrightarrow \forall\mathbf{y}\forall\mathbf{x}\mathbf{A}$.

c) $\vdash \exists\mathbf{x}\forall\mathbf{y}\mathbf{A} \rightarrow \forall\mathbf{y}\exists\mathbf{x}\mathbf{A}$.

d) Give an example of a formula of N of the form $\forall\mathbf{x}\exists\mathbf{y}\mathbf{A} \rightarrow \exists\mathbf{y}\forall\mathbf{x}\mathbf{A}$ which is not valid in $\mathfrak{N}$.

7. a) Let **A′** be obtained from **A** by replacing some occurrences of **B** by **B′**. Let $\mathbf{x}_1, \ldots, \mathbf{x}_n$ include all those variables which have an occurrence within these occurrences of **B** and **B′** which is free in **B** or **B′** and bound in **A** or **A′**. Show that

$$\vdash \forall\mathbf{x}_1 \ldots \forall\mathbf{x}_n(\mathbf{B} \leftrightarrow \mathbf{B}') \rightarrow (\mathbf{A} \leftrightarrow \mathbf{A}').$$

[Use the method of §3.3 and the equivalence theorem.]

b) Give an example of formulas **A**, **A′**, **B**, and **B′** of N satisfying the conditions of (a) such that $(\mathbf{B} \leftrightarrow \mathbf{B}') \rightarrow (\mathbf{A} \leftrightarrow \mathbf{A}')$ is not valid in $\mathfrak{N}$.

8. a) Let **A′** be obtained from **A** by replacing some occurrences of **a** not immediately following $\exists$ or $\forall$ by **a′**. Let $\mathbf{x}_1, \ldots, \mathbf{x}_n$ include all those variables which have an occurrence within these occurrences of **a** and **a′** which is bound in **A** or **A′**. Show that

$$\vdash \forall \mathbf{x}_1 \ldots \forall \mathbf{x}_n(\mathbf{a} = \mathbf{a}') \rightarrow (\mathbf{A} \leftrightarrow \mathbf{A}').$$

[Similar to 7(a).]

b) Give an example of terms **a** and **a′** and formulas **A** and **A′** of N satisfying the conditions of (a) such that $\mathbf{a} = \mathbf{a}' \rightarrow (\mathbf{A} \leftrightarrow \mathbf{A}')$ is not valid in $\mathfrak{N}$.

9. Let T be a theory. Let T' be the formal system obtained from T by omitting the equality axioms and adding as new axioms all formulas $\mathbf{x} = \mathbf{y} \rightarrow \mathbf{A} \rightarrow \mathbf{A}_\mathbf{x}[\mathbf{y}]$ for **A** atomic. Show that T and T' have the same theorems.

10. a) Show that if **x** does not occur in **a**, then

$$\vdash \mathbf{A}_\mathbf{x}[\mathbf{a}] \leftrightarrow \forall \mathbf{x}(\mathbf{x} = \mathbf{a} \rightarrow \mathbf{A}).$$

b) Give examples of formulas of N of the forms

$$\mathbf{A}_\mathbf{x}[\mathbf{a}] \leftrightarrow \exists \mathbf{x}(\mathbf{x} = \mathbf{a} \mathbin{\&} \mathbf{A}) \quad \text{and} \quad \mathbf{A}_\mathbf{x}[\mathbf{a}] \leftrightarrow \forall \mathbf{x}(\mathbf{x} = \mathbf{a} \rightarrow \mathbf{A})$$

which are not valid in $\mathfrak{N}$.

11. a) A formula is in *disjunctive form* if it is a disjunction of conjunctions of formulas which are either elementary or negations of elementary formulas. Show that for every formula **A**, there is a formula **A′** in disjunctive form such that $\mathbf{A} \leftrightarrow \mathbf{A}'$ is a tautology.

b) A formula is in *conjunctive form* if it is a conjunction of disjunctions of formulas which are either elementary or negations of elementary formulas. Show that for every formula **A**, there is a formula **A′** in conjunctive form such that $\mathbf{A} \leftrightarrow \mathbf{A}'$ is a tautology. [Start from a formula **B** in disjunctive form such that $\neg \mathbf{A} \leftrightarrow \mathbf{B}$ is a tautology.]

12. Let **A** be a formula possibly containing the defined symbols & and $\forall$. Let $\mathbf{A}^\circ$ be obtained from **A** as follows: for every occurrence of an atomic formula **B** in **A**, if **B** is immediately preceded by $\neg$, omit the $\neg$; otherwise, insert a $\neg$ before **B**. Let $\mathbf{A}^*$ be obtained from $\mathbf{A}^\circ$ by replacing $\vee$, &, $\exists$, and $\forall$ everywhere by &, $\vee$, $\forall$, and $\exists$ respectively. Show that $\vdash \mathbf{A}^* \leftrightarrow \neg \mathbf{A}$. [Use induction on the length of **A**.]

CHAPTER 4

THE CHARACTERIZATION PROBLEM

4.1 THE REDUCTION THEOREM

The primary object of a formal system is to provide a framework for proving theorems. Hence a particularly important problem for any formal system F is: find a necessary and sufficient condition that a formula of F be a theorem of F. This is called the *characterization problem* for F. We propose to study the characterization problem for theories.

There is a trivial solution to the characterization problem for a theory T: a formula is a theorem iff it has a proof. This is an unsatisfactory solution because the condition for **A** to be a theorem depends upon all formulas which might appear in a proof of **A**. In a satisfactory solution, the condition must depend only upon **A** and formulas closely related to **A**.

If we are looking for a solution of the characterization problem which works for all theories, we must modify this requirement a little. Clearly whether or not **A** is a theorem of T depends strongly on what the nonlogical axioms of T are. Hence we must expect the condition for **A** to be a theorem of T to refer not only to **A**, but also to the nonlogical axioms of T. If these nonlogical axioms are sufficiently simple, this will not be a disadvantage. For theories with complicated nonlogical axioms, it is necessary to abandon general solutions and seek a solution adapted to the particular theory.

There are some simple results which relate the characterization problems for different theories. To state them, we introduce some concepts which are often used in the study of theories.

The first-order language L' is an *extension* of the first-order language L if every nonlogical symbol of L is a nonlogical symbol of L'. (This is subject to the convention of §2.4 concerning the use of nonlogical symbols in two different languages.) A theory T' is an *extension* of a theory T if $L(T')$ is an extension of $L(T)$ and every theorem of T is a theorem of T'. For the latter, it is clearly necessary and sufficient that every nonlogical axiom of T be a theorem of T'; but it is *not* necessary that every nonlogical axiom of T be a nonlogical axiom of T'.

A *conservative* extension of T is an extension T' of T such that every formula of T which is a theorem of T' is also a theorem of T. For example, if we obtain T' from T by adding some new constants, then T' is a conservative extension of T by the theorem on constants.

The theories T and T' are *equivalent* if each is an extension of the other, i.e., if they have the same language and the same theorems. This implies that they are conservative extensions of each other. If T' is a (conservative) extension of T, then any theory equivalent to T' is a (conservative) extension of any theory equivalent to T.

If T' is a conservative extension of T, then a formula of T is a theorem of T iff it is a theorem of T'; so a solution of the characterization problem for T' gives a solution for T. In particular, if T and T' are equivalent, then the characterization problems for T and T' are equivalent.

If Γ is a set of formulas in the theory T, then $T[\Gamma]$ is the theory obtained from T by adding all of the formulas in Γ as new nonlogical axioms.

Reduction Theorem. Let Γ be a set of formulas in the theory T, and let $\mathbf{A}$ be a formula of T. Then $\mathbf{A}$ is a theorem of $T[\Gamma]$ iff there is a theorem of T of the form $\mathbf{B}_1 \rightarrow \cdots \rightarrow \mathbf{B}_n \rightarrow \mathbf{A}$, where each $\mathbf{B}_i$ is the closure of a formula in Γ.

Proof. If such a theorem of T exists, then $\mathbf{B}_1, \ldots, \mathbf{B}_n$, and $\mathbf{B}_1 \rightarrow \cdots \rightarrow \mathbf{B}_n \rightarrow \mathbf{A}$ are all theorems of $T[\Gamma]$ by the closure theorem; so $\mathbf{A}$ is a theorem of $T[\Gamma]$ by the detachment rule. Now suppose that $\mathbf{A}$ has a proof in $T[\Gamma]$, and let $\mathbf{B}_1, \ldots, \mathbf{B}_n$ be the closures of the formulas in Γ which are used as nonlogical axioms in the proof. Using the closure theorem again, we see that $\mathbf{A}$ is a theorem of $T[\mathbf{B}_1, \ldots, \mathbf{B}_n]$; so $\mathbf{B}_1 \rightarrow \cdots \rightarrow \mathbf{B}_n \rightarrow \mathbf{A}$ is a theorem of T by the deduction theorem.

The reduction theorem reduces the characterization problem for $T[\Gamma]$ to that for T. Now any theory T' is $T[\Gamma]$, where T is obtained from T' by omitting all nonlogical axioms and Γ is the set of nonlogical axioms of T'. Hence to solve the characterization problem for all theories, it suffices to solve it for theories with no nonlogical axioms. Of course, the condition in the solution for T' then depends upon the nonlogical axioms of T'; but we have seen that this is inevitable.

A theory T is *inconsistent* if every formula of T is a theorem of T; otherwise, T is *consistent*. The problem of whether or not T is consistent is a special case of the characterization problem for T. It is important because, as we shall see in the next section, T is consistent iff it has a model.

If for some formula $\mathbf{A}$, both $\mathbf{A}$ and $\neg\mathbf{A}$ are theorems of T, then T is inconsistent; for every formula is a tautological consequence of $\mathbf{A}$ and $\neg\mathbf{A}$. We often use this fact tacitly in discussions of consistency.

If T' is an extension of T, and T' is consistent, then T is consistent; for if $\mathbf{A}$ is any formula of T, then one of $\mathbf{A}$ and $\neg\mathbf{A}$ is not a theorem of T' and hence not a theorem of T. If T' is a conservative extension of T, then T' is consistent iff T is consistent. For if T is consistent and $\mathbf{A}$ is a formula of T, then one of $\mathbf{A}$ and $\neg\mathbf{A}$ is not a theorem of T and hence not a theorem of T'.

We now reformulate the reduction theorem to apply to consistency.

Reduction Theorem for Consistency. Let Γ be a nonempty set of formulas in the theory T. Then $T[\Gamma]$ is inconsistent iff there is a theorem of T which is a disjunction of negations of closures of distinct formulas in Γ.

Proof. If such a formula $\neg \mathbf{A}_1 \vee \cdots \vee \neg \mathbf{A}_n$ exists, then each of $\mathbf{A}_1, \ldots, \mathbf{A}_n$, $\neg \mathbf{A}_1 \vee \cdots \vee \neg \mathbf{A}_n$ is a theorem of $T[\Gamma]$; so by the tautology theorem, every formula is a theorem of $T[\Gamma]$. Now suppose that $T[\Gamma]$ is inconsistent, and let **B** be any formula in T. Then $\mathbf{B}$ & $\neg \mathbf{B}$ is a theorem of $T[\Gamma]$; so by the reduction theorem, $\vdash_T \mathbf{A}_1 \rightarrow \cdots \rightarrow \mathbf{A}_n \rightarrow \mathbf{B}$ & $\neg \mathbf{B}$, where each $\mathbf{A}_i$ is the closure of a formula in Γ. We may clearly suppose that $n > 0$ and that $\mathbf{A}_i$ are distinct. By the tautology theorem, $\vdash_T \neg \mathbf{A}_1 \vee \cdots \vee \neg \mathbf{A}_n$.

Corollary. Let $\mathbf{A}'$ be the closure of **A**. Then **A** is a theorem of T iff $T[\neg \mathbf{A}']$ is inconsistent.

Proof. By the theorem, $T[\neg \mathbf{A}']$ is inconsistent iff $\vdash_T \neg\neg \mathbf{A}'$. By the tautology theorem and the closure theorem, this holds iff $\vdash_T \mathbf{A}$.

4.2 THE COMPLETENESS THEOREM

Our first solution to the characterization problem is a result already mentioned in Chapter 2.

Completeness Theorem, First Form (*Gödel*). A formula **A** of a theory T is a theorem of T iff it is valid in T.

This theorem has a second form, which concerns consistency.

Completeness Theorem, Second Form. A theory T is consistent iff it has a model.

We first show that the second form of the completeness theorem implies the first. In view of the closure theorem and its corollary, it suffices to prove the first form for a closed formula **A**. By the corollary to the reduction theorem for consistency, **A** is a theorem of T iff $T[\neg \mathbf{A}]$ is inconsistent. By the second form of the completeness theorem, this holds iff $T[\neg \mathbf{A}]$ has no model. Now since **A** is closed, a model of $T[\neg \mathbf{A}]$ is simply a model of T in which **A** is not valid. Hence **A** is a theorem of T iff **A** is valid in every model of T.

In one direction, the second form follows from the validity theorem. Suppose that T has a model $\mathcal{A}$. If **A** is a closed formula of T, $\mathcal{A}(\mathbf{A}$ & $\neg \mathbf{A}) = \mathsf{F}$; so **A** & $\neg$**A** is not valid in $\mathcal{A}$ and hence not a theorem of T. Thus T is consistent.

We shall now give Henkin's proof of the other half of the second form of the completeness theorem. We begin with some observations on structures and extensions.

Suppose that the first-order language L' is an extension of the first-order language L, and let $\mathcal{A}'$ be a structure for L'. By omitting certain of the functions and predicates of $\mathcal{A}'$, we obtain a structure $\mathcal{A}$ for L. We call $\mathcal{A}$ the *restriction* of $\mathcal{A}'$ to L, and designate it by $\mathcal{A}' \mid L$. We also say that $\mathcal{A}'$ is an *expansion* of $\mathcal{A}$ to L'.

If $\mathcal{A}'$ is an expansion of $\mathcal{A}$ to L', then $\mathcal{A}$ and $\mathcal{A}'$ have the same individuals. We therefore use the same constant as a name for an individual a in $L(\mathcal{A})$ and $L'(\mathcal{A}')$. It is then easy to verify that $\mathcal{A}(\mathbf{A}) = \mathcal{A}'(\mathbf{A})$ for every closed formula **A** of $L(\mathcal{A})$. From this and the corollary to the closure theorem, we see that the same formulas of L are valid in $\mathfrak{A}$ as in $\mathfrak{A}'$. From this and the validity theorem, we obtain the following result.

Lemma 1. If T' is an extension of T, and $\mathfrak{A}'$ is a model of T', then the restriction of $\mathfrak{A}'$ to $L(T)$ is a model of T.

Returning to the completeness theorem, we are faced with the following problem: given a consistent theory T, how shall we find a model of T? Since only the theory is given, we must build our model from syntactical materials. Now the expressions of T which designate particular individuals are the variable-free terms. The basic idea is to take these terms as individuals, and let the theorems of T tell us what is to be true of these individuals. Actually, the individuals will be sets of variable-free terms; this is necessary because the theorems of T may force two variable-free terms to represent the same individual.

Let T be a theory containing a constant. We shall define a structure $\mathfrak{A}$ which we call the *canonical structure* for T. If $\mathbf{a}$ and $\mathbf{b}$ are variable-free terms of T, then we define $\mathbf{a} \sim \mathbf{b}$ to mean $\vdash_T \mathbf{a} = \mathbf{b}$. Now $\vdash_T \mathbf{a} = \mathbf{a}$ and

$$\vdash_T \mathbf{a} = \mathbf{b} \rightarrow (\mathbf{a} = \mathbf{c} \leftrightarrow \mathbf{b} = \mathbf{c})$$

by the identity axioms and the equality theorem. Hence

$$\mathbf{a} \sim \mathbf{a} \quad \text{and} \quad \mathbf{a} \sim \mathbf{b} \rightarrow (\mathbf{a} \sim \mathbf{c} \leftrightarrow \mathbf{b} \sim \mathbf{c}).$$

Taking $\mathbf{c}$ to be $\mathbf{a}$ in the latter and using the former, we get

$$\mathbf{a} \sim \mathbf{b} \rightarrow \mathbf{b} \sim \mathbf{a}.$$

Hence $\sim$ is an equivalence relation.

We let $|\mathfrak{A}|$ be the set of all equivalence classes of $\sim$. The equivalence class of $\mathbf{a}$ is designated by $\mathbf{a}^\circ$. We complete the definition of $\mathfrak{A}$ by setting

$$\mathbf{f}_{\mathfrak{A}}(\mathbf{a}_1^\circ, \ldots, \mathbf{a}_n^\circ) = (\mathbf{f}\mathbf{a}_1 \ldots \mathbf{a}_n)^\circ,$$

$$\mathbf{p}_{\mathfrak{A}}(\mathbf{a}_1^\circ, \ldots, \mathbf{a}_n^\circ) \quad \text{iff} \quad \vdash_T \mathbf{p}\mathbf{a}_1 \ldots \mathbf{a}_n.$$

We must show that the right-hand sides depend only on the $\mathbf{a}_i^\circ$ and not on the $\mathbf{a}_i$. For this, suppose that $\mathbf{a}_i^\circ = \mathbf{b}_i^\circ$ for $i = 1, \ldots, n$. Then $\vdash_T \mathbf{a}_i = \mathbf{b}_i$; so by the equality theorem,

$$\vdash_T \mathbf{f}\mathbf{a}_1 \ldots \mathbf{a}_n = \mathbf{f}\mathbf{b}_1 \ldots \mathbf{b}_n,$$

$$\vdash_T \mathbf{p}\mathbf{a}_1 \ldots \mathbf{a}_n \leftrightarrow \mathbf{p}\mathbf{b}_1 \ldots \mathbf{b}_n.$$

Hence

$$(\mathbf{f}\mathbf{a}_1 \ldots \mathbf{a}_n)^\circ = (\mathbf{f}\mathbf{b}_1 \ldots \mathbf{b}_n)^\circ$$

and

$$\vdash_T \mathbf{p}\mathbf{a}_1 \ldots \mathbf{a}_n \quad \text{iff} \quad \vdash_T \mathbf{p}\mathbf{b}_1 \ldots \mathbf{b}_n.$$

This is just what we wanted to prove.

Next we show that

$$\mathfrak{A}(\mathbf{a}) = \mathbf{a}^\circ \tag{1}$$

for every variable-free term **a** of T. We use induction on the length of **a**. Since **a** is variable-free, it must have the form $\mathbf{f}\mathbf{a}_1 \ldots \mathbf{a}_n$. Hence, using the induction hypothesis, we have

$$\begin{aligned}\mathcal{A}(\mathbf{a}) &= \mathbf{f}_{\mathcal{A}}(\mathcal{A}(\mathbf{a}_1), \ldots, \mathcal{A}(\mathbf{a}_n)) \\ &= \mathbf{f}_{\mathcal{A}}(\mathbf{a}_1^\circ, \ldots, \mathbf{a}_n^\circ) = \mathbf{a}^\circ.\end{aligned}$$

It follows that if **A** is a variable-free atomic formula, then $\mathcal{A}(\mathbf{A}) = \mathsf{T}$ iff $\vdash_T \mathbf{A}$. For suppose that **A** is $\mathbf{p}\mathbf{a}_1 \ldots \mathbf{a}_n$, where **p** is not $=$. Then

$$\begin{aligned}\mathcal{A}(\mathbf{A}) = \mathsf{T} &\leftrightarrow \mathbf{p}_{\mathcal{A}}(\mathcal{A}(\mathbf{a}_1), \ldots, \mathcal{A}(\mathbf{a}_n)) \\ &\leftrightarrow \mathbf{p}_{\mathcal{A}}(\mathbf{a}_1^\circ, \ldots, \mathbf{a}_n^\circ) \\ &\leftrightarrow \vdash_T \mathbf{A}.\end{aligned}$$

Now suppose that **A** is $\mathbf{a} = \mathbf{b}$. Then

$$\begin{aligned}\mathcal{A}(\mathbf{A}) = \mathsf{T} &\leftrightarrow \mathcal{A}(\mathbf{a}) = \mathcal{A}(\mathbf{b}) \\ &\leftrightarrow \mathbf{a}^\circ = \mathbf{b}^\circ \\ &\leftrightarrow \mathbf{a} \sim \mathbf{b} \\ &\leftrightarrow \vdash_T \mathbf{A}.\end{aligned}$$

We have shown that a variable-free atomic formula is valid in $\mathcal{A}$ iff it is a theorem. This is not necessarily true for all closed formulas for two reasons. First, there may not be enough variable-free terms; there may be a theorem which asserts that some individual has a certain property without there being a variable-free term which represents such an individual. Second, the theorems may not determine the truth or falsity of all closed formulas; there may be a closed formula **A** such that neither **A** nor $\neg$**A** is a theorem. We shall now introduce hypotheses about T which eliminate these difficulties.

A theory T is a *Henkin theory* if for every closed instantiation $\exists\mathbf{x}\mathbf{A}$ of T, there is a constant **e** such that $\vdash_T \exists\mathbf{x}\mathbf{A} \rightarrow \mathbf{A}_\mathbf{x}[\mathbf{e}]$.

A formula **A** of T is *undecidable* in T if neither **A** nor $\neg$**A** is a theorem of T; otherwise, **A** is *decidable* in T. A theory T is *complete* if it is consistent and if every closed formula in T is decidable in T. This may be restated: T is complete if for every closed formula **A** in T, exactly one of **A** and $\neg$**A** is a theorem of T.

We shall study complete theories in detail later. For the moment, we only note that it would be unreasonable to require that *every* formula be decidable. For example, neither $x = 0$ nor $x \neq 0$ is true for all meanings of the variable; so we should not expect either of them to be a theorem.

Lemma 2. Let T be a complete Henkin theory; $\mathcal{A}$ the canonical structure for T; **A** a closed formula of T. Then $\mathcal{A}(\mathbf{A}) = \mathsf{T}$ iff **A** is a theorem of T.

Proof. We use induction on the height of **A**. For **A** atomic, the result has already been proved. Suppose that **A** is $\neg$**B**. Then $\mathcal{A}(\mathbf{A}) = \mathsf{T}$ iff $\mathcal{A}(\mathbf{B}) = \mathsf{F}$, hence, by induction hypothesis, iff **B** is not a theorem of T. Since T is complete, this holds iff **A** is a theorem of T.

Suppose that **A** is **B** $\vee$ **C**. If $\mathfrak{A}(\mathbf{A}) = \mathbf{T}$, then either $\mathfrak{A}(\mathbf{B}) = \mathbf{T}$ or $\mathfrak{A}(\mathbf{C}) = \mathbf{T}$; so, by induction hypothesis, one of **B** and **C** is a theorem of T. By the tautology theorem, **A** is a theorem of T. Now suppose that $\mathfrak{A}(\mathbf{A}) = \mathbf{F}$. Then

$$\mathfrak{A}(\mathbf{B}) = \mathfrak{A}(\mathbf{C}) = \mathbf{F};$$

so neither **B** nor **C** is a theorem of T. By completeness, $\vdash_T \neg\mathbf{B}$ and $\vdash_T \neg\mathbf{C}$; so $\vdash_T \neg\mathbf{A}$ by the tautology theorem. By the consistency of T, **A** is not a theorem of T.

Suppose that **A** is $\exists\mathbf{xB}$. Then $\mathfrak{A}(\mathbf{A}) = \mathbf{T}$ iff $\mathfrak{A}(\mathbf{B_x[i]}) = \mathbf{T}$ for some **i**. Now each **i** is the name of an individual $\mathbf{a}^\circ$. Since $\mathfrak{A}(\mathbf{a}) = \mathbf{a}^\circ$, it follows from the lemma of §2.5 that $\mathfrak{A}(\mathbf{B_x[i]}) = \mathfrak{A}(\mathbf{B_x[a]})$. Hence $\mathfrak{A}(\mathbf{A}) = \mathbf{T}$ iff $\mathfrak{A}(\mathbf{B_x[a]}) = \mathbf{T}$ for some variable-free term **a**. By induction hypothesis, this holds iff $\vdash_T \mathbf{B_x[a]}$ for some variable-free term **a**. We must therefore show that this last condition is equivalent to $\vdash_T \mathbf{A}$. Now $\mathbf{B_x[a]} \to \mathbf{A}$ is a substitution axiom; so if $\vdash \mathbf{B_x[a]}$ for some **a**, then $\vdash\mathbf{A}$ by the detachment rule. Now suppose that $\vdash\mathbf{A}$. Since T is a Henkin theory, there is an **e** such that $\vdash \mathbf{A} \to \mathbf{B_x[e]}$; so $\vdash\mathbf{B_x[e]}$ by the detachment rule.

Corollary. If T is a complete Henkin theory, then the canonical structure for T is a model of T.

Proof. Let **A** be a nonlogical axiom of T, and let $\mathbf{A}'$ be the closure of **A**. Then $\vdash_T \mathbf{A}'$; so $\mathbf{A}'$ is valid in $\mathfrak{A}$ by the lemma; so **A** is valid in $\mathfrak{A}$.

Our next problem is to see how to obtain Henkin theories. For this purpose, we introduce some definitions.

Let L be a first-order language. We shall define the *special constants* of *level* n by induction on n. Suppose that the special constants of all levels less than n have been defined. Let $\exists\mathbf{xA}$ be a closed instantiation formed with these constants and the symbols of L. If $n > 0$, suppose also that $\exists\mathbf{xA}$ contains at least one special constant of level $n - 1$. Then the symbol consisting of the letter c with the subscript $\exists\mathbf{xA}$ is a special constant of level n, called the special constant *for* $\exists\mathbf{xA}$.

The language obtained from L by adding all the special constants of all levels is designated by L_c. If $\exists\mathbf{xA}$ is a closed instantiation of L_c, then there is in L_c a unique special constant for $\exists\mathbf{xA}$; its level is the least natural number which is greater than the levels of all the special constants occurring in $\exists\mathbf{xA}$. We use **r**, **s**, and **t** as syntactical variables which vary through special constants. If **r** is the special constant for $\exists\mathbf{xA}$, then the formula $\exists\mathbf{xA} \to \mathbf{A_x[r]}$ is called the *special axiom for* **r**.

Now let T be a theory with the language L. Then T_c is the theory whose language is L_c and whose nonlogical axioms are the nonlogical axioms of T and the special axioms for the special constants of L_c. It is obvious that T_c is a Henkin theory.

Lemma 3. T_c is a conservative extension of T.

Proof. Let T' be obtained from T by adding the special constants (but no new axioms). By the theorem on constants, T' is a conservative extension of T; so it will suffice to show that every formula **A** of T which is a theorem of T_c is a theorem of T'. By the reduction theorem, we have $\vdash_{T'} \mathbf{B}_1 \to \cdots \to \mathbf{B}_k \to \mathbf{A}$, where $\mathbf{B}_1, \ldots, \mathbf{B}_k$ are distinct special axioms. We now use induction on k. If $k = 0$, there is nothing to prove. Let $k > 0$. We may suppose that the level of the special constant **r** for which $\mathbf{B}_1$ is the special axiom is at least as great as the levels of the special constants for which $\mathbf{B}_2, \ldots, \mathbf{B}_k$ are the special axioms. Then **r** does not occur in $\mathbf{B}_2, \ldots, \mathbf{B}_k$; and it certainly does not occur in **A**. It follows by the theorem on constants that if $\mathbf{B}_1$ is $\exists \mathbf{x}\mathbf{C} \to \mathbf{C}_\mathbf{x}[\mathbf{r}]$ and **y** is a new variable, then

$$\vdash_{T'} (\exists \mathbf{x}\mathbf{C} \to \mathbf{C}_\mathbf{x}[\mathbf{y}]) \to \mathbf{B}_2 \to \cdots \to \mathbf{B}_k \to \mathbf{A}.$$

Hence by the $\exists$-introduction rule,

$$\vdash_{T'} \exists \mathbf{y}(\exists \mathbf{x}\mathbf{C} \to \mathbf{C}_\mathbf{x}[\mathbf{y}]) \to \mathbf{B}_2 \to \cdots \to \mathbf{B}_k \to \mathbf{A}.$$

Now $\vdash_{T'} \exists \mathbf{x}\mathbf{C} \to \exists \mathbf{y}\mathbf{C}_\mathbf{x}[\mathbf{y}]$ by the variant theorem; so $\vdash_{T'} \exists \mathbf{y}(\exists \mathbf{x}\mathbf{C} \to \mathbf{C}_\mathbf{x}[\mathbf{y}])$ by the prenex operations. Hence $\vdash_{T'} \mathbf{B}_2 \to \cdots \to \mathbf{B}_k \to \mathbf{A}$ by the detachment rule; so $\vdash_{T'} \mathbf{A}$ by the induction hypothesis.

Next we need a method for obtaining complete theories. For this, we shall need a result from set theory, which we state without proof. (For an outline of the proof, see Problem 4 of Chapter 9.)

Let E be a set, and let J be a class of subsets of E. We say that J *has finite character* if for every subset A of E, A is in J iff every finite subset of A is in J. A set A in J is a *maximal element* of J if A is not a subset of any other member of J.

Teichmüller-Tukey Lemma. If J is a nonempty class of subsets of E which is of finite character, then J contains a maximal element.

An extension T' of T is a *simple* extension if T and T' have the same language.

Lindenbaum's Theorem. If T is a consistent theory, then T has a complete simple extension.

Proof. Let E be the set of formulas in T. Let J be the class of all subsets A of E such that $T[A]$ is consistent. We show that J is of finite character. If A is in J, then every subset A' of A is in J; for $T[A]$ is an extension of $T[A']$. Suppose that A is not in J, so that $T[A]$ is inconsistent. Select a formula **A** of T and proofs of **A** and $\neg$**A** in $T[A]$. Let A' be the set of formulas in A which are used as nonlogical axioms in these proofs. Then A' is a finite subset of A; and $T[A']$ is inconsistent, so that A' is not in J.

The empty set is in J; so by the Teichmüller-Tukey lemma, there is a maximal element A in J. Clearly $T[A]$ is a consistent simple extension of T; we show that it is complete. We must show that if **A** is a closed formula of T which is not a theorem of $T[A]$, then $\neg$**A** is a theorem of $T[A]$. Let A' be obtained by adding $\neg$**A** to A. By the corollary to the reduction theorem for consistency, $T[A']$ is consistent; so A' is in J. By the maximality of A, this implies that $A = A'$; so $\neg$**A** is in A. Hence $\neg$**A** is a theorem of $T[A]$.

The proof of the completeness theorem is now easy. Suppose that T is a consistent theory. Then by Lemma 3, T_c is a consistent theory. By Lindenbaum's theorem, there is a complete simple extension U of T_c. Since U is a simple extension of a Henkin theory, it is a Henkin theory. Hence by the corollary to Lemma 2, U has a model. Since U is an extension of T, T has a model by Lemma 1.

We can also combine our lemmas to obtain a result which will be useful later.

Lemma 4. Let T be a theory, and let U be a consistent simple extension of T_c. Then U has a model $\mathcal{A}$ such that each individual of $\mathcal{A}$ is $\mathcal{A}(\mathbf{r})$ for infinitely many special constants $\mathbf{r}$.

Proof. By Lindenbaum's theorem, there is a complete simple extension U' of U. Since U' is a Henkin theory, the canonical structure $\mathcal{A}$ for U' is a model of U', and hence, by Lemma 1, of U. By (1), every individual a of $\mathcal{A}$ is $\mathcal{A}(\mathbf{a})$ for some variable-free term $\mathbf{a}$. By the identity axioms and the substitution theorem, $\vdash_U \exists\mathbf{x}(\mathbf{x} = \mathbf{a})$; so if $\mathbf{r}$ is the special constant for $\exists\mathbf{x}(\mathbf{x} = \mathbf{a})$, then $\vdash_U \mathbf{r} = \mathbf{a}$. It follows that $\mathcal{A}(\mathbf{r}) = \mathcal{A}(\mathbf{a}) = a$. By replacing $\mathbf{x}$ by other variables, we can find infinitely many other special constants with the same property.

Each form of the completeness theorem establishes the equivalence between a syntactical concept and a semantical concept. Many other such equivalences can be derived from these. We give one example.

Corollary. Let T and T' be theories with the same language. Then T' is an extension of T iff every model of T' is a model of T. Hence T and T' are equivalent iff they have the same models.

Proof. The condition is necessary by Lemma 1. If the condition holds, then every formula valid in T is valid in T'; so every theorem of T is a theorem of T' by the first form of the completeness theorem.

Here is an application of the corollary. As is well known, there are many sets of axioms for groups other than the one which we formalized in G. Suppose that we formalize one of them in a theory G' with the same language as G. Then G and G' have the same models (viz., the groups) and hence are equivalent.

4.3 THE CONSISTENCY THEOREM

The characterization problem deals entirely with concrete objects. The solution given by the completeness theorem, however, deals with models, which are abstract objects. It is therefore natural to seek a finitary solution of the characterization problem for theories. We shall give such a solution in this and the next section.

A theory is *open* if all of its nonlogical axioms are open. In this section we consider a special case of the characterization problem for theories: the problem of the consistency of open theories.

Let $\mathbf{r}$ be the special constant for $\exists\mathbf{x}\mathbf{A}$. A formula *belongs* to $\mathbf{r}$ if it is either the special axiom for $\mathbf{r}$ or a closed substitution axiom $L(T_c)$ of the form

$\mathbf{A_x[a]} \to \exists\mathbf{xA}$. We designate by $\Delta(T)$ the set of formulas in T_c which either belong to some special constant or are closed instances of identity axioms, equality axioms, or nonlogical axioms of T.

Lemma 1. If $\vdash_T \mathbf{A}$, and $\mathbf{A}'$ is a closed instance of $\mathbf{A}$ in $L(T_c)$, then $\mathbf{A}'$ is a tautological consequence of formulas in $\Delta(T)$.

Proof. We use induction on theorems (in the form described in §3.1). If $\mathbf{A}$ is a substitution axiom, then $\mathbf{A}'$ is a closed substitution axiom and hence in $\Delta(T)$. If $\mathbf{A}$ is an identity axiom, an equality axiom, or a nonlogical axiom, then clearly $\mathbf{A}'$ is in $\Delta(T)$. If $\mathbf{A}$ is a tautological consequence of $\mathbf{B}_1, \ldots, \mathbf{B}_n$, then $\mathbf{A}'$ is a tautological consequence of closed instances $\mathbf{B}_1', \ldots, \mathbf{B}_n'$ of $\mathbf{B}_1, \ldots, \mathbf{B}_n$ respectively. By induction hypothesis, $\mathbf{B}_1', \ldots, \mathbf{B}_n'$ are tautological consequences of formulas in $\Delta(T)$; so $\mathbf{A}'$ is also. Finally, suppose that $\mathbf{A}$ is $\exists\mathbf{xB} \to \mathbf{C}$, and is inferred from $\mathbf{B} \to \mathbf{C}$ by the $\exists$-introduction rule. Then $\mathbf{A}'$ is $\exists\mathbf{xB}' \to \mathbf{C}'$. If $\mathbf{r}$ is the special constant for $\exists\mathbf{xB}'$, then $\mathbf{B}_\mathbf{x}'[\mathbf{r}] \to \mathbf{C}'$ is a closed instance of $\mathbf{B} \to \mathbf{C}$, and hence, by induction hypothesis, a tautological consequence of formulas in $\Delta(T)$. Since $\mathbf{A}'$ is a tautological consequence of $\mathbf{B}_\mathbf{x}'[\mathbf{r}] \to \mathbf{C}'$ and the special axiom $\exists\mathbf{xB}' \to \mathbf{B}_\mathbf{x}'[\mathbf{r}]$, it is a tautological consequence of formulas in $\Delta(T)$.

A formula is a *quasi-tautology* if it is a tautological consequence of instances of identity axioms and equality axioms.

Consistency Theorem (*Hilbert-Ackermann*). An open theory T is inconsistent iff there is a quasi-tautology which is a disjunction of negations of instances of nonlogical axioms of T.

Proof. Suppose that $\neg\mathbf{A}_1 \vee \cdots \vee \neg\mathbf{A}_n$ is such a quasi-tautology. Then $\mathbf{A}_1, \ldots, \mathbf{A}_n$, and $\neg\mathbf{A}_1 \vee \cdots \vee \neg\mathbf{A}_n$ are all theorems of T; so, by the tautology theorem, T is inconsistent.

Now suppose that T is inconsistent. Let $\mathbf{r}$ be any special constant in T_c. Then $\mathbf{r} \neq \mathbf{r}$ is an instance of the theorem $x \neq x$ of T; so by Lemma 1, there are formulas $\mathbf{A}_1, \ldots, \mathbf{A}_k$ in $\Delta(T)$ such that $\mathbf{A}_1 \to \cdots \to \mathbf{A}_k \to \mathbf{r} \neq \mathbf{r}$ is a tautology. Since $\mathbf{r} = \mathbf{r}$ is in $\Delta(T)$, we may suppose it is one of the $\mathbf{A}_i$. Then $\neg\mathbf{A}_1 \vee \cdots \vee \neg\mathbf{A}_k$ is a tautology.

The *rank* of the special constant for $\exists\mathbf{xA}$ is the number of occurrences of $\exists$ in $\exists\mathbf{xA}$; it is at least 1. We let $\Delta_n(T)$ be the set obtained from $\Delta(T)$ by omitting the formulas belonging to special constants of rank greater than n. Thus $\Delta_0(T)$ consists of the variable-free instances in $L(T_c)$ of the identity axioms, equality axioms, and nonlogical axioms of T.

We call $\mathbf{A}_1, \ldots, \mathbf{A}_k$ a *special sequence* if $\neg\mathbf{A}_1 \vee \cdots \vee \neg\mathbf{A}_k$ is a tautology; i.e., if there is no truth valuation V such that $V(\mathbf{A}_i) = \mathbf{T}$ for all i. We have seen that there is a special sequence whose formulas are all in $\Delta(T)$ and hence are all in $\Delta_n(T)$ for some n. Suppose that we have a special sequence consisting of formulas in $\Delta_0(T)$. By replacing each special constant by a new variable, we obtain a special sequence whose formulas are instances in $L(T)$ of identity axioms, equality axioms, and nonlogical axioms of T. If $\mathbf{A}_1, \ldots, \mathbf{A}_k$ are the instances of nonlogical axioms occurring in this sequence, then clearly $\neg\mathbf{A}_1 \vee \cdots \vee \neg\mathbf{A}_k$ is a quasi-tautology. It follows that the proof of the theorem reduces to the proof of the following lemma.

Lemma 2. If $n > 0$, and there is a special sequence consisting of formulas in $\Delta_n(T)$, then there is a special sequence consisting of formulas in $\Delta_{n-1}(T)$.

Proof. By hypothesis, there is a special sequence consisting of formulas in $\Delta_{n-1}(T)$ and formulas belonging to special constants of rank n. We prove the conclusion by induction on the number of these constants of rank n. If there are none, there is nothing to prove; so we suppose that there is at least one. Let $\mathbf{r}$ be one which has as high a level as possible, and let $\mathbf{r}_1, \ldots, \mathbf{r}_s$ be the remaining ones. We shall construct a special sequence consisting of formulas in $\Delta_{n-1}(T)$ and formulas belonging to $\mathbf{r}_1, \ldots, \mathbf{r}_s$; in view of the induction hypothesis, this will complete the proof.

Let $\mathbf{A}_1, \ldots, \mathbf{A}_r$ be the formulas in the given special sequence which are in $\Delta_{n-1}(T)$ or belong to one of $\mathbf{r}_1, \ldots, \mathbf{r}_s$. Let the remaining formulas in the given special sequence be the special axiom $\exists\mathbf{x}\mathbf{B} \to \mathbf{B}_\mathbf{x}[\mathbf{r}]$ for $\mathbf{r}$ and the formulas

$$\mathbf{B}_\mathbf{x}[\mathbf{a}_i] \to \exists\mathbf{x}\mathbf{B}, \qquad i = 1, \ldots, p.$$

We show that no $\mathbf{A}_i$ contains an occurrence of $\exists\mathbf{x}\mathbf{B}$. This is immediate if $\mathbf{A}_i$ is an instance of an identity, equality, or nonlogical axiom; for then $\mathbf{A}_i$ is open. Suppose that $\mathbf{A}_i$ is a formula $\exists\mathbf{y}\mathbf{C} \to \mathbf{C}_\mathbf{y}[\mathbf{s}]$ or $\mathbf{C}_\mathbf{y}[\mathbf{a}] \to \exists\mathbf{y}\mathbf{C}$ belonging to $\mathbf{s}$. Since the rank of $\mathbf{r}$ is as great as the rank of $\mathbf{s}$, $\exists\mathbf{x}\mathbf{B}$ contains as many occurrences of $\exists$ as $\exists\mathbf{y}\mathbf{C}$. Thus $\exists\mathbf{x}\mathbf{B}$ cannot occur in $\mathbf{C}_\mathbf{y}[\mathbf{s}]$ or $\mathbf{C}_\mathbf{y}[\mathbf{a}]$. It cannot occur in $\exists\mathbf{y}\mathbf{C}$ either, unless $\exists\mathbf{y}\mathbf{C}$ is the same as $\exists\mathbf{x}\mathbf{B}$. But this is impossible, since $\mathbf{s}$ is distinct from $\mathbf{r}$.

If in our given special sequence, we replace every occurrence of $\exists\mathbf{x}\mathbf{B}$ by $\mathbf{B}_\mathbf{x}[\mathbf{r}]$, we obtain a new special sequence. (This follows from a remark in §3.1.) These replacements do not affect the $\mathbf{A}_i$. They change $\exists\mathbf{x}\mathbf{B} \to \mathbf{B}_\mathbf{x}[\mathbf{r}]$ into the tautology $\mathbf{B}_\mathbf{x}[\mathbf{r}] \to \mathbf{B}_\mathbf{x}[\mathbf{r}]$, and change $\mathbf{B}_\mathbf{x}[\mathbf{a}_i] \to \exists\mathbf{x}\mathbf{B}$ into $\mathbf{B}_\mathbf{x}[\mathbf{a}_i] \to \mathbf{B}_\mathbf{x}[\mathbf{r}]$. It follows that

$$\mathbf{A}_1, \ldots, \mathbf{A}_r, \mathbf{B}_\mathbf{x}[\mathbf{a}_1] \to \mathbf{B}_\mathbf{x}[\mathbf{r}], \ldots, \mathbf{B}_\mathbf{x}[\mathbf{a}_p] \to \mathbf{B}_\mathbf{x}[\mathbf{r}] \tag{1}$$

is a special sequence.

For each expression $\mathbf{u}$ of $L(T_c)$, let $\mathbf{u}^{(i)}$ be the expression obtained from $\mathbf{u}$ by replacing $\mathbf{r}$ everywhere (including within subscripts of special constants) by $\mathbf{a}_i$. If $\mathbf{C}_1, \ldots, \mathbf{C}_k$ is a special sequence, then so is $\mathbf{C}_1{}^{(i)}, \ldots, \mathbf{C}_k{}^{(i)}$ (by the same remark of §3.1). Applying this to (1), and noting that $\mathbf{B}_\mathbf{x}[\mathbf{r}]^{(i)}$ is $\mathbf{B}_\mathbf{x}[\mathbf{a}_i]$ because $\mathbf{r}$ does not appear in $\mathbf{B}$, we obtain the special sequence

$$\mathbf{A}_1{}^{(i)}, \ldots, \mathbf{A}_r{}^{(i)}, \mathbf{B}_\mathbf{x}[\mathbf{a}_1]^{(i)} \to \mathbf{B}_\mathbf{x}[\mathbf{a}_i], \ldots, \mathbf{B}_\mathbf{x}[\mathbf{a}_p]^{(i)} \to \mathbf{B}_\mathbf{x}[\mathbf{a}_i]. \tag{2}$$

We now claim that the sequence consisting of all the $\mathbf{A}_i$ and all the $\mathbf{A}_i{}^{(j)}$ $(i = 1, \ldots, r, j = 1, \ldots, p)$ is a special sequence. For suppose that there were a truth valuation V assigning $\mathbf{T}$ to all of these formulas. According to (2), V would, for each i, have to assign $\mathbf{F}$ to some $\mathbf{B}_\mathbf{x}[\mathbf{a}_j]^{(i)} \to \mathbf{B}_\mathbf{x}[\mathbf{a}_i]$. Thus $V(\mathbf{B}_\mathbf{x}[\mathbf{a}_i]) = \mathbf{F}$ for all i; so V assigns $\mathbf{T}$ to each $\mathbf{B}_\mathbf{x}[\mathbf{a}_i] \to \mathbf{B}_\mathbf{x}[\mathbf{r}]$. This is impossible by (1).

It will now suffice to show that each $\mathbf{A}_i^{(j)}$ either is in $\Delta_{n-1}(T)$ or belongs to one of $\mathbf{r}_1, \ldots, \mathbf{r}_s$. If $\mathbf{A}_i$ is an instance of an identity, equality, or nonlogical axiom of T, then $\mathbf{A}_i^{(j)}$ is also. Now suppose that $\mathbf{A}_i$ is a sentence $\exists \mathbf{y}\mathbf{C} \rightarrow \mathbf{C}_{\mathbf{y}}[\mathbf{s}]$ or $\mathbf{C}_{\mathbf{y}}[\mathbf{a}] \rightarrow \exists \mathbf{y}\mathbf{C}$ belonging to $\mathbf{s}$. Then $\mathbf{A}_i^{(j)}$ is

$$\exists \mathbf{y}\mathbf{C}^{(j)} \rightarrow \mathbf{C}_{\mathbf{y}}^{(j)}[\mathbf{s}^{(j)}]$$

or

$$\mathbf{C}_{\mathbf{y}}^{(j)}[\mathbf{a}^{(j)}] \rightarrow \exists \mathbf{y}\mathbf{C}^{(j)}.$$

Since $\mathbf{s}$ is not $\mathbf{r}$, it is clear that $\mathbf{s}^{(j)}$ is the special constant for $\exists \mathbf{y}\mathbf{C}^{(j)}$. Hence $\mathbf{A}_i^{(j)}$ is a sentence belonging to $\mathbf{s}^{(j)}$. Clearly $\mathbf{s}$ and $\mathbf{s}^{(j)}$ have the same rank. Hence if $\mathbf{A}_i$ belongs to $\Delta_{n-1}(T)$, then $\mathbf{A}_i^{(j)}$ does also. Suppose that $\mathbf{s}$ is one of $\mathbf{r}_1, \ldots, \mathbf{r}_s$. Since the level of $\mathbf{r}$ is then as great as the level of $\mathbf{s}$, $\mathbf{r}$ does not appear in $\exists \mathbf{y}\mathbf{C}$ and hence not in $\mathbf{s}$. Thus $\mathbf{s}^{(j)}$ is $\mathbf{s}$, and $\mathbf{A}_i^{(j)}$ belongs to one of $\mathbf{r}_1, \ldots, \mathbf{r}_s$.

By combining the fact that $\mathfrak{N}$ is a model of N with the completeness theorem, we get a proof of the consistency of N. We now indicate how we can convert this into a finitary proof of the consistency of N. First we replace the individuals of $\mathfrak{N}$ by concrete objects. For this purpose, it suffices to replace the natural number n by the expression consisting of n strokes. Next we note that if we are given a variable-free term $\mathbf{a}$ or formula $\mathbf{A}$, we can actually compute $\mathfrak{N}(\mathbf{a})$ or $\mathfrak{N}(\mathbf{A})$. It follows that in certain cases, we can give a finitary proof that an open formula $\mathbf{A}$ of $L(N)$ is valid in $\mathfrak{N}$. In particular, we can prove that every instance of a nonlogical axiom of N is valid in $\mathfrak{N}$ and that every open quasi-tautology is valid in $\mathfrak{N}$. Now it is clearly impossible to have open formulas $\mathbf{A}_1, \ldots, \mathbf{A}_n$ such that $\mathbf{A}_1, \ldots, \mathbf{A}_n$, and $\neg \mathbf{A}_1 \vee \cdots \vee \neg \mathbf{A}_n$ are all valid in $\mathfrak{N}$; so by the consistency theorem, N is consistent.

We thus have two proofs of the consistency of N, one finitary and one nonfinitary. Since the finitary proof is longer, it is natural to ask what additional benefits we can expect from a finitary proof.

In the first place, as previously mentioned, finitary proofs can throw light on the nature of the concrete and the abstract. For example, the finitary proof of the consistency of N shows that Hilbert's view of abstract mathematics discussed in §1.2 is tenable for N. We can think of the variable-free formulas of N as expressing concrete results, and of the quantifiers as being abstract notions introduced to prove concrete results. Our consistency proof then shows that any such concrete result is correct, i.e., that any variable-free formula $\mathbf{A}$ provable in N is valid in $\mathfrak{N}$. For otherwise, $\neg \mathbf{A}$ would be valid in $\mathfrak{N}$; and the above proof would then show that $N[\neg \mathbf{A}]$ is consistent, which contradicts the corollary to the reduction theorem for consistency.

In addition, a finitary proof often gives more information than an abstract proof, simply because the restriction to finitary methods often requires us to prove more than is stated in order to obtain the desired result. For example, our abstract consistency proof for N proceeds by showing that every theorem of N is true. Since the truth of a formula of N is not a finitary concept, the finitary proof must prove something stronger about the theorems of N. If we analyze the proof more closely, we find that we can actually produce a formula which is true but does not have the stronger property. Thus we obtain a true formula of N which is not a theorem of N. We shall not carry this construction through, since a rather similar result about another theory will be considered in Chapter 8.

4.4 HERBRAND'S THEOREM

We now turn to the finitary solution of the characterization problem for theories. Since the proof of the reduction theorem is finitary, it will suffice to give a solution for theories with no nonlogical axioms. In view of the results of §3.5, it suffices to give a solution for formulas in prenex form. We may even restrict ourselves to closed formulas in prenex form; this follows from the closure theorem and the fact that the closure of a formula in prenex form is in prenex form.

A formula in prenex form is *existential* if all of the quantifiers in its prefix are existential. We begin with a solution for closed existential formulas.

Lemma 1. Let T be a theory with no nonlogical axioms. A closed existential formula $\mathbf{A}$ is a theorem of T iff there is a quasi-tautology which is a disjunction of instances of the matrix of $\mathbf{A}$.

Proof. Suppose that $\mathbf{A}$ is $\exists \mathbf{x}_1 \ldots \exists \mathbf{x}_n \mathbf{B}$ with $\mathbf{B}$ open. By the corollary to the reduction theorem for consistency, $\mathbf{A}$ is a theorem iff $T[\neg \mathbf{A}]$ is inconsistent. By the prenex operations and the closure theorem, $T[\neg \mathbf{A}]$ is equivalent to $T[\neg \mathbf{B}]$. Thus $\mathbf{A}$ is a theorem iff $T[\neg \mathbf{B}]$ is inconsistent. By the consistency theorem, this holds iff there is a quasi-tautology $\neg\neg \mathbf{B}_1 \vee \cdots \vee \neg\neg \mathbf{B}_n$, where each $\mathbf{B}_i$ is an instance of $\mathbf{B}$. Since $\neg\neg \mathbf{B}_1 \vee \cdots \vee \neg\neg \mathbf{B}_n$ is a quasi-tautology iff $\mathbf{B}_1 \vee \cdots \vee \mathbf{B}_n$ is a quasi-tautology, we get the lemma.

Before completing the solution, we introduce an extension T_c' of T_c. By a *special equality axiom*, we mean a formula

$$\forall \mathbf{x}(\mathbf{A} \leftrightarrow \mathbf{B}) \rightarrow \mathbf{r} = \mathbf{s},$$

where $\mathbf{r}$ and $\mathbf{s}$ are the special constants for $\exists \mathbf{x} \mathbf{A}$ and $\exists \mathbf{x} \mathbf{B}$ respectively. We obtain T_c' from T_c by adding all the special equality axioms as new nonlogical axioms.

Lemma 2. T_c' is a conservative extension of T.

Proof. We let $T[\mathbf{r}_1, \ldots, \mathbf{r}_n]$ be the theory obtained from T by adding the constants $\mathbf{r}_1, \ldots, \mathbf{r}_n$ and the special axioms and special equality axioms which contain only these special constants. As in the proof of Lemma 3 of §4.2, we are reduced to proving the following: if level $(\mathbf{r}_i) \leqslant$ level $(\mathbf{r})$ for $i = 1, \ldots, n$, then $T[\mathbf{r}_1, \ldots, \mathbf{r}_n, \mathbf{r}]$ is a conservative extension of $T[\mathbf{r}_1, \ldots, \mathbf{r}_n]$. We shall show that if a formula $\mathbf{A}$ of $T[\mathbf{r}_1, \ldots, \mathbf{r}_n]$ has a proof in $T[\mathbf{r}_1, \ldots, \mathbf{r}_n, \mathbf{r}]$, then it has such a proof which uses no special equality axioms containing $\mathbf{r}$; we can then complete the proof as in Lemma 3 of §4.2.

We note that for each special constant appearing on the left-hand side of a special equality axiom, there is a special constant of higher level appearing on the right-hand side. It follows that the special equality axioms in our given proof of **A** can contain **r** only on the right side. We may suppose that all of these axioms have the form $\forall \mathbf{x}(\mathbf{B} \leftrightarrow \mathbf{C}) \to \mathbf{r} = \mathbf{s}$; for $\forall \mathbf{x}(\mathbf{C} \leftrightarrow \mathbf{B}) \to \mathbf{s} = \mathbf{r}$ can be derived from $\forall \mathbf{x}(\mathbf{B} \leftrightarrow \mathbf{C}) \to \mathbf{r} = \mathbf{s}$ by the equivalence theorem. We may also suppose that none of them have $\mathbf{r} = \mathbf{r}$ as the right-hand side; for such an axiom could be derived from the identity axioms.

Let $\forall \mathbf{x}(\mathbf{B} \leftrightarrow \mathbf{C}) \to \mathbf{r} = \mathbf{s}$ be one of these special equality axioms in the given proof of **A**. Let T' be the theory obtained from $T[\mathbf{r}_1, \ldots, \mathbf{r}_n]$ by adding the constant **r** and the two axioms $\mathbf{r} = \mathbf{s}$ and $\forall \mathbf{x}(\mathbf{B} \leftrightarrow \mathbf{C})$. We show that **A** is a theorem of T'. For this, it will suffice to prove in T' all the nonlogical axioms in the given proof of **A** which contain **r**. First of all, we can derive $\mathbf{B} \leftrightarrow \mathbf{C}$ and hence $\mathbf{B_x[r]} \leftrightarrow \mathbf{C_x[r]}$ from the axiom $\forall \mathbf{x}(\mathbf{B} \leftrightarrow \mathbf{C})$. From $\mathbf{B_x[r]} \leftrightarrow \mathbf{C_x[r]}$ and $\mathbf{r} = \mathbf{s}$ we derive $\mathbf{B_x[r]} \leftrightarrow \mathbf{C_x[s]}$ by the equality theorem. We can then derive the special axiom $\exists \mathbf{xB} \to \mathbf{B_x[r]}$ for **r** from the special axiom $\exists \mathbf{xC} \to \mathbf{C_x[s]}$ for **s** by the equivalence theorem. Now consider a special equality axiom $\forall \mathbf{x}(\mathbf{B} \leftrightarrow \mathbf{D}) \to \mathbf{r} = \mathbf{t}$ occurring in the given proof of **A**. We can derive this from the special equality axiom $\forall \mathbf{x}(\mathbf{C} \leftrightarrow \mathbf{D}) \to \mathbf{s} = \mathbf{t}$ (which does not contain **r**) by the equivalence and equality theorems.

Since $\vdash_{T'} \mathbf{A}$, it follows from the deduction theorem and the theorem on constants that

$$\mathbf{y} = \mathbf{s} \to \forall \mathbf{x}(\mathbf{B} \leftrightarrow \mathbf{C}) \to \mathbf{A}$$

is a theorem of $T[\mathbf{r}_1, \ldots, \mathbf{r}_n]$. Substituting **s** for **y** and using the identity axioms, it follows that $\forall \mathbf{x}(\mathbf{B} \leftrightarrow \mathbf{C}) \to \mathbf{A}$ is a theorem of $T[\mathbf{r}_1, \ldots, \mathbf{r}_n]$. It follows by the tautology theorem that $\neg(\forall \mathbf{x}(\mathbf{B} \leftrightarrow \mathbf{C}) \to \mathbf{r} = \mathbf{s}) \to \mathbf{A}$ is provable in $T[\mathbf{r}_1, \ldots, \mathbf{r}_n, \mathbf{r}]$ without using any nonlogical axioms containing **r**.

Now let $\mathbf{D}_1, \ldots, \mathbf{D}_k$ be the special equality axioms containing **r** which are used in the given proof of **A**. By the deduction theorem, $\mathbf{D}_1 \to \cdots \to \mathbf{D}_k \to \mathbf{A}$ has a proof not using special equality axioms containing **r**. We have just shown that each $\neg \mathbf{D}_i \to \mathbf{A}$ also has such a proof. Thus **A** has such a proof by the tautology theorem.

We add the following remark about T'_c. Let **r** be the special constant for $\exists \mathbf{x} \neg \mathbf{A}$. Then $\exists \mathbf{x} \neg \mathbf{A} \to \neg \mathbf{A_x[r]}$ is an axiom of T'_c. Bringing the left-hand side to prenex form and using the tautology theorem,

$$\vdash_{T'_c} \mathbf{A_x[r]} \to \forall \mathbf{xA}. \qquad (1)$$

We now return to the characterization problem. We shall associate a closed existential formula $\mathbf{A}_H$ with each closed formula **A** in prenex form. If **A** is existential, then $\mathbf{A}_H$ is **A**. If not, then **A** has the form $\exists \mathbf{x}_1 \ldots \exists \mathbf{x}_n \forall \mathbf{yB}$ ($n \geqslant 0$). We introduce a new n-ary function symbol **f** and let $\mathbf{A}^*$ be $\exists \mathbf{x}_1 \ldots \exists \mathbf{x}_n \mathbf{B_y}[\mathbf{fx}_1 \ldots \mathbf{x}_n]$. Then $\mathbf{A}^*$ has one less universal quantifier than **A**. If $\mathbf{A}^*$ is not existential, then we form $\mathbf{A}^{**}$, $\mathbf{A}^{***}$, etc., until we come to an existential formula. This existential formula is $\mathbf{A}_H$.

Herbrand's Theorem. Let T be a theory with no nonlogical axioms, and let **A** be a closed formula in prenex form in T. Then **A** is a theorem of T iff there is a quasi-tautology which is a disjunction of instances of the matrix of $\mathbf{A}_H$.

Proof. Let T' be obtained from T by adding the new function symbols of $\mathbf{A}_H$. We shall show that $\vdash_T \mathbf{A}$ iff $\vdash_{T'} \mathbf{A}_H$; the theorem will then follow immediately by Lemma 1.

Using the notation of the definition of $\mathbf{A}_H$, we have $\vdash_{T'} \forall \mathbf{y}\mathbf{B} \rightarrow \mathbf{B}_{\mathbf{y}}[\mathbf{f}\mathbf{x}_1 \ldots \mathbf{x}_n]$ by the substitution theorem, and hence $\vdash_{T'} \mathbf{A} \rightarrow \mathbf{A}^*$ by the distribution rule. Similarly $\vdash_{T'} \mathbf{A}^* \rightarrow \mathbf{A}^{**}$, $\vdash_{T'} \mathbf{A}^{**} \rightarrow \mathbf{A}^{***}$, etc.; so $\vdash_{T'} \mathbf{A} \rightarrow \mathbf{A}_H$. If $\vdash_T \mathbf{A}$, then $\vdash_{T'} \mathbf{A}$; so $\vdash_{T'} \mathbf{A}_H$ by the detachment rule.

In proving the converse we shall suppose, to simplify the notation, that **A** is $\exists \mathbf{x} \forall \mathbf{y} \exists \mathbf{z} \forall \mathbf{w}\mathbf{B}$ with **B** open. (The method of proof will be perfectly general, however.) When **B** is followed by square brackets, the subscripted variables are understood to be **x**, **y**, **z**, **w**. Hence we may rewrite **A** as

$$\exists \mathbf{x} \forall \mathbf{y} \exists \mathbf{z} \forall \mathbf{w}\mathbf{B}[\mathbf{x}, \mathbf{y}, \mathbf{z}, \mathbf{w}],$$

and $\mathbf{A}_H$ is

$$\exists \mathbf{x} \exists \mathbf{z}\mathbf{B}[\mathbf{x}, \mathbf{f}\mathbf{x}, \mathbf{z}, \mathbf{g}\mathbf{x}\mathbf{z}].$$

We now introduce the following notation: if **a** and **b** are variable-free terms of T'_c, then $\mathbf{r}(\mathbf{a})$ is the special constant for

$$\exists \mathbf{y} \neg \exists \mathbf{z} \forall \mathbf{w}\mathbf{B}[\mathbf{a}, \mathbf{y}, \mathbf{z}, \mathbf{w}]$$

and $\mathbf{s}(\mathbf{a}, \mathbf{b})$ is the special constant for

$$\exists \mathbf{w} \neg \mathbf{B}[\mathbf{a}, \mathbf{r}(\mathbf{a}), \mathbf{b}, \mathbf{w}].$$

From (1) and the substitution theorem, we have in T_c:

$$\begin{aligned}
&\vdash \mathbf{B}[\mathbf{a}, \mathbf{r}(\mathbf{a}), \mathbf{b}, \mathbf{s}(\mathbf{a}, \mathbf{b})] \rightarrow \forall \mathbf{w}\mathbf{B}[\mathbf{a}, \mathbf{r}(\mathbf{a}), \mathbf{b}, \mathbf{w}],\\
&\vdash \forall \mathbf{w}\mathbf{B}[\mathbf{a}, \mathbf{r}(\mathbf{a}), \mathbf{b}, \mathbf{w}] \rightarrow \exists \mathbf{z} \forall \mathbf{w}\mathbf{B}[\mathbf{a}, \mathbf{r}(\mathbf{a}), \mathbf{z}, \mathbf{w}],\\
&\vdash \exists \mathbf{z} \forall \mathbf{w}\mathbf{B}[\mathbf{a}, \mathbf{r}(\mathbf{a}), \mathbf{z}, \mathbf{w}] \rightarrow \forall \mathbf{y} \exists \mathbf{z} \forall \mathbf{w}\mathbf{B}[\mathbf{a}, \mathbf{y}, \mathbf{z}, \mathbf{w}],\\
&\vdash \forall \mathbf{y} \exists \mathbf{z} \forall \mathbf{w}\mathbf{B}[\mathbf{a}, \mathbf{y}, \mathbf{z}, \mathbf{w}] \rightarrow \exists \mathbf{x} \forall \mathbf{y} \exists \mathbf{z} \forall \mathbf{w}\mathbf{B}[\mathbf{x}, \mathbf{y}, \mathbf{z}, \mathbf{w}].
\end{aligned}$$

Thus

$$\vdash \mathbf{B}[\mathbf{a}, \mathbf{r}(\mathbf{a}), \mathbf{b}, \mathbf{s}(\mathbf{a}, \mathbf{b})] \rightarrow \mathbf{A}. \tag{2}$$

We also have

$$\vdash \mathbf{a} = \mathbf{a}' \rightarrow \mathbf{r}(\mathbf{a}) = \mathbf{r}(\mathbf{a}') \tag{3}$$

and

$$\vdash \mathbf{a} = \mathbf{a}' \rightarrow \mathbf{b} = \mathbf{b}' \rightarrow \mathbf{s}(\mathbf{a}, \mathbf{b}) = \mathbf{s}(\mathbf{a}', \mathbf{b}'). \tag{4}$$

To prove (3), we use the special equality axiom

$$\forall \mathbf{y}(\neg \exists \mathbf{z} \forall \mathbf{w}\mathbf{B}[\mathbf{a}, \mathbf{y}, \mathbf{z}, \mathbf{w}] \leftrightarrow \neg \exists \mathbf{z} \forall \mathbf{w}\mathbf{B}[\mathbf{a}', \mathbf{y}, \mathbf{z}, \mathbf{w}]) \rightarrow \mathbf{r}(\mathbf{a}) = \mathbf{r}(\mathbf{a}')$$

and the equality theorem. To prove (4), we use the special equality axiom

$$\forall \mathbf{w}(\neg \mathbf{B}[\mathbf{a}, \mathbf{r}(\mathbf{a}), \mathbf{b}, \mathbf{w}] \leftrightarrow \neg \mathbf{B}[\mathbf{a}', \mathbf{r}(\mathbf{a}'), \mathbf{b}', \mathbf{w}]) \rightarrow \mathbf{s}(\mathbf{a}, \mathbf{b}) = \mathbf{s}(\mathbf{a}', \mathbf{b}')$$

with (3) and the equality theorem.

Now assume that $\vdash_{T'} \mathbf{A}_H$. By Lemma 1, there is a quasi-tautology

$$\mathbf{B}[\mathbf{a}_1, \mathbf{fa}_1, \mathbf{b}_1, \mathbf{ga}_1\mathbf{b}_1] \vee \cdots \vee \mathbf{B}[\mathbf{a}_n, \mathbf{fa}_n, \mathbf{b}_n, \mathbf{ga}_n\mathbf{b}_n], \tag{5}$$

where $\mathbf{a}_i$ and $\mathbf{b}_i$ are terms of $L(T')$. We now modify (5) as follows. We first replace each variable by some special constant. Next we choose a part of the form **fa** or **gab**, where **a** and **b** do not contain **f** or **g**. If the part selected is **fa**, we replace it everywhere by **r(a)**; if the part selected is **gab**, we replace it everywhere by **s(a, b)**. We continue to make such replacements until we have eliminated all occurrences of **f** and **g**. There results a formula

$$\mathbf{B}[\mathbf{a}_1', \mathbf{r}(\mathbf{a}_1'), \mathbf{b}_1', \mathbf{s}(\mathbf{a}_1', \mathbf{b}_1')] \vee \cdots \vee \mathbf{B}[\mathbf{a}_n', \mathbf{r}(\mathbf{a}_n'), \mathbf{b}_n', \mathbf{s}(\mathbf{a}_n', \mathbf{b}_n')] \tag{6}$$

of T_c'. We will show that (6) is a theorem of T_c'. It will then follow from (2) and the tautology theorem that **A** is a theorem of T_c' and hence, by Lemma 2, a theorem of T.

The formula (5) is a tautological consequence of instances $\mathbf{C}_1, \ldots, \mathbf{C}_r$ of identity and equality axioms. If we make the same transformations on $\mathbf{C}_1, \ldots, \mathbf{C}_r$ which we made to get (6) from (5), we obtain formulas $\mathbf{C}_1', \ldots, \mathbf{C}_r'$. By a remark in §3.1, (6) is a tautological consequence of $\mathbf{C}_1', \ldots, \mathbf{C}_r'$; so we need only prove the $\mathbf{C}_i'$ in T_c'. But $\mathbf{C}_i'$ is again an instance of an identity or an equality axiom unless $\mathbf{C}_i$ is an instance of an equality axiom

$$\mathbf{x} = \mathbf{x}' \rightarrow \mathbf{fx} = \mathbf{fx}' \quad \text{or} \quad \mathbf{x} = \mathbf{x}' \rightarrow \mathbf{y} = \mathbf{y}' \rightarrow \mathbf{gxy} = \mathbf{gx}'\mathbf{y}'.$$

In this case, $\mathbf{C}_i'$ is a theorem by (3) or (4). This completes the proof of Herbrand's theorem.

4.5 ADDITION OF FUNCTION SYMBOLS

There is a type of reasoning frequently used in mathematics which we have not yet considered. To illustrate it, suppose that we are discussing natural numbers and have proved that for every x, there is a prime y such that $y > x$. In the course of a later proof we might say: let y be a prime such that $y > x$. We would then have to keep in mind through the rest of the proof that y depends upon x. If we wished to indicate this by the notation, we would say instead: for each x, let $f(x)$ be a prime greater than x. Of course f would be a new symbol which does not appear in the result we are trying to prove. Our next result shows that an analogous method applied in a theory T does not lead to any results which cannot be proved in T.

Theorem on Functional Extensions. Let $\mathbf{x}, \mathbf{y}_1, \ldots, \mathbf{y}_n$ be distinct variables, and let $\exists \mathbf{x}\mathbf{A}$ be a theorem of T in which no variable other than $\mathbf{y}_1, \ldots, \mathbf{y}_n$

is free. Let T' be the theory obtained from T by adding a new n-ary function symbol $\mathbf{f}$ and a new nonlogical axiom $\mathbf{A_x[fy_1 \ldots y_n]}$. Then T' is a conservative extension of T.

Proof. In view of the closure theorem, it is sufficient to prove that every closed formula $\mathbf{B}$ of T which is a theorem of T' is a theorem of T. By the reduction theorem, there is a proof using no nonlogical axioms of a formula

$$\forall \mathbf{y}_1 \ldots \forall \mathbf{y}_n \mathbf{A_x[fy_1 \ldots y_n]} \rightarrow \mathbf{C}_1 \rightarrow \cdots \rightarrow \mathbf{C}_k \rightarrow \mathbf{B}, \qquad (1)$$

where each $\mathbf{C}_i$ is the closure of a nonlogical axiom of T. Let $\mathbf{C}$ be a prenex form of $\mathbf{A} \rightarrow \mathbf{C}_1 \rightarrow \cdots \rightarrow \mathbf{C}_k \rightarrow \mathbf{B}$. Then $\exists \mathbf{y}_1 \ldots \exists \mathbf{y}_n \mathbf{C_x[fy_1 \ldots y_n]}$ is a prenex form of (1), and hence is provable without nonlogical axioms.

Let $\mathbf{D}$ be $\exists \mathbf{y}_1 \ldots \exists \mathbf{y}_n \forall \mathbf{x} \mathbf{C}$. Then $\mathbf{D}$ does not contain $\mathbf{f}$. If we use $\mathbf{f}$ as the new function symbol in forming the $\mathbf{D}^*$ of the last section, then $\mathbf{D}^*$ is

$$\exists \mathbf{y}_1 \ldots \exists \mathbf{y}_n \mathbf{C_x[fy_1 \ldots y_n]};$$

so $\mathbf{D}^*$ is provable without nonlogical axioms. Since $\mathbf{D}_H$ is the same as $\mathbf{D}^*_H$, it follows from Herbrand's theorem that $\mathbf{D}$ is also provable without nonlogical axioms, and hence is a theorem of T. From this and the equivalence theorem,

$$\vdash_T \exists \mathbf{y}_1 \ldots \exists \mathbf{y}_n \forall \mathbf{x}(\mathbf{A} \rightarrow \mathbf{C}_1 \rightarrow \cdots \rightarrow \mathbf{C}_k \rightarrow \mathbf{B});$$

so by the prenex operations,

$$\vdash_T \forall \mathbf{y}_1 \ldots \forall \mathbf{y}_n \exists \mathbf{x} \mathbf{A} \rightarrow \mathbf{C}_1 \rightarrow \cdots \rightarrow \mathbf{C}_k \rightarrow \mathbf{B}.$$

Since $\forall \mathbf{y}_1 \ldots \forall \mathbf{y}_n \exists \mathbf{x} \mathbf{A}, \mathbf{C}_1, \ldots, \mathbf{C}_k$ are all theorems of T, it follows by the detachment rule that $\mathbf{B}$ is a theorem of T.

A formula is *universal* if it is in prenex form and all of the quantifiers in its prefix are universal. We shall associate a closed universal formula $\mathbf{A}_S$ with each closed formula $\mathbf{A}$ in prenex form. If $\mathbf{A}$ is universal, then $\mathbf{A}_S$ is $\mathbf{A}$. If not, then $\mathbf{A}$ has the form $\forall \mathbf{x}_1 \ldots \forall \mathbf{x}_n \exists \mathbf{y} \mathbf{B}$ ($n \geqslant 0$). We introduce a new n-ary function symbol $\mathbf{f}$ and let $\mathbf{A}^\circ$ be $\forall \mathbf{x}_1 \ldots \forall \mathbf{x}_n \mathbf{B_y[fx_1 \ldots x_n]}$. Then $\mathbf{A}^\circ$ has one less existential quantifier than $\mathbf{A}$. If $\mathbf{A}^\circ$ is not universal, we form $\mathbf{A}^{\circ\circ}$, $\mathbf{A}^{\circ\circ\circ}$, etc., until we come to a universal formula. This universal formula is $\mathbf{A}_S$.

By repeated use of the theorem on functional extensions, we see that if $\mathbf{A}$ is a closed formula in prenex form which is a theorem of T, and if T' is obtained from T by adding the function symbols of $\mathbf{A}_S$ and the new nonlogical axiom $\mathbf{A}_S$, then T' is a conservative extension of T. We also note that $\mathbf{A}_S \rightarrow \mathbf{A}$ is provable without nonlogical axioms. It will obviously suffice to show that $\mathbf{A}^\circ \rightarrow \mathbf{A}$ can be so proved. Now in the above notation, $\mathbf{B_y[fx_1 \ldots x_n]} \rightarrow \exists \mathbf{y} \mathbf{B}$ is a substitution axiom; so we get $\mathbf{A}^\circ \rightarrow \mathbf{A}$ by the distribution rule.

Skolem's Theorem. Every theory has an open conservative extension.

Proof. Let T be a theory. Let T_1 be obtained from T by replacing each nonlogical axiom by the closure of one of its prenex forms. By §3.5 and the closure theorem, T_1 is equivalent to T. Now obtain T_2 from T_1 as follows: for each nonlogical axiom **A** of T_1, add the function symbols of $\mathbf{A}_S$ and add $\mathbf{A}_S$ as a new axiom. If we use the above results together with the fact that a proof in T_2 can use only a finite number of the new function symbols and axioms, we see that T_2 is a conservative extension of T_1. Now obtain T_3 from T_2 by omitting the nonlogical axioms of T_1. Since $\mathbf{A}_S \rightarrow \mathbf{A}$ is provable without nonlogical axioms, these omitted axioms are provable in T_3; so T_3 is equivalent to T_2. The nonlogical axioms of T_3 are then universal. We obtain T_4 from T_3 by replacing each nonlogical axiom by its matrix. Then T_4 is open; and by the closure theorem, T_4 is equivalent to T_3, and hence is a conservative extension of T.

One use of Skolem's theorem is to give finitary consistency proofs for theories which are not open. Given such a theory T, we construct a conservative open extension T' of T and then prove that T' is consistent by the method of §4.3. If the nonlogical axioms of T are sufficiently simple, this method is quite effective. However, if T has complicated nonlogical axioms, then T' may be so complicated that there is no model of T' which can be described in a finitary manner; or it may be that all such models are too complicated to be handled conveniently.

4.6 EXTENSIONS BY DEFINITIONS

We now turn to the problem of defining new function and predicate symbols in a theory. To illustrate the problem, suppose that we want to introduce $\leqslant$ in N. We can do this by agreeing that $\mathbf{a} \leqslant \mathbf{b}$ is to be an abbreviation for $\mathbf{a} < \mathbf{b} \vee \mathbf{a} = \mathbf{b}$. The difficulty with this is that $\leqslant$ is then a defined symbol and hence not a predicate symbol at all.

A more satisfactory procedure is to form an extension N' of N by adding a new binary predicate symbol $\leqslant$ and a new axiom

$$x \leqslant y \leftrightarrow x < y \vee x = y.$$

From this new axiom we can prove

$$\mathbf{a} \leqslant \mathbf{b} \leftrightarrow \mathbf{a} < \mathbf{b} \vee \mathbf{a} = \mathbf{b}$$

by the substitution rule; and we can then use the equivalence theorem to replace $\mathbf{a} \leqslant \mathbf{b}$ by $\mathbf{a} < \mathbf{b} \vee \mathbf{a} = \mathbf{b}$ and vice versa.

We wish to show that passing from N to N' does not really do any more than introducing the defined symbol $\leqslant$ in N. To make this precise, let **A** be a formula of N'. If we regard $\leqslant$ as a defined symbol in N, then **A** is a defined formula in N which abbreviates a formula $\mathbf{A}^*$ in N. We wish to show that $\vdash_{N'} \mathbf{A}$ iff $\vdash_N \mathbf{A}^*$.

Now let us look at the general situation. We have a theory T; distinct variables $\mathbf{x}_1, \ldots, \mathbf{x}_n$; and a formula **D** of T in which no variable other than $\mathbf{x}_1, \ldots, \mathbf{x}_n$ is free. We form T' from T by adding a new n-ary predicate symbol **p** and a new nonlogical axiom $\mathbf{p}\mathbf{x}_1 \ldots \mathbf{x}_n \leftrightarrow \mathbf{D}$, which we call the *defining axiom* of **p**.

Given a formula **A** of T', we obtain a formula **A*** of T, called the *translation of* **A** *into* T, as follows. We select a variant **D′** of **D** in which no variable of **A** is bound, and we replace each part $\mathbf{pa}_1 \dots \mathbf{a}_n$ of **A** by

$$\mathbf{D}'_{\mathbf{x}_1,\dots,\mathbf{x}_n}[\mathbf{a}_1, \dots, \mathbf{a}_n].$$

The freedom of choice in **D′** is unimportant; different choices of **D′** give different answers for **A***, but these answers are variants of one another.

We shall now show that $\vdash_{T'} \mathbf{A}$ iff $\vdash_T \mathbf{A}^*$. It will suffice to prove:

i) $\vdash_{T'} \mathbf{A} \leftrightarrow \mathbf{A}^*$;

ii) T' is a conservative extension of T.

For then $\vdash_{T'} \mathbf{A}$ iff $\vdash_{T'} \mathbf{A}^*$ by (i), and $\vdash_{T'} \mathbf{A}^*$ iff $\vdash_T \mathbf{A}^*$ by (ii).

We first prove (i). In view of the equivalence theorem, it will suffice to show that (in the above notation)

$$\vdash_{T'} \mathbf{pa}_1 \dots \mathbf{a}_n \leftrightarrow \mathbf{D}'_{\mathbf{x}_1,\dots,\mathbf{x}_n}[\mathbf{a}_1, \dots, \mathbf{a}_n].$$

This follows from the defining axiom of **p** by the variant theorem and the substitution rule.

If **A** is a formula of T, then **A*** is **A**. Hence to prove (ii), it suffices to prove that $\vdash_T \mathbf{A}^*$ for every theorem **A** of T'. We do this by induction on theorems (in the form described in §3.1).

Suppose that **A** is a substitution axiom $\mathbf{B}_\mathbf{x}[\mathbf{a}] \rightarrow \exists\mathbf{x}\mathbf{B}$. As we easily prove by induction on the length of **B**, $\mathbf{B}_\mathbf{x}[\mathbf{a}]^*$ is a variant of $\mathbf{B}^*_\mathbf{x}[\mathbf{a}]$. Hence **A*** is a variant of the substitution axiom $\mathbf{B}^*_\mathbf{x}[\mathbf{a}] \rightarrow \exists\mathbf{x}\mathbf{B}^*$. If **A** is an identity axiom or an equality axiom not containing **p**, then **A*** is **A**. If **A** is an equality axiom,

$$\mathbf{y}_1 = \mathbf{y}'_1 \rightarrow \dots \rightarrow \mathbf{y}_n = \mathbf{y}'_n \rightarrow \mathbf{py}_1 \dots \mathbf{y}_n \rightarrow \mathbf{py}'_1 \dots \mathbf{y}'_n,$$

then **A*** is

$$\mathbf{y}_1 = \mathbf{y}'_1 \rightarrow \dots \rightarrow \mathbf{y}_n = \mathbf{y}'_n \rightarrow \mathbf{D}'[\mathbf{y}_1, \dots, \mathbf{y}_n] \rightarrow \mathbf{D}'[\mathbf{y}'_1, \dots, \mathbf{y}'_n].$$

This follows from Corollary 2 to the equality theorem. If **A** is a nonlogical axiom of T, then **A*** is **A**. If **A** is the defining axiom of **p**, then **A*** is $\mathbf{D}' \leftrightarrow \mathbf{D}$, which is a theorem by the variant theorem.

If **A** is a tautological consequence of $\mathbf{B}_1, \dots, \mathbf{B}_n$, then **A*** is a tautological consequence of $\mathbf{B}_1^*, \dots, \mathbf{B}_n^*$; so **A*** is a theorem by the induction hypothesis and the tautology theorem. If **A** is $\exists\mathbf{x}\mathbf{B} \rightarrow \mathbf{C}$ and is inferred from $\mathbf{B} \rightarrow \mathbf{C}$ by the $\exists$-introduction rule, then **A*** is $\exists\mathbf{x}\mathbf{B}^* \rightarrow \mathbf{C}^*$. By induction hypothesis, $\vdash \mathbf{B}^* \rightarrow \mathbf{C}^*$; and, since **x** is not free in **C**, it is not free in **C***. Hence $\vdash \exists\mathbf{x}\mathbf{B}^* \rightarrow \mathbf{C}^*$ by the $\exists$-introduction rule.

Now let us consider the analogous problem for function symbols. As an example, suppose that we are in a theory containing $\cdot$ in which the individuals are the positive real numbers, and that we wish to define the square root function

$\sqrt{\ }$. We would add $\sqrt{\ }$ as a new unary function symbol and add a new axiom

$$y = \sqrt{x} \leftrightarrow y \cdot y = x.$$

Before we do this, we should be able to prove in our theory that every individual has one and only one square root; i.e., we should be able to prove

$$\exists y(y \cdot y = x)$$

and

$$y \cdot y = x \;\&\; y' \cdot y' = x \rightarrow y = y'.$$

Assuming that all of this is done, how can we translate a formula $..\sqrt{x}..$ containing the new symbol back into the original theory? One way is to translate it as

$$\exists y(y \cdot y = x \;\&\; ..y..).$$

We can now describe the general situation. We have a theory T; distinct variables $\mathbf{x}_1, \ldots, \mathbf{x}_n$, $\mathbf{y}$, $\mathbf{y}'$; and a formula $\mathbf{D}$ in which no variable other than $\mathbf{x}_1, \ldots, \mathbf{x}_n$, $\mathbf{y}$ is free. We have

$$\vdash_T \exists \mathbf{y}\mathbf{D} \tag{1}$$

and

$$\vdash_T \mathbf{D} \;\&\; \mathbf{D_y}[\mathbf{y}'] \rightarrow \mathbf{y} = \mathbf{y}'. \tag{2}$$

We form T' from T by adding a new n-ary function symbol $\mathbf{f}$ and a new nonlogical axiom $\mathbf{y} = \mathbf{f}\mathbf{x}_1 \ldots \mathbf{x}_n \leftrightarrow \mathbf{D}$, which we call the *defining axiom* of $\mathbf{f}$. We call (1) the *existence condition* and (2) the *uniqueness condition* for $\mathbf{f}$.

We now define the formula $\mathbf{A}^*$ of T corresponding to a formula $\mathbf{A}$ of T'. We do this only for atomic $\mathbf{A}$; in the general case, $\mathbf{A}^*$ is obtained by replacing each atomic part $\mathbf{B}$ of $\mathbf{A}$ by $\mathbf{B}^*$. Our definition is by induction on the number of occurrences of $\mathbf{f}$ in $\mathbf{A}$. If there are no such occurrences, then $\mathbf{A}^*$ is $\mathbf{A}$. Otherwise, $\mathbf{A}$ can be written as $\mathbf{B_z}[\mathbf{f}\mathbf{a}_1 \ldots \mathbf{a}_n]$, where $\mathbf{a}_1, \ldots, \mathbf{a}_n$ do not contain $\mathbf{f}$ and $\mathbf{B}$ is an atomic formula containing one less occurrence of $\mathbf{f}$ than $\mathbf{A}$. We choose a variant $\mathbf{D}'$ of $\mathbf{D}$ in which no variable of $\mathbf{A}$ is bound, and let $\mathbf{A}^*$ be

$$\exists \mathbf{z}(\mathbf{D}'_{\mathbf{x}_1,\ldots,\mathbf{x}_n,\mathbf{y}}[\mathbf{a}_1, \ldots, \mathbf{a}_n, \mathbf{z}] \;\&\; \mathbf{B}^*).$$

We again wish to show that $\vdash_{T'} \mathbf{A}$ iff $\vdash_T \mathbf{A}^*$. Again it suffices to prove:

i) $\vdash_{T'} \mathbf{A} \leftrightarrow \mathbf{A}^*$;

ii) T' is a conservative extension of T.

It suffices to prove (i) for $\mathbf{A}$ atomic. We do this by induction on the number of occurrences of $\mathbf{f}$ in $\mathbf{A}$. If there are none, $\mathbf{A} \leftrightarrow \mathbf{A}^*$ is the tautology $\mathbf{A} \leftrightarrow \mathbf{A}$. Now suppose that $\mathbf{f}$ occurs in $\mathbf{A}$, and use the above notation. From the defining axiom of $\mathbf{f}$, the variant theorem, and the substitution rule,

$$\vdash \mathbf{z} = \mathbf{f}\mathbf{a}_1 \ldots \mathbf{a}_n \leftrightarrow \mathbf{D}'[\mathbf{a}_1, \ldots, \mathbf{a}_n, \mathbf{z}];$$

so by the equivalence theorem

$$\vdash \exists \mathbf{z}(\mathbf{z} = \mathbf{fa}_1 \ldots \mathbf{a}_n \ \& \ \mathbf{B}^*) \leftrightarrow \mathbf{A}^*.$$

Since $\vdash \mathbf{B} \leftrightarrow \mathbf{B}^*$ by induction hypothesis,

$$\vdash \exists \mathbf{z}(\mathbf{z} = \mathbf{fa}_1 \ldots \mathbf{a}_n \ \& \ \mathbf{B}) \leftrightarrow \mathbf{A}^*;$$

so by Corollary 3 to the equality theorem,

$$\vdash \mathbf{B}_\mathbf{z}[\mathbf{fa}_1 \ldots \mathbf{a}_n] \leftrightarrow \mathbf{A}^*,$$

that is, $\vdash \mathbf{A} \leftrightarrow \mathbf{A}^*$.

To prove (ii), let T'' be obtained from T by adding **f** and the new axiom

$$\mathbf{D}_\mathbf{y}[\mathbf{fx}_1 \ldots \mathbf{x}_n]. \tag{3}$$

By (1) and the theorem on functional extensions, T'' is a conservative extension of T. Hence it will suffice to show that T' and T'' are equivalent. We can prove (3) in T' by substituting $\mathbf{fx}_1 \ldots \mathbf{x}_n$ for **y** in the defining axiom of **f** and using the identity axioms. Now by the equality theorem,

$$\vdash_{T''} \mathbf{y} = \mathbf{fx}_1 \ldots \mathbf{x}_n \rightarrow (\mathbf{D} \leftrightarrow \mathbf{D}_\mathbf{y}[\mathbf{fx}_1 \ldots \mathbf{x}_n]) \tag{4}$$

while by (2) and the substitution rule,

$$\vdash_{T''} \mathbf{D} \ \& \ \mathbf{D}_\mathbf{y}[\mathbf{fx}_1 \ldots \mathbf{x}_n] \rightarrow \mathbf{y} = \mathbf{fx}_1 \ldots \mathbf{x}_n. \tag{5}$$

From (3), (4), and (5) we can derive the defining axiom of **f** by the tautology theorem.

Remark. The equivalence of T' and T'' shows that we could equally well have adopted (3) as our new axiom. We shall therefore sometimes also call (3) the defining axiom of **f**.

A special case which occurs frequently is that **D** is $\mathbf{y} = \mathbf{a}$, where **a** is a term containing no variable other than $\mathbf{x}_1, \ldots, \mathbf{x}_n$. The defining axiom of **f** (in the form (3)) then becomes $\mathbf{fx}_1 \ldots \mathbf{x}_n = \mathbf{a}$. The existence and uniqueness conditions are always provable in this case. For the existence condition $\exists \mathbf{y}(\mathbf{y} = \mathbf{a})$ follows from $\mathbf{a} = \mathbf{a}$ and the substitution axioms; and the uniqueness condition

$$\mathbf{y} = \mathbf{a} \ \& \ \mathbf{y}' = \mathbf{a} \rightarrow \mathbf{y} = \mathbf{y}'$$

follows from

$$\mathbf{y}' = \mathbf{a} \rightarrow (\mathbf{y} = \mathbf{y}' \leftrightarrow \mathbf{y} = \mathbf{a}),$$

which is a case of the equality theorem.

We say that T' is an *extension by definitions* of T if T' is obtained from T by a finite number of extensions of the two types which we have described. When this is the case, we have for each formula **A** of T' a formula $\mathbf{A}^*$ of T, the translation of **A** into T, such that $\vdash_{T'} \mathbf{A}$ iff $\vdash_T \mathbf{A}^*$. Moreover, T' is a conservative extension of T; so T' is consistent iff T is consistent.

Now suppose that T' is an extension by definitions of T and that $\mathcal{A}$ is a model of T. We claim that there is a unique expansion $\mathcal{A}'$ of $\mathcal{A}$ which is a model of T'. It is clearly sufficient to verify this when T' contains only one nonlogical symbol not in T. If this symbol is a predicate symbol $\mathbf{p}$ with the defining axiom

$$\mathbf{p}\mathbf{x}_1 \ldots \mathbf{x}_n \leftrightarrow \mathbf{D},$$

then

$$\mathbf{p}_{\mathcal{A}'}(a_1, \ldots, a_n) \leftrightarrow \mathcal{A}(\mathbf{D}_{\mathbf{x}_1,\ldots,\mathbf{x}_n}[\mathbf{i}_1, \ldots, \mathbf{i}_n]) = \mathbf{T},$$

where $\mathbf{i}_1, \ldots, \mathbf{i}_n$ are the names of $a_1, \ldots, a_n$. If the new symbol is a function symbol $\mathbf{f}$ with the defining axiom $\mathbf{f}\mathbf{x}_1 \ldots \mathbf{x}_n = \mathbf{y} \leftrightarrow \mathbf{D}$, then $\mathbf{f}_{\mathcal{A}'}(a_1, \ldots, a_n)$ is the unique b such that

$$\mathcal{A}(\mathbf{D}_{\mathbf{x}_1,\ldots,\mathbf{x}_n,\mathbf{y}}[\mathbf{i}_1, \ldots, \mathbf{i}_n, \mathbf{j}]) = \mathbf{T},$$

where $\mathbf{i}_1, \ldots, \mathbf{i}_n, \mathbf{j}$ are the names of $a_1, \ldots, a_n, b$. The fact that such a b exists and is unique follows from the fact that the existence and uniqueness conditions for $\mathbf{f}$ are valid in $\mathcal{A}$.

4.7 INTERPRETATIONS

We have so far discussed structures in English. We could, of course, translate the entire discussion into any language in which there is sufficient set-theoretic notation to discuss functions, predicates, etc.

If we only wish to discuss a particular structure $\mathcal{A}$ for L, it is not necessary to have so rich a language. It suffices to have a language L' which has a symbol for the universe of $\mathcal{A}$ and a symbol for each function and predicate of $\mathcal{A}$. This raises a slight difficulty: the functions and predicates designated by symbols of L' take all individuals of L' as arguments, while the functions and predicates of $\mathcal{A}$ take only the individuals of $\mathcal{A}$ as arguments. This is overcome by allowing the symbol of L' to designate any function or predicate which is an extension of the function or predicate of $\mathcal{A}$.

Let L and L' be first-order languages. An *interpretation* I of L in L' consists of:

i) a unary predicate symbol U_I of L', called the *universe* of I;

ii) for each n-ary function symbol $\mathbf{f}$ of L, an n-ary function symbol $\mathbf{f}_I$ of L';

iii) for each n-ary predicate symbol $\mathbf{p}$ of L other than $=$, an n-ary predicate symbol $\mathbf{p}_I$ of L'.

An *interpretation* of L in a theory T' is an interpretation I of L in $L(T')$ such that

$$\vdash_{T'} \exists \mathbf{x} U_I \mathbf{x} \tag{1}$$

and

$$\vdash_{T'} U_I\mathbf{x}_1 \rightarrow \cdots \rightarrow U_I\mathbf{x}_n \rightarrow U_I\mathbf{f}_I\mathbf{x}_1 \ldots \mathbf{x}_n \tag{2}$$

for each $\mathbf{f}$ in L. The first condition requires that the universe be nonempty; the second requires that $\mathbf{f}_I$ represent a function whose restriction to the universe of I

takes values in the universe of I. (Both (1) and (2) are to hold for all choices of the variables; but it is clear that if (1) holds for one $\mathbf{x}$, and, for each $\mathbf{f}$, (2) holds for one set $\mathbf{x}_1, \ldots, \mathbf{x}_n$ of distinct variables, then (1) and (2) hold for all choices of the variables.)

Let I be an interpretation of L in L'. We will define for each formula $\mathbf{A}$ of L a formula $\mathbf{A}^{(I)}$ of L', called the *interpretation* of $\mathbf{A}$ by I, whose meaning is that $\mathbf{A}$ is valid in the structure being discussed. We let $\mathbf{A}_I$ be the formula of L' obtained from $\mathbf{A}$ by the two following steps:

a) replace each nonlogical symbol $\mathbf{u}$ by $\mathbf{u}_I$;

b) replace each part $\exists\mathbf{x}\mathbf{B}$ by $\exists\mathbf{x}(U_I\mathbf{x} \,\&\, \mathbf{B})$.

Then $\mathbf{A}^{(I)}$ is

$$U_I\mathbf{x}_1 \rightarrow \cdots \rightarrow U_I\mathbf{x}_n \rightarrow \mathbf{A}_I,$$

where $\mathbf{x}_1, \ldots, \mathbf{x}_n$ are the variables free in $\mathbf{A}$ (and hence in $\mathbf{A}_I$) in alphabetical order. If $\mathbf{a}$ is a term of L, $\mathbf{a}_I$ designates the term obtained from $\mathbf{a}$ by step (a) above.

We can now define the notion which corresponds to a model in the present situation. An *interpretation* of a theory T in a theory T' is an interpretation I of $L(T)$ in T' such that $\vdash_{T'} \mathbf{A}^{(I)}$ for every nonlogical axiom $\mathbf{A}$ of T.

Our next result is the analogue of the validity theorem in the present situation.

Interpretation Theorem. If I is an interpretation of T in T', then $\vdash_{T'} \mathbf{A}^{(I)}$ for every theorem $\mathbf{A}$ of T.

Proof. We need two preliminary results. First, if $\mathbf{a}$ is a term of T and $\mathbf{x}_1, \ldots, \mathbf{x}_n$ are all the variables in $\mathbf{a}$, then

$$\vdash U_I\mathbf{x}_1 \rightarrow \cdots \rightarrow U_I\mathbf{x}_n \rightarrow U_I\mathbf{a}_I. \tag{3}$$

(Here and in what follows, $\vdash$ means $\vdash_{T'}$.) The proof of (3) is by induction on the length of $\mathbf{a}$, and uses (2). We leave the details to the reader.

The second preliminary result is that if $\mathbf{x}_1, \ldots, \mathbf{x}_n$ include the variables free in $\mathbf{A}$, and if

$$\vdash U_I\mathbf{x}_1 \rightarrow \cdots \rightarrow U_I\mathbf{x}_n \rightarrow \mathbf{A}_I, \tag{4}$$

then $\vdash \mathbf{A}^{(I)}$. From (4) by the tautology theorem, we obtain

$$\vdash U_I\mathbf{y}_1 \rightarrow \cdots \rightarrow U_I\mathbf{y}_k \rightarrow \mathbf{A}^{(I)}, \tag{5}$$

where $\mathbf{y}_1, \ldots, \mathbf{y}_k$ are distinct and are not free in $\mathbf{A}$ and hence not free in $\mathbf{A}^{(I)}$. Applying the $\exists$-introduction rule to (5), and then using (1) and the detachment rule, we have

$$\vdash U_I\mathbf{y}_2 \rightarrow \cdots \rightarrow U_I\mathbf{y}_k \rightarrow \mathbf{A}^{(I)}.$$

We have now merely to repeat this step $k - 1$ more times.

We now prove the theorem by induction on theorems in T (in the form described in §3.1). Suppose that $\mathbf{A}$ is a substitution axiom $\mathbf{B}_\mathbf{x}[\mathbf{a}] \rightarrow \exists\mathbf{x}\mathbf{B}$, and let

$\mathbf{y}_1, \ldots, \mathbf{y}_n$ be the variables other than $\mathbf{x}$ free in $\mathbf{B}$ and $\mathbf{a}$. By (3),

$$\vdash U_I\mathbf{y}_1 \to \cdots \to U_I\mathbf{y}_n \to U_I\mathbf{a}_I.$$

By the substitution theorem,

$$\vdash(U_I\mathbf{a}_I \mathbin{\&} (\mathbf{B}_I)_\mathbf{x}[\mathbf{a}_I]) \to \exists\mathbf{x}(U_I\mathbf{x} \mathbin{\&} \mathbf{B}_I).$$

From these results by the tautology theorem,

$$\vdash U_I\mathbf{y}_1 \to \cdots \to U_I\mathbf{y}_n \to (\mathbf{B}_I)_\mathbf{x}[\mathbf{a}_I] \to \exists\mathbf{x}(U_I\mathbf{x} \mathbin{\&} \mathbf{B}_I),$$

that is,

$$\vdash U_I\mathbf{y}_1 \to \cdots \to U_I\mathbf{y}_n \to \mathbf{A}_I.$$

Hence $\vdash\mathbf{A}^{(I)}$ by the second preliminary result.

If $\mathbf{A}$ is an identity or an equality axiom, then so is $\mathbf{A}_I$; so $\vdash\mathbf{A}_I$ and hence $\vdash\mathbf{A}^{(I)}$. If $\mathbf{A}$ is a nonlogical axiom, then $\vdash\mathbf{A}^{(I)}$ by hypothesis.

Suppose that $\mathbf{A}$ is a tautological consequence of $\mathbf{B}_1, \ldots, \mathbf{B}_k$. By a remark in §3.1, $\mathbf{A}_I$ is a tautological consequence of $(\mathbf{B}_1)_I, \ldots, (\mathbf{B}_k)_I$. It follows that if $\mathbf{x}_1, \ldots, \mathbf{x}_n$ are all the variables free in $\mathbf{A}, \mathbf{B}_1, \ldots, \mathbf{B}_k$, then (4) is a tautological consequence of $\mathbf{B}_1^{(I)}, \ldots, \mathbf{B}_k^{(I)}$. Hence $\vdash\mathbf{A}^{(I)}$ by the induction hypothesis, the tautology theorem, and the second preliminary result.

Suppose that $\mathbf{A}$ is $\exists\mathbf{x}\mathbf{B} \to \mathbf{C}$, and is inferred from $\mathbf{B} \to \mathbf{C}$ by the $\exists$-introduction rule. Let $\mathbf{x}_1, \ldots, \mathbf{x}_n$ be the variables free in $\mathbf{A}$. Then $\mathbf{x}$ is not an $\mathbf{x}_i$. By induction hypothesis and the tautology theorem,

$$\vdash U_I\mathbf{x}_1 \to \cdots \to U_I\mathbf{x}_n \to U_I\mathbf{x} \to \mathbf{B}_I \to \mathbf{C}_I.$$

Hence by the tautology theorem and the $\exists$-introduction rule,

$$\vdash\exists\mathbf{x}(U_I\mathbf{x} \mathbin{\&} \mathbf{B}_I) \to U_I\mathbf{x}_1 \to \cdots \to U_I\mathbf{x}_n \to \mathbf{C}_I,$$

from which $\vdash\mathbf{A}^{(I)}$ by the tautology theorem.

Corollary. If I is an interpretation of T in T', and T' is consistent, then T is consistent.

Proof. Suppose that T is inconsistent, and let $\mathbf{A}$ be a closed formula of T. Then $\vdash_T \mathbf{A}$ and $\vdash_T \neg\mathbf{A}$. Since $(\neg\mathbf{A})^{(I)}$ is $\neg\mathbf{A}^{(I)}$, $\mathbf{A}^{(I)}$ and $\neg\mathbf{A}^{(I)}$ are theorems of T', contradicting the consistency of T'.

Since the proof of the interpretation theorem is finitary, we may use the corollary in giving finitary proofs of consistency.

In forming $\mathbf{A}_I$ and $\mathbf{A}^{(I)}$, we are supposed to eliminate defined symbols. However, it is clearly unnecessary to eliminate $\to$, &, and $\leftrightarrow$. If $\mathbf{A}$ contains a part $\forall\mathbf{x}\mathbf{B}$, then this becomes $\neg\exists\mathbf{x}\neg\mathbf{B}$ upon elimination of $\forall$, and thus gives rise to a part

$$\neg\exists\mathbf{x}(U_I\mathbf{x} \mathbin{\&} \neg\mathbf{B})$$

in $\mathbf{A}_I$ and $\mathbf{A}^{(I)}$. By the tautology theorem and the equivalence theorem, this part is equivalent to

$$\neg\exists \mathbf{x} \neg(U_I\mathbf{x} \rightarrow \mathbf{B}),$$

that is, to $\forall\mathbf{x}(U_I\mathbf{x} \rightarrow \mathbf{B})$. Hence we may simply replace a part $\forall\mathbf{x}\mathbf{B}$ by $\forall\mathbf{x}(U_I\mathbf{x} \rightarrow \mathbf{B})$.

Let I be an interpretation of T in T', and let U be an extension by definitions of T. We are going to show that, under suitable hypotheses, I can be extended to an interpretation of U in an extension by definitions of T'.

First suppose that U is obtained from T by adding a new predicate symbol $\mathbf{p}$ with the defining axiom

$$\mathbf{p}\mathbf{x}_1 \ldots \mathbf{x}_n \leftrightarrow \mathbf{D}. \tag{6}$$

We form an extension by definitions U' of T' by adding a new predicate symbol $\mathbf{p}'$ with the defining axiom

$$\mathbf{p}'\mathbf{x}_1 \ldots \mathbf{x}_n \leftrightarrow \mathbf{D}_I. \tag{7}$$

We then extend I by letting $\mathbf{p}_I$ be $\mathbf{p}'$. To see that I is an interpretation of U in U', we must see that the interpretation of (6) is provable in U'. This interpretation is

$$U_I\mathbf{x}_1 \rightarrow \cdots \rightarrow U_I\mathbf{x}_n \rightarrow (\mathbf{p}'\mathbf{x}_1 \ldots \mathbf{x}_n \leftrightarrow \mathbf{D}_I)$$

and hence is provable from (7).

In treating function symbols, there is a new problem. If $\mathbf{f}$ is a new function symbol in an extension by definitions of T, then $\mathbf{f}$ designates a unique function on the universe of our model; and we have to extend this function to the universe of T'. Our solution is to pick an individual of T' and assign it as the value for all new sets of arguments.

We therefore suppose that a constant $\mathbf{e}$ in T' is fixed. Suppose that U is obtained from T by adding a new function symbol $\mathbf{f}$ with the defining axiom

$$\mathbf{f}\mathbf{x}_1 \ldots \mathbf{x}_n = \mathbf{y} \leftrightarrow \mathbf{D}. \tag{8}$$

We form an extension by definitions U' of T' by adding a new function symbol $\mathbf{f}'$ with the defining axiom

$$\begin{aligned}\mathbf{f}'\mathbf{x}_1 \ldots \mathbf{x}_n = \mathbf{y} \leftrightarrow & (U_I\mathbf{x}_1 \;\&\; \cdots \;\&\; U_I\mathbf{x}_n \;\&\; \mathbf{D}_I \;\&\; U_I\mathbf{y}) \\ & \vee (\neg(U_I\mathbf{x}_1 \;\&\; \cdots \;\&\; U_I\mathbf{x}_n) \;\&\; \mathbf{y} = \mathbf{e}).\end{aligned} \tag{9}$$

It is easy to verify that the interpretations of the existence and uniqueness conditions for $\mathbf{f}$ imply the existence and uniqueness conditions for $\mathbf{f}'$. Again we extend I by letting $\mathbf{f}_I$ be $\mathbf{f}'$. We must then prove

$$U_I\mathbf{x}_1 \rightarrow \cdots \rightarrow U_I\mathbf{x}_n \rightarrow U_I\mathbf{f}'\mathbf{x}_1 \ldots \mathbf{x}_n.$$

This follows from (9). The interpretation of (8) is

$$U_I\mathbf{x}_1 \rightarrow \cdots \rightarrow U_I\mathbf{x}_n \rightarrow U_I\mathbf{y} \rightarrow (\mathbf{f}'\mathbf{x}_1 \ldots \mathbf{x}_n = \mathbf{y} \leftrightarrow \mathbf{D}_I),$$

which also follows from (9).

We say that T is *interpretable* in T' if there is an interpretation I of T in an extension by definitions of T'. From the corollary to the interpretation theorem, we see that if T is interpretable in T' and T' is consistent, then T is consistent. (The proof of this is entirely finitary.) By the above discussion, we see that if T is interpretable in T', and if there is a constant in T' or in an extension by definitions of T', then every extension by definitions of T is interpretable in T'.

PROBLEMS

1. Prove Lindenbaum's theorem for a theory T with only countably many nonlogical symbols without using the Teichmüller-Tukey lemma. [Show that there are only countably many formulas in T. If $\mathbf{A}_1, \mathbf{A}_2, \ldots$ are all the closed formulas, choose $\mathbf{B}_i$ inductively so that $\mathbf{B}_i$ is either $\mathbf{A}_i$ or $\neg\mathbf{A}_i$ and $T[\mathbf{B}_1, \ldots, \mathbf{B}_i]$ is consistent. Add the $\mathbf{B}_i$ to T as new axioms.]

2. Let L be a first-order language containing a constant. A set Γ of variable-free formulas of L is *tautologically inconsistent* if there is a quasi-tautology which is a disjunction of negations of formulas in Γ; otherwise, Γ is *tautologically consistent*. If Γ is tautologically consistent, and if for every variable-free formula $\mathbf{A}$, either $\mathbf{A}$ or $\neg\mathbf{A}$ is in Γ, then Γ is *tautologically complete*.

a) Show that every tautologically consistent set is a subset of a tautologically complete set. [Show that a maximal tautologically consistent set is tautologically complete, and use the Teichmüller-Tukey lemma.]

b) Show that if Γ is tautologically complete, then there is a structure $\mathcal{A}$ for L such that for every open formula $\mathbf{A}$, $\mathbf{A}$ is valid in $\mathcal{A}$ iff every variable-free instance of $\mathbf{A}$ is in Γ. [Choose $\mathcal{A}$ similar to the canonical structure for a theory, using Γ to replace the set of theorems of the theory.]

c) Use (a), (b), and Skolem's theorem to give a new proof of the completeness theorem.

d) Use (a), (b), and the completeness theorem to give a new proof of the consistency theorem.

3. a) Let $L(T')$ be an extension of $L(T)$. Show that T' is an extension of T iff the restriction to $L(T)$ of every model of T' is a model of T.

b) Show that if T' is an extension of T, and every model of T has an expansion which is a model of T', then T' is a conservative extension of T.

4. a) Prove the theorem on functional extensions by means of models. [Use 3(b).]

b) Use Lemma 1 of §4.4 and (a) to give a new proof of Herbrand's theorem.

5. Let Γ be a set of formulas in L, and let $\mathfrak{T}(\Gamma)$ be the set of mappings from Γ to the set of truth values. We consider $\mathfrak{T}(\Gamma)$ as the product space $\Pi_{\mathbf{A}\in\Gamma}T_{\mathbf{A}}$, where each $T_{\mathbf{A}}$ is the set of truth values. If we give each $T_{\mathbf{A}}$ the discrete topology, $\mathfrak{T}(\Gamma)$ becomes a compact Hausdorff space by Tychonoff's theorem. Let $\mathfrak{TV}(\Gamma)$ be the set of V in $\mathfrak{T}(\Gamma)$ such that $V(\neg\mathbf{A}) = H_\neg(V(\mathbf{A}))$ whenever $\mathbf{A}$ and $\neg\mathbf{A}$ are in Γ, and $V(\mathbf{A} \vee \mathbf{B}) = H_\vee(V(\mathbf{A}), V(\mathbf{B}))$ whenever $\mathbf{A}$, $\mathbf{B}$, and $\mathbf{A} \vee \mathbf{B}$ are in Γ.

a) Show that for any formula $\mathbf{A}$ in Γ, the set of V in $\mathfrak{T}(\Gamma)$ such that $V(\mathbf{A}) = \mathbf{T}$ is open and closed.

b) Show that the space $\mathfrak{T}\mathfrak{B}(\Gamma)$ is closed in $\mathfrak{T}(\Gamma)$ and hence is compact. [Use (a).]

c) Use (a) and (b) to give a new solution of 2(a). [Make use of the finite intersection property.]

6. The *disjuncts* of a formula **A** are defined by induction on the length of **A** as follows: if **A** is not a disjunction, the only disjunct of **A** is **A**; if **A** is **B** $\vee$ **C**, then the disjuncts of **A** are the disjuncts of **B** and the disjuncts of **C**.

Let T be a theory, and let T' be the following formal system. The language of T' is the language of T. The axioms of T' are the identity axioms, the equality axioms, and the nonlogical axioms of T and all formulas $\neg$**A** $\vee$ **A** with **A** atomic. The rules of T' are:

i) $\vee$*-rule:* infer **B** from **A** if every disjunct of **A** is a disjunct of **B**;

ii) $\neg\neg$*-rule:* infer $\neg\neg$**A** $\vee$ **B** from **A** $\vee$ **B**;

iii) $\neg\vee$*-rule:* infer $\neg$(**A** $\vee$ **B**) $\vee$ **C** from $\neg$**B** $\vee$ **C** and $\neg$**A** $\vee$ **C**;

iv) $\exists$*-rule:* infer $\exists$**xA** $\vee$ **B** from $\mathbf{A_x[a]}$ $\vee$ **B**;

v) $\neg\exists$*-rule:* infer $\neg\exists$**xA** $\vee$ **B** from $\neg$**A** $\vee$ **B** provided that **x** is not free in **B**;

vi) the cut rule.

Show that T and T' have the same theorems.

7. Let T and T' be as in Problem 6. Let L' be the language obtained from the language of T by adding an infinite number of new constants. Let U be the following formal system. The language of U is L'. The axioms of U are the closed instances in L' of axioms of T'. The rules of U are the same as the rules of T', except that in each rule, the hypotheses and the conclusion are required to be closed, and that the $\neg\exists$-rule is amended to read: infer $\neg\exists$**xA** $\vee$ **B** from $\neg\mathbf{A_x[e]}$ $\vee$ **B**, provided that **e** is a new constant which does not appear in $\neg\exists$**xA** $\vee$ **B**. Show that every closed instance in L' of a theorem of T is a theorem of U.

8. Let T be an open theory, and let U be as in Problem 7. Obtain U' from U by replacing the cut rule by the *weak cut rule*: infer **B** $\vee$ **C** from **A** $\vee$ **B** and $\neg$**A** $\vee$ **C** provided that **A** is variable-free (and **B** and **C** are closed). A formula **A** of U' is a *cut* formula if for each pair of closed formulas **B** and **C** such that $\vdash_{U'}$ **A** $\vee$ **B** and $\vdash_{U'}$ $\neg$**A** $\vee$ **C**, we have $\vdash_{U'}$ **B** $\vee$ **C**.

a) Show that if $\vdash_{U'}$ **A** and $\mathbf{e}_1, \ldots, \mathbf{e}_n$ are new constants, then there is a proof of **A** in U' such that:

i) each formula in the proof except the last is used exactly once as a hypothesis to a rule to infer a later formula;

ii) if the $\neg\exists$-rule is used to infer $\neg\exists$**xB** $\vee$ **C** from $\neg\mathbf{B_x[e]}$ $\vee$ **C**, then **e** is not one of $\mathbf{e}_1, \ldots, \mathbf{e}_n$.

b) Show that if $\vdash_{U'}$ **A**, and **A**$'$ is obtained from **A** by replacing a new constant **e** everywhere by a variable-free term **a**, then $\vdash_{U'}$ **A**$'$. [Use (a).]

c) Show that if **A** is not variable-free and $\vdash_{U'}$ $\neg\neg$**A** $\vee$ **B**, then $\vdash_{U'}$ **A** $\vee$ **B**. [Start with a proof of $\neg\neg$**A** $\vee$ **B** given by (a). In each formula of this proof, replace $\neg\neg$**A** at some places where it occurs as a disjunct by **A**.] Conclude that if **A** is a cut formula, then $\neg$**A** is a cut formula.

d) Show that if $\mathbf{A} \vee \mathbf{B}$ is not variable-free, and $\vdash_{U'} \neg(\mathbf{A} \vee \mathbf{B}) \vee \mathbf{C}$, then $\vdash_{U'} \neg\mathbf{A} \vee \mathbf{C}$ and $\vdash_{U'} \neg\mathbf{B} \vee \mathbf{C}$. [Similar to (c).] Conclude that if **A** and **B** are cut formulas, then $\mathbf{A} \vee \mathbf{B}$ is a cut formula.

e) Show that if $\vdash_{U'} \neg\exists\mathbf{x}\mathbf{A} \vee \mathbf{C}$ and **e** is a new constant not appearing in $\exists\mathbf{x}\mathbf{A} \vee \mathbf{C}$, then $\vdash_{U'} \neg\mathbf{A}_{\mathbf{x}}[\mathbf{e}] \vee \mathbf{C}$. [Similar to (c).]

f) Show that if $\mathbf{A}_{\mathbf{x}}[\mathbf{a}]$ is a cut formula for every variable-free term **a**, then $\exists\mathbf{x}\mathbf{A}$ is a cut formula. [Suppose that $\vdash_{U'} \exists\mathbf{x}\mathbf{A} \vee \mathbf{B}$ and $\vdash_{U'} \neg\exists\mathbf{x}\mathbf{A} \vee \mathbf{C}$. Start with a proof of $\exists\mathbf{x}\mathbf{A} \vee \mathbf{B}$ as described in (a), and replace suitable occurrences of $\exists\mathbf{x}\mathbf{A}$ by **C**. Use the fact that if $\vdash_{U'} \mathbf{A}_{\mathbf{x}}[\mathbf{a}] \vee \mathbf{D}$, then $\vdash_{U'} \mathbf{C} \vee \mathbf{D}$ by (e) and (b).]

g) Show that every theorem of U is a theorem of U'. [Use (c), (d), and (f) to prove that every closed formula is a cut formula.]

h) Use (g) and Problem 7 to give a new proof of the consistency theorem. [Note that a proof of a variable-free formula in U' satisfying (a) can contain no quantifiers.]

9. Let T be an open theory, and let **A** be a closed formula in T in prenex form. Show that $\vdash_T \mathbf{A}$ iff there is a disjunction of instances of the matrix of $\mathbf{A}_H$ which is a tautological consequence of instances of identity axioms, equality axioms, and nonlogical axioms of T. [Use the reduction theorem and Herbrand's theorem.]

10. Show that if **A** is a theorem of T, then there is a proof of **A** in which no nonlogical symbols appear except those in **A** and in the nonlogical axioms of T. [Use the reduction theorem and Herbrand's theorem.]

11. Let **A** be an open formula, and let n be the number of terms occurring in **A**. Show that **A** is a quasi-tautology iff **A** is valid in every structure having n or less individuals. [Show that we may assume that **A** is variable-free. If **A** is not a quasi-tautology, obtain a structure $\mathcal{A}$ in which **A** is not valid by a construction like that of the canonical structure for a theory, replacing the theorems of the theory by the formulas to which a suitable truth valuation assigns **T**.]

12. The theories T and T' are *weakly equivalent* if some extension by definitions of T is equivalent to some extension by definitions of T'.

a) Show that for every theory T, there is a theory T' containing no function symbols which is weakly equivalent to T. [Obtain an extension by definitions U of T by adding for each **f** in T a predicate symbol **p** with a defining axiom $\mathbf{p}\mathbf{x}_1 \ldots \mathbf{x}_n\mathbf{y} \leftrightarrow \mathbf{f}\mathbf{x}_1 \ldots \mathbf{x}_n = \mathbf{y}$. Choose T' in such a way that $\mathbf{y} = \mathbf{f}\mathbf{x}_1 \ldots \mathbf{x}_n \leftrightarrow \mathbf{p}\mathbf{x}_1 \ldots \mathbf{x}_n\mathbf{y}$ can be used as a defining axiom for **f**.]

b) Show that if T is an open theory, then there is a theory T' weakly equivalent to T such that T' contains no function symbols and such that the nonlogical axioms of T' are existential. [Similar to (a).] If T has only one nonlogical axiom and only a finite number of function symbols, show that T' may be chosen to have only one axiom.

c) A formula **A** is *satisfiable* in A if it is valid in some model having A as its universe. A formula is in *Skolem form* if it is in prenex form and contains no function symbols and all of its universal quantifiers precede all of its existential quantifiers. Show that, given a formula **A**, we may construct a closed formula **B** in Skolem form such that for every A, **A** is satisfiable in A iff **B** is satisfiable in A. [Use (b).]

13. Let L be a first-order language.

a) Let $\mathfrak{A}$ be a structure for L having a finite number of individuals, and let Q be a finite set of nonlogical symbols of L. Suppose that we are given $\mathfrak{A}(\mathbf{a})$ for every $\mathbf{a}$ of the form $\mathbf{f}\mathbf{i}_1 \ldots \mathbf{i}_n$ with $\mathbf{f}$ in Q and $\mathfrak{A}(\mathbf{A})$ for every $\mathbf{A}$ of the form $\mathbf{p}\mathbf{i}_1 \ldots \mathbf{i}_n$ with $\mathbf{p}$ in Q. Show that we may compute $\mathfrak{A}(\mathbf{A})$ for every closed formula of $L(\mathfrak{A})$ all of whose nonlogical symbols are in Q.

b) Show that, given a formula $\mathbf{A}$ of L and a number n, we may decide whether $\mathbf{A}$ is valid in every structure for L having n individuals.

c) Suppose that $\mathbf{A}$ is an existential formula of L containing no function symbols other than constants. Show that if $\mathbf{A}$ is not logically valid, then there is a structure in which $\mathbf{A}$ is not valid having n or fewer individuals, where n is the number of variables and constants in $\mathbf{A}$.

d) Show that given an existential formula $\mathbf{A}$ containing no function symbols other than constants, we may decide whether or not $\mathbf{A}$ is logically valid. [Use (b) and (c).]

CHAPTER 5

THE THEORY OF MODELS

5.1 THE COMPACTNESS THEOREM

The purpose of the completeness theorem is to show that every logical consequence of a set of nonlogical axioms can be proved from these nonlogical axioms by means of the logical axioms and rules. It might therefore seem that the consequences of the theorem would be tied to our choice of the logical axioms and rules, and hence that the theorem would not tell us anything about the nature of models. This is not so because the logical axioms and rules which we have chosen have two simple properties; and the fact that a set of logical axioms and rules with these properties suffices to obtain all logical consequences of every set of nonlogical axioms has important consequences.

The first property is expressed in the following statement: if we have a method of deciding whether or not a formula is a nonlogical axiom, then we have a method for deciding whether or not a sequence of formulas is a proof. We will study this property and its consequences in the next chapter.

The second property is that the logical rules are finite. We are going to derive an important consequence of this fact.

A theory T is a *part* of a theory T' if T and T' have the same language and every nonlogical axiom of T is a nonlogical axiom of T'. A theory T is *finitely axiomatized* if it has only a finite number of nonlogical axioms.

Compactness Theorem. A formula $\mathbf{A}$ in a theory T is valid in T iff it is valid in some finitely axiomatized part of T.

Proof. In view of the completeness theorem, we have only to show that a formula is a theorem of T iff it is a theorem in some finitely axiomatized part of T. This is obvious, since only a finite number of nonlogical axioms can be used in a proof.

Corollary. A theory T has a model iff every finitely axiomatized part of T has a model.

Proof. Take the formula of the theorem to be $x \neq x$, noting that $x \neq x$ is not valid in any structure.

The compactness theorem does not depend on the logical axioms and rules, since it does not say anything about them. This is perhaps clearer if we state the theorem in a different way: a formula $\mathbf{A}$ is a logical consequence of a set Γ of formulas iff it is a logical consequence of a finite subset of Γ. It is possible to give a proof which does not utilize logical axioms and rules (see Problem 30); but such a proof is by no means trivial.

We shall give some applications of the compactness theorem to the *elementary theory of fields*. This theory, which we designate by *FL*, has as nonlogical symbols the constants 0, 1, and -1 and the binary function symbols $+$ and $\cdot$. The nonlogical axioms of *FL* are:

FL1. $(x + y) + z = x + (y + z)$.
FL2. $x + 0 = x$.
FL3. $x + (-1 \cdot x) = 0$.
FL4. $x + y = y + x$.
FL5. $(x \cdot y) \cdot z = x \cdot (y \cdot z)$.
FL6. $x \cdot 1 = x$.
FL7. $x \neq 0 \rightarrow \exists y(x \cdot y = 1)$.
FL8. $x \cdot y = y \cdot x$.
FL9. $x \cdot (y + z) = (x \cdot y) + (x \cdot z)$.
FL10. $0 \neq 1$.

This theory has the same relation to fields that *G* has to groups. Briefly stated, the models of *FL* are just the fields.

We can get the elementary theories of certain special types of fields by adding further nonlogical axioms. Thus let $\mathbf{A}_n$ be the formula

$$1 + 1 + \cdots + 1 = 0,$$

where there are *n* occurrences of 1 on the left (and we are using the convention of association to the right). By adding the nonlogical axioms

$$\neg\mathbf{A}_2,\ \neg\mathbf{A}_3, \ldots,\ \neg\mathbf{A}_{n-1},\ \mathbf{A}_n,$$

we get the elementary theory *FL*(n) of fields of characteristic n ($n \geqslant 2$). To get the elementary theory *FL*(0) of fields of characteristic 0, we add all of the $\neg\mathbf{A}_n$ as nonlogical axioms. Since every field has characteristic 0 or a prime, it follows from the completeness theorem that *FL*(n) is inconsistent if n is composite.

We shall now show that if **A** is valid in *FL*(0), then there is an n_0 such that **A** is valid in *FL*(n) for every $n \geqslant n_0$. By the compactness theorem, **A** is valid in some finite part *T* of *FL*(0); and we have only to choose n_0 larger than all the n such that $\neg\mathbf{A}_n$ is a nonlogical axiom of *T*. A consequence of this result is that we cannot replace the infinite number of axioms we added to *FL* to get *FL*(0) by a finite number. If we could, the conjunction of these axioms would be valid in fields of characteristic zero but in no other fields. Choosing n_0 as above, we would conclude that there are no fields of characteristic greater than n_0, which is absurd.

We might also try to form the elementary theory of finite fields; that is, we might look for a simple extension *T* of *FL* whose models are just the finite fields.

However, this cannot be done. To see this, let $\mathbf{B}_n$ be a formula which asserts that there are at least n individuals. (For example, $\mathbf{B}_3$ is

$$\exists x \exists y \exists z(x \neq y \ \& \ x \neq z \ \& \ y \neq z).)$$

Suppose that we had such a theory T, and let T' be obtained from T by adding all the $\mathbf{B}_n$ as nonlogical axioms. Then T' has no model; so some finitely axiomatized part T'' of T' has no model. Choose n_0 larger than all the n such that $\mathbf{B}_n$ is a nonlogical axiom of T'', and choose a finite field $\mathcal{A}$ having more than n_0 elements. Then $\mathcal{A}$ is a model of T'', which is a contradiction.

5.2 ISOMORPHISMS AND SUBSTRUCTURES

Let $\mathcal{A}$ be a structure. We can obtain structures similar to $\mathcal{A}$ by the following process; we replace each individual of $\mathcal{A}$ by a new individual (replacing distinct individuals by distinct individuals), but otherwise leave the functions and predicates of $\mathcal{A}$ unchanged. We shall now describe this process more exactly.

Let ϕ be a mapping from A to B. We say that ϕ is *injective* (or *one-one*) if for every a and a' in A, $\phi(a) = \phi(a')$ implies $a = a'$. We say that ϕ is *surjective* (or *onto*) if for every b in B, there is an a in A such that $\phi(a) = b$. We say that ϕ is *bijective* if it is both injective and surjective.

Now let $\mathcal{A}$ be a structure for L, and let ϕ be a bijective mapping from $|\mathcal{A}|$ to B. We define a structure $\mathcal{B}$ for L with universe B as follows. If $b_1, \ldots, b_n$ are in B, then there are uniquely determined individuals $a_1, \ldots, a_n$ of $\mathcal{A}$ such that $\phi(a_1) = b_1, \ldots, \phi(a_n) = b_n$. We set

$$\mathbf{f}_{\mathcal{B}}(b_1, \ldots, b_n) = \phi(\mathbf{f}_{\mathcal{A}}(a_1, \ldots, a_n)),$$
$$\mathbf{p}_{\mathcal{B}}(b_1, \ldots, b_n) \leftrightarrow \mathbf{p}_{\mathcal{A}}(a_1, \ldots, a_n).$$

If $\mathcal{B}$ is constructed from $\mathcal{A}$ in the manner just described, we say that $\mathcal{A}$ is *isomorphic* to $\mathcal{B}$ and that ϕ is an *isomorphism* of $\mathcal{A}$ and $\mathcal{B}$. Put in another way, an isomorphism of $\mathcal{A}$ and $\mathcal{B}$ is a bijective mapping ϕ from $|\mathcal{A}|$ to $|\mathcal{B}|$ such that for $a_1, \ldots, a_n$ in $|\mathcal{A}|$,

$$\mathbf{f}_{\mathcal{B}}(\phi(a_1), \ldots, \phi(a_n)) = \phi(\mathbf{f}_{\mathcal{A}}(a_1, \ldots, a_n)) \tag{1}$$

and

$$\mathbf{p}_{\mathcal{B}}(\phi(a_1), \ldots, \phi(a_n)) \leftrightarrow \mathbf{p}_{\mathcal{A}}(a_1, \ldots, a_n). \tag{2}$$

Example. If $\mathcal{A}$ and $\mathcal{B}$ are groups, considered as models of G, then an isomorphism of $\mathcal{A}$ and $\mathcal{B}$ in the above sense is just an isomorphism of $\mathcal{A}$ and $\mathcal{B}$ in the usual sense of group theory.

The identity mapping from $|\mathcal{A}|$ to $|\mathcal{A}|$ (i.e., the mapping ϕ such that $\phi(a) = a$ for all a in $|\mathcal{A}|$) is an isomorphism of $\mathcal{A}$ and $\mathcal{A}$. If ϕ is an isomorphism of $\mathcal{A}$ and $\mathcal{B}$, then the inverse mapping of ϕ is an isomorphism of $\mathcal{B}$ and $\mathcal{A}$. If ϕ is an isomorphism of $\mathcal{A}$ and $\mathcal{B}$ and ψ is an isomorphism of $\mathcal{B}$ and $\mathcal{C}$, then the composite mapping $\psi\phi$ is an isomorphism of $\mathcal{A}$ and $\mathcal{C}$. These facts imply that being isomorphic is an equivalence relation among the structures for L.

Let $\mathcal{A}$ and $\mathcal{B}$ be structures for L, and let ϕ be a mapping from $|\mathcal{A}|$ to $|\mathcal{B}|$. If $\mathbf{i}$ is the name of an individual a of $\mathcal{A}$, we use $\mathbf{i}^\phi$ to designate the name of the individual $\phi(a)$ of $\mathcal{B}$. If $\mathbf{u}$ is an expression of $L(\mathcal{A})$, $\mathbf{u}^\phi$ is the expression obtained from $\mathbf{u}$ by replacing each name $\mathbf{i}$ by $\mathbf{i}^\phi$.

Lemma 1. Let ϕ be an isomorphism of $\mathcal{A}$ and $\mathcal{B}$. Then $\phi(\mathcal{A}(\mathbf{a})) = \mathcal{B}(\mathbf{a}^\phi)$ for every variable-free term $\mathbf{a}$ of $L(\mathcal{A})$, and $\mathcal{A}(\mathbf{A}) = \mathcal{B}(\mathbf{A}^\phi)$ for every closed formula $\mathbf{A}$ of $L(\mathcal{A})$.

Proof. We prove the first part by induction on the length of $\mathbf{a}$. If $\mathbf{a}$ is a name, the result is clear. Otherwise, $\mathbf{a}$ is $\mathbf{f}\mathbf{a}_1 \ldots \mathbf{a}_n$ with $\mathbf{f}$ in L. Using (1) and the induction hypothesis, we have

$$\begin{aligned}\phi(\mathcal{A}(\mathbf{a})) &= \phi(\mathbf{f}_\mathcal{A}(\mathcal{A}(\mathbf{a}_1), \ldots, \mathcal{A}(\mathbf{a}_n)))\\ &= \mathbf{f}_\mathcal{B}(\phi(\mathcal{A}(\mathbf{a}_1)), \ldots, \phi(\mathcal{A}(\mathbf{a}_n)))\\ &= \mathbf{f}_\mathcal{B}(\mathcal{B}(\mathbf{a}_1^\phi), \ldots, \mathcal{B}(\mathbf{a}_n^\phi)) = \mathcal{B}(\mathbf{a}^\phi).\end{aligned}$$

We now prove the second part by induction on the length of $\mathbf{A}$. If $\mathbf{A}$ is an atomic formula $\mathbf{p}\mathbf{a}_1 \ldots \mathbf{a}_n$ where $\mathbf{p}$ is not $=$, then by (2) and the first part,

$$\begin{aligned}\mathcal{A}(\mathbf{A}) = \mathbf{T} &\leftrightarrow \mathbf{p}_\mathcal{A}(\mathcal{A}(\mathbf{a}_1), \ldots, \mathcal{A}(\mathbf{a}_n))\\ &\leftrightarrow \mathbf{p}_\mathcal{B}(\phi(\mathcal{A}(\mathbf{a}_1)), \ldots, \phi(\mathcal{A}(\mathbf{a}_n)))\\ &\leftrightarrow \mathbf{p}_\mathcal{B}(\mathcal{B}(\mathbf{a}_1^\phi), \ldots, \mathcal{B}(\mathbf{a}_n^\phi))\\ &\leftrightarrow \mathcal{B}(\mathbf{A}^\phi) = \mathbf{T}.\end{aligned}$$

If $\mathbf{A}$ is $\mathbf{a} = \mathbf{b}$, then, since ϕ is injective,

$$\begin{aligned}\mathcal{A}(\mathbf{A}) = \mathbf{T} &\leftrightarrow \mathcal{A}(\mathbf{a}) = \mathcal{A}(\mathbf{b})\\ &\leftrightarrow \phi(\mathcal{A}(\mathbf{a})) = \phi(\mathcal{A}(\mathbf{b}))\\ &\leftrightarrow \mathcal{B}(\mathbf{a}^\phi) = \mathcal{B}(\mathbf{b}^\phi)\\ &\leftrightarrow \mathcal{B}(\mathbf{A}^\phi) = \mathbf{T}.\end{aligned}$$

If $\mathbf{A}$ is a negation or a disjunction, the result follows easily from the induction hypothesis. Now suppose that $\mathbf{A}$ is $\exists\mathbf{x}\mathbf{B}$. Since ϕ is surjective, every $\mathbf{j}$ in $L(\mathcal{B})$ is $\mathbf{i}^\phi$ for some $\mathbf{i}$ in $L(\mathcal{A})$. Hence by the induction hypothesis,

$$\begin{aligned}\mathcal{A}(\mathbf{A}) = \mathbf{T} &\leftrightarrow \mathcal{A}(\mathbf{B}_\mathbf{x}[\mathbf{i}]) = \mathbf{T} \quad \text{for some } \mathbf{i} \text{ in } L(\mathcal{A})\\ &\leftrightarrow \mathcal{B}(\mathbf{B}^\phi{}_\mathbf{x}[\mathbf{i}^\phi]) = \mathbf{T} \quad \text{for some } \mathbf{i} \text{ in } L(\mathcal{A})\\ &\leftrightarrow \mathcal{B}(\mathbf{B}^\phi{}_\mathbf{x}[\mathbf{j}]) = \mathbf{T} \quad \text{for some } \mathbf{j} \text{ in } L(\mathcal{B})\\ &\leftrightarrow \mathcal{B}(\mathbf{A}^\phi) = \mathbf{T}.\end{aligned}$$

Two structures $\mathcal{A}$ and $\mathcal{B}$ for L are *elementarily equivalent* if the same formulas of L are valid in $\mathcal{A}$ and $\mathcal{B}$. This clearly implies that $\mathcal{A}$ and $\mathcal{B}$ are models of the same theories.

The corollary to the closure theorem shows that if the same closed formulas are valid in $\mathcal{A}$ and $\mathcal{B}$, then $\mathcal{A}$ and $\mathcal{B}$ are elementarily equivalent. Hence, by Lemma 1, isomorphic structures are elementarily equivalent.

Let $\mathcal{A}$ and $\mathcal{B}$ be structures for L. An *embedding* of $\mathcal{A}$ in $\mathcal{B}$ is an injective mapping ϕ from $|\mathcal{A}|$ to $|\mathcal{B}|$ such that (1) and (2) hold for all nonlogical symbols $\mathbf{f}$ and $\mathbf{p}$

of L and all $a_1, \ldots, a_n$ in $|\mathfrak{A}|$. If $|\mathfrak{A}|$ is a subset of $|\mathfrak{B}|$ and the identity mapping from $|\mathfrak{A}|$ to $|\mathfrak{B}|$ is an embedding of $\mathfrak{A}$ in $\mathfrak{B}$, then $\mathfrak{A}$ is a *substructure* of $\mathfrak{B}$ and $\mathfrak{B}$ is an *extension* of $\mathfrak{A}$. For this case, (1) and (2) become

$$\mathbf{f}_{\mathfrak{A}}(a_1, \ldots, a_n) = \mathbf{f}_{\mathfrak{B}}(a_1, \ldots, a_n), \tag{3}$$

$$\mathbf{p}_{\mathfrak{A}}(a_1, \ldots, a_n) \leftrightarrow \mathbf{p}_{\mathfrak{B}}(a_1, \ldots, a_n). \tag{4}$$

If $\mathfrak{A}$ and $\mathfrak{B}$ are models of some theory, we sometimes say *submodel* for substructure.

Example. If $\mathfrak{A}$ and $\mathfrak{B}$ are groups (considered as models of G), then $\mathfrak{A}$ is a submodel of $\mathfrak{B}$ iff $\mathfrak{A}$ is a subgroup of $\mathfrak{B}$.

Let $\mathfrak{B}$ be a structure for L, and let A be a nonempty subset of $|\mathfrak{B}|$. If there is a substructure $\mathfrak{A}$ of $\mathfrak{B}$ with universe A, it is unique; for it is defined by (3) and (4). It is clear that (3) and (4) define a structure $\mathfrak{A}$ iff A satisfies the following condition: if $a_1, \ldots, a_n$ are in A and $\mathbf{f}$ is a function symbol of L, then $\mathbf{f}_{\mathfrak{B}}(a_1, \ldots, a_n)$ is in A.

Let $\mathfrak{A}$ and $\mathfrak{B}$ be structures for L such that $|\mathfrak{A}|$ is a subset of $|\mathfrak{B}|$. If a is an individual of $\mathfrak{A}$, we use the same name for a in $L(\mathfrak{A})$ and in $L(\mathfrak{B})$. Hence $L(\mathfrak{B})$ is an extension of $L(\mathfrak{A})$.

Lemma 2. Let $\mathfrak{A}$ and $\mathfrak{B}$ be structures for L, and let ϕ be a mapping from $|\mathfrak{A}|$ to $|\mathfrak{B}|$. Then ϕ is an embedding of $\mathfrak{A}$ in $\mathfrak{B}$ iff $\mathfrak{A}(\mathbf{A}) = \mathfrak{B}(\mathbf{A}^{\phi})$ for every variable-free formula $\mathbf{A}$ of $L(\mathfrak{A})$.

Proof. The proof that the condition holds if ϕ is an embedding is just like the proof of Lemma 1. Now suppose that the condition holds. Let $a_1, \ldots, a_n$ be individuals of $\mathfrak{A}$, and let $\mathbf{i}_1, \ldots, \mathbf{i}_n$ be the names of $a_1, \ldots, a_n$ respectively. Then $\mathfrak{A}(\mathbf{pi}_1 \ldots \mathbf{i}_n) = \mathfrak{B}(\mathbf{pi}_1^{\phi} \ldots \mathbf{i}_n^{\phi})$. This implies (2). If $\mathbf{j}$ is the name of $\mathbf{f}_{\mathfrak{A}}(a_1, \ldots, a_n)$, then

$$\mathfrak{B}(\mathbf{fi}_1^{\phi} \ldots \mathbf{i}_n^{\phi} = \mathbf{j}^{\phi}) = \mathfrak{A}(\mathbf{fi}_1 \ldots \mathbf{i}_n = \mathbf{j}) = \mathbf{T}.$$

This implies (1). Finally, if $\phi(a_1) = \phi(a_2)$, then

$$\mathfrak{A}(\mathbf{i}_1 = \mathbf{i}_2) = \mathfrak{B}(\mathbf{i}_1^{\phi} = \mathbf{i}_2^{\phi}) = \mathbf{T},$$

so $a_1 = a_2$. Thus ϕ is injective.

Corollary. Let $\mathfrak{A}$ and $\mathfrak{B}$ be structures for L such that $|\mathfrak{A}|$ is a subset of $|\mathfrak{B}|$. Then $\mathfrak{A}$ is a substructure of $\mathfrak{B}$ iff $\mathfrak{A}(\mathbf{A}) = \mathfrak{B}(\mathbf{A})$ for every variable-free formula $\mathbf{A}$ of $L(\mathfrak{A})$.

Let Γ be a set of formulas in L, and let $\mathfrak{A}$ be a structure for L. Then $\Gamma(\mathfrak{A})$ designates the set of $\mathfrak{A}$-instances of formulas in Γ. Thus if Γ is the set of all open formulas in L, then $\Gamma(\mathfrak{A})$ is the set of all variable-free formulas in $L(\mathfrak{A})$; if Γ is the set of all formulas in L, then $\Gamma(\mathfrak{A})$ is the set of all closed formulas in $L(\mathfrak{A})$.

Let $\mathfrak{A}$ and $\mathfrak{B}$ be structures for L such that $|\mathfrak{A}|$ is a subset of $|\mathfrak{B}|$, and let Γ be a set of formulas in L. We say that $\mathfrak{A}$ is a Γ-*substructure* of $\mathfrak{B}$ and that $\mathfrak{B}$ is a Γ-*extension* of $\mathfrak{A}$ if for every formula $\mathbf{A}$ in $\Gamma(\mathfrak{A})$, $\mathfrak{A}(\mathbf{A}) = \mathbf{T}$ implies $\mathfrak{B}(\mathbf{A}) = \mathbf{T}$.

If the negation of every formula in Γ is in Γ, then the same is true of $\Gamma(\mathfrak{A})$. It follows that if $\mathfrak{A}$ is a Γ-substructure of $\mathfrak{B}$, then $\mathfrak{A}(\mathbf{A}) = \mathfrak{B}(\mathbf{A})$ for every formula $\mathbf{A}$ in $\Gamma(\mathfrak{A})$. For if $\mathfrak{A}(\mathbf{A}) = \mathbf{F}$, then $\mathfrak{A}(\neg\mathbf{A}) = \mathbf{T}$; so $\mathfrak{B}(\neg\mathbf{A}) = \mathbf{T}$; so $\mathfrak{B}(\mathbf{A}) = \mathbf{F}$.

From the remark just made and the corollary to Lemma 2, we see that if Γ is the set of open formulas of L, then a Γ-substructure is simply a substructure and a Γ-extension is simply an extension. If Γ is the set of all formulas in L, then we say *elementary substructure* for Γ-substructure and *elementary extension* for Γ-extension. Then if $\mathfrak{A}$ is an elementary substructure of $\mathfrak{B}$, we have $\mathfrak{A}(\mathbf{A}) = \mathfrak{B}(\mathbf{A})$ for every closed formula $\mathbf{A}$ in L; so $\mathfrak{A}$ and $\mathfrak{B}$ are elementarily equivalent.

If $\mathfrak{A}$ is a Γ-substructure of $\mathfrak{B}$, then $\mathfrak{A}$ is a Δ-substructure of $\mathfrak{B}$ for every subset Δ of Γ. If the subset Δ consists of formulas in a language L' having L as an extension, then $\mathfrak{A}|L'$ is a Δ-substructure of $\mathfrak{B}|L'$.

Lemma 3. If $\mathbf{x} = \mathbf{e}$ is in Γ and $\mathfrak{A}$ is a Γ-substructure of $\mathfrak{B}$, then $\mathfrak{A}(\mathbf{e}) = \mathfrak{B}(\mathbf{e})$.

Proof. If $\mathbf{i}$ is the name of the individual $\mathfrak{A}(\mathbf{e})$, then $\mathbf{i} = \mathbf{e}$ is in $\Gamma(\mathfrak{A})$ and $\mathfrak{A}(\mathbf{i} = \mathbf{e}) = \mathbf{T}$. Hence $\mathfrak{B}(\mathbf{i} = \mathbf{e}) = \mathbf{T}$; so $\mathfrak{B}(\mathbf{e}) = \mathfrak{B}(\mathbf{i}) = \mathfrak{A}(\mathbf{i}) = \mathfrak{A}(\mathbf{e})$.

Let $\mathfrak{A}$ and $\mathfrak{B}$ be structures for L such that $|\mathfrak{A}|$ is a subset of $|\mathfrak{B}|$. We expand $\mathfrak{B}$ to a structure $\mathfrak{B}_{\mathfrak{A}}$ for $L(\mathfrak{A})$ as follows: if $\mathbf{i}$ is the name of an individual a of $\mathfrak{A}$, then $\mathfrak{B}_{\mathfrak{A}}$ assigns a to $\mathbf{i}$.

If $L' = L(\mathfrak{A})$ is the language of $\mathfrak{B}_{\mathfrak{A}}$, then a name $\mathbf{i}$ in L' serves two functions in $L'(\mathfrak{B}_{\mathfrak{A}})$; it is a constant of L' and a name of an individual of $\mathfrak{B}_{\mathfrak{A}}$. We claim that this makes no difference, at least for the purpose of obtaining truth values $\mathfrak{B}_{\mathfrak{A}}(\mathbf{A})$. To see this, let us temporarily regard any such name $\mathbf{i}$ as a constant of L', and use $\mathbf{i}'$ as the name of the corresponding individual of $\mathfrak{B}_{\mathfrak{A}}$. Then $\mathbf{i}'$ is the name of $\mathfrak{B}_{\mathfrak{A}}(\mathbf{i})$. If $\mathbf{A}'$ results from $\mathbf{A}$ by changing $\mathbf{i}$ to $\mathbf{i}'$, then $\mathfrak{B}_{\mathfrak{A}}(\mathbf{A}) = \mathfrak{B}_{\mathfrak{A}}(\mathbf{A}')$ by the lemma of §2.5.

Let Γ be a set of formulas in L, and let $\mathfrak{A}$ be a structure for L. The Γ-*diagram* of $\mathfrak{A}$, designated by $D_{\Gamma}(\mathfrak{A})$, is the theory whose language is $L(\mathfrak{A})$ and whose nonlogical axioms are the formulas $\mathbf{A}$ in $\Gamma(\mathfrak{A})$ such that $\mathfrak{A}(\mathbf{A}) = \mathbf{T}$. If Γ is the set of open formulas, we write $D(\mathfrak{A})$ for $D_{\Gamma}(\mathfrak{A})$; if Γ is the set of all formulas, we write $D_e(\mathfrak{A})$ for $D_{\Gamma}(\mathfrak{A})$.

Diagram Lemma. Let Γ be a set of formulas in L, and let $\mathfrak{A}$ and $\mathfrak{B}$ be structures for L such that $|\mathfrak{A}|$ is a subset of $|\mathfrak{B}|$. Then $\mathfrak{A}$ is a Γ-substructure of $\mathfrak{B}$ iff $\mathfrak{B}_{\mathfrak{A}}$ is a model of $D_{\Gamma}(\mathfrak{A})$.

Proof. For $\mathbf{A}$ a closed formula of $L(\mathfrak{A})$, we have $\mathfrak{B}_{\mathfrak{A}}(\mathbf{A}) = \mathfrak{B}(\mathbf{A})$ because $\mathfrak{B}_{\mathfrak{A}}$ is an expansion of $\mathfrak{B}$. The lemma then follows from the definitions of Γ-substructure and Γ-diagram.

A set Γ of formulas of L is *regular* if every formula of the form $\mathbf{x} = \mathbf{y}$ or $\mathbf{x} \neq \mathbf{y}$ is in Γ, and if for every formula $\mathbf{A}$ in Γ, every formula of the form $\mathbf{A}[\mathbf{x}_1, \ldots, \mathbf{x}_n]$ is in Γ.

Model Extension Theorem (*Keisler*). Let $\mathfrak{A}$ be a structure for L, T a theory with language L, Γ a regular set of formulas in L. Then $\mathfrak{A}$ has a Γ-extension which is a model of T iff every theorem of T which is a disjunction of negations of formulas in Γ is valid in $\mathfrak{A}$.

Proof. Suppose that such a Γ-extension $\mathfrak{B}$ exists. We must show that if

$$\vdash_T \neg \mathbf{A}_1 \vee \cdots \vee \neg \mathbf{A}_n,$$

where each $\mathbf{A}_i$ is in Γ, then $\neg \mathbf{A}_1 \vee \cdots \vee \neg \mathbf{A}_n$ is valid in $\mathfrak{A}$. If not, there is an $\mathfrak{A}$-instance $\neg \mathbf{A}'_1 \vee \cdots \vee \neg \mathbf{A}'_n$ of $\neg \mathbf{A}_1 \vee \cdots \vee \neg \mathbf{A}_n$ such that $\mathfrak{A}(\mathbf{A}'_i) = \mathbf{T}$ for $i = 1, \ldots, n$. Since $\mathfrak{B}$ is a Γ-extension of $\mathfrak{A}$ and $\mathbf{A}'_i$ is in $\Gamma(\mathfrak{A})$, $\mathfrak{B}(\mathbf{A}'_i) = \mathbf{T}$. Hence $\mathfrak{B}(\neg \mathbf{A}'_1 \vee \cdots \vee \neg \mathbf{A}'_n) = \mathbf{F}$. This is impossible, since $\neg \mathbf{A}'_1 \vee \cdots \vee \neg \mathbf{A}'_n$ is a $\mathfrak{B}$-instance of a theorem of T and $\mathfrak{B}$ is a model of T.

Now suppose that the condition holds. We form T' from T by adding all the names of $L(\mathfrak{A})$ as new constants; and we form T'' from T' by adding all the nonlogical axioms of $D_\Gamma(\mathfrak{A})$ as new axioms. We shall show that T'' is consistent. If not, then by the reduction theorem for consistency we have

$$\vdash_{T'} \neg \mathbf{A}'_1 \vee \cdots \vee \neg \mathbf{A}'_n,$$

where each $\mathbf{A}'_i$ is a formula in $\Gamma(\mathfrak{A})$ such that $\mathfrak{A}(\mathbf{A}'_i) = \mathbf{T}$. Hence by the theorem on constants,

$$\vdash_T \neg \mathbf{A}_1 \vee \cdots \vee \neg \mathbf{A}_n,$$

where $\mathbf{A}_i$ results from $\mathbf{A}'_i$ by replacing the names by new variables. Then $\mathbf{A}_i$ results from a formula in Γ by a substitution of variables, and hence, by the regularity of Γ, is in Γ. Hence by the hypothesis, $\neg \mathbf{A}_1 \vee \cdots \vee \neg \mathbf{A}_n$ is valid in $\mathfrak{A}$. This implies that $\mathfrak{A}(\neg \mathbf{A}'_1 \vee \cdots \vee \neg \mathbf{A}'_n) = \mathbf{T}$; so $\mathfrak{A}(\mathbf{A}'_i) = \mathbf{F}$ for some i, a contradiction.

By the completeness theorem, T'' has a model $\mathfrak{B}'$. If $\mathbf{i}$ and $\mathbf{j}$ are the names of distinct individuals of $\mathfrak{A}$, then $\mathbf{i} \neq \mathbf{j}$ is a formula in $\Gamma(\mathfrak{A})$ (since Γ is regular) and $\mathfrak{A}(\mathbf{i} \neq \mathbf{j}) = \mathbf{T}$. Hence $\mathbf{i} \neq \mathbf{j}$ is an axiom of $D_\Gamma(\mathfrak{A})$; so $\mathfrak{B}'(\mathbf{i} \neq \mathbf{j}) = \mathbf{T}$ and hence $\mathfrak{B}'(\mathbf{i}) \neq \mathfrak{B}'(\mathbf{j})$. It readily follows that, by replacing $\mathfrak{B}'$ by an isomorphic structure, we may suppose that for each name $\mathbf{i}$ in $L(\mathfrak{A})$, $\mathfrak{B}'(\mathbf{i})$ is the individual whose name is $\mathbf{i}$. This means that if $\mathfrak{B}$ is the restriction of $\mathfrak{B}'$ to L, then $\mathfrak{B}_\mathfrak{A} = \mathfrak{B}'$. Since $\mathfrak{B}'$ is a model of $D_\Gamma(\mathfrak{A})$, it follows by the diagram lemma that $\mathfrak{B}$ is a Γ-extension of $\mathfrak{A}$, while $\mathfrak{B}$ is a model of T by Lemma 1 of §4.2.

Corollary. Let Γ be a regular set of formulas in L, and let Δ be a set of formulas containing every formula $\forall \mathbf{x}_1 \ldots \forall \mathbf{x}_n \mathbf{A}$, where $\mathbf{A}$ is a disjunction of negations of formulas in Γ. If $\mathfrak{A}$ is a structure for L and $\mathfrak{B}$ is a Δ-extension of $\mathfrak{A}$, then there is a Γ-extension $\mathfrak{C}$ of $\mathfrak{B}$ which is an elementary extension of $\mathfrak{A}$.

Proof. Let Γ' be the set of formulas $\mathbf{A}[\mathbf{i}_1, \ldots, \mathbf{i}_k]$, where $\mathbf{A}$ is in Γ and $\mathbf{i}_1, \ldots, \mathbf{i}_k$ are names in $L(\mathfrak{A})$. We show that there is a Γ'-extension of $\mathfrak{B}_\mathfrak{A}$ which is a model of $D_e(\mathfrak{A})$. By the theorem, we need only show that if $\mathbf{A}_1, \ldots, \mathbf{A}_n$ are in Γ' and

$\neg \mathbf{A}_1 \vee \cdots \vee \neg \mathbf{A}_n$ is a theorem of $D_e(\mathcal{A})$, then $\neg \mathbf{A}_1 \vee \cdots \vee \neg \mathbf{A}_n$ is valid in $\mathcal{B}_{\mathcal{A}}$. The closure $\mathbf{B}$ of $\neg \mathbf{A}_1 \vee \cdots \vee \neg \mathbf{A}_n$ is a theorem of $D_e(\mathcal{A})$; so by the diagram lemma, $\mathcal{A}(\mathbf{B}) = \mathcal{A}_{\mathcal{A}}(\mathbf{B}) = \mathbf{T}$. Since $\mathbf{B}$ is in $\Delta(\mathcal{A})$, it is then an axiom of $D_\Delta(\mathcal{A})$; so by the diagram lemma again, $\mathcal{B}_{\mathcal{A}}(\mathbf{B}) = \mathbf{T}$. Hence $\neg \mathbf{A}_1 \vee \cdots \vee \neg \mathbf{A}_n$ is valid in $\mathcal{B}_{\mathcal{A}}$.

Let $\mathcal{C}'$ be a Γ'-extension of $\mathcal{B}_{\mathcal{A}}$ which is a model of $D_e(\mathcal{A})$, and let $\mathcal{C} = \mathcal{C}'|L$. Since Γ is a subset of Γ', $\mathcal{C}$ is a Γ-extension of $\mathcal{B}$. If $\mathbf{i}$ is the name of an individual a of $\mathcal{A}$, $\mathcal{C}'(\mathbf{i}) = \mathcal{B}_{\mathcal{A}}(\mathbf{i}) = a$ by Lemma 3. This implies that $\mathcal{C}' = \mathcal{C}_{\mathcal{A}}$; so $\mathcal{C}$ is an elementary extension of $\mathcal{A}$ by the diagram lemma.

We shall apply our results to the following problem: under what conditions on the models of T is T equivalent to a theory whose nonlogical axioms are in Γ? We shall solve this problem when Γ is the set of open formulas and when Γ is the set of existential formulas.

Lemma 4. Let Γ be a set of formulas in $L(T)$, and let Γ' be the set of formulas in Γ which are theorems of T. If every structure for $L(T)$ in which all the formulas of Γ' are valid is a model of T, then T is equivalent to a theory whose nonlogical axioms are in Γ.

Proof. Let T' be the theory with language $L(T)$ whose nonlogical axioms are the formulas in Γ'. Clearly T is an extension of T'. By the hypothesis, every model of T' is a model of T; so by the corollary to the completeness theorem, T' is an extension of T. Hence T is equivalent to T'.

Łoś-Tarski Theorem. A theory T is equivalent to an open theory iff every substructure of a model of T is a model of T.

Proof. Suppose that T is equivalent to the open theory T'. By the corollary to the completeness theorem, it suffices to show that every substructure $\mathcal{A}$ of a model $\mathcal{B}$ of T' is a model of T'. By the corollary to Lemma 2, every open formula valid in $\mathcal{B}$ is valid in $\mathcal{A}$; so $\mathcal{A}$ is a model of T'.

Suppose that the condition of the theorem holds. By Lemma 4, it suffices to show that if every open theorem of T is valid in $\mathcal{A}$, then $\mathcal{A}$ is a model of T. By the model extension theorem, $\mathcal{A}$ has an extension which is a model of T; so by the condition of the theorem, $\mathcal{A}$ is a model of T.

A sequence $\mathcal{A}_1, \mathcal{A}_2, \ldots$ of structures for L is a *chain* if for each n, $\mathcal{A}_{n+1}$ is an extension of $\mathcal{A}_n$. Given such a chain, we define a structure $\mathcal{A}$ which we call the *union* of the chain. The universe of $\mathcal{A}$ is the union of the universes of the $\mathcal{A}_n$. If $a_1, \ldots, a_k$ are in this union, then there is an n such that all of $a_1, \ldots, a_k$ are individuals of $\mathcal{A}_n$. We then set

$$\mathbf{f}_{\mathcal{A}}(a_1, \ldots, a_k) = \mathbf{f}_{\mathcal{A}_n}(a_1, \ldots, a_k),$$
$$\mathbf{p}_{\mathcal{A}}(a_1, \ldots, a_k) \leftrightarrow \mathbf{p}_{\mathcal{A}_n}(a_1, \ldots, a_k).$$

Using the definition of a chain, we easily verify that this definition is independent of the choice of n and that $\mathcal{A}$ is an extension of each $\mathcal{A}_n$.

An *elementary chain* is a chain $\mathcal{A}_1, \mathcal{A}_2, \ldots$ such that for each n, $\mathcal{A}_{n+1}$ is an elementary extension of $\mathcal{A}_n$.

Tarski's Lemma. If $\mathcal{A}_1, \mathcal{A}_2, \ldots$ is an elementary chain, then the union $\mathcal{A}$ of the chain is an elementary extension of each $\mathcal{A}_n$.

Proof. We must show that if $\mathbf{A}$ is a closed formula in $L(\mathcal{A}_n)$, then $\mathcal{A}_n(\mathbf{A}) = \mathcal{A}(\mathbf{A})$. We use induction on the length of $\mathbf{A}$. If $\mathbf{A}$ is atomic, $\mathcal{A}_n(\mathbf{A}) = \mathcal{A}(\mathbf{A})$ by the corollary to Lemma 2. If $\mathbf{A}$ is a negation or a disjunction, the result follows immediately from the induction hypothesis. Now suppose that $\mathbf{A}$ is $\exists\mathbf{x}\mathbf{B}$. If $\mathcal{A}(\mathbf{A}) = \mathsf{F}$, then $\mathcal{A}(\mathbf{B_x[i]}) = \mathsf{F}$ for all $\mathbf{i}$ in $L(\mathcal{A})$. By induction hypothesis, $\mathcal{A}_n(\mathbf{B_x[i]}) = \mathsf{F}$ for all $\mathbf{i}$ in $L(\mathcal{A}_n)$; so $\mathcal{A}_n(\mathbf{A}) = \mathsf{F}$. If $\mathcal{A}(\mathbf{A}) = \mathsf{T}$, then $\mathcal{A}(\mathbf{B_x[i]}) = \mathsf{T}$ for some $\mathbf{i}$ in $L(\mathcal{A})$. Choose k so that $k > n$ and $\mathbf{i}$ is a name in $L(\mathcal{A}_k)$. By induction hypothesis, $\mathcal{A}_k(\mathbf{B_x[i]}) = \mathsf{T}$; so $\mathcal{A}_k(\mathbf{A}) = \mathsf{T}$. Since $\mathcal{A}_k$ is an elementary extension of $\mathcal{A}_n$, it follows that $\mathcal{A}_n(\mathbf{A}) = \mathsf{T}$.

Chang-Łoś-Suszko Theorem. A theory T is equivalent to a theory whose nonlogical axioms are existential iff every union of a chain of models of T is a model of T.

Proof. Suppose that T is equivalent to a theory T' whose nonlogical axioms are existential. By the corollary to the completeness theorem, it suffices to prove that the union $\mathcal{A}$ of a chain $\mathcal{A}_1, \mathcal{A}_2, \ldots$ of models of T' is a model of T'. Let $\exists\mathbf{x}_1 \ldots \exists\mathbf{x}_n\mathbf{A}$ be an $\mathcal{A}$-instance of a nonlogical axiom of T'. For large enough k, $\exists\mathbf{x}_1 \ldots \exists\mathbf{x}_n\mathbf{A}$ is an $\mathcal{A}_k$-instance of this axiom; so $\mathcal{A}_k(\exists\mathbf{x}_1 \ldots \exists\mathbf{x}_n\mathbf{A}) = \mathsf{T}$. It follows that

$$\mathcal{A}_k(\mathbf{A}[\mathbf{i}_1, \ldots, \mathbf{i}_n]) = \mathsf{T}$$

for some $\mathbf{i}_1, \ldots, \mathbf{i}_n$ in $L(\mathcal{A}_k)$. By the corollary to Lemma 2,

$$\mathcal{A}(\mathbf{A}[\mathbf{i}_1, \ldots, \mathbf{i}_n]) = \mathsf{T};$$

so

$$\mathcal{A}(\exists\mathbf{x}_1 \ldots \exists\mathbf{x}_n\mathbf{A}) = \mathsf{T}.$$

We have shown that $\mathcal{A}$ is a model of T'.

Now suppose that every union of a chain of models of T is a model of T. By Lemma 4, it suffices to show that if $\mathcal{A}$ is a structure in which every existential theorem of T is valid, then $\mathcal{A}$ is a model of T. We shall construct a chain $\mathcal{A}_1, \mathcal{A}_2, \ldots$ such that $\mathcal{A}_1 = \mathcal{A}$, $\mathcal{A}_{2n}$ is a model of T, and $\mathcal{A}_{2n+3}$ is an elementary extension of $\mathcal{A}_{2n+1}$. Assume that this is done, and let $\mathcal{B}$ be the union of the chain. Then $\mathcal{B}$ is the union of the chain $\mathcal{A}_2, \mathcal{A}_4, \ldots$ of models of T, and hence is a model of T. But $\mathcal{B}$ is also the union of the elementary chain $\mathcal{A}_1, \mathcal{A}_3, \ldots$. Hence by Tarski's lemma, $\mathcal{B}$ is an elementary extension of $\mathcal{A}_1 = \mathcal{A}$ and therefore is elementarily equivalent to $\mathcal{A}$. It follows that $\mathcal{A}$ is a model of T.

We now define the $\mathcal{A}_n$. Suppose that $\mathcal{A}_{2n-1}$ is defined and is an elementary extension of $\mathcal{A}_1 = \mathcal{A}$; we shall construct $\mathcal{A}_{2n}$ and $\mathcal{A}_{2n+1}$. Let Γ be the set of universal formulas in L; we show that there is a Γ-extension $\mathcal{A}_{2n}$ of $\mathcal{A}_{2n-1}$ which is a model of T. By the model extension theorem, it suffices to show that if $\vdash_T \mathbf{A}$, where $\mathbf{A}$ is a disjunction of negations of universal formulas, then $\mathbf{A}$ is valid in

$\mathfrak{A}_{2n-1}$. Now **A** has a prenex form **B** which is existential. By hypothesis, **B** is valid in $\mathfrak{A}$ and hence in $\mathfrak{A}_{2n-1}$. Thus $\mathfrak{A}_{2n-1}$ is a model for the theory with **B** as its only nonlogical axiom. Since **A** is a theorem of this theory, **A** is valid in $\mathfrak{A}_{2n-1}$.

Since every open formula is in Γ, $\mathfrak{A}_{2n}$ is an extension of $\mathfrak{A}_{2n-1}$. By the corollary to the model extension theorem, there is an extension $\mathfrak{A}_{2n+1}$ of $\mathfrak{A}_{2n}$ which is an elementary extension of $\mathfrak{A}_{2n-1}$. This completes the proof.

If we apply this theorem to G, and use the well-known fact that the union of a chain of groups is a group, we conclude that G is equivalent to a theory whose nonlogical axioms are existential. Axioms for such a theory are also well known; they consist of the associative law and the axioms $\exists x(x \cdot y = z)$ and $\exists x(y \cdot x = z)$.

5.3 CARDINALITY OF MODELS

In this section, we shall assume some elementary results on cardinals. Proofs can be found in Chapter 9 or in any text on set theory.

By the *cardinal* of a structure $\mathfrak{A}$, we mean the cardinal of its universe $|\mathfrak{A}|$. We shall say that $\mathfrak{A}$ is finite or infinite, countable or uncountable if $|\mathfrak{A}|$ has the corresponding property.

Let $\mathfrak{m}$ be an infinite cardinal. A first-order language L is an $\mathfrak{m}$-*language* if the set of nonlogical symbols of L has cardinal $\leqslant \mathfrak{m}$. A theory T is an $\mathfrak{m}$-*theory* if $L(T)$ is an $\mathfrak{m}$-language. We say *countable language* and *countable theory* for $\aleph_0$-language and $\aleph_0$-theory.

Lemma. If $\mathfrak{m}$ is an infinite cardinal and L is an $\mathfrak{m}$-language, then L_c contains at most $\mathfrak{m}$ special constants.

Proof. We show by induction on n that there are at most $\mathfrak{m}$ special constants of level n; it will follow that there are at most $\aleph_0 \cdot \mathfrak{m} = \mathfrak{m}$ special constants. Since L contains only countably many logical symbols, it contains at most $\aleph_0 + \mathfrak{m} = \mathfrak{m}$ symbols. If Q is the set of all symbols of L and all special constants of levels less than n, then by induction hypothesis, the cardinal of Q is at most $\mathfrak{m} + n \cdot \mathfrak{m} = \mathfrak{m}$. Hence for each k, the number of expressions of length k formed with symbols in Q is at most $\mathfrak{m}^k = \mathfrak{m}$; so the total number of expressions formed with symbols of Q is at most $\aleph_0 \cdot \mathfrak{m} = \mathfrak{m}$. It follows immediately that the number of special constants of level n is at most $\mathfrak{m}$.

Cardinality Theorem (*Tarski*). Let $\mathfrak{m}$ be an infinite cardinal, and let T be an $\mathfrak{m}$-theory having an infinite model. Then T has a model of cardinal $\mathfrak{m}$.

Proof. We form U from T as follows: we add a set of new constants of cardinal $\mathfrak{m}$, and if **e** and **e**′ are distinct new constants, we add an axiom $\mathbf{e} \neq \mathbf{e}'$. We shall show that U has a model. By the corollary to the compactness theorem, it suffices to show that every finite part U' of U has a model. Let $\mathbf{e}_1, \ldots, \mathbf{e}_k$ be the new constants appearing in the nonlogical axioms of U'. Let $\mathfrak{A}$ be an infinite model of T, and let $a_1, \ldots, a_k$ be distinct individuals of $\mathfrak{A}$. Expand $\mathfrak{A}$ to a structure $\mathfrak{A}'$ for U'

by assigning the individual a_i to $\mathbf{e}_i$ and assigning any individual to new constants other than $\mathbf{e}_1, \ldots, \mathbf{e}_k$. Clearly $\mathfrak{A}'$ is a model of U'.

The number of nonlogical constants in U is at most $\mathfrak{m} + \mathfrak{m} = \mathfrak{m}$; so U is an $\mathfrak{m}$-theory. By the lemma, U_c contains at most $\mathfrak{m}$ special constants. By the above and Lemma 3 of §4.2, U_c is consistent. Hence by Lemma 4 of §4.2, U_c has a model $\mathfrak{A}$ having cardinal $\leqslant \mathfrak{m}$. The axioms of U guarantee that $\mathfrak{A}(\mathbf{e}) \neq \mathfrak{A}(\mathbf{e}')$ if $\mathbf{e}$ and $\mathbf{e}'$ are distinct new constants; so the cardinal of $\mathfrak{A}$ is exactly $\mathfrak{m}$. The restriction of $\mathfrak{A}$ to $L(T)$ is then a model of T of cardinal $\mathfrak{m}$.

Corollary (*Löwenheim-Skolem*). If T is a countable theory having a model, then T has a countable model.

There are some paradoxes resulting from the cardinality theorem and the Löwenheim-Skolem theorem which were first pointed out by Skolem. We can certainly formalize enough mathematics in a countable theory to prove that the set of real numbers is uncountable. How can such a theory have a countable model? The explanation is this. The set of real numbers in the model is indeed countable, and therefore there is a bijective mapping from it to the set of natural numbers. But this mapping is *not* in the model; so it does not make invalid the theorem of the theory which states that there is no bijective mapping from the set of real numbers to the set of natural numbers.

Another paradox arises from the Peano axioms. There is a well-known proof that any set of objects satisfying these axioms is isomorphic to the set of natural numbers and hence countable. If we formalize the Peano axioms in $L(N)$, however, the cardinality theorem shows that the resulting theory has uncountable models. The difficulty here is that we cannot fully express the induction axiom (which is one of the Peano axioms) in $L(N)$. We shall discuss this situation more fully in Chapter 8.

5.4 JOINT CONSISTENCY

Let T and T' be theories. The *union* of T and T', designated by $T \cup T'$, is the theory whose nonlogical symbols are the nonlogical symbols of T and the nonlogical symbols of T', and whose nonlogical axioms are the nonlogical axioms of T and the nonlogical axioms of T'.

The theory $T \cup T'$ may be inconsistent, even if both T and T' are consistent; for there may be a formula $\mathbf{A}$ such that $\vdash_T \mathbf{A}$ and $\vdash_{T'} \neg\mathbf{A}$. We shall show that if there is no such $\mathbf{A}$, then $T \cup T'$ is consistent. This shows that any inconsistency in $T \cup T'$ can be "localized" in a formula $\mathbf{A}$ which is a formula in both T and T'.

Joint Consistency Theorem (*Craig-Robinson*). Let T and T' be theories. Then $T \cup T'$ is inconsistent iff there is a closed formula $\mathbf{A}$ such that $\vdash_T \mathbf{A}$ and $\vdash_{T'} \neg\mathbf{A}$.

Proof. If such a formula $\mathbf{A}$ exists, then $\mathbf{A}$ and $\neg\mathbf{A}$ are theorems of $T \cup T'$; so $T \cup T'$ is inconsistent. We shall now suppose that there is no such $\mathbf{A}$ and prove that $T \cup T'$ is consistent.

Let Γ be the set of closed formulas in $L(T)$ which are theorems of T'. Then $T[\Gamma]$ is consistent. For otherwise it would follow from the reduction theorem for consistency that there is a theorem **A** of T which is a disjunction of negations of closed theorems of T'. Then $\vdash_T \mathbf{A}$ and $\vdash_{T'} \neg\mathbf{A}$, contradicting our hypothesis.

Let L be the first-order language whose nonlogical symbols are the nonlogical symbols common to $L(T)$ and $L(T')$. We shall construct an elementary chain $\mathcal{A}_1, \mathcal{A}_2, \ldots$ of models of T and an elementary chain $\mathcal{A}'_1, \mathcal{A}'_2, \ldots$ of models of T' such that $\mathcal{A}_1|L, \mathcal{A}'_1|L, \mathcal{A}_2|L, \mathcal{A}'_2|L, \ldots$ is an elementary chain.

Let $\mathcal{A}_1$ be any model of $T[\Gamma]$. Let Δ be the set of all formulas of L. Then Δ is regular. Let $\mathcal{B}$ be an expansion to $L(T')$ of $\mathcal{A}_1|L$. Then the same formulas of L are valid in $\mathcal{A}_1$ and $\mathcal{B}$. If a formula in L is a theorem of T', its closure is in Γ and hence is valid in $\mathcal{A}_1$; so the formula itself is valid in $\mathcal{A}_1$ and hence in $\mathcal{B}$. It follows by the model extension theorem that there is a model $\mathcal{A}'_1$ of T' which is a Δ-extension of $\mathcal{B}$. Then $\mathcal{A}'_1|L$ is an elementary extension of $\mathcal{B}|L = \mathcal{A}_1|L$.

We now describe the construction of $\mathcal{A}_n$ for $n > 1$; the construction of $\mathcal{A}'_n$ is similar. Let $\mathcal{C}$ be an expansion to $L(T)$ of $\mathcal{A}'_{n-1}|L$. Then $\mathcal{C}|L$ is an elementary extension of $\mathcal{A}_{n-1}|L$; so $\mathcal{C}$ is a Δ-extension of $\mathcal{A}_{n-1}$. It follows by the corollary to the model extension theorem that there is a Δ-extension $\mathcal{A}_n$ of $\mathcal{C}$ which is an elementary extension of $\mathcal{A}_{n-1}$. Then $\mathcal{A}_n$ is an elementary extension of $\mathcal{A}_1$ and hence a model of T. Clearly $\mathcal{A}_n|L$ is an elementary extension of $\mathcal{C}|L = \mathcal{A}'_{n-1}|L$.

Let $\mathcal{A}$ be the union of $\mathcal{A}_1, \mathcal{A}_2, \ldots$ and let $\mathcal{A}'$ be the union of $\mathcal{A}'_1, \mathcal{A}'_2, \ldots$. By Tarski's lemma, $\mathcal{A}$ is an elementary extension of $\mathcal{A}_1$ and hence a model of T. Similarly, $\mathcal{A}'$ is a model of T'. Now $\mathcal{A}|L = \mathcal{A}'|L$, since both are the union of the chain $\mathcal{A}_1|L, \mathcal{A}'_1|L, \mathcal{A}_2|L, \mathcal{A}'_2|L, \ldots$. From this we see easily that there is a structure $\mathcal{B}$ for $T \cup T'$ such that $\mathcal{B}|L(T) = \mathcal{A}$ and $\mathcal{B}|L(T') = \mathcal{A}'$. Then $\mathcal{B}$ is a model of $T \cup T'$; so $T \cup T'$ is consistent.

Corollary (*Craig Interpolation Lemma*). Let T and T' be theories, and let $\mathbf{A} \to \mathbf{B}$ be a theorem of $T \cup T'$ such that **A** is a formula of T and **B** is a formula of T'. Then there is a formula **C** such that $\vdash_T \mathbf{A} \to \mathbf{C}$ and $\vdash_{T'} \mathbf{C} \to \mathbf{B}$.

Proof. First suppose that **A** and **B** are closed. In the theory $T[\mathbf{A}] \cup T'[\neg\mathbf{B}]$, we can prove **A**, $\neg\mathbf{B}$, and $\mathbf{A} \to \mathbf{B}$; so by the tautology theorem, $T[\mathbf{A}] \cup T'[\neg\mathbf{B}]$ is inconsistent. By the theorem, there is a closed formula **C** such that **C** is a theorem of $T[\mathbf{A}]$ and $\neg\mathbf{C}$ is a theorem of $T'[\neg\mathbf{B}]$. By the deduction theorem, $\vdash_T \mathbf{A} \to \mathbf{C}$ and $\vdash_{T'} \neg\mathbf{B} \to \neg\mathbf{C}$; so by the tautology theorem, $\vdash_{T'} \mathbf{C} \to \mathbf{B}$.

In the general case, we substitute a new constant for each variable free in **A** or **B**, obtaining formulas **A**′ and **B**′. Then the above tells us that for a suitable **C**′, $\vdash_U \mathbf{A}' \to \mathbf{C}'$ and $\vdash_{U'} \mathbf{C}' \to \mathbf{B}'$, where U and U' result from T and T' respectively by adding constants. By the variant theorem, **C**′ may be chosen so that none of the variables bound in **C**′ is free in **A** or **B**. By replacing the new constants by the original variables, we obtain a formula **C** which has the required properties by the theorem on constants.

We shall give an application of the interpolation lemma. Let Q be a set of nonlogical symbols in the theory T. A predicate symbol **p** not in Q is *definable in* *terms of Q in T* if there is a formula **A** containing no nonlogical symbols not in Q such that $\vdash_T \mathbf{px}_1 \ldots \mathbf{x}_n \leftrightarrow \mathbf{A}$ (where $\mathbf{x}_1, \ldots, \mathbf{x}_n$ are distinct). A function symbol **f** not in Q is *definable in terms of Q in T* if there is a formula **A** containing no nonlogical symbols not in Q such that $\vdash_T \mathbf{y} = \mathbf{fx}_1 \ldots \mathbf{x}_n \leftrightarrow \mathbf{A}$ (where $\mathbf{x}_1, \ldots, \mathbf{x}_n, \mathbf{y}$ are distinct).

Let $\mathcal{A}$ and $\mathcal{B}$ be structures for L, **u** a nonlogical symbol of L, and ϕ a bijective mapping from $|\mathcal{A}|$ to $|\mathcal{B}|$. We say that ϕ is a **u**-*isomorphism* of $\mathcal{A}$ and $\mathcal{B}$ if ϕ is an isomorphism of the restrictions of $\mathcal{A}$ and $\mathcal{B}$ to the language whose only nonlogical symbol is **u**.

Definability Theorem (*Beth*). Let Q be a set of nonlogical symbols in T, and let **u** be a nonlogical symbol of T which is not in Q. Then **u** is definable in terms of Q in T iff for every pair of models $\mathcal{A}$ and $\mathcal{B}$ of T and every bijective mapping ϕ from $|\mathcal{A}|$ to $|\mathcal{B}|$ which is a **v**-isomorphism for every **v** in Q, ϕ is a **u**-isomorphism.

Proof. We suppose that **u** is a predicate symbol **p**; if **u** is a function symbol, the proof is essentially the same. Suppose that we have $\vdash_T \mathbf{px}_1 \ldots \mathbf{x}_n \leftrightarrow \mathbf{A}$, where **A** contains no nonlogical symbols not in Q. Let $\mathcal{A}$, $\mathcal{B}$, and ϕ be as in the theorem. If $\mathbf{i}_1, \ldots, \mathbf{i}_n$ are names in $L(\mathcal{A})$ and **B** is $\mathbf{A}_{\mathbf{x}_1,\ldots,\mathbf{x}_n}[\mathbf{i}_1, \ldots, \mathbf{i}_n]$, then

$$\mathcal{A}(\mathbf{pi}_1 \ldots \mathbf{i}_n \leftrightarrow \mathbf{B}) = \mathbf{T},$$
$$\mathcal{B}(\mathbf{pi}_1^{\phi} \ldots \mathbf{i}_n^{\phi} \leftrightarrow \mathbf{B}^{\phi}) = \mathbf{T},$$
$$\mathcal{A}(\mathbf{B}) = \mathcal{B}(\mathbf{B}^{\phi}).$$

Hence

$$\mathcal{A}(\mathbf{pi}_1 \ldots \mathbf{i}_n) = \mathcal{B}(\mathbf{pi}_1^{\phi} \ldots \mathbf{i}_n^{\phi}).$$

It follows that

$$\mathbf{p}_{\mathcal{A}}(a_1, \ldots, a_n) \leftrightarrow \mathbf{p}_{\mathcal{B}}(\phi(a_1), \ldots, \phi(a_n))$$

for $a_1, \ldots, a_n$ in $|\mathcal{A}|$; so ϕ is a **p**-isomorphism.

Now suppose the condition of the theorem holds. For each nonlogical n-ary function or predicate symbol **v** of T which is not in Q, introduce a new n-ary function or predicate symbol $\mathbf{v}'$. We obtain T' from T by replacing each **v** by $\mathbf{v}'$ (leaving the symbols in Q unchanged). We will show that $\mathbf{px}_1 \ldots \mathbf{x}_n \rightarrow \mathbf{p}'\mathbf{x}_1 \ldots \mathbf{x}_n$ is a theorem of $T \cup T'$. By the completeness theorem, we need only show that it is valid in every model $\mathcal{A}$ of $T \cup T'$. Now $\mathcal{A} \mid L(T)$ and $\mathcal{A} \mid L(T')$ are models of T and T' respectively. Construct a structure $\mathcal{B}$ for $L(T)$ by taking $|\mathcal{B}| = |\mathcal{A}|$, $\mathbf{v}_{\mathcal{B}} = \mathbf{v}_{\mathcal{A}}$ for **v** in Q, and $\mathbf{v}_{\mathcal{B}} = \mathbf{v}'_{\mathcal{A}}$ for **v** not in Q. Since $\mathcal{A} \mid L(T')$ is a model of T', it is evident that $\mathcal{B}$ is a model of T. Obviously the identity mapping from $\mathcal{A} \mid L(T)$ to $\mathcal{B}$ is a **v**-isomorphism for **v** in Q; so it is a **p**-isomorphism. It follows that $\mathbf{p}_{\mathcal{A}} = \mathbf{p}_{\mathcal{B}} = \mathbf{p}'_{\mathcal{A}}$; and this implies that $\mathbf{px}_1 \ldots \mathbf{x}_n \rightarrow \mathbf{p}'\mathbf{x}_1 \ldots \mathbf{x}_n$ is valid in $\mathcal{A}$.

We apply the Craig interpolation lemma to the theorem

$$\mathbf{px}_1 \ldots \mathbf{x}_n \rightarrow \mathbf{p}'\mathbf{x}_1 \ldots \mathbf{x}_n$$

of $T \cup T'$. We obtain a formula **A** such that

$$\vdash_T \mathbf{px}_1 \ldots \mathbf{x}_n \rightarrow \mathbf{A} \quad \text{and} \quad \vdash_{T'} \mathbf{A} \rightarrow \mathbf{p}'\mathbf{x}_1 \ldots \mathbf{x}_n.$$

Since **A** is in T and T', it contains no nonlogical symbols not in Q. From the choice of T',

$$\vdash_{T'} \mathbf{A} \rightarrow \mathbf{p}'\mathbf{x}_1 \ldots \mathbf{x}_n$$

implies

$$\vdash_T \mathbf{A} \rightarrow \mathbf{p}\mathbf{x}_1 \ldots \mathbf{x}_n;$$

so

$$\vdash_T \mathbf{p}\mathbf{x}_1 \ldots \mathbf{x}_n \leftrightarrow \mathbf{A}$$

by the tautology theorem.

5.5 COMPLETE THEORIES

We have used the notion of a complete theory in proving the completeness theorem. In addition to this, there are several important applications of complete theories.

For the first application, suppose that we are given a structure $\mathfrak{A}$ for L. The *theory of* $\mathfrak{A}$, designated by $Th(\mathfrak{A})$, is the theory whose language is L and whose nonlogical axioms are the formulas of L which are valid in $\mathfrak{A}$. Clearly $\mathfrak{A}$ is a model of $Th(\mathfrak{A})$; so by the validity theorem, the theorems of $Th(\mathfrak{A})$ are just the formulas valid in $\mathfrak{A}$.

The theory $Th(\mathfrak{A})$ is usually unmanageable because we do not know what its nonlogical axioms are. It is therefore desirable to find a simple axiomatization of $Th(\mathfrak{A})$, that is, to find a theory T equivalent to $Th(\mathfrak{A})$ whose nonlogical axioms are simple. We will, of course, only consider axioms which are valid in $\mathfrak{A}$; that is, we will only consider theories T having $\mathfrak{A}$ as a model. The following lemma shows that any such theory which is complete is equivalent to $Th(\mathfrak{A})$.

Lemma 1. For a consistent theory T, the following are equivalent:

a) T is complete;

b) every two models of T are elementarily equivalent;

c) for every model $\mathfrak{A}$ of T, T is equivalent to $Th(\mathfrak{A})$.

Proof. We first show that (a) implies (b). We must prove that if $\mathfrak{A}$ and $\mathfrak{B}$ are models of T, then every closed formula **A** is valid in both $\mathfrak{A}$ and $\mathfrak{B}$ or neither. But the former holds if $\vdash_T \mathbf{A}$ and the latter holds if $\vdash_T \neg\mathbf{A}$; so one or the other holds by (a).

We now show that (b) implies (c). Clearly $Th(\mathfrak{A})$ is an extension of T. By (b), every model of T is elementarily equivalent to $\mathfrak{A}$ and hence is a model of $Th(\mathfrak{A})$; so T is an extension of $Th(\mathfrak{A})$ by the corollary to the completeness theorem.

We now show that (c) implies (a). Let $\mathfrak{A}$ be a model of T. If **A** is a closed formula, then either **A** or $\neg$**A** is valid in $\mathfrak{A}$ and hence a theorem of $Th(\mathfrak{A})$. Thus $Th(\mathfrak{A})$ is complete; so T is complete by (c).

Our second application also stems from Lemma 1. Suppose that T is a complete theory, and suppose that we have shown that **A** is valid in some model of T. By (b) of Lemma 1, we can conclude that **A** is valid in every model of T.

A third application of complete theories will be discussed in Chapter 6. These applications suggest the importance of finding methods for proving that specific theories are complete. We present one such method here; another will be given in the next section.

We say that **A** is *equivalent to* **B** *in* T if $\vdash_T \mathbf{A} \leftrightarrow \mathbf{B}$. We say that T *admits elimination of quantifiers* if every formula in T is equivalent in T to an open formula. It is clear that if T admits elimination of quantifiers, then so does every simple extension of T.

Lemma 2. Suppose that T is consistent; that T admits elimination of quantifiers; that T contains a constant; and that every variable-free formula of T is decidable in T. Then T is complete.

Proof. If **A** is a closed formula of T, then $\vdash_T \mathbf{A} \leftrightarrow \mathbf{B}$ for some open formula **B**. Since we may substitute a constant for the variables free in **B**, we may suppose that **B** is variable-free. Then **B** is decidable in T; so **A** is decidable in T.

In practice, the difficult hypothesis to verify in Lemma 2 is that T admits elimination of quantifiers. We shall prove a theorem which gives a sufficient condition for this to be the case.

A formula is *simply existential* if it is of the form $\exists \mathbf{x}\mathbf{A}$ with **A** open.

Lemma 3. If every simply existential formula is equivalent in T to an open formula, then T admits elimination of quantifiers.

Proof. We prove by induction on the length of **A** that **A** is equivalent to an open formula. If **A** is atomic, this is clear. If **A** is a negation or a disjunction, then the result follows from the induction hypothesis and the equivalence theorem. Now suppose that **A** is $\exists \mathbf{x}\mathbf{B}$. By induction hypothesis, **B** is equivalent to an open formula $\mathbf{B}'$; so by the equivalence theorem, **A** is equivalent to the simply existential formula $\exists \mathbf{x}\mathbf{B}'$. Since $\exists \mathbf{x}\mathbf{B}'$ is equivalent to an open formula, **A** is also.

Lemma 4. Let **A** be a closed formula in T. Suppose that T contains a constant and that for every two models $\mathcal{A}$ and $\mathcal{A}'$ of T such that $\mathcal{A}(\mathbf{B}) = \mathcal{A}'(\mathbf{B})$ for every variable-free formula **B** in T, we have $\mathcal{A}(\mathbf{A}) = \mathcal{A}'(\mathbf{A})$. Then **A** is equivalent in T to a variable-free formula.

Proof. Let Γ be the set of variable-free theorems of $T[\mathbf{A}]$. It will suffice to show that **A** is a theorem of $T[\Gamma]$. For this implies by the reduction theorem that $\vdash_T \mathbf{B}_1 \rightarrow \cdots \rightarrow \mathbf{B}_n \rightarrow \mathbf{A}$, where each $\mathbf{B}_i$ is in Γ. Since $\mathbf{B}_i$ is in Γ, the deduction theorem shows that $\vdash_T \mathbf{A} \rightarrow \mathbf{B}_i$. Then by the tautology theorem,

$$\vdash_T \mathbf{A} \leftrightarrow (\mathbf{B}_1 \,\&\, \cdots \,\&\, \mathbf{B}_n);$$

and $\mathbf{B}_1 \,\&\, \cdots \,\&\, \mathbf{B}_n$ is variable-free.

Suppose that **A** is not a theorem of $T[\Gamma]$. By the completeness theorem, there is a model $\mathcal{A}$ of $T[\Gamma]$ such that $\mathcal{A}(\mathbf{A}) = \mathbf{F}$. Let Δ be the set of variable-free formulas which are valid in $\mathcal{A}$. Let $\mathcal{A}'$ be any model of $T(\Delta)$. For every variable-free formula **B**, $\mathcal{A}(\mathbf{B}) = \mathcal{A}'(\mathbf{B})$. For if $\mathcal{A}(\mathbf{B}) = \mathbf{T}$, then **B** is in Δ and hence $\mathcal{A}'(\mathbf{B}) = \mathbf{T}$; while if

$\mathfrak{A}(\mathbf{B}) = \mathbf{F}$, then $\neg\mathbf{B}$ is in Δ and hence $\mathfrak{A}'(\mathbf{B}) = \mathbf{F}$. It follows that $\mathfrak{A}'(\mathbf{A}) = \mathfrak{A}(\mathbf{A}) = \mathbf{F}$; so $\mathfrak{A}'(\neg\mathbf{A}) = \mathbf{T}$. We have shown that $\neg\mathbf{A}$ is valid in $T[\Delta]$. By the completeness theorem, $\neg\mathbf{A}$ is a theorem of $T[\Delta]$; so by the reduction theorem,

$$\vdash_T \mathbf{C}_1 \rightarrow \cdots \rightarrow \mathbf{C}_m \rightarrow \neg\mathbf{A}$$

with $\mathbf{C}_1, \ldots, \mathbf{C}_n$ in Δ. Then by the tautology theorem,

$$\vdash_T \mathbf{A} \rightarrow \neg(\mathbf{C}_1 \ \& \cdots \& \ \mathbf{C}_m);$$

so $\neg(\mathbf{C}_1 \ \& \cdots \& \ \mathbf{C}_m)$ is in Γ. We thus conclude that

$$\mathbf{C}_1, \ldots, \mathbf{C}_m \quad \text{and} \quad \neg(\mathbf{C}_1 \ \& \cdots \& \ \mathbf{C}_m)$$

are all valid in $\mathfrak{A}$; and this is impossible.

We say that T satisfies the *isomorphism condition* if for every two models $\mathfrak{A}$ and $\mathfrak{A}'$ of T and every isomorphism ϕ of a substructure of $\mathfrak{A}$ and a substructure of $\mathfrak{A}'$, there is an extension of ϕ which is an isomorphism of a submodel of $\mathfrak{A}$ and a submodel of $\mathfrak{A}'$. We say that T satisfies the *submodel condition* if for every model $\mathfrak{B}$ of T, every submodel $\mathfrak{A}$ of $\mathfrak{B}$, and every closed simply existential formula $\mathbf{A}$ of $L(\mathfrak{A})$, we have $\mathfrak{A}(\mathbf{A}) = \mathfrak{B}(\mathbf{A})$.

Lemma 5. Let T' be obtained from T by adding a new constant $\mathbf{e}$. If T satisfies the isomorphism (submodel) condition, then T' does also.

Proof. Suppose that T satisfies the isomorphism condition. Let $\mathfrak{A}$ and $\mathfrak{A}'$ be models of T', and let ϕ be an isomorphism of the substructure $\mathfrak{B}$ of $\mathfrak{A}$ and the substructure $\mathfrak{B}'$ of $\mathfrak{A}'$. Then ϕ can be extended to an isomorphism ϕ' of a submodel $\mathfrak{C}$ of $\mathfrak{A} \mid L(T)$ and a submodel $\mathfrak{C}'$ of $\mathfrak{A}' \mid L(T)$. We expand $\mathfrak{C}$ to a structure for $L(T')$ by setting $\mathbf{e}_{\mathfrak{C}} = \mathbf{e}_{\mathfrak{A}}$; this gives a submodel of $\mathfrak{A}$. We obtain a submodel of $\mathfrak{A}'$ similarly. Since $\phi(\mathbf{e}_{\mathfrak{A}}) = \mathbf{e}_{\mathfrak{A}'}$, ϕ' is an isomorphism of these submodels.

Suppose that T satisfies the submodel condition. Let $\mathfrak{B}$ be a model of T', $\mathfrak{A}$ a submodel of $\mathfrak{B}$, and $\mathbf{A}$ a closed simply existential formula of $L(\mathfrak{A})$. Let $\mathbf{A}'$ be obtained from $\mathbf{A}$ by substituting the name of $\mathfrak{A}(\mathbf{e})$ for $\mathbf{e}$. By the lemma of §2.5,

$$\mathfrak{A}(\mathbf{A}') = \mathfrak{A}(\mathbf{A}) \quad \text{and} \quad \mathfrak{B}(\mathbf{A}') = \mathfrak{B}(\mathbf{A}).$$

Applying the submodel condition in T to $\mathfrak{A} \mid L(T)$ and $\mathfrak{B} \mid L(T)$, we find that $\mathfrak{A}(\mathbf{A}') = \mathfrak{B}(\mathbf{A}')$; so $\mathfrak{A}(\mathbf{A}) = \mathfrak{B}(\mathbf{A})$.

Lemma 6. If T satisfies the isomorphism condition and the submodel condition and contains a constant, then every closed simply existential formula in T is equivalent in T to a variable-free formula.

Proof. Let $\mathbf{A}$ be closed and simply existential. By Lemma 4, it suffices to verify that if $\mathfrak{A}$ and $\mathfrak{A}'$ are models of T such that $\mathfrak{A}(\mathbf{B}) = \mathfrak{A}'(\mathbf{B})$ for every variable-free $\mathbf{B}$, then $\mathfrak{A}(\mathbf{A}) = \mathfrak{A}'(\mathbf{A})$. Let B be the set of all $\mathfrak{A}(\mathbf{a})$ for $\mathbf{a}$ variable-free. Since T contains a constant, B is nonempty; and clearly $\mathbf{f}_{\mathfrak{A}}(a_1, \ldots, a_n)$ is in B whenever $a_1, \ldots, a_n$ are in B. It follows that B is the universe of a substructure $\mathfrak{B}$ of $\mathfrak{A}$.

Let $\mathcal{B}'$ be the corresponding substructure of $\mathcal{A}'$. We claim that there is an isomorphism ϕ of $\mathcal{B}$ and $\mathcal{B}'$ defined by $\phi(\mathcal{A}(\mathbf{a})) = \mathcal{A}'(\mathbf{a})$. The fact that ϕ is well defined and bijective follows from $\mathcal{A}(\mathbf{a} = \mathbf{b}) = \mathcal{A}'(\mathbf{a} = \mathbf{b})$. The fact that ϕ is an isomorphism then follows from $\mathcal{A}(\mathbf{fa}_1 \ldots \mathbf{a}_n = \mathbf{b}) = \mathcal{A}'(\mathbf{fa}_1 \ldots \mathbf{a}_n = \mathbf{b})$ and $\mathcal{A}(\mathbf{pa}_1 \ldots \mathbf{a}_n) = \mathcal{A}'(\mathbf{pa}_1 \ldots \mathbf{a}_n)$.

By the isomorphism condition, ϕ can be extended to an isomorphism of a submodel $\mathcal{C}$ of $\mathcal{A}$ and a submodel $\mathcal{C}'$ of $\mathcal{A}'$. Since $\mathbf{A}$ is simply existential, the submodel condition implies that $\mathcal{A}(\mathbf{A}) = \mathcal{C}(\mathbf{A})$ and $\mathcal{A}'(\mathbf{A}) = \mathcal{C}'(\mathbf{A})$. But $\mathcal{C}(\mathbf{A}) = \mathcal{C}'(\mathbf{A})$, since $\mathcal{C}$ and $\mathcal{C}'$ are isomorphic; so $\mathcal{A}(\mathbf{A}) = \mathcal{A}'(\mathbf{A})$.

Quantifier Elimination Theorem. If T satisfies the isomorphism condition and the submodel condition, then T admits elimination of quantifiers.

Proof. In view of Lemma 3, it suffices to prove that every simply existential formula $\mathbf{A}$ in T is equivalent in T to an open formula. Let $\mathbf{A}'$ be obtained from $\mathbf{A}$ by replacing each variable free in $\mathbf{A}$ by a new constant; and let T' be obtained from T by adding these constants (or by adding one new constant if $\mathbf{A}$ is closed). From Lemmas 5 and 6, $\mathbf{A}'$ is equivalent in T' to a variable-free formula; so by the theorem on constants, $\mathbf{A}$ is equivalent in T to an open formula.

To verify that T satisfies the submodel condition, it suffices to verify that the following holds for every model $\mathcal{B}$ of T and every submodel $\mathcal{A}$ of $\mathcal{B}$: if $\mathbf{A}_1, \ldots, \mathbf{A}_n$ are atomic formulas of $L(\mathcal{A})$ in which no variable except $\mathbf{x}$ is free, and $\mathbf{j}$ is a name in $L(\mathcal{B})$, then there is a name $\mathbf{i}$ in $L(\mathcal{A})$ such that

$$\mathcal{B}(\mathbf{A}_i[\mathbf{i}]) = \mathcal{B}(\mathbf{A}_i[\mathbf{j}]) \qquad \text{for} \quad i = 1, \ldots, n.$$

For suppose this verified, and let $\exists\mathbf{x}\mathbf{A}$ be a closed simply existential formula in $L(\mathcal{A})$. Obviously $\mathcal{A}(\exists\mathbf{x}\mathbf{A}) = \mathbf{T}$ implies $\mathcal{B}(\exists\mathbf{x}\mathbf{A}) = \mathbf{T}$; we must prove the converse. Suppose that $\mathcal{B}(\exists\mathbf{x}\mathbf{A}) = \mathbf{T}$, and choose $\mathbf{j}$ in $L(\mathcal{B})$ so that $\mathcal{B}(\mathbf{A}_\mathbf{x}[\mathbf{j}]) = \mathbf{T}$. Choose $\mathbf{i}$ as above, with $\mathbf{A}_1, \ldots, \mathbf{A}_n$ the atomic formulas occurring in $\mathbf{A}$. Then

$$\mathcal{A}(\mathbf{A}[\mathbf{i}]) = \mathcal{B}(\mathbf{A}[\mathbf{i}]) = \mathcal{B}(\mathbf{A}[\mathbf{j}]) = \mathbf{T};$$

so

$$\mathcal{A}(\exists\mathbf{x}\mathbf{A}) = \mathbf{T}.$$

We shall apply our results to find axiomatizations for the field of complex numbers and the field of real numbers. (These axiomatizations are due to Tarski.)

The field of complex numbers is *algebraically closed*; that is, every nonconstant polynomial with coefficients in the field has a root in the field. We obtain the elementary theory *ACF* of algebraically closed fields from *FL* by adding for each $n \geqslant 1$ an axiom stating that every polynomial of degree n has a root. For example, the axiom for $n = 2$ is

$$y \neq 0 \rightarrow \exists x(y \cdot x \cdot x + z \cdot x + w = 0).$$

We now use the quantifier elimination theorem to show that *ACF* admits elimination of quantifiers. A substructure of a field contains 0, 1, and -1 and is closed under addition and multiplication; so it is a subring containing 1. Thus the isomorphism condition for *ACF* amounts to the following: if $\mathfrak{A}$ and $\mathfrak{A}'$ are algebraically closed fields, and ϕ is an isomorphism of a subring of $\mathfrak{A}$ containing 1 and a subring of $\mathfrak{A}'$ containing 1, then ϕ can be extended to an isomorphism of an algebraically closed subfield of $\mathfrak{A}$ and an algebraically closed subfield of $\mathfrak{A}'$. To obtain this extension, we first extend to an isomorphism of the smallest subfields of $\mathfrak{A}$ and $\mathfrak{A}'$ including these subrings, and then extend to an isomorphism of the smallest algebraically closed subfields including these subfields. The fact that these two extensions can be made is proved in standard texts on algebra.

Suppose that $\mathfrak{A}$ is a field and that **a** is a term of $L(\mathfrak{A})$ containing no variable except **x**. Then there is a polynomial $p_{\mathbf{a}}$ with coefficients in $\mathfrak{A}$ which represents **a** in the following sense: if $\mathfrak{B}$ is an extension of $\mathfrak{A}$, and **j** is the name of the individual b of $\mathfrak{B}$, then $\mathfrak{B}(\mathbf{a}_{\mathbf{x}}[\mathbf{j}]) = p_{\mathbf{a}}(b)$. The proof is by induction on the length of **a**; we leave the details to the reader. Now suppose that **b** is another such term, and let $r = p_{\mathbf{a}} - p_{\mathbf{b}}$. If **A** is the atomic formula $\mathbf{a} = \mathbf{b}$, then $\mathfrak{B}(\mathbf{A}_{\mathbf{x}}[\mathbf{j}]) = \mathbf{T}$ iff $r(b) = 0$.

If we combine these remarks with the remarks following the quantifier elimination theorem, we see that to prove that *ACF* satisfies the submodel condition, it suffices to prove the following result: if $\mathfrak{B}$ is an algebraically closed field, $\mathfrak{A}$ is an algebraically closed subfield of $\mathfrak{B}$, $r_1, \ldots, r_n$ are polynomials with coefficients in $\mathfrak{A}$, and b is in $\mathfrak{B}$, then there is an a in $\mathfrak{A}$ such that $r_i(a) = 0 \leftrightarrow r_i(b) = 0$ for all i. We may suppose that no r_i is the constant polynomial 0, since such an r_i could be dropped from the list. If some $r_i(b)$ is 0, then b is algebraic over $\mathfrak{A}$ and hence belongs to $\mathfrak{A}$; so we may take $a = b$. Otherwise, we have to find an a in $\mathfrak{A}$ such that $r_i(a) \neq 0$ for all i. Since a polynomial has only finitely many roots, it suffices to show that $\mathfrak{A}$ is infinite. But if $a_1, \ldots, a_k$ were all the elements of $\mathfrak{A}$, then the polynomial

$$(X - a_1) \cdot \ldots \cdot (X - a_k) + 1$$

would have no root in $\mathfrak{A}$.

We let $ACF(n)$ be the theory $ACF \cup FL(n)$. We show that if n is 0 or a prime, then $ACF(n)$ is complete. Since $ACF(n)$ has a model, it is consistent; and since it is a simple extension of *ACF*, it admits elimination of quantifiers. Hence by Lemma 2, we need only prove that every variable-free formula of $ACF(n)$ is decidable in $ACF(n)$. In view of the completeness theorem, it suffices to prove that if **A** is a variable-free formula of $ACF(n)$, then $\mathfrak{A}(\mathbf{A})$ is the same for all models $\mathfrak{A}$ of $ACF(n)$. If $n = 0$, let $\mathfrak{B}$ be the field of rational numbers; if n is a prime, let $\mathfrak{B}$ be the field of integers modulo n. Every model $\mathfrak{A}$ of $ACF(n)$ has a substructure isomorphic to $\mathfrak{B}$; so $\mathfrak{A}(\mathbf{A}) = \mathfrak{B}(\mathbf{A})$ for all models $\mathfrak{A}$ of $ACF(n)$ by the corollary to Lemma 2 of §5.2.

It follows that $ACF(0)$ is equivalent to the theory of the field of complex numbers. By our second application of complete theories, this implies that every formula of *FL* which is valid in the field of complex numbers is valid in every algebraically closed field of characteristic 0. The value of this is that we have many tools (such as contour integration) for proving a formula valid in the field of complex numbers which are not available in other fields.

In dealing with the field of real numbers, it is best to consider the predicate $<$. We therefore introduce the elementary theory *OF* of ordered fields. This is obtained from *FL* by adding the predicate symbol $<$ and the axioms

OF1. $\neg(x < x)$,

OF2. $x < y \to y < z \to x < z$,

OF3. $x < y \vee x = y \vee y < x$,

OF4. $x < y \to x + z < y + z$,

OF5. $0 < x \to 0 < y \to 0 < x \cdot y$.

We shall assume the basic theory of ordered fields in the following.

The ordered field of real numbers is *real closed*. This means that every positive element has a square root and that every polynomial of odd degree has a root. We obtain the elementary theory *RCF* of real closed fields from *OF* by adding an axiom

$$0 < x \to \exists y(y \cdot y = x),$$

and, for each odd n, an axiom like that added to obtain *ACF*.

We now use the quantifier elimination theorem to show that *RCF* admits elimination of quantifiers. The isomorphism condition now says the following: if $\mathfrak{A}$ and $\mathfrak{A}'$ are real closed fields, and ϕ is an isomorphism of a subring of $\mathfrak{A}$ containing 1 and a subring of $\mathfrak{A}'$ containing 1, then ϕ can be extended to an isomorphism of a real closed subfield of $\mathfrak{A}$ and a real closed subfield of $\mathfrak{A}'$. (Of course, isomorphisms here must preserve the order.) Again we first extend to the smallest subfields including the subrings, and then to the smallest real closed subfields including these subfields.

Much as above, we see that the verification of the submodel condition reduces to proving the following: if $\mathfrak{B}$ is a real closed field, $\mathfrak{A}$ is a real closed subfield of $\mathfrak{B}$, $r_1, \ldots, r_n$ are polynomials with coefficients in $\mathfrak{A}$, and b is in $\mathfrak{B}$, then there is an a in $\mathfrak{A}$ such that

$$r_i(a) = 0 \leftrightarrow r_i(b) = 0 \qquad \text{and} \qquad r_i(a) < 0 \leftrightarrow r_i(b) < 0.$$

Again we may suppose no r_i is the constant polynomial 0. If $r_i(b) = 0$, then b is algebraic over $\mathfrak{A}$ and hence belongs to $\mathfrak{A}$; so we may take $b = a$. We may therefore suppose that $r_i(b) \neq 0$ for all i.

Suppose first that there is a root c of some r_i and a root d of some r_j such that $c < b < d$. Since a polynomial has only finitely many roots, we may suppose that none of $r_1, \ldots, r_n$ has a root between c and d. Then each of $r_1, \ldots, r_n$ has the same sign at every point of $\mathfrak{B}$ between c and d. (This is because a polynomial in a real closed field which is positive at some point and negative at some other point has a root between these two points.) By the argument used above, c and d are in $\mathfrak{A}$; so we may take $a = \frac{1}{2}(c + d)$.

Now suppose that no root of any r_i is greater than b. Then the sign of r_i at all points greater than b is the same as its sign at b. Now the sign of r_i for large values of the argument is determined by the sign of the highest coefficient of r_i, and hence is the same in $\mathfrak{A}$ and in $\mathfrak{B}$. Thus if we choose a sufficiently large in $\mathfrak{A}$, each $r_i(a)$ will have the same sign as $r_i(b)$. A similar proof holds if no root of any r_i is less than b.

Since an ordered field always has an ordered subfield isomorphic to the ordered field of rational numbers, we can show as above that every variable-free formula in *RCF* is decidable in *RCF*, and hence that *RCF* is complete. Hence *RCF* is equivalent to the theory of the ordered field of real numbers. We may easily obtain from this an axiomatization of the field of real numbers (see Problem 17).

5.6 CATEGORICITY

We say that a theory T is *categorical* if every two models of T are isomorphic. For example, let T be the theory with no nonlogical symbols and the single nonlogical axiom $x = y$. Then every model of T contains just one individual, and it is easy to see that any two such models are isomorphic.

One can give somewhat more complicated examples than this; but all of them have only finite models. For if T has an infinite model, then the cardinality theorem shows that T has models of many different cardinalities; and two models with different cardinals cannot be isomorphic.

This suggests an extension of our definition. Let $\mathfrak{m}$ be an infinite cardinal. A theory T is $\mathfrak{m}$-*categorical* (or *categorical in power* $\mathfrak{m}$) if every two models of T of cardinal $\mathfrak{m}$ are isomorphic. We then have the following possibilities.

i) T is $\mathfrak{m}$-categorical for every $\mathfrak{m}$. An example is the theory with no nonlogical symbols and no nonlogical axioms.

ii) T is not $\mathfrak{m}$-categorical for any $\mathfrak{m}$. An example is the theory with no nonlogical axioms and a unary predicate symbol as its only nonlogical symbol. It is also known that *RCF* has this property.

iii) T is $\aleph_0$-categorical but not $\mathfrak{m}$-categorical for any uncountable $\mathfrak{m}$. For example, let the only nonlogical symbol of T be a unary predicate symbol P. Let the nonlogical axioms of T assert that for each natural number k, there are at least k individuals in the set P and at least k individuals not in the set P. Another example is discussed in Problem 23.

iv) T is $\mathfrak{m}$-categorical for every uncountable $\mathfrak{m}$ but not $\aleph_0$-categorical. For example, let the nonlogical symbols of T be an infinite sequence $\mathbf{e}_1, \mathbf{e}_2, \ldots$ of constants, and let the nonlogical axioms be the $\mathbf{e}_i \neq \mathbf{e}_j$ for $i \neq j$. It is also known that *ACF* (0) has this property.

For countable theories, these are the only possibilities; for a theorem of Morley states that if a countable theory is $\mathfrak{m}$-categorical for one uncountable $\mathfrak{m}$, then it is $\mathfrak{m}$-categorical for all uncountable $\mathfrak{m}$. Since the proof is quite long, we shall not give it here.

Part of the usefulness of categoricity comes from the following theorem.

Łoś-Vaught Theorem. Let $\mathfrak{m}$ be an infinite cardinal. If T is a consistent $\mathfrak{m}$-theory having only infinite models, and T is $\mathfrak{m}$-categorical, then T is complete.

Proof. Suppose that there is a closed formula $\mathbf{A}$ which is not decidable in T. By the corollary to the reduction theorem for consistency, both $T[\neg\mathbf{A}]$ and $T[\neg\neg\mathbf{A}]$ are consistent. Thus both have models; and these models are infinite because every model of T is infinite. By the cardinality theorem, there is a model $\mathcal{A}$ of $T[\neg\mathbf{A}]$ of cardinal $\mathfrak{m}$ and a model $\mathcal{B}$ of $T[\neg\neg\mathbf{A}]$ of cardinal $\mathfrak{m}$. Since T is $\mathfrak{m}$-categorical, $\mathcal{A}$ and $\mathcal{B}$ are isomorphic. This is impossible, since $\mathcal{A}(\mathbf{A}) = \mathsf{F}$ and $\mathcal{B}(\mathbf{A}) = \mathsf{T}$.

With the result mentioned in (iv), this gives a new proof of the completeness of $ACF(0)$.

We shall now investigate $\aleph_0$-categoricity. In the remainder of the section, $\mathbf{z}_1, \mathbf{z}_2, \ldots$ will be the variables in alphabetical order. We write $\mathbf{A}[\mathbf{a}_1, \ldots, \mathbf{a}_n]$ for $\mathbf{A}_{\mathbf{z}_1,\ldots,\mathbf{z}_n}[\mathbf{a}_1, \ldots, \mathbf{a}_n]$. We designate by $S_n(L)$ the set of formulas in L in which no variable other than $\mathbf{z}_1, \ldots, \mathbf{z}_n$ is free.

Let $\mathcal{A}$ be a structure for L, and let $a_1, \ldots, a_n$ be individuals in $\mathcal{A}$. The *type* of the n-tuple $(a_1, \ldots, a_n)$ is the set of all formulas $\mathbf{A}$ in $S_n(L)$ such that

$$\mathcal{A}(\mathbf{A}[\mathbf{i}_1, \ldots, \mathbf{i}_n]) = \mathsf{T},$$

where $\mathbf{i}_1, \ldots, \mathbf{i}_n$ are the names of $a_1, \ldots, a_n$ respectively. An *n-type in $\mathcal{A}$* is a type of an n-tuple of individuals of $\mathcal{A}$. Thus the unique 0-type in $\mathcal{A}$ is the set of closed formulas valid in $\mathcal{A}$. If T is a theory, we write $S_n(T)$ for $S_n(L(T))$, and call each n-type in a model of T an *n-type in T.*

We note some simple facts about types. If Γ is an n-type in a structure for L and $\mathbf{A}$ is in $S_n(L)$, then exactly one of $\mathbf{A}$ and $\neg\mathbf{A}$ belongs to Γ. It follows that if one n-type is included in another, then the two n-types are identical. If

$$\vdash_T \mathbf{A}_1 \vee \cdots \vee \mathbf{A}_k,$$

where $\mathbf{A}_1, \ldots, \mathbf{A}_k$ are in $S_n(T)$, then every n-type in T contains at least one of the $\mathbf{A}_i$. It follows that if $\vdash_T \mathbf{A}_1 \rightarrow \cdots \rightarrow \mathbf{A}_k \rightarrow \mathbf{B}$, where $\mathbf{A}_1, \ldots, \mathbf{A}_k, \mathbf{B}$ are in $S_n(T)$, then any n-type in T which contains $\mathbf{A}_1, \ldots, \mathbf{A}_k$ also contains $\mathbf{B}$.

Lemma. Let T be a countable theory, and let Γ be a nonempty set of formulas in $S_n(T)$ such that no disjunction of negations of formulas in Γ is a theorem of T. Then there is an n-type in a countable model of T which includes Γ.

Proof. Form T' from T by adding n new constants $\mathbf{e}_1, \ldots, \mathbf{e}_n$ and adding $\mathbf{A}[\mathbf{e}_1, \ldots, \mathbf{e}_n]$ as an axiom for each $\mathbf{A}$ in Γ. By the hypothesis, the reduction theorem for consistency, and the theorem on constants, T' is consistent. By the completeness theorem and the Löwenheim-Skolem theorem, T' has a countable model $\mathcal{A}$. Let $a_i = \mathcal{A}(\mathbf{e}_i)$. Then $\mathcal{A} \mid L(T)$ is a countable model of T, and the type of $(a_1, \ldots, a_n)$ in this model includes Γ.

Corollary. If T is a countable theory and Γ is an n-type in T, then Γ is an n-type in a countable model of T.

Proof. If $\mathbf{A}_1, \ldots, \mathbf{A}_k$ are in Γ, we cannot have $\vdash_T \neg\mathbf{A}_1 \vee \cdots \vee \neg\mathbf{A}_k$; for this would imply that some $\neg\mathbf{A}_i$ is in Γ. By the lemma, some n-type in a countable model of T includes Γ and therefore is equal to Γ.

Let Γ be a set of formulas in $S_n(T)$. A formula $\mathbf{A}$ of $S_n(T)$ is a *generator* of Γ if $\neg\mathbf{A}$ is not a theorem of T and $\vdash_T \mathbf{A} \rightarrow \mathbf{B}$ for every formula $\mathbf{B}$ in Γ. If Γ has a generator, we say that Γ is *principal.*

If $\mathbf{A}$ is a generator of an n-type Γ, then $\mathbf{A}$ is in Γ. For otherwise, $\neg\mathbf{A}$ is in Γ; so $\vdash_T \mathbf{A} \rightarrow \neg\mathbf{A}$; so $\vdash_T \neg\mathbf{A}$ by the tautology theorem. It follows that Γ is the set of all formulas $\mathbf{B}$ in $S_n(T)$ such that $\vdash_T \mathbf{A} \rightarrow \mathbf{B}$. This shows that a principal n-type is determined by any of its generators.

Ehrenfeucht's Theorem. Let T be a countable consistent theory, and let Γ be a subset of $S_n(T)$ which is not principal. Then there is a countable model $\mathfrak{A}$ of T such that no n-type in $\mathfrak{A}$ includes Γ.

Proof. By the lemma of §5.3, T_c contains only countably many special constants. Hence we may arrange the set of n-tuples of distinct special constants of T_c in a sequence.

We define inductively a sequence $\mathbf{A}_1, \mathbf{A}_2, \ldots$ of closed formulas of T_c so that:

a) $T_k = T_c[\mathbf{A}_1, \ldots, \mathbf{A}_k]$ is consistent;

b) $\mathbf{A}_k$ is $\neg\mathbf{A}[\mathbf{r}_1, \ldots, \mathbf{r}_n]$, where $\mathbf{A}$ is in Γ and $(\mathbf{r}_1, \ldots, \mathbf{r}_n)$ is the kth n-tuple in the above sequence.

First of all, T is consistent; so $T_0 = T_c$ is consistent by Lemma 3 of §4.2.

Now suppose that $\mathbf{A}_1, \ldots, \mathbf{A}_{k-1}$ have been selected, and let Δ be the set of formulas $\mathbf{A}$ in $S_n(T)$ such that $\mathbf{A}[\mathbf{r}_1, \ldots, \mathbf{r}_n]$ is a theorem of T_{k-1}. For any such $\mathbf{A}$,

$$\mathbf{A}_1 \rightarrow \cdots \rightarrow \mathbf{A}_{k-1} \rightarrow \mathbf{A}[\mathbf{r}_1, \ldots, \mathbf{r}_n]$$

is a theorem of T_c by the deduction theorem. From the proof of Lemma 3 of §4.2, we see that it has a proof in T_c which uses special axioms only for $\mathbf{r}_1, \ldots, \mathbf{r}_n$, and the special constants occurring in $\mathbf{A}_1, \ldots, \mathbf{A}_{k-1}$. It follows that if $\mathbf{B}$ is the conjunction of these special axioms and $\mathbf{A}_1, \ldots, \mathbf{A}_{k-1}$, then $\mathbf{B} \rightarrow \mathbf{A}[\mathbf{r}_1, \ldots, \mathbf{r}_n]$ is provable in T_c without special axioms for every $\mathbf{A}$ in Δ.

We may write $\mathbf{B}$ as $\mathbf{C}[\mathbf{r}_1, \ldots, \mathbf{r}_m]$, where $m \geqslant n$, $\mathbf{r}_1, \ldots, \mathbf{r}_m$ are distinct, and $\mathbf{C}$ is in $S_m(T)$. By the theorem on constants, $\vdash_T \mathbf{C} \rightarrow \mathbf{A}$ for every $\mathbf{A}$ in Δ. If $\mathbf{C}'$ is the formula $\exists \mathbf{z}_{n+1} \ldots \exists \mathbf{z}_m \mathbf{C}$ of $S_n(T)$, then $\vdash_T \mathbf{C}' \rightarrow \mathbf{A}$ for every $\mathbf{A}$ in Δ by the $\exists$-introduction rule.

Since $\mathbf{C}[\mathbf{r}_1, \ldots, \mathbf{r}_m]$ is a theorem of T_{k-1}, $\mathbf{C}'[\mathbf{r}_1, \ldots, \mathbf{r}_n]$ is a theorem of T_{k-1} by the substitution theorem and the detachment rule. Since T_{k-1} is consistent, $\neg\mathbf{C}'[\mathbf{r}_1, \ldots, \mathbf{r}_n]$ is not a theorem of T_{k-1}; so $\neg\mathbf{C}'$ is not a theorem of T. But $\mathbf{C}'$ is not a generator of Γ; so there must be a formula $\mathbf{A}$ in Γ such that $\mathbf{C}' \rightarrow \mathbf{A}$

is not a theorem of T. Then $\mathbf{A}$ is not in Δ. Let $\mathbf{A}_k$ be $\neg\mathbf{A}[\mathbf{r}_1, \ldots, \mathbf{r}_n]$. Then T_k is consistent by the corollary to the reduction theorem for consistency.

Form T' by adding all of $\mathbf{A}_1, \mathbf{A}_2, \ldots$ as new axioms to T_c. Every finite part of T' has some T_k as an extension and therefore is consistent. Hence by the compactness theorem and the completeness theorem, T' is consistent. It follows by Lemma 4 of §4.2 that T' has a model $\mathcal{A}$ such that every individual of $\mathcal{A}$ is $\mathcal{A}(\mathbf{r})$ for infinitely many $\mathbf{r}$. Such a model is certainly countable.

Let $a_1, \ldots, a_n$ be individuals of $\mathcal{A}$. We can then find distinct special constants $\mathbf{r}_1, \ldots, \mathbf{r}_n$ such that $\mathcal{A}(\mathbf{r}_1) = a_1, \ldots, \mathcal{A}(\mathbf{r}_n) = a_n$. For some $\mathbf{A}$ in Γ, $\neg\mathbf{A}[\mathbf{r}_1, \ldots, \mathbf{r}_n]$ is an axiom of T' and hence is valid in $\mathcal{A}$. Using the lemma of §2.5, we conclude that $\neg\mathbf{A}$ is in the type of $(a_1, \ldots, a_n)$; so Γ is not included in the type of $(a_1, \ldots, a_n)$. This shows that Γ is not included in an n-type in $\mathcal{A}$.

Ryll-Nardzewski's Theorem. Let T be a complete countable theory having only infinite models. Then the following are equivalent:

a) T is $\aleph_0$-categorical;

b) for every n, T has only finitely many n-types;

c) for every n, every n-type in T is principal.

Proof. Suppose that T has an n-type Γ which is not principal; we shall show that (a) and (b) are false. By Ehrenfeucht's theorem, there is a countable model $\mathcal{A}$ of T in which Γ is not an n-type; while by the corollary to the lemma, there is a countable model $\mathcal{B}$ of T in which Γ is an n-type. Since $\mathcal{A}$ and $\mathcal{B}$ are not isomorphic, (a) is false.

Now we show that (b) is false. Since Γ is not principal, and since a conjunction of formulas in Γ is in Γ, we may choose inductively formulas $\mathbf{A}_1, \mathbf{A}_2, \ldots$ in Γ such that for each k, $(\mathbf{A}_1 \,\&\, \cdots \,\&\, \mathbf{A}_{k-1}) \rightarrow \mathbf{A}_k$ is not a theorem of T. Then $\neg\mathbf{A}_1 \vee \cdots \vee \neg\mathbf{A}_{k-1} \vee \neg\neg\mathbf{A}_k$ is not a theorem of T. Hence by the lemma there is an n-type Γ_k in T which contains $\mathbf{A}_1, \ldots, \mathbf{A}_{k-1}, \neg\mathbf{A}_k$. Clearly the Γ_k are distinct n-types.

We now assume that (c) is true and prove (a) and (b). We begin with (b). Fix n. For each n-type in T choose a generator, and let these generators be $\mathbf{A}_1, \mathbf{A}_2, \ldots$. Then no n-type can contain all of $\neg\mathbf{A}_1, \neg\mathbf{A}_2, \ldots$. Hence by the lemma and the tautology theorem, there is a k such that $\mathbf{A}_1 \vee \cdots \vee \mathbf{A}_k$ is a theorem of T. Then each n-type in T contains an $\mathbf{A}_i$. But if an n-type contains $\mathbf{A}_i$, it contains the n-type with generator $\mathbf{A}_i$ and hence is identical with that n-type. Thus the only n-types are those with generators $\mathbf{A}_1, \ldots, \mathbf{A}_k$.

Now we prove (a). Let $\mathcal{A}$ and $\mathcal{B}$ be models of T with cardinal $\aleph_0$. Arrange each of $|\mathcal{A}|$ and $|\mathcal{B}|$ in a sequence of distinct elements. We shall arrange them in new sequences $a_1, a_2, \ldots$ and $b_1, b_2, \ldots$ so that for each n, $(a_1, \ldots, a_n)$ and $(b_1, \ldots, b_n)$ have the same type. We must first show that this holds for $n = 0$, that is, that the empty sequence has the same type in $\mathcal{A}$ and in $\mathcal{B}$. This means that the same closed formulas are valid in $\mathcal{A}$ and $\mathcal{B}$; so it holds by Lemma 1 of §5.5.

Now suppose that $a_1, \ldots, a_{n-1}, b_1, \ldots, b_{n-1}$ have been chosen. First suppose that n is even. Let a_n be the first individual in the old sequence of individuals of $\mathfrak{A}$ which is not among $a_1, \ldots, a_{n-1}$. Let Γ be the type of $(a_1, \ldots, a_n)$, and let **A** be a generator of Γ. Then $\exists \mathbf{z}_n \mathbf{A}$ is in the type of $(a_1, \ldots, a_{n-1})$ and hence in the type of $(b_1, \ldots, b_{n-1})$. It follows that we may choose b_n in $|\mathfrak{B}|$ so that **A** is in the type of $(b_1, \ldots, b_n)$. The type of $(b_1, \ldots, b_n)$ then includes Γ and hence must be Γ. Moreover, $b_n \neq b_i$ for $i < n$; for $a_n \neq a_i$, and hence $\mathbf{z}_n \neq \mathbf{z}_i$ is in Γ.

If n is odd, we let b_n be the first individual in the old sequence of individuals of $\mathfrak{B}$ which is not among $b_1, \ldots, b_{n-1}$. We can then choose a_n as above. The first n members of the old sequence of elements of $\mathfrak{A}$ appear among $a_1, \ldots, a_{2n}$; so every individual of $\mathfrak{A}$ is an a_i. Similarly, every individual of $\mathfrak{B}$ is a b_i. We can therefore define a bijective mapping ϕ from $|\mathfrak{A}|$ to $|\mathfrak{B}|$ by $\phi(a_i) = b_i$. By Lemma 2 of §5.2, ϕ is an isomorphism.

As an application of Ryll-Nardzewski's theorem, we can show that $Th(\mathfrak{N})$ is not $\aleph_0$-categorical. In fact, it is easy to see that if i and j are distinct natural numbers, then the type of i is different from the type of j. Thus $Th(\mathfrak{N})$ has countable models which are not isomorphic to $\mathfrak{N}$.

PROBLEMS

1. Let Γ be a set of closed formulas in L. Let $\mathfrak{T}(\Gamma)$ be as in Problem 5 of Chapter 4, and let $\mathfrak{T}\mathfrak{S}(\Gamma)$ be the set of V in $\mathfrak{T}(\Gamma)$ such that for some structure $\mathfrak{A}$ for L, $V(\mathbf{A}) = \mathfrak{A}(\mathbf{A})$ for all **A** in Γ. Show that $\mathfrak{T}\mathfrak{S}(\Gamma)$ is closed in $\mathfrak{T}(\Gamma)$, and hence is compact. [Use the compactness theorem.]

2. Let $\mathfrak{J}$ be a class of structures for L. If there is a formula **A** of L such that $\mathfrak{J}$ is the class of structures in which **A** is valid, then $\mathfrak{J}$ is an *elementary class*. If there is a theory T with language L such that $\mathfrak{J}$ is the class of models of T, then $\mathfrak{J}$ is a *generalized elementary class*.

a) Show that a class of structures for L is an elementary class iff it is the class of all models of a finitely axiomatized theory with language L.

b) Show that a class of structures for L is a generalized elementary class iff it is the intersection of a family of elementary classes.

c) Let A be a collection of generalized elementary classes, and let $\mathfrak{J}$ be an elementary class which includes the intersection of the members of A. Show that there is a finite subcollection A' of A such that $\mathfrak{J}$ includes the intersection of the members of A'. [Use the compactness theorem.]

d) Let $\mathfrak{J}_1, \mathfrak{J}_2, \ldots$ be a sequence of elementary classes such that for all n, $\mathfrak{J}_{n+1}$ is a proper subclass of $\mathfrak{J}_n$. Show that the intersection of the $\mathfrak{J}_n$ is not an elementary class. [Use (c).]

e) Let $\mathfrak{J}$ be a generalized elementary class, $\mathfrak{J}'$ the class of structures for L which are not in $\mathfrak{J}$. Show that $\mathfrak{J}$ is an elementary class iff $\mathfrak{J}'$ is a generalized elementary class. [Use (b), (c), and (a).]

3. a) Let $\mathfrak{A}$ be a structure for L, and let B be a nonempty subset of $|\mathfrak{A}|$. Show that there is a smallest subset C of $|\mathfrak{A}|$ which includes B and is the universe of a substructure of $\mathfrak{A}$. We call the substructure with universe C the substructure *generated* by B.

b) Let $\mathfrak{B}$ be the substructure of $\mathfrak{A}$ generated by B. Show that if $\mathfrak{m}$ is the largest of $\aleph_0$, the cardinal of B, and the cardinal of the set of function symbols of L, then the cardinal of $\mathfrak{B}$ is at most $\mathfrak{m}$. In particular, if L is a countable language and B is countable, then $\mathfrak{B}$ is countable.

c) A structure $\mathfrak{A}$ is *finitely generated* if there is a finite subset B of $|\mathfrak{A}|$ such that the substructure of $\mathfrak{A}$ generated by B is $\mathfrak{A}$. Show that a structure $\mathfrak{A}$ has an extension which is a model of T iff every finitely generated substructure of $\mathfrak{A}$ has an extension which is a model of T. [If $\mathfrak{A}$ has no such extension, then, by the model extension theorem, there is an open theorem **A** of T not valid in $\mathfrak{A}$. Show that **A** is not valid in some finitely generated substructure of $\mathfrak{A}$.]

d) Show that a structure $\mathfrak{A}$ for an open theory T is a model of T iff every finitely generated substructure of $\mathfrak{A}$ is a model of T. [Use (c) and the Łoś-Tarski theorem.]

e) A group $\mathfrak{A}$ is *divisible* if for every positive integer n and every individual a of $\mathfrak{A}$, there is an individual b of $\mathfrak{A}$ such that $b^n = a$. Show that every Abelian group is isomorphic to a subgroup of a divisible Abelian group. [A cyclic group is isomorphic to a subgroup of the multiplicative group of nonzero complex numbers, which is divisible. Extend the result to finitely generated Abelian groups and then use (c).]

4. a) Let Γ be a regular set of formulas in L, and let $\mathfrak{A}$ and $\mathfrak{B}$ be structures for L such that every disjunction of negations of formulas in Γ which is valid in $\mathfrak{A}$ is valid in $\mathfrak{B}$. Show that there is a Γ-extension of $\mathfrak{B}$ which is isomorphic to an elementary extension of $\mathfrak{A}$. [Expand $\mathfrak{B}$ to a structure $\mathfrak{B}'$ for $L(\mathfrak{A})$. Find a Γ-extension of $\mathfrak{B}'$ which is a model of $D_e(\mathfrak{A})$.]

b) Show that if $\mathfrak{A}$ and $\mathfrak{B}$ are structures for L, then $\mathfrak{A}$ and $\mathfrak{B}$ are elementarily equivalent iff they have isomorphic elementary extensions. [Use (a).]

5. Let $\mathfrak{A}$ and $\mathfrak{B}$ be structures for L, and let Γ be a set of formulas in L. A Γ-*morphism* from $\mathfrak{A}$ to $\mathfrak{B}$ is a mapping ϕ from $|\mathfrak{A}|$ to $|\mathfrak{B}|$ such that for every formula **A** in $\Gamma(\mathfrak{A})$, $\mathfrak{A}(\mathbf{A}) = \mathbf{T}$ implies that $\mathfrak{B}(\mathbf{A}^\phi) = \mathbf{T}$.

a) Show that if $|\mathfrak{A}|$ is a subset of $|\mathfrak{B}|$, then $\mathfrak{A}$ is a Γ-substructure of $\mathfrak{B}$ iff the identity mapping from $|\mathfrak{A}|$ to $|\mathfrak{B}|$ is a Γ-morphism.

b) Show that if $\mathbf{x} = \mathbf{e}$ is in Γ and ϕ is a Γ-morphism from $\mathfrak{A}$ to $\mathfrak{B}$, then $\phi(\mathfrak{A}(\mathbf{e})) = \mathfrak{B}(\mathbf{e})$.

c) If ϕ is a mapping from $|\mathfrak{A}|$ to $|\mathfrak{B}|$, we expand $\mathfrak{B}$ to a structure $\mathfrak{B}_\phi$ for $L(\mathfrak{A})$ by assigning $\phi(a)$ to the name of a. Show that ϕ is a Γ-morphism iff $\mathfrak{B}_\phi$ is a model of $D_\Gamma(\mathfrak{A})$. [First show that $\mathfrak{B}_\phi(\mathbf{A}) = \mathfrak{B}_\phi(\mathbf{A}^\phi) = \mathfrak{B}(\mathbf{A}^\phi)$ for **A** a closed formula of $L(\mathfrak{A})$.]

d) A non-empty set Γ of formulas is *invariant* if for each formula **A** in Γ, every formula $\mathbf{A}[\mathbf{x}_1, \ldots, \mathbf{x}_n]$ is in Γ. Show that if Γ is invariant and $L(T) = L$, then there is a Γ-morphism from $\mathfrak{A}$ to a model of T iff every theorem of T which is a disjunction of negations of formulas in Γ is valid in $\mathfrak{A}$. [Like the proof of the model extension theorem.]

e) Let Γ be an invariant set of formulas containing $x = y$; Δ a set of formulas containing every formula $\forall \mathbf{x}_1 \ldots \forall \mathbf{x}_n \mathbf{A}$, where **A** is a disjunction of negations of formulas in Γ; and ϕ a Δ-morphism from $\mathfrak{A}$ to $\mathfrak{B}$. Show that there is a Γ-morphism ψ from $\mathfrak{B}$ to an elementary extension $\mathfrak{C}$ of $\mathfrak{A}$ such that $\psi\phi$ is the identity mapping from $|\mathfrak{A}|$ to $|\mathfrak{C}|$. [Like the proof of the corollary to the model extension theorem.]

6. A formula is *positive* if it is in prenex form, and its matrix is built from atomic formulas by repeatedly taking disjunctions and conjunctions. A formula is *negative* if it is the negation of a positive formula. If Γ is the set of atomic (positive) (negative) formulas, we say *homomorphism* (*positive homomorphism*) (*negative homomorphism*) for Γ-morphism.

a) Show that a mapping ϕ from $|\mathfrak{A}|$ to $|\mathfrak{B}|$ is a homomorphism iff for all nonlogical symbols $\mathbf{f}$ and $\mathbf{p}$ and all $a_1, \ldots, a_n$ in $|\mathfrak{A}|$,

$$\phi(\mathbf{f}_{\mathfrak{A}}(a_1, \ldots, a_n)) = \mathbf{f}_{\mathfrak{B}}(\phi(a_1), \ldots, \phi(a_n))$$

and

$$\mathbf{p}_{\mathfrak{A}}(a_1, \ldots, a_n) \rightarrow \mathbf{p}_{\mathfrak{B}}(\phi(a_1), \ldots, \phi(a_n)).$$

Show that in this case, $\phi(\mathfrak{A}(\mathbf{a})) = \mathfrak{B}(\mathbf{a}^{\phi})$ for every variable-free term $\mathbf{a}$ in $L(\mathfrak{A})$.

b) Show that a surjective homomorphism is a positive homomorphism.

c) Show that if ϕ is a negative homomorphism from $\mathfrak{A}$ to $\mathfrak{B}$, then there is a positive homomorphism ψ from $\mathfrak{B}$ to an elementary extension $\mathfrak{C}$ of $\mathfrak{A}$ such that $\psi\phi$ is the identity mapping from $|\mathfrak{A}|$ to $|\mathfrak{C}|$. [Use 5(e).]

d) Let ϕ be a positive homomorphism from $\mathfrak{A}$ to $\mathfrak{B}$. Show that there is an elementary extension $\mathfrak{A}'$ of $\mathfrak{A}$ and a negative homomorphism ψ from $\mathfrak{B}_{\phi}$ to $\mathfrak{A}'_{\mathfrak{A}}$. [Use 5(d) to get a negative homomorphism from $\mathfrak{B}_{\phi}$ to a model of $D_e(\mathfrak{A})$.]

e) Let ϕ be a positive homomorphism from $\mathfrak{A}$ to $\mathfrak{B}$. Show that there are elementary extensions $\mathfrak{A}'$ and $\mathfrak{B}'$ of $\mathfrak{A}$ and $\mathfrak{B}$ respectively and a positive homomorphism ϕ' from $\mathfrak{A}'$ to $\mathfrak{B}'$ which is an extension of ϕ such that $\phi'(|\mathfrak{A}'|)$ includes $|\mathfrak{B}|$. [Choose $\mathfrak{A}'$ as in (d), and use (c) to find a positive homomorphism ϕ' from $\mathfrak{A}'_{\mathfrak{A}}$ to an elementary extension $\mathfrak{C}$ of $\mathfrak{B}_{\phi}$. Let $\mathfrak{B}' = \mathfrak{C} \mid L$.]

f) We say that $\mathfrak{B}$ is a *homomorphic image* of $\mathfrak{A}$ if there is a surjective homomorphism from $\mathfrak{A}$ to $\mathfrak{B}$. Show that if there is a positive homomorphism from $\mathfrak{A}'$ to $\mathfrak{B}'$, then some elementary extension $\mathfrak{B}$ of $\mathfrak{B}'$ is a homomorphic image of some elementary extension $\mathfrak{A}$ of $\mathfrak{A}'$. [Take $\mathfrak{A}$ and $\mathfrak{B}$ as unions of elementary chains constructed by using (e).]

g) Show that a theory T is equivalent to a theory whose nonlogical axioms are positive iff every homomorphic image of a model of T is a model of T. [For the "only if" part use (b). Suppose that $\mathfrak{B}$ is a structure in which every positive theorem of T is valid. Show that if Γ is the set of negative formulas valid in $\mathfrak{B}$, then $T[\Gamma]$ is consistent. Use 5(d) to find a positive homomorphism from a model $\mathfrak{A}$ of $T[\Gamma]$ to an elementary extension of $\mathfrak{B}$. Then use (f) to show that $\mathfrak{B}$ is a model of T.]

7. For each i in the nonempty set I, let $\mathfrak{A}_i$ be a structure for L. The *direct product* $\mathfrak{A} = \prod_{i \in I} \mathfrak{A}_i$ is the structure for L defined as follows. The universe of $\mathfrak{A}$ is $\prod_{i \in I} |\mathfrak{A}_i|$; and

$$(\mathbf{f}_{\mathfrak{A}}(a_1, \ldots, a_n))_i = \mathbf{f}_{\mathfrak{A}_i}((a_1)_i, \ldots, (a_n)_i),$$
$$\mathbf{p}_{\mathfrak{A}}(a_1, \ldots, a_n) \leftrightarrow \mathbf{p}_{\mathfrak{A}_i}((a_1)_i, \ldots, (a_n)_i) \text{ for all } i.$$

a) Let $\pi_i(a) = (a)_i$ for a in $|\mathfrak{A}|$. Show that π_i is a surjective homomorphism from $\mathfrak{A}$ to $\mathfrak{A}_i$.

b) Show that if $\mathbf{A}$ is a variable-free atomic formula in $L(\mathfrak{A})$, then

$$\mathfrak{A}(\mathbf{A}) = \mathbf{T} \quad \text{iff} \quad \mathfrak{A}_i(\mathbf{A}^{\pi_i}) = \mathbf{T} \text{ for all } i \text{ in } I.$$

[Use (a) and 6(a).]

c) A formula is a *McKinsey* formula if it is a disjunction of formulas, each of which is either atomic or the negation of an atomic formula and at most one of which is atomic. Show that if $\mathbf{A}$ is a closed McKinsey formula in $L(\mathfrak{A})$ such that $\mathfrak{A}_i(\mathbf{A}^{\pi_i}) = \mathbf{T}$ for all i in I, then $\mathfrak{A}(\mathbf{A}) = \mathbf{T}$. [Assume that $\mathfrak{A}(\mathbf{A}) = \mathbf{F}$ and use (b).]

d) A formula is a *Horn* formula if it is in prenex form and its matrix is a conjunction of McKinsey formulas. Show that the result of (c) extends to Horn formulas. [Use induction on the number of quantifiers.] Conclude that if a Horn formula is valid in each $\mathfrak{A}_i$, then it is valid in $\mathfrak{A}$.

e) Let T be a theory such that every direct product of models of T is a model of T. Let $\neg\mathbf{A}_1 \vee \cdots \vee \neg\mathbf{A}_n \vee \mathbf{B}_1 \vee \cdots \vee \mathbf{B}_m$ be a theorem of T, where $m > 0$ and the $\mathbf{A}_i$ and $\mathbf{B}_i$ are atomic. Show that for some j, $\neg\mathbf{A}_1 \vee \cdots \vee \neg\mathbf{A}_n \vee \mathbf{B}_j$ is a theorem of T. [Assume that there is no such j. Find a model $\mathfrak{A}_i$ of T such that $\neg\mathbf{A}_1 \vee \cdots \vee \neg\mathbf{A}_n \vee \mathbf{B}_i$ is not valid in $\mathfrak{A}_i$, and show that $\neg\mathbf{A}_1 \vee \cdots \vee \neg\mathbf{A}_n \vee \mathbf{B}_1 \vee \cdots \vee \mathbf{B}_m$ is not valid in the direct product of the $\mathfrak{A}_i$.]

f) Show that a theory T is equivalent to a theory whose nonlogical axioms are McKinsey formulas iff every substructure of a model of T is a model of T and every direct product of models of T is a model of T. [Use the Łoś-Tarski theorem, (c), Problem 11(b) of Chapter 3, and (e).]

8. A class $\mathfrak{J}$ of structures for L is *closed* if every substructure of a structure in $\mathfrak{J}$ is in $\mathfrak{J}$, every homomorphic image of a structure in $\mathfrak{J}$ is in $\mathfrak{J}$, and every direct product of structures in $\mathfrak{J}$ is in $\mathfrak{J}$. We write $At(\mathfrak{J})$ for the set of atomic formulas which are valid in every structure in $\mathfrak{J}$.

a) Let $\mathfrak{J}$ be a closed class of structures for L. Let L' be obtained from L by adding a nonempty set of new constants, and let $\mathfrak{J}'$ be the set of expansions to L' of structures in I. Show that $\mathfrak{J}'$ is closed and that $At(\mathfrak{J})$ is a subset of $At(\mathfrak{J}')$.

b) Let the notation be as in (a). Let $\mathbf{A}$ be a variable-free atomic formula in L' which is not an instance of a formula in $At(\mathfrak{J})$. Show that there is an $\mathfrak{A}$ in $\mathfrak{J}'$ such that $\mathfrak{A}(\mathbf{A}) = \mathbf{F}$ and such that every individual of $\mathfrak{A}$ is $\mathfrak{A}(\mathbf{a})$ for a variable-free term $\mathbf{a}$ in L'. [Choose $\mathfrak{B}$ in $\mathfrak{J}'$ such that $\mathfrak{B}(\mathbf{A}) = \mathbf{F}$ and take a substructure.]

c) Let the notation be as in (a). Show that there is a structure $\mathfrak{A}$ in $\mathfrak{J}'$ such that:

i) every variable-free atomic formula of L' which is valid in $\mathfrak{A}$ is an instance of a formula in $At(\mathfrak{J})$;

ii) every individual of $\mathfrak{A}$ is $\mathfrak{A}(\mathbf{a})$ for a variable-free term $\mathbf{a}$ in L'.

[Use (b) and 7(b).]

d) Let $\mathfrak{J}$ be a closed class of structures for L. Show that if $\mathfrak{B}$ is a structure for L in which every formula of $At(\mathfrak{J})$ is valid, then $\mathfrak{B}$ is in $\mathfrak{J}$. [Let L' be $L(\mathfrak{B})$, and let $\mathfrak{A}$ be as in (c). If $\mathbf{A}$ is a variable-free atomic formula of L', then $\mathfrak{A}(\mathbf{A}) = \mathbf{T}$ implies $\mathfrak{B}(\mathbf{A}) = \mathbf{T}$. Define a surjective homomorphism ϕ from $\mathfrak{A}$ to $\mathfrak{B}_{\mathfrak{B}}$ by setting $\phi(\mathfrak{A}(\mathbf{a})) = \mathfrak{B}_{\mathfrak{B}}(\mathbf{a})$.]

e) Show that a class $\mathfrak{J}$ of structures for L is closed iff it is the class of all models of a theory with language L whose nonlogical axioms are atomic. [Use (d) for the "only if" part.]

9. If T is a theory whose language is an extension of L, the *restriction* of T to L is the theory whose language is L and whose nonlogical axioms are the formulas of L which are theorems of T.

a) Let T be the restriction of T' to L. Show that a structure $\mathfrak{A}$ for L is a model of T iff some expansion of an elementary extension of $\mathfrak{A}$ is a model of T'. [If $\mathfrak{A}$ is a model of T, use the joint consistency theorem to show that $T' \cup D_e(\mathfrak{A})$ is consistent.]

b) Give an example of a theory T', a restriction T of T', and a model $\mathfrak{A}$ of T which has no expansion which is a model of T'. [Let T' have as nonlogical symbols the constants $\mathbf{e}_1, \mathbf{e}_2, \ldots$ and as nonlogical axioms all formulas $\mathbf{e}_i \neq \mathbf{e}_j$ for $i \neq j$. Obtain T by omitting $\mathbf{e}_1$.]

c) Let T' be an extension of T. Show that T' is a conservative extension of T iff for every model $\mathfrak{A}$ of T, some expansion of an elementary extension of $\mathfrak{A}$ is a model of T'. [Use (a).]

d) Let T be a theory whose language is an extension of L, and let $\mathfrak{J}$ be the class of restrictions to L of models of T. Suppose that every substructure of a structure in $\mathfrak{J}$ is in $\mathfrak{J}$. Show that $\mathfrak{J}$ is the class of models of the restriction T' of T to L and that T' is equivalent to an open theory. [Use (a) and the Łoś-Tarski theorem.]

e) Let T be an open theory whose language is obtained from L by adding predicate symbols. Let $\mathfrak{A}$ be a structure for L. Show that $\mathfrak{A}$ has an expansion which is a model of T iff every finitely generated substructure of $\mathfrak{A}$ has an expansion which is a model of T. [If $\mathfrak{J}$ is as in (d), show that every substructure of a structure in $\mathfrak{J}$ is in $\mathfrak{J}$. Then use (d) and 3(c).]

f) The *four-color conjecture* states that any map can be colored with four colors so that no two adjacent countries have the same color. Show that if the four-color conjecture holds for maps with finitely many countries, then it holds for all maps. [Consider a map as a structure with the countries as individuals and a binary predicate *is adjacent to*. Show that a map can be colored with four colors iff it has an expansion to a model of a suitable open theory containing four unary predicates. Then use (e).]

10. We write $\exists\Gamma$ for the set of all formulas $\exists \mathbf{x}_1 \ldots \exists \mathbf{x}_n \mathbf{A}$ with $\mathbf{A}$ in Γ, and $\forall\Gamma$ for the set of all formulas $\forall \mathbf{x}_1 \ldots \forall \mathbf{x}_n \mathbf{A}$ with $\mathbf{A}$ in Γ. We define $\exists_n$ and $\forall_n$ inductively as follows: $\exists_0$ and $\forall_0$ are both the set of all open formulas; $\exists_{n+1} = \exists\forall_n$; and $\forall_{n+1} = \forall\exists_n$. We let B_n be the set of formulas obtained from formulas in $\exists_n$ and $\forall_n$ by repeatedly taking negations and disjunctions.

a) Let Γ be a set of formulas in L, $\mathfrak{B}$ a structure for L, $\mathfrak{A}$ a substructure of $\mathfrak{B}$. Show that if some elementary extension of $\mathfrak{A}$ is a Γ-extension of $\mathfrak{B}$, then $\mathfrak{B}$ is a $\forall\Gamma'$-extension of $\mathfrak{A}$ where Γ' is the set of negations of formulas in Γ''.

b) Show that every formula in B_n has a prenex form which is in $\exists_{n+1}$ and a prenex form which is in $\forall_{n+1}$. Show that the negation of a formula in $\exists_n$ ($\forall_n$) has a prenex form which is in $\forall_n$ ($\exists_n$).

c) An *n-sandwich* for L is a sequence $\mathfrak{A}_0, \ldots, \mathfrak{A}_n$ of structures for L such that $\mathfrak{A}_{i+1}$ is an extension of $\mathfrak{A}_i$ for $i < n$ and $\mathfrak{A}_{i+2}$ is an elementary extension of $\mathfrak{A}_i$ for $i < n - 1$. Let $\mathfrak{B}$ be an extension of $\mathfrak{A}$. Show that $\mathfrak{B}$ is a $\forall_n$-extension of $\mathfrak{A}$ iff there is an $(n + 1)$-sandwich whose first two structures are $\mathfrak{A}$ and $\mathfrak{B}$. [Use (a), (b), and the corollary to the model extension theorem.]

d) Show that a theory T is equivalent to a theory whose nonlogical axioms are in $\exists_n$ iff for every $(n + 1)$-sandwich for $L(T)$, if the second structure in the sandwich is a model of T, then the first structure in the sandwich is a model of T. [For the "only if" part, use (b) and (c). If the condition holds, and $\mathfrak{A}$ is a structure in which all the theorems of T in $\exists_n$ are valid, use (b) and the model extension theorem to get a $\forall_n$-extension of $\mathfrak{A}$ which is a model of T. Then use (c).]

11. a) Let $\mathfrak{A}_1, \mathfrak{A}_2, \ldots$ be a chain of structures for L such that for each k, $\mathfrak{A}_{k+1}$ is a $\forall_n$-extension of $\mathfrak{A}_k$. Show that the union $\mathfrak{A}$ of the chain is a $\forall_n$-extension of each $\mathfrak{A}_k$. [Like the proof of Tarski's lemma.]

b) Let T and T' be theories with the same language whose non-logical axioms are in $\exists_n$. Show that, for $n \geqslant 1$, $T \cup T'$ is inconsistent iff there is a closed formula **A** in B_n such that $\vdash_T$ **A** and $\vdash_{T'} \neg$**A**. [Suppose that no such **A** exists. Let Γ be the set of closed formulas in B_n which are theorems of T, and define Γ' similarly. Show that $T[\Gamma']$ and $T'[\Gamma]$ are consistent and that the same formulas in B_{n-1} are provable in $T[\Gamma']$ and $T'[\Gamma]$. Using the model extension theorem, find a chain $\mathfrak{A}_1, \mathfrak{A}_2, \ldots$ such that $\mathfrak{A}_{k+1}$ is a B_{n-1}-extension of $\mathfrak{A}_k$, $\mathfrak{A}_{2k+1}$ is a model of $T[\Gamma']$, and $\mathfrak{A}_{2k}$ is a model of $T'[\Gamma]$. Use (a) to show that the union of the chain is a $\forall_{n-1}$-extension of each $\mathfrak{A}_k$ and hence a model of $T \cup T'$.] If $n = 0$, the same result holds, provided that we omit the requirement that **A** be closed. [Use the consistency theorem.]

c) Let $\vdash_T$ **A** $\rightarrow$ **B**, where **A** is in $\forall_{n+1}$ and **B** is in $\exists_{n+1}$. Suppose that all the nonlogical axioms of T are in $\exists_n$. Show that there is a formula **C** in B_n such that $\vdash_T$ **A** $\rightarrow$ **C** and $\vdash_T$ **C** $\rightarrow$ **B**. [Like the proof of the Craig interpolation lemma, using (b).]

d) Let T be a theory whose nonlogical axioms are in $\exists_n$. Show that if **A** is equivalent in T to a formula in $\exists_{n+1}$ and is equivalent in T to a formula in $\forall_{n+1}$, then **A** is equivalent in T to a formula in B_n. [Use (c).]

12. a) Let $\mathfrak{B}$ be an extension of $\mathfrak{A}$ such that for every closed formula $\exists$**xA** in $L(\mathfrak{A})$ such that $\mathfrak{B}(\exists\mathbf{xA}) = \mathbf{T}$, there is an **i** in $L(\mathfrak{A})$ such that $\mathfrak{B}(\mathbf{A_x[i]}) = \mathbf{T}$. Show that $\mathfrak{B}$ is an elementary extension of $\mathfrak{A}$.

b) Let $\mathfrak{m}$ be an infinite cardinal, L an $\mathfrak{m}$-langauge, $\mathfrak{A}$ a structure for L, and B a subset of $|\mathfrak{A}|$ having cardinal $\leqslant \mathfrak{m}$. Show that there is an elementary substructure $\mathfrak{B}$ of $\mathfrak{A}$ having cardinal $\mathfrak{m}$ such that B is a subset of $|\mathfrak{B}|$. [Define inductively a sequence $B_0, B_1, \ldots$ of subsets of $|\mathfrak{A}|$ such that (i) $B_0 = B$; (ii) if $a_1, \ldots, a_n$ are in B_k, then $\mathbf{f}_A(a_1, \ldots, a_n)$ is in B_{k+1} for all **f**; (iii) if $\mathfrak{A}(\exists\mathbf{xA}[\mathbf{i}_1, \ldots, \mathbf{i}_n]) = \mathbf{T}$, where **A** is a formula of L and $\mathbf{i}_1, \ldots, \mathbf{i}_n$ are names of individuals in B_k, then $\mathfrak{A}(\mathbf{A}[\mathbf{j}, \mathbf{i}_1, \ldots, \mathbf{i}_n]) = \mathbf{T}$ for some name **j** of an individual of B_{k+1}; (iv) B_k has cardinal $\mathfrak{m}$ for $k > 0$. Let $|\mathfrak{B}|$ be the union of the B_k and use (a).]

c) Let $\mathfrak{m}$ be an infinite cardinal, L an $\mathfrak{m}$-language, and $\mathfrak{A}$ a structure for L whose cardinal is $\leqslant \mathfrak{m}$. Show that $\mathfrak{A}$ has an elementary extension whose cardinal is $\mathfrak{m}$. [Apply the cardinality theorem to $D_e(\mathfrak{A})$.]

13. Let $\mathfrak{A}$ be a structure for L, A_1 and A_2 subsets of $|\mathfrak{A}|$. An *isomorphism of* A_1 *and* A_2 *in* $\mathfrak{A}$ is a bijective mapping ϕ from A_1 to A_2 such that $\mathfrak{A}(\mathbf{A}) = \mathfrak{A}(\mathbf{A}^{\phi})$ for every closed formula **A** of $L(\mathfrak{A})$ such that all of the names in **A** are names of individuals in A_1. An *automorphism* of $\mathfrak{A}$ is an isomorphism of $\mathfrak{A}$ and $\mathfrak{A}$.

a) Show that a bijective mapping from $|\mathfrak{A}|$ to $|\mathfrak{A}|$ is an automorphism of $\mathfrak{A}$ iff it is an isomorphism of $|\mathfrak{A}|$ and $|\mathfrak{A}|$ in $\mathfrak{A}$.

b) Let ϕ be an isomorphism of A_1 and A_2 in $\mathfrak{A}$. Show that there is an elementary extension $\mathfrak{B}$ of $\mathfrak{A}$ and an isomorphism ϕ' of $|\mathfrak{A}|$ and a subset of $|\mathfrak{B}|$ in $\mathfrak{B}$ which extends ϕ. [For each name **i** in $L(\mathfrak{A})$, introduce a new name **i**$'$; and for each **A** in $L(\mathfrak{A})$, let **A**$'$ be obtained from $\mathbf{A}$ by replacing each $\mathbf{i}$ by $\mathbf{i}'$. Obtain T' from $D_e(\mathfrak{A})$ by replacing each $\mathbf{i}$ by $\mathbf{i}'$. Obtain T from $D_e(\mathfrak{A})$ as follows: if $\mathbf{i}$ is the name of an individual in A_1, add the constant $\mathbf{i}'$ and the axiom $\mathbf{i}' = \mathbf{i}^{\phi}$. Expand $\mathfrak{A}_{\mathfrak{A}}$ to a structure $\mathfrak{A}'$ for T by setting $\mathfrak{A}'(\mathbf{i}') = \mathfrak{A}(\mathbf{i}^{\phi})$. Show that $\mathfrak{A}'$ is a model of T, and that if $\mathbf{A}'$ is a closed formula of T, then

$$\mathfrak{A}'(\mathbf{A}') = \mathfrak{A}'(\mathbf{A}^{\phi}) = \mathfrak{A}(\mathbf{A}).$$

Conclude that if $\vdash_T \mathbf{A}'$, then $\mathbf{A}$ is valid in $\mathfrak{A}$. Use this and the joint consistency theorem to show that $T \cup T'$ is consistent. Let $\mathfrak{C}$ be a model of $T \cup T'$ such that $\mathfrak{C}(\mathbf{i}) = \mathfrak{A}(\mathbf{i})$ for $\mathbf{i}$ in $L(\mathfrak{A})$. Let $\mathfrak{B}$ be $\mathfrak{C} \mid L$ and let $\phi'(\mathfrak{A}(\mathbf{i})) = \mathfrak{C}(\mathbf{i}')$.]

c) Let ϕ be an isomorphism of A_1 and A_2 in $\mathfrak{A}$. Show that there is an elementary extension $\mathfrak{B}$ of $\mathfrak{A}$ and an isomorphism ϕ' of a subset of $|\mathfrak{B}|$ and $|\mathfrak{A}|$ in $\mathfrak{B}$ which extends ϕ. [Apply (b) to the inverse of ϕ.]

d) Show that every isomorphism of subsets of $|\mathfrak{A}|$ in $\mathfrak{A}$ can be extended to an automorphism of an elementary extension of $\mathfrak{A}$. [Form an elementary chain beginning with $\mathfrak{A}$ in which the terms are obtained alternatively by (b) and (c).]

e) Show that if T has an infinite model, then it has a model in which there are two distinct individuals having the same type. [It is sufficient to prove consistent the theory T' obtained from T by adding two constants $\mathbf{e}$ and $\mathbf{e}'$; the axiom $\mathbf{e} \neq \mathbf{e}'$; and the axiom $\mathbf{A}[\mathbf{e}] \leftrightarrow \mathbf{A}[\mathbf{e}']$ for each $\mathbf{A}$ in $S_1(T)$. If it is inconsistent, then

$$\vdash_T (\mathbf{A}_1 \leftrightarrow \mathbf{A}_1[\mathbf{y}]) \rightarrow \cdots \rightarrow (\mathbf{A}_n \leftrightarrow \mathbf{A}_n[\mathbf{y}]) \rightarrow \mathbf{z}_1 = \mathbf{y}$$

with $\mathbf{A}_1, \ldots, \mathbf{A}_n$ in $S_1(T)$. Conclude that no model of T has more than 2^n individuals.]

f) Show that if T has an infinite model, then T has a model which has an automorphism other than the identity mapping. [Use (e) and (d).]

14. Let T be a theory, $\mathbf{p}$ a predicate symbol of T, and Q a set of nonlogical symbols of T not containing $\mathbf{p}$. We say that $\mathbf{p}$ is *disjunctively definable* in terms of Q in T if there is a theorem of T which is a disjunction of closures of formulas of the form $\mathbf{px}_1 \ldots \mathbf{x}_n \leftrightarrow \mathbf{A}$, where $\mathbf{x}_1, \ldots, \mathbf{x}_n$ are distinct and $\mathbf{A}$ contains no nonlogical symbol not in Q.

a) Show that if $\mathbf{p}$ is not disjunctively definable in terms of Q, then there is a model $\mathfrak{A}$ of T such that for every formula $\mathbf{A}$ which contains no nonlogical symbol not in Q, $\mathbf{px}_1 \ldots \mathbf{x}_n \leftrightarrow \mathbf{A}$ is not valid in $\mathfrak{A}$. [Let Γ be the set of negations of closures of such sentences, and show that $T[\Gamma]$ is consistent.]

b) Show that $\mathbf{p}$ is disjunctively definable in terms of Q iff for every model $\mathfrak{A}$ of T and every bijective mapping ϕ from $|\mathfrak{A}|$ to $|\mathfrak{A}|$ which is a $\mathbf{u}$-isomorphism for every $\mathbf{u}$ in Q, ϕ is a $\mathbf{p}$-isomorphism. [If $\mathbf{p}$ is not disjunctively definable in terms of Q, take $\mathfrak{A}$ as in (a) and use the definability theorem to get models $\mathfrak{B}$ and $\mathfrak{C}$ of $Th(\mathfrak{A})$ and a bijective mapping ϕ from $|\mathfrak{B}|$ to $|\mathfrak{C}|$ which is a $\mathbf{u}$-isomorphism for all $\mathbf{u}$ in Q but not a $\mathbf{p}$-isomorphism. After replacing $\mathfrak{C}$ by an isomorphic model, use 4(b) to obtain a common elementary extension of $\mathfrak{B}$ and $\mathfrak{C}$, and apply a suitable extension of 13(d).]

c) Extend the results of this problem to function symbols.

15. Assume the following theorem: if $\mathfrak{A}$ is a field and p is a nonconstant polynomial with coefficients in $\mathfrak{A}$, then there is an extension of $\mathfrak{A}$ which is a field in which p has a root.

a) Show that for every field $\mathfrak{A}$, there is an extension $\mathfrak{B}$ of $\mathfrak{A}$ which is a field in which every nonconstant polynomial with coefficients in $\mathfrak{A}$ has a root. [Obtain T from $FL \cup D(\mathfrak{A})$

by adding for each nonconstant polynomial p with coefficients in $\mathcal{A}$ an axiom stating that p has a root. Use the compactness theorem to show that T has a model.]

b) Show that every field $\mathcal{A}$ is a subfield of an algebraically closed field. [Form a chain of fields beginning with $\mathcal{A}$ by (a), and apply the Chang-Łoś-Suszko theorem.]

16. A theory T is *model-complete* if for every model $\mathcal{B}$ of T, every submodel of $\mathcal{B}$ is an elementary submodel of $\mathcal{B}$.

a) Show that if T admits elimination of quantifiers, then T is model-complete. Show that if T is a model-complete open theory, then T admits elimination of quantifiers. [Use the quantifier elimination theorem.]

b) Suppose that for every model $\mathcal{B}$ of T, every submodel of $\mathcal{B}$ is a $\forall_1$-submodel of $\mathcal{B}$. Show that T is model-complete. [Let $\mathcal{A}$ be a submodel of $\mathcal{B}$. Using induction on n and 10(c), show that there is an n-sandwich whose first two structures are $\mathcal{A}$ and $\mathcal{B}$. Then apply 10(c) again.]

c) Show that if $\mathcal{A}$ and $\mathcal{B}$ are structures for L such that $|\mathcal{A}|$ is a subset of $|\mathcal{B}|$, then $\mathcal{A}$ is an elementary substructure of $\mathcal{B}$ iff $\mathcal{A}_{\mathcal{A}}$ is elementarily equivalent to $\mathcal{B}_{\mathcal{A}}$.

d) Show that a theory T is model-complete iff for every model $\mathcal{A}$ of T, $T \cup D(\mathcal{A})$ is complete. [Use (c), the diagram lemma, and Lemma 1 of §5.5.]

e) A model $\mathcal{A}$ of T is *prime* if every model of T has a submodel which is isomorphic to $\mathcal{A}$. Show that if T is model-complete and has a prime model, then T is complete.

17. Let T be obtained from FL by adding the axioms

$$\exists y(y \cdot y = x \vee y \cdot y = -x),$$
$$x \cdot x \neq -1,$$
$$\exists z(z \cdot z = x \cdot x + y \cdot y),$$

and, for each odd n, an axiom stating that every polynomial of degree n has a root.

a) Show that if $\mathcal{A}$ is a model of T, and if for a and b in $|\mathcal{A}|$ we define $a < b$ to mean that $a = b + c^2$ for some nonzero c in $|\mathcal{A}|$, then $\mathcal{A}$ becomes a real closed field.

b) Show that some extension by definitions of T is equivalent to RCF. [Use (a), the corollary to the completeness theorem, and the completeness of RCF.] Conclude that T is complete and hence is equivalent to the theory of the field of real numbers.

c) Show that T does not admit elimination of quantifiers. [Show that if **A** is an open formula of T in which no variable except **x** is free, and P is the set of real numbers a such that **A** is true when **x** designates a, then either P is finite or all but a finite number of real numbers belong to P. Conclude that $\exists y(x = y \cdot y)$ is not equivalent in T to an open formula.]

18. Let $\mathcal{A}$ be a field, and let $\mathcal{B}$ be the ring of polynomials in n indeterminants with coefficients in $\mathcal{A}$.

a) Let $f_1, \ldots, f_k$ be polynomials in $\mathcal{B}$ which are simultaneously zero for some set of arguments in some extension field of $\mathcal{A}$. Show that $f_1, \ldots, f_k$ are simultaneously zero for some set of arguments in each algebraically closed extension of $\mathcal{A}$. [Show that we may take $\mathcal{A}$ algebraically closed. Use 16(a) and 16(d) to show that $ACF \cup D(\mathcal{A})$ is complete; then use Lemma 1 of §5.5 and the diagram lemma.]

b) If I is a proper ideal in $\mathcal{B}$, then all the polynomials in I are simultaneously zero for some set of arguments in some extension field $\mathcal{C}$ of $\mathcal{A}$. [Use Zorn's lemma to find a maximal ideal J including I. Note that $\mathcal{B}/J$ is a field and that the natural mapping from $\mathcal{B}$ to $\mathcal{B}/J$ is an isomorphism on $\mathcal{A}$. Take $\mathcal{C}$ isomorphic to $\mathcal{B}/J$.]

c) Let $\mathcal{C}$ be an algebraically closed extension field of $\mathcal{A}$, and let $f_1, \ldots, f_k$ be polynomials in $\mathcal{B}$ which are not simultaneously zero for any set of arguments in $\mathcal{C}$. Show that the ideal in $\mathcal{B}$ generated by $f_1, \ldots, f_k$ is $\mathcal{B}$. [Use (a) and (b).]

d) Let $f, g_1, \ldots, g_k$ be polynomials in $\mathcal{B}$. Suppose that there is an algebraically closed extension field $\mathcal{C}$ of $\mathcal{A}$ such that the common roots of $g_1, \ldots, g_k$ in $\mathcal{C}$ are all roots of f. Show that some power of f belongs to the ideal in $\mathcal{B}$ generated by $g_1, \ldots, g_k$ (Hilbert's *Nullstellensatz*). [Let Z be a new indeterminant. Conclude from (c) that we have $1 = h_1g_1 + \cdots + h_kg_k + h(1 - Zf)$ for suitable polynomials $h_1, \ldots, h_k, h$. Substitute $1/f$ for Z and clear of fractions.]

e) If $f, g_1, \ldots, g_k$ are as in (d), we have $f^m = h_1g_1 + \cdots + h_kg_k$ for suitable $m, h_1, \ldots, h_k$. Show that there are bounds for m and the degrees of the h_i which depend only upon n and the degrees of f and the g_i. [Given n and the degrees of $f, g_1, \ldots, g_k$, construct a formula $\mathbf{A}$ of FL such that if $\mathbf{x}_1, \ldots, \mathbf{x}_s$ are given the values of the coefficients of $f, g_1, \ldots, g_k$, then $\mathbf{A}$ is true in an extension $\mathcal{C}$ of $\mathcal{A}$ iff the common roots of $g_1, \ldots, g_k$ in $\mathcal{C}$ are all roots of f. Let $\mathbf{B}$ be an open formula which is equivalent to $\mathbf{A}$ in ACF. Let $\mathbf{C}_r$ be a formula which, under the same meaning of $\mathbf{x}_1, \ldots, \mathbf{x}_s$, is true iff

$$f^m = h_1g_1 + \cdots + h_kg_k$$

for some $m \leqslant r$ and some polynomials $h_1, \ldots, h_k$ of degrees $\leqslant r$. Then if $\mathbf{e}_1, \ldots, \mathbf{e}_s$ are new constants, $\neg\mathbf{B}[\mathbf{e}_1, \ldots, \mathbf{e}_s]$ is a logical consequence of the nonlogical axioms of FL and the formulas $\neg\mathbf{C}_r[\mathbf{e}_1, \ldots, \mathbf{e}_s]$. Apply the compactness theorem.]

19. Let $\mathcal{A}$ be an ordered field.

a) Let $f_1, \ldots, f_k$ be polynomials in n indeterminants with coefficients in $\mathcal{A}$ which are simultaneously zero for some set of arguments in some ordered field which is an extension of $\mathcal{A}$. Show that $f_1, \ldots, f_k$ are simultaneously zero for some set of arguments in each real closed extension of $\mathcal{A}$. [Like 18(a).]

b) A polynomial f in n indeterminants with coefficients in $\mathcal{A}$ is *positive* in an extension $\mathcal{B}$ of $\mathcal{A}$ if it assumes only nonnegative values for arguments in $\mathcal{B}$. Show that if f is positive in some real closed extension of $\mathcal{A}$, then it is positive in every real closed extension of $\mathcal{A}$. [Like 18(a).]

c) The Artin-Schreier theorem states that if $\mathcal{B}$ is an extension of the field $\mathcal{A}$, and b is an individual of $\mathcal{B}$ which cannot be put in the form $c_1a_1^2 + \cdots + c_ka_k^2$ with the c_i positive individuals of $\mathcal{A}$, then there is an ordering of $\mathcal{B}$ which make $\mathcal{B}$ into an ordered field, extends the ordering of $\mathcal{A}$, and makes $b < 0$. Assume this, and prove the following theorem of Artin. Let f be a polynomial in n indeterminants with coefficients in $\mathcal{A}$ which is positive in some real closed extension of $\mathcal{A}$. Then $f = c_1g_1^2 + \cdots + c_kg_k^2$ where the c_i are positive elements of $\mathcal{A}$ and the g_i are rational functions with coefficients in $\mathcal{A}$. Moreover, if $\mathcal{A}$ is real closed or is the field of rational numbers, then the c_i may all be taken to be 1. [Assume that f cannot be written in this form. Use the Artin-Schreier theorem to order the field of rational functions so that $f < 0$, and show that this contradicts (b).]

d) Show that in Artin's theorem, there is a bound on k and the degrees of the numerators and denominators of the g_i which depends only upon n and the degree of f. [Like 18(e).]

20. Let $\mathfrak{A}$ and $\mathfrak{B}$ be structures for L, $a_1, \ldots, a_k$ individuals of $\mathfrak{A}$, and $b_1, \ldots, b_k$ individuals of $\mathfrak{B}$. We define the *n-equivalence* of $(a_1, \ldots, a_k)$ and $(b_1, \ldots, b_k)$ by induction on n as follows. We say $(a_1, \ldots, a_k)$ and $(b_1, \ldots, b_k)$ are 0-equivalent if the types of $(a_1, \ldots, a_k)$ and $(b_1, \ldots, b_k)$ contain the same atomic formulas. We say $(a_1, \ldots, a_k)$ and $(b_1, \ldots, b_k)$ are $(n + 1)$-equivalent if for each a in $|\mathfrak{A}|$ there is a b in $|\mathfrak{B}|$ such that $(a_1, \ldots, a_k, a)$ and $(b_1, \ldots, b_k, b)$ are n-equivalent, and for each b in $|\mathfrak{B}|$ there is an a in $|\mathfrak{A}|$ such that $(a_1, \ldots, a_k, a)$ and $(b_1, \ldots, b_k, b)$ are n-equivalent. Show that if $(a_1, \ldots, a_k)$ and $(b_1, \ldots, b_k)$ are n-equivalent, then the types of $(a_1, \ldots, a_k)$ and $(b_1, \ldots, b_k)$ contain the same formulas of height n. [Use induction on n.] Conclude that if $(a_1, \ldots, a_k)$ and $(b_1, \ldots, b_k)$ are n-equivalent for every n, then they have the same type.

21. Let EQ be the theory whose only nonlogical symbol is the binary predicate symbol $\sim$, and whose nonlogical axioms are

$$x \sim x,$$

$$x \sim y \rightarrow y \sim x,$$

$$x \sim y \rightarrow y \sim z \rightarrow x \sim z.$$

Then a model $\mathfrak{A}$ for EQ consists of a nonempty set $|\mathfrak{A}|$ and an equivalence relation $\sim_{\mathfrak{A}}$ on $|\mathfrak{A}|$.

a) For each n and k, show that there is a closed formula $\mathbf{A}_{n,k}$ which is valid in a model $\mathfrak{A}$ of EQ iff $\mathfrak{A}$ has at least n equivalence classes having k members and a closed formula $\mathbf{B}_{n,k}$ which is valid in a model $\mathfrak{A}$ of EQ iff $\mathfrak{A}$ has at least n equivalence classes having at least k members.

b) If $\mathfrak{A}$ and $\mathfrak{B}$ are models of EQ, we write $\mathfrak{A} \equiv \mathfrak{B}$ if

$$\mathfrak{A}(\mathbf{A}_{n,k}) = \mathfrak{B}(\mathbf{A}_{n,k}) \quad \text{and} \quad \mathfrak{A}(\mathbf{B}_{n,k}) = \mathfrak{B}(\mathbf{B}_{n,k}) \quad \text{for all } n \text{ and } k.$$

Suppose that $\mathfrak{A} \equiv \mathfrak{B}$, that $(a_1, \ldots, a_k)$ is a k-tuple in $|\mathfrak{A}|$ which is 0-equivalent to the k-tuple $(b_1, \ldots, b_k)$ in $|\mathfrak{B}|$, and that for $i = 1, \ldots, k$, either the equivalence classes of a_i and b_i have the same finite number of members, or both these equivalence classes have more than $n + k$ members. Show that $(a_1, \ldots, a_k)$ is n-equivalent to $(b_1, \ldots, b_k)$. [Use induction on n.]

c) Show that if $\mathfrak{A}$ and $\mathfrak{B}$ are models of EQ, then $\mathfrak{A} \equiv \mathfrak{B}$ iff $\mathfrak{A}$ is elementarily equivalent to $\mathfrak{B}$. [Use (b) and 20.]

22. Let L be a language whose only nonlogical symbols are the unary predicate symbols $\mathbf{p}_1, \ldots, \mathbf{p}_k$.

a) Show that for each subset J of $\{1, \ldots, k\}$ and each n, there is a closed formula $\mathbf{A}_{J,n}$ which is valid in a structure $\mathfrak{A}$ for L iff there are at least n individuals a in $\mathfrak{A}$ such that $[i \mid (\mathbf{p}_i)_{\mathfrak{A}}(a)] = J$.

b) If $\mathfrak{A}$ and $\mathfrak{B}$ are structures for L, we write $\mathfrak{A} \equiv \mathfrak{B}$ if $\mathfrak{A}(\mathbf{A}_{J,n}) = \mathfrak{B}(\mathbf{A}_{J,n})$ for all J and n. Show that if $\mathfrak{A} \equiv \mathfrak{B}$, then any r-tuple in $\mathfrak{A}$ and any r-tuple in $\mathfrak{B}$ which are

0-equivalent are n-equivalent for all n. Conclude that $\mathfrak{A} \equiv \mathfrak{B}$ iff $\mathfrak{A}$ and $\mathfrak{B}$ are elementarily equivalent. [Use 20.]

23. Let DO be the theory whose only nonlogical symbol is $<$, and whose nonlogical axioms are the axioms OF1 through OF3 of OF and

$$x < y \rightarrow \exists z(x < z \,\&\, z < y),$$
$$\exists x(x < y),$$
$$\exists x(y < x).$$

a) Let $\mathfrak{A}$ and $\mathfrak{B}$ be models of DO, and let $(a_1, \ldots, a_k)$ be a k-tuple in $|\mathfrak{A}|$ which is 0-equivalent to the k-tuple $(b_1, \ldots, b_k)$ in $|\mathfrak{B}|$. Show that $(a_1, \ldots, a_k)$ and $(b_1, \ldots, b_k)$ have the same type. [Show that they are n-equivalent for every n and use 20.]

b) Show that DO is complete. [Use (a) and Lemma 1 of §5.5.]

c) Show that DO is $\aleph_0$-categorical. [Use (a), (b), and Ryll-Nardzewski's theorem.]

d) Prove (c) without using (a) or (b). [Given models $\mathfrak{A}$ and $\mathfrak{B}$ of cardinal $\aleph_0$, choose sequences $a_1, a_2, \ldots$ and $b_1, b_2, \ldots$ as in the proof of Ryll-Nardzewski's theorem so that $a_i < a_j$ iff $b_i < b_j$.] Obtain a new proof of (b).

e) Show that if $\mathfrak{m}$ is the cardinal of the set of real numbers, then DO is not $\mathfrak{m}$-categorical. [Construct a model of DO by starting with the rational numbers and replacing each element by a linearly ordered set isomorphic to the set of real numbers. Show that this model is not isomorphic to the set of real numbers.]

24. Let the nonlogical symbols of T be a unary predicate symbol $\mathbf{p}$ and two infinite sequences of constants $\mathbf{e}_1, \mathbf{e}_2, \ldots$ and $\mathbf{e}'_1, \mathbf{e}'_2, \ldots$. Let the nonlogical axioms of T be the $\mathbf{p}(\mathbf{e}_i)$ and the $\neg\mathbf{p}(\mathbf{e}'_i)$ and all $\mathbf{e} \neq \mathbf{e}'$ for $\mathbf{e}$ and $\mathbf{e}'$ distinct constants. Show that T is complete but is not $\mathfrak{m}$-categorical for any infinite cardinal $\mathfrak{m}$. [Use the method of §5.5.]

25. If T is complete and has a finite model, then T is categorical. [Suppose that $\mathfrak{A}$ is a finite model and that $\mathfrak{B}$ is a model not isomorphic to $\mathfrak{A}$. Let $\mathbf{i}_1, \ldots, \mathbf{i}_n$ be the names of individuals of $\mathfrak{A}$. If ϕ is a bijective mapping from $|\mathfrak{A}|$ to $|\mathfrak{B}|$, there is a sentence $\mathbf{A}_\phi$ in $S_n(T)$ such that

$$\mathfrak{A}(\mathbf{A}_\phi[\mathbf{i}_1, \ldots, \mathbf{i}_n]) = \mathsf{T} \quad \text{and} \quad \mathfrak{B}(\mathbf{A}_\phi[\mathbf{i}_1^\phi, \ldots, \mathbf{i}_n^\phi]) = \mathsf{F}.$$

Let $\mathbf{B}$ be the conjunction of the $\mathbf{A}_\phi$, the formulas $\mathbf{z}_i \neq \mathbf{z}_j$ for $1 \leqslant i < j \leqslant n$, and the formula

$$\forall \mathbf{z}_{n+1}(\mathbf{z}_{n+1} = \mathbf{z}_1 \vee \cdots \vee \mathbf{z}_{n+1} = \mathbf{z}_n).$$

Then $\mathfrak{A}(\exists \mathbf{z}_1 \ldots \exists \mathbf{z}_n \mathbf{B}) = \mathsf{T}$ and $\mathfrak{B}(\exists \mathbf{z}_1 \ldots \exists \mathbf{z}_n \mathbf{B}) = \mathsf{F}$.]

26. Let T be a countable complete theory having only infinite models and let $\mathfrak{A}$ be a countable model of T. We say $\mathfrak{A}$ is *weakly saturated* if for each n, every n-type in T is an n-type in $\mathfrak{A}$. We say $\mathfrak{A}$ is *saturated* if for every $a_1, \ldots, a_{n-1}$ in $|\mathfrak{A}|$ and every n-type Γ in T which includes the type of $(a_1, \ldots, a_{n-1})$, there is an a_n in $|\mathfrak{A}|$ such that Γ is the type of $(a_1, \ldots, a_n)$. We say $\mathfrak{A}$ is *homogeneous* if whenever $a_1, \ldots, a_n, b_1, \ldots, b_n$ are individuals of $\mathfrak{A}$ such that the types of $(a_1, \ldots, a_n)$ and $(b_1, \ldots, b_n)$ are the same, then there is an automorphism ϕ of $\mathfrak{A}$ such that $\phi(a_1) = b_1, \ldots, \phi(a_n) = b_n$. We say $\mathfrak{A}$ is *universal* if every countable model of T is isomorphic to an elementary substructure of $\mathfrak{A}$.

a) Show that a saturated model $\mathfrak{A}$ of T is universal. [Let $\mathfrak{B}$ be a countable model with individuals $b_1, b_2, \ldots$. Choose individuals $a_1, a_2, \ldots$ of $\mathfrak{A}$ inductively so that $(a_1, \ldots, a_n)$ and $(b_1, \ldots, b_n)$ have the same type. Show that the a_i are the individuals of an elementary substructure of $\mathfrak{A}$ isomorphic to $\mathfrak{B}$.]

b) Let $\mathfrak{A}$ and $\mathfrak{B}$ be saturated models of T. Let $a_1, \ldots, a_n$ be individuals of $\mathfrak{A}$, and let $b_1, \ldots, b_n$ be individuals of $\mathfrak{B}$ such that $(a_1, \ldots, a_n)$ and $(b_1, \ldots, b_n)$ have the same type. Show that there is an isomorphism ϕ of $\mathfrak{A}$ and $\mathfrak{B}$ such that

$$\phi(a_1) = b_1, \ldots, \phi(a_n) = b_n.$$

[Like the proof of Ryll-Nardzewski's theorem.] Conclude that any two saturated models of T are isomorphic, and that a saturated model is homogeneous.

c) Show that a universal model is weakly saturated. [Use the lemma of §5.6.]

d) Show that a weakly saturated homogeneous model is saturated.

e) Let $\mathfrak{A}$ be a countable model of T; $a_1, \ldots, a_{n-1}$ individuals of $\mathfrak{A}$; Γ an n-type in T which includes the type of $(a_1, \ldots, a_{n-1})$. Show that there is a countable elementary extension $\mathfrak{B}$ of $\mathfrak{A}$ and an element a_n of $|\mathfrak{B}|$ such that Γ is the type of $(a_1, \ldots, a_n)$. [Let $\mathbf{i}_1, \ldots, \mathbf{i}_{n-1}$ be the names of $a_1, \ldots, a_{n-1}$. Obtain T' from $D_e(\mathfrak{A})$ by adding a new constant $\mathbf{e}$, and, for each $\mathbf{A}$ in Γ, a new axiom $\mathbf{A}[\mathbf{i}_1, \ldots, \mathbf{i}_{n-1}, \mathbf{e}]$. If $\mathbf{A}_1, \ldots, \mathbf{A}_k$ are in Γ, then $\exists \mathbf{z}_n(\mathbf{A}_1 \ \& \cdots \& \ \mathbf{A}_k)$ is in the type of $(a_1, \ldots, a_{n-1})$, since its negation cannot be in that type. Conclude that T' is consistent, and let $\mathfrak{B}$ be a restriction of a model of T'.]

f) Assume that for every n, T has only countably many n-types. Show that T has a saturated model. [If $\mathfrak{A}$ is any countable model, apply (e) and Tarski's lemma to obtain an elementary extension $\mathfrak{B}$ of $\mathfrak{A}$ such that the conclusion of (e) holds for every choice of Γ and $a_1, \ldots, a_{n-1}$. Combine this result with Tarski's lemma.]

g) Show that the following are equivalent:

i) for each n, T has only countably many n-types;

ii) T has a saturated model;

iii) T has a universal model;

iv) T has a weakly saturated model.

[Use (f), (a), and (c).]

27. Let T be a countable complete theory. Using the notation of Problem 5 of Chapter 4, identify each subset Γ of $S_n(T)$ with the element V of $\mathfrak{T}(S_n(T))$ such that $V(\mathbf{A}) = \mathbf{T}$ iff $\mathbf{A}$ is in Γ. The set of n-types in T is then a subspace of $\mathfrak{T}(S_n(T))$; it is designated by $\mathfrak{T}\mathfrak{y}_n(T)$.

a) Show that an element V of $\mathfrak{T}(S_n(T))$ is in $\mathfrak{T}\mathfrak{y}_n(T)$ iff V is in $\mathfrak{T}\mathfrak{V}(S_n(T))$ and $V(\mathbf{A}) = \mathbf{T}$ for every formula $\mathbf{A}$ in $S_n(T)$ which is a theorem of T. [Use the lemma of §5.6.] Conclude that $\mathfrak{T}\mathfrak{y}_n(T)$ is closed in $\mathfrak{T}\mathfrak{V}(S_n(T))$ and hence is compact.

b) For $\mathbf{A}$ in $S_n(T)$, let $\Phi_\mathbf{A}$ be the set of V in $\mathfrak{T}\mathfrak{y}_n(T)$ such that $V(\mathbf{A}) = \mathbf{T}$. Show that the $\Phi_\mathbf{A}$ form a base for $\mathfrak{T}\mathfrak{y}_n(T)$, and that $\Phi_\mathbf{A}$ is nonempty iff $\vdash_T \exists \mathbf{z}_1 \ldots \exists \mathbf{z}_n \mathbf{A}$.

c) Let $\Gamma \subset S_n(T)$, and let Φ be the set of n-types in T which include Γ. Show that a formula $\mathbf{A}$ in $S_n(T)$ is a generator of Γ iff $\Phi_\mathbf{A}$ is a nonempty subset of Φ. Conclude that an n-type is principal iff it is an isolated point of $\mathfrak{T}\mathfrak{y}_n(T)$.

d) Show that if $\mathfrak{A}$ is a model of T, then the n-types in $\mathfrak{A}$ form a dense subset of $\mathfrak{Ty}_n(T)$. [Use (b).] Conclude that an n-type in T is principal iff it is an n-type in every model of T. [Use (c) and Ehrenfeucht's theorem.]

e) Let R be a set of types in T such that for each n, $R \cap \mathfrak{Ty}_n(T)$ is nowhere dense in $\mathfrak{Ty}_n(T)$. Show that there is a countable model $\mathfrak{A}$ of T such that no type in $\mathfrak{A}$ is in R. [Like Ehrenfeucht's theorem, using (c).]

f) Let $\mathfrak{A}$ and $\mathfrak{B}$ be countable infinite models of T, and suppose that every n-type in $\mathfrak{A}$ is principal. Show that $\mathfrak{A}$ is isomorphic to an elementary submodel of $\mathfrak{B}$. [Let $a_1, a_2, \ldots$ be the individuals of $\mathfrak{A}$. As in the proof of Ryll-Nardzewski's theorem, choose individuals $b_1, b_2, \ldots$ of $\mathfrak{B}$ such that for each n, $(a_1, \ldots, a_n)$ and $(b_1, \ldots, b_n)$ have the same type. Then proceed as in 26(a).] Show that if, in addition, every type in $\mathfrak{B}$ is principal, then $\mathfrak{A}$ and $\mathfrak{B}$ are isomorphic. [Like the proof of Ryll-Nardzewski's theorem.]

g) A model $\mathfrak{A}$ of T is *elementarily prime* if every model of T is isomorphic to an elementary extension of $\mathfrak{A}$. Show that a model $\mathfrak{A}$ of T is elementarily prime iff $\mathfrak{A}$ is countable and every type in $\mathfrak{A}$ is principal. [Use the Löwenheim-Skolem theorem, Ehrenfeucht's theorem, Problem 25, and (f).]

h) Show that any two elementarily prime models of T are isomorphic. [Use (f) and (g).]

i) Show that T has an elementarily prime model iff for every n, the set of principal n-types is dense in $\mathfrak{Ty}_n(T)$. [Use (g), (d), (e), and (c).]

28. Let $\mathfrak{U}$ be a class of subsets of a nonempty space I. We say that $\mathfrak{U}$ is an *ultrafilter* on I if

i) $\mathfrak{U}$ satisfies the finite intersection property;

ii) for every subset J of I, either J or J^c (the complement of J in I) is in $\mathfrak{U}$.

a) Show that if $\mathfrak{B}$ is a class of subsets of I satisfying the finite intersection property, then $\mathfrak{B}$ is included in an ultrafilter on I. [Use the Teichmüller-Tukey lemma.]

b) Let $\mathfrak{U}$ be an ultrafilter on I. Show that if $J \in \mathfrak{U}$ and $J \subset K$, then $K \in \mathfrak{U}$. Show that the intersection of two members of $\mathfrak{U}$ is a member of $\mathfrak{U}$.

29. Let $\mathfrak{U}$ be an ultrafilter on a nonempty space I. For each i in I, let $\mathfrak{A}_i$ be a structure for L, and let $\mathfrak{B} = \prod_{i \in I} \mathfrak{A}_i$. For a and b in $|\mathfrak{B}|$, let $a \sim b$ mean that $[i \mid (a)_i = (b)_i] \in \mathfrak{U}$.

a) Show that $\sim$ is an equivalence relation.

b) Let A be the set of equivalence classes of $\sim$, and let $\phi(b)$ be the equivalence class of b. Show that we may define a structure $\mathfrak{A}$ with universe A by

$$\mathbf{f}_{\mathfrak{A}}(\phi(a_1), \ldots, \phi(a_n)) = \phi(\mathbf{f}_{\mathfrak{B}}(a_1, \ldots, a_n)),$$
$$\mathbf{p}_{\mathfrak{A}}(\phi(a_1), \ldots, \phi(a_n)) \leftrightarrow [i \mid \mathbf{p}_{\mathfrak{A}_i}((a_1)_i, \ldots, (a_n)_i)] \in \mathfrak{U}.$$

We call $\mathfrak{A}$ an *ultraproduct* of the $\mathfrak{A}_i$, and designate it by $\prod_{i \in I} \mathfrak{A}_i / \mathfrak{U}$.

c) If $\mathbf{a}$ is a variable-free term in $L(\mathfrak{B})$, show that $\mathfrak{A}(\mathbf{a}^{\phi}) = \phi(\mathfrak{B}(\mathbf{a}))$.

d) If $\mathbf{A}$ is a closed formula of $L(\mathfrak{B})$, and π_i is defined by $\pi_i(b) = (b)_i$, show that

$$\mathfrak{A}(\mathbf{A}^{\phi}) = \mathbf{T} \quad \text{iff} \quad [i \mid \mathfrak{A}_i(\mathbf{A}^{\pi_i}) = \mathbf{T}] \in \mathfrak{U}.$$

[Use induction on the length of $\mathbf{A}$ and 28(b).] In particular, a closed formula $\mathbf{A}$ of L

is valid in $\mathfrak{A}$ iff $[i \mid \mathfrak{A}_i(\mathbf{A}) = \mathbf{T}] \in \mathfrak{U}$. Conclude that if each $\mathfrak{A}_i$ is a model of the theory T, then $\mathfrak{A}$ is a model of T.

e) If all of the $\mathfrak{A}_i$ are equal to $\mathfrak{C}$, then $\mathfrak{A}$ is called an *ultrapower* of $\mathfrak{C}$. Show that in this case, $\mathfrak{C}$ is isomorphic to an elementary substructure of $\mathfrak{A}$. [Map an individual c of $\mathfrak{C}$ into the equivalence class of the element b of $|\mathfrak{B}|$ which has $(b)_i = c$ for all i, and use (d).]

30. a) Use ultraproducts to prove the compactness theorem without using the completeness theorem. [Suppose that the closed formula $\mathbf{A}$ is not valid in any finitely axiomatized part of T. Let I be the class of formulas $\neg\mathbf{A} \,\&\, \mathbf{A}_1 \,\&\, \cdots \,\&\, \mathbf{A}_n$ with $\mathbf{A}_1, \ldots, \mathbf{A}_n$ nonlogical axioms of T. For $\mathbf{B}$ in I, let $\mathfrak{A}_\mathbf{B}$ be a structure in which $\mathbf{B}$ is valid, and let $J_\mathbf{B}$ be the set of $\mathbf{B}'$ in I such that $\mathbf{B}$ is valid in $\mathfrak{A}_{\mathbf{B}'}$. Use 28(a) to find an ultrafilter $\mathfrak{U}$ on I containing each $J_\mathbf{B}$. By 29(d), each $\mathbf{B}$ in I is valid in $\prod \mathfrak{A}_\mathbf{B}/\mathfrak{U}$.]

b) Let $\mathfrak{A}$ and $\mathfrak{B}$ be structures for L. Show that $\mathfrak{A}$ is elementarily equivalent to $\mathfrak{B}$ iff $\mathfrak{A}$ is isomorphic to an elementary substructure of an ultrapower of $\mathfrak{B}$. [The "if" part follows from 29(e). Suppose that $\mathfrak{A}$ and $\mathfrak{B}$ are elementarily equivalent. Let I be the set of closed formulas $\mathbf{A}$ in $L(\mathfrak{A})$ such that $\mathfrak{A}(\mathbf{A}) = \mathbf{T}$. If $\mathbf{A}$ is in I, we may define a mapping $\phi_\mathbf{A}$ from $|\mathfrak{A}|$ to $|\mathfrak{B}|$ such that $\mathfrak{B}(\mathbf{A}^{\phi_\mathbf{A}}) = \mathbf{T}$. Let $J_\mathbf{A} = [\mathbf{B} \mid \mathfrak{B}(\mathbf{A}^{\phi_\mathbf{B}}) = \mathbf{T}]$. Find an ultrafilter $\mathfrak{U}$ containing all the $J_\mathbf{A}$. Let $\mathfrak{B}_\mathbf{A} = \mathfrak{B}$, $\mathfrak{C} = \prod \mathfrak{B}_\mathbf{A}/\mathfrak{U}$. For a in $|\mathfrak{A}|$, let $\phi(a)$ be the equivalence class in $|\mathfrak{C}|$ of the element b such that $b_\mathbf{A} = \phi_\mathbf{A}(a)$ for $\mathbf{A}$ in I. Use 29(d) to show that $\mathfrak{C}(\mathbf{A}^\phi) = \mathbf{T}$ for $\mathbf{A}$ in I.]

c) Show that a class $\mathfrak{J}$ of structures for L is a generalized elementary class iff every ultraproduct of structures in $\mathfrak{J}$ is in $\mathfrak{J}$ and every structure elementarily equivalent to a structure in $\mathfrak{J}$ is in $\mathfrak{J}$. [For the "only if" part, use 29(d). Suppose that the conditions hold. Let the nonlogical axioms of T be all closed formulas valid in every structure of $\mathfrak{J}$. Let $\mathfrak{A}$ be a model of T, and let I be the class of closed formulas valid in $\mathfrak{A}$. For each $\mathbf{A}$ in I, choose a $\mathfrak{B}_\mathbf{A}$ in $\mathfrak{J}$ so that $\mathfrak{B}_\mathbf{A}(\mathbf{A}) = \mathbf{T}$. Let $Q_\mathbf{A}$ be the set of $\mathbf{B}$ in I such that $\mathfrak{B}_\mathbf{B}(\mathbf{A}) = \mathbf{T}$, and let $\mathfrak{U}$ be an ultrafilter containing all the $Q_\mathbf{A}$. Show that $\mathfrak{A}$ is elementarily equivalent to $\prod \mathfrak{B}_\mathbf{B}/\mathfrak{U}$.]

CHAPTER 6

INCOMPLETENESS AND UNDECIDABILITY

6.1 CALCULABILITY

A *decision method* for a formal system F is a method by which, given a formula of F, we can decide in a finite number of steps whether or not it is a theorem of F. The *decision problem* for F is the following: find a decision method for F or prove that no such method exists.

Although a solution of the decision problem for F gives a solution of the characterization problem, the converse is not always true. For example, if T is a theory with no nonlogical axioms, Herbrand's theorem gives no solution to the decision problem. To decide by Herbrand's theorem whether or not a given formula is a theorem, we must test infinitely many formulas to see whether they are quasi-tautologies; and this cannot be done in a finite number of steps. Of course, the completeness theorem gives no solution to the decision problem either; for we have no way of deciding whether or not a given formula is valid in T.

We can abstract a more general problem from the decision problem for formal systems. Suppose that A is a subset of E. A *decision method* for A in E is a method by which, given an element a of E, we can decide in a finite number of steps whether or not a is in A. The *decision problem* for A in E is the following: find a decision method for A in E or prove that no such method exists.

The decision problem for a formal system F is the special case in which E is the set of formulas in F and A is the set of theorems of F. Many other examples have arisen in mathematics. An example of a decision problem which is still unsolved is Hilbert's tenth problem: find a method for deciding if a given Diophantine equation has a solution. Here E is the set of Diophantine equations, and A is the set of Diophantine equations having a solution. Another example is discussed in the Appendix.

For our definition of a decision method for A in E to make sense, each element of E must be such that it can be given to us in a single step. This means that the elements of E must be concrete objects. In the examples given above, the elements of E were expressions in some language; and we can be given such an expression by having it written down for us. If E is the set of natural numbers, we can, as in §4.3, replace the natural number n by the symbol consisting of n strokes, and then proceed in the same way. It would not make any difference if we were given the natural number n by having its decimal representation written down for us; for we have a method of converting this into the expression consisting of n strokes.

We can state a similar problem for a mapping F from a set A to a set B. A *decision method* for F is a method by which, given an element a of A, we can obtain $F(a)$ in a finite number of steps. Here both the elements of A and the elements of B must be concrete objects. The *decision problem* for F is: find a decision method for F or prove that no such method exists.

The decision problem for functions is a generalization of the decision problem for sets. For suppose that A is a subset of E. Define a mapping F from E to the set of natural numbers by letting $F(a) = 0$ if a is in A and letting $F(a) = 1$ if a is not in A. Then a decision method for F would provide a decision method for A in E and vice versa; so the decision problem for F is equivalent to the decision problem for A in E.

Our definitions are still very imprecise in one respect: we have not specified exactly what a *method* is. As a step toward explaining this, we remark that a method must be *mechanical*. Perhaps the best way to elucidate this remark is to give some examples of methods which are excluded by it. First, methods which involve chance procedures are excluded; we cannot decide whether or not a is in A by tossing a coin. Second, methods which involve magic are excluded; we cannot decide whether or not a is in A by asking a fortune teller. Third, methods which require insight are excluded; we cannot use a method which requires us to solve a mathematical problem unless the method provides instructions for solving that problem. These exclusions are clearly necessary if we wish to be able to give negative solutions of the decision problem. For example, we cannot give a mathematical proof that a fortune teller is unable to always tell whether or not a given a is in A.

In a more positive direction, a mechanical method is one which could be carried out by a suitably designed machine. Of course, we have in mind an ideal machine, not limited, as real machines are, by problems of size, mechanical breakdown, etc. A machine for computing F will have an input device into which we can feed the argument a; it will then compute $F(a)$. Of course, the machine itself must be independent of a. This is implicit in the notion of a method; if we compute $F(a)$ differently for each different a, we do not have a method, but only madness.

We have still not given a precise definition of a *method*. Indeed, it seems quite hopeless to describe, say, all possible mechanical methods for a mapping from the natural numbers to the natural numbers. We claim, however, that this is not necessary for solving decision problems.

First suppose that we want to give a positive solution of a decision problem. We then simply give the decision method, and verify that it is mechanical and that it always leads to the correct answers.

A set or mapping is *calculable* if it has a decision method. Thus a negative solution of a decision problem consists of a proof that a set or mapping is not calculable. For this, it is only necessary to give a precise definition of *calculable*.

This may not seem of much help, since it is not apparent that we can define *calculable* without defining *method* first. However, we shall see that this can be done in at least some cases.

Our procedure in the rest of the chapter is as follows. We introduce a class of functions from natural numbers to natural numbers. After some study of this class, we shall give arguments to show that this class is just the class of calculable functions from the natural numbers to the natural numbers. We shall then use this to obtain a precise version of the decision problem for theories. Finally, we shall obtain methods for giving solutions to decision problems for theories.

6.2 RECURSIVE FUNCTIONS

We adopt some conventions which will shorten the statement of our results considerably. In this chapter, unless otherwise stated, *number* means *natural number*; *set* means *set of natural numbers*; *function* means *function from the set of natural numbers to the set of natural numbers*; and *predicate* means *predicate in the set of natural numbers.*

We use small Latin letters to designate natural numbers. We use capital Latin letters to designate functions and predicates; generally F, G, and H for functions and P, Q, and R for predicates. The symbols of N will be used informally in our discussions with their usual meaning. Thus we write

$$\forall x P(a, x) \vee F(a, k) = 2$$

to mean *either* $P(a, x)$ *for all numbers* x *or* $F(a, k) = 2$. The notion of free and bound will be used in the general sense explained in §2.3; an occurrence of x is free if the meaning of the expression depends on the value of x.

We shall use small German letters to stand for finite sequences of distinct Latin letters. Thus we might write $F(\mathfrak{a})$ instead of $F(a_1, \ldots, a_n)$. If two distinct German letters, say $\mathfrak{a}$ and $\mathfrak{b}$, appear in the same context, it is understood that the letters in the sequence abbreviated by $\mathfrak{a}$ are all distinct from the letters in the sequence abbreviated by $\mathfrak{b}$. If a German letter appears as an argument to a function or predicate, it is assumed that the abbreviated sequence has the correct number of letters. Thus if F is n-ary and we write $F(\mathfrak{a})$, then we assume that $\mathfrak{a}$ is a sequence of n letters. If $\mathfrak{a}$ stands for $a_1, \ldots, a_n$, then we let $\exists\mathfrak{a}$ stand for $\exists a_1 \ldots \exists a_n$ and $\forall\mathfrak{a}$ stand for $\forall a_1 \ldots \forall a_n$.

If P is an n-ary predicate, we define an n-ary function K_P by

$$\begin{aligned} K_P(\mathfrak{a}) &= 0 \quad \text{if } P(\mathfrak{a}), \\ &= 1 \quad \text{if } \neg P(\mathfrak{a}). \end{aligned}$$

We call K_P the *representing function* of P. As we noted in the last section, a predicate is calculable iff its representing function is calculable.

We now give some examples of calculable functions. If $1 \leqslant i \leqslant n$, we define the function I_i^n by

$$I_i^n(a_1, \ldots, a_n) = a_i.$$

Clearly I_i^n is calculable. The binary functions $+$ and $\cdot$ are calculable; decision methods for these functions are taught in elementary school arithmetic. The binary predicate $<$ is calculable; so its representing function $K_<$ is calculable.

Next we give two methods for obtaining calculable functions from other calculable functions. First, suppose that we define a function F by

$$F(\mathfrak{a}) = G(H_1(\mathfrak{a}), \ldots, H_k(\mathfrak{a})),$$

where $G, H_1, \ldots, H_k$ are calculable functions. Then F is calculable. In fact, $F(\mathfrak{a})$ may be calculated by first calculating the values $b_1, \ldots, b_k$ of $H_1(\mathfrak{a}), \ldots, H_k(\mathfrak{a})$, and then calculating $G(b_1, \ldots, b_k)$.

To explain the second method, we need some notation. If $\ldots x \ldots$ is a sentence which is true for some x, then $\mu x(\ldots x \ldots)$ denotes the smallest x for which $\ldots x \ldots$ is true. For example, $\mu x(x = a) = a$. As this example shows, the value of $\mu x(\ldots x \ldots)$ does not depend on the value of x, that is, the occurrences of x in $\mu x(\ldots x \ldots)$ are bound. We call μx a *μ-operator*.

Now suppose that we define F by

$$F(\mathfrak{a}) = \mu x(G(\mathfrak{a}, x) = 0),$$

where G is a calculable function such that for each $\mathfrak{a}$, there is an x such that $G(\mathfrak{a}, x) = 0$. (This last condition needed to ensure that $F(\mathfrak{a})$ is defined for all $\mathfrak{a}$.) Then F is calculable. In fact, we can calculate $F(\mathfrak{a})$ by successively calculating $G(\mathfrak{a}, 0), G(\mathfrak{a}, 1), \ldots$ until we obtain a zero value.

We now define the *recursive* functions by a generalized inductive definition consisting of three rules R1 through R3.

R1. The I_i^n, $+$, $\cdot$, and $K_<$ are recursive.

R2. If $G, H_1, \ldots, H_k$ are recursive, and F is defined by

$$F(\mathfrak{a}) = G(H_1(\mathfrak{a}), \ldots, H_k(\mathfrak{a})),$$

then F is recursive.

R3. If G is recursive and $\forall \mathfrak{a}\, \exists x(G(\mathfrak{a}, x) = 0)$, and F is defined by

$$F(\mathfrak{a}) = \mu x(G(\mathfrak{a}, x) = 0),$$

then F is recursive.

To prove that every recursive function has some property P, it suffices to prove that R1 through R3 remain true when *recursive function* is replaced by *function having property P*. Such a proof is called a *proof by induction on recursive functions*. Using the above discussion, we can prove by induction on recursive functions that every recursive function is calculable. The converse is by no means evident; we shall return to it in §6.5.

A predicate is *recursive* if its representing function is recursive. It follows from the above that every recursive predicate is calculable. Again we postpone discussion of the converse until §6.5.

6.3 EXPLICIT DEFINITIONS

We shall continue the list R1 through R3 with further rules for obtaining recursive functions and predicates.

R4. If $Q, H_1, \ldots, H_k$ are recursive, and P is defined by

$$P(\mathfrak{a}) \leftrightarrow Q(H_1(\mathfrak{a}), \ldots, H_k(\mathfrak{a})),$$

then P is recursive.

Proof. We have

$$K_P(\mathfrak{a}) = K_Q(H_1(\mathfrak{a}), \ldots, H_k(\mathfrak{a})).$$

Hence P is recursive by R2 and the definition of a recursive predicate.

R5. If P is recursive and $\forall\mathfrak{a}\,\exists x P(\mathfrak{a}, x)$, and F is defined by

$$F(\mathfrak{a}) = \mu x P(\mathfrak{a}, x),$$

then F is recursive.

Proof. Since

$$F(\mathfrak{a}) = \mu x(K_P(\mathfrak{a}, x) = 0),$$

F is recursive by R3.

We have already met definitions of the form $F(\mathfrak{a}) = \ldots$ or $P(\mathfrak{a}) \leftrightarrow$ ___, where ... and ___ contain only previously defined symbols. Such a definition is called an *explicit definition*. By using R1 through R5, we can show that functions and predicates defined by certain explicit definitions are recursive. We give some illustrations of this.

Suppose that F is defined by

$$F(a, b, c) = G(H(b, c), K(G(b, c, c), a), c),$$

where G, H, and K are previously defined recursive functions. We shall successively prove that larger and larger parts of the right-hand side are recursive functions of a, b, c. For this purpose, we define

$$\begin{aligned}
F_1(a, b, c) &= a,\\
F_2(a, b, c) &= b,\\
F_3(a, b, c) &= c,\\
F_4(a, b, c) &= H(b, c),\\
F_5(a, b, c) &= G(b, c, c),\\
F_6(a, b, c) &= K(G(b, c, c), a).
\end{aligned}$$

Then F_1 is I_1^3 and hence is recursive by R1. Similarly, F_2 and F_3 are recursive. Now

$$F_4(a, b, c) = H(F_2(a, b, c), F_3(a, b, c));$$

so F_4 is recursive by R2. Similarly, F_5 is recursive. Since

$$F_6(a, b, c) = K(F_5(a, b, c), F_1(a, b, c)),$$

F_6 is recursive by R2. Finally,

$$F(a, b, c) = G(F_4(a, b, c), F_6(a, b, c), F_3(a, b, c));$$

so F is recursive by R2.

A similar technique applies to predicates. If bound occurrences of a variable appear on the right-hand side, some of the functions and predicates used in the proof will have this variable as an argument and some will not. Thus suppose that P is defined by

$$P(a, b) \leftrightarrow Q(b, \mu x R(x, F(b, a))),$$

where Q, R, and F are recursive and are such that $\mu x R(x, F(b, a))$ is defined for all a and b. We then define

$$\begin{aligned}
&F_1(a, b, x) = F(b, a) = F(I_2^3(a, b, x), I_1^3(a, b, x)),\\
&P_1(a, b, x) \leftrightarrow R(x, F(b, a)) \leftrightarrow R(I_3^3(a, b, x), F_1(a, b, x)),\\
&F_2(a, b) = \mu x R(x, F(b, a)) = \mu x P_1(a, b, x),\\
&P(a, b) \leftrightarrow Q(I_2^2(a, b), F_2(a, b)).
\end{aligned}$$

We then use R1 through R5 to show that all of these functions and predicates are recursive.

We may summarize our conclusion as follows: if a function or predicate has an explicit definition using only variables, symbols for recursive functions and predicates, and μ-operators, then it is recursive. (It is understood that μ-operators are to be used only when they are defined for all values of the variables.) The remaining results of this section will enable us to expand the class of symbols which may be used in such definitions.

R6. Every constant function is recursive.

Proof. Let F_k be the n-ary function with the constant value k; we show by induction on k that F_k is recursive. For $k = 0$ we have the explicit definition

$$F_0(\mathfrak{a}) = \mu x(I_{n+1}^{n+1}(\mathfrak{a}, x) = 0);$$

while for $k = r + 1$, we have the explicit definition

$$F_k(\mathfrak{a}) = \mu x(F_r(\mathfrak{a}) < x).$$

Note that the last definition is permissible because $<$ is recursive by R1.

It follows from R6 that we may use constants in explicit definitions of recursive functions and predicates.

We let $\neg P$ be the predicate defined by $(\neg P)(\mathfrak{a}) \leftrightarrow \neg P(\mathfrak{a})$. We define $P \vee Q$ by

$$(P \vee Q)(\mathfrak{a}) \leftrightarrow P(\mathfrak{a}) \vee Q(\mathfrak{a}),$$

and we define $P \rightarrow Q$, $P \mathbin{\&} Q$, and $P \leftrightarrow Q$ similarly.

R7. If P is recursive, then $\neg P$ is recursive. If P and Q are recursive, then $P \vee Q$, $P \rightarrow Q$, $P \mathbin{\&} Q$, and $P \leftrightarrow Q$ are recursive.

Proof. We have the explicit definitions

$$K_{\neg P}(\mathfrak{a}) = K_<(0, K_P(\mathfrak{a})), \qquad K_{P \vee Q}(\mathfrak{a}) = K_P(\mathfrak{a}) \cdot K_Q(\mathfrak{a}).$$

From these and the definition of a recursive predicate, we see that $\neg P$ and $P \vee Q$ are recursive. To treat the remaining cases, we use the fact that $P \rightarrow Q$ is $\neg P \vee Q$, $P \,\&\, Q$ is $\neg(P \rightarrow \neg Q)$, and $P \leftrightarrow Q$ is $(P \rightarrow Q) \,\&\, (Q \rightarrow P)$.

It follows from R7 that we may use $\neg$, $\vee$, $\rightarrow$, &, and $\leftrightarrow$ in explicit definitions of recursive functions and predicates.

R8. The predicates $<$, $\leqslant$, $>$, $\geqslant$, and $=$ are recursive.

Proof. By R1, $<$ is recursive. The others have the explicit definitions

$$a \leqslant b \leftrightarrow \neg(b < a),$$
$$a > b \leftrightarrow b < a,$$
$$a \geqslant b \leftrightarrow b \leqslant a,$$
$$a = b \leftrightarrow a \leqslant b \,\&\, b \leqslant a.$$

We are now going to define a modified type of μ-operator which does not have the disadvantage of sometimes being undefined. Suppose that $\_\_x\_\_$ is a formula and that ... is an expression not containing x which represents a number. We then define

$$\mu x_{x<\ldots}(\_\_x\_\_) = \mu x(\_\_x\_\_ \vee x = \ldots).$$

It is clear that the right-hand side is defined. The value of $\mu x_{x<\ldots}(\_\_x\_\_)$ is the smallest x less than ... which makes $\_\_x\_\_$ true, provided there is such an x; if there is no such x, $\mu x_{x<\ldots}(\_\_x\_\_) = \ldots$. Note that the occurrences of x in $\mu x_{x<\ldots}(\_\_x\_\_)$ are bound. We call $\mu x_{x<\ldots}$ a *bounded μ-operator,* in contrast to the *unbounded μ-operator* μx. (However, *μ-operator* continues to mean *unbounded μ-operator.*)

From the definition of the bounded μ-operator and previous results:

R9. If P is recursive, and F is defined by

$$F(a, \mathfrak{a}) = \mu x_{x<a} P(\mathfrak{a}, x),$$

then F is recursive.

It follows that we may use bounded μ-operators in explicit definitions of recursive functions and predicates.

We shall see later that we may define nonrecursive predicates explicitly by using quantifiers. We shall therefore introduce modified quantifiers which can be used in explicit definitions of recursive functions and predicates.

Let $\_\_x\_\_$ and ... be as above. We define

$$\exists x_{x<\ldots}(\_\_x\_\_) \leftrightarrow \mu x_{x<\ldots}(\_\_x\_\_) < a$$

and

$$\forall x_{x<\ldots}(\_\_x\_\_) \leftrightarrow \neg\exists x_{x<\ldots} \neg(\_\_x\_\_).$$

Then $\exists x_{x<\ldots}(\_\_x\_\_)$ is true iff $\_\_x\_\_$ is true for some x less than ..., and $\forall x_{x<\ldots}(\_\_x\_\_)$ is true iff $\_\_x\_\_$ is true for every x less than .... We may thus think of $\exists x_{x<\ldots}$ as an existential quantifier on a variable which varies through the numbers less than ... ; and we may explain $\forall x_{x<\ldots}$ similarly. We call $\exists x_{x<\ldots}$ and $\forall x_{x<\ldots}$ *bounded quantifiers*; the former is a *bounded existential quantifier* and the latter is a *bounded universal quantifier*. For contrast, $\exists x$ and $\forall x$ are called *unbounded quantifiers*. (However, *quantifier* continues to mean *unbounded quantifier*.)

From the definitions of the bounded quantifiers and our previous results:

R10. If R is recursive, and P and Q are defined by

$$P(a, \mathfrak{a}) \leftrightarrow \exists x_{x<a} R(\mathfrak{a}, x)$$

and

$$Q(a, \mathfrak{a}) \leftrightarrow \forall x_{x<a} R(\mathfrak{a}, x),$$

then P and Q are recursive.

It follows that we may use bounded quantifiers in explicit definitions of recursive functions and predicates.

We shall also allow $x \leqslant \ldots$ to occur as a subscript to μx, $\exists x$, or $\forall x$. It is then to be understood as an abbreviation of $x < \ldots + 1$. By the above, we may use this in explicit definitions.

The ordinary subtraction function is not a function in our sense, since its values are not all natural numbers. We therefore define a modified subtraction function $\dot{-}$ as follows:

$$a \dot{-} b = a - b \quad \text{if} \quad a \geqslant b,$$

and $a \dot{-} b = 0$ otherwise.

R11. The function $\dot{-}$ is recursive.

Proof. We have the explicit definition

$$a \dot{-} b = \mu x(b + x = a \vee a < b).$$

A slight generalization of explicit definition is *definition by cases*. Here the value specified for the function or predicate is different in different cases. The definition of $\dot{-}$ given above is an example.

R12. Let $G_1, \ldots, G_k$ be recursive functions, and let $R_1, \ldots, R_k$ be recursive predicates such that for each $\mathfrak{a}$, exactly one of $R_1(\mathfrak{a}), \ldots, R_k(\mathfrak{a})$ holds. If F is defined by

$$\begin{aligned} F(\mathfrak{a}) &= G_1(\mathfrak{a}) \quad &\textit{if} \quad R_1(\mathfrak{a}),\\ &\vdots \\ &= G_k(\mathfrak{a}) \quad &\textit{if} \quad R_k(\mathfrak{a}), \end{aligned}$$

then F is recursive.

Proof. We have the explicit definition

$$F(\mathfrak{a}) = G_1(\mathfrak{a}) \cdot K_{\neg R_1}(\mathfrak{a}) + \cdots + G_k(\mathfrak{a}) \cdot K_{\neg R_k}(\mathfrak{a}).$$

R13. Let $Q_1, \ldots, Q_k$ be recursive predicates, and let $R_1, \ldots, R_k$ be recursive predicates such that for each $\mathfrak{a}$, exactly one of $R_1(\mathfrak{a}), \ldots, R_k(\mathfrak{a})$ holds. If P is defined by

$$\begin{aligned} P(\mathfrak{a}) &\leftrightarrow Q_1(\mathfrak{a}) \quad && \textit{if} \ \ R_1(\mathfrak{a}), \\ &\ \ \vdots \\ &\leftrightarrow Q_k(\mathfrak{a}) \quad && \textit{if} \ \ R_k(\mathfrak{a}), \end{aligned}$$

then P is recursive.

Proof. We may define K_P by

$$\begin{aligned} K_P(\mathfrak{a}) &= K_{Q_1}(\mathfrak{a}) \quad && \textit{if} \ \ R_1(\mathfrak{a}), \\ &\ \ \vdots \\ &= K_{Q_k}(\mathfrak{a}) \quad && \textit{if} \ \ R_k(\mathfrak{a}). \end{aligned}$$

Hence P is recursive by R12.

In actual practice, the $G_1, \ldots, G_k, R_1, \ldots, R_k$ of R12 and the $Q_1, \ldots, Q_k, R_1, \ldots, R_k$ of R13 are replaced by explicit definitions of these functions and predicates. Since R_k must be $\neg(R_1 \vee \cdots \vee R_{k-1})$, we sometimes just write *otherwise* for $R_k(\mathfrak{a})$. Thus a typical definition to which R12 applies is

$$\begin{aligned} F(a, b) &= a \quad && \textit{if} \ \ a < b, \\ &= b + 2 \quad && \textit{if} \ \ b \leqslant a \ \& \ a = 4, \\ &= 2 \quad && \textit{otherwise}. \end{aligned}$$

We can, of course, apply the above results to sets (i.e., to unary predicates). Thus by R7, the union and intersection of two recursive sets are recursive, and the complement of a recursive set is recursive. Moreover, every finite set A is recursive. For if A is empty, it has the explicit definition

$$A(a) \leftrightarrow a < a,$$

while if A has the members $k_1, \ldots, k_n$, then it has the explicit definition

$$A(a) \leftrightarrow a = k_1 \vee \cdots \vee a = k_n.$$

We insert here a warning about the use of dots in explicit definitions. In the above example, the expression represented by the dots depends upon A; if we knew what A was, we could write it out in full. However, we must not use dots when the expression which they represent depends on the value of an argument. Thus

$$P(a, b) \leftrightarrow a = F(0) \vee a = F(1) \vee \cdots \vee a = F(b)$$

is not a legitimate explicit definition, since the expression represented by the dots depends upon the value of b.

6.4 SEQUENCE NUMBERS

Our next object is to assign a number to each finite sequence of numbers in such a way that the associated functions and predicates are recursive. This will depend on the following result.

Lemma (*Gödel*). There is a binary recursive function β such that

$$\beta(a, i) \leqslant a \dot{-} 1$$

for all a and i, and such that for any numbers $a_0, a_1, \ldots, a_{n-1}$, there is a number a such that $\beta(a, i) = a_i$ for all $i < n$.

To prove the lemma, we shall need a few elementary results from number theory. Since we shall later want to see that these results are provable in a certain theory, we shall give rather detailed proofs.

We write $Div(a, b)$ if a is divisible by b, that is, if $\exists x(a = b \cdot x)$. If a and b are not 0 and $\forall x\big(Div(ax, b) \rightarrow Div(x, b)\big)$, we say that a and b are *relatively prime*, and write $RP(a, b)$. Then

$$RP(a, b) \rightarrow RP(b, a). \tag{1}$$

For assume that $RP(a, b)$ and that $Div(bx, a)$. Then $bx = ay$ for some y. Hence $Div(ay, b)$; so $Div(y, b)$; so $y = bz$ for some z. From this, $bx = abz$ and hence $x = az$; so $Div(x, a)$.

Suppose that $a_1, \ldots, a_n, b_1, \ldots, b_m$ are different from 0 and 1 and that $RP(a_i, b_j)$ for all i and j. Then there is a number c which is divisible by all of the a_i and none of the b_j. We prove this by induction on n. If $n = 0$, take $c = 1$. If $n \neq 0$, there is a c' which is divisible by $a_1, \ldots, a_{n-1}$ and not by $b_1, \ldots, b_m$; we then take $c = a_n c'$.

Next we show that

$$k \neq 0 \;\&\; z \neq 0 \;\&\; Div(z, k) \rightarrow RP\big(1 + (j + k)z, 1 + jz\big). \tag{2}$$

First, we have

$$Div(x + xjz, z) \rightarrow Div(x, z) \qquad \text{for all } x;$$

so $RP(1 + jz, z)$. By (1), $RP(z, 1 + jz)$. Now suppose that

$$Div\big(x + x(j + k)z, 1 + jz\big).$$

Then clearly $Div(xkz, 1 + jz)$. Since $RP(z, 1 + jz)$, we have

$$Div(xk, 1 + jz).$$

Since $Div(z, k)$, it follows that $Div(xz, 1 + jz)$. Using $RP(z, 1 + jz)$ again, we have $Div(x, 1 + jz)$. This proves (2).

We define a function OP by

$$OP(a, b) = (a + b) \cdot (a + b) + a + 1. \tag{3}$$

Then

$$OP(a, b) = OP(a', b') \rightarrow a = a' \;\&\; b = b'. \tag{4}$$

For assume the left-hand side. If $a + b < a' + b'$, then

$$OP(a, b) \leqslant (a + b + 1)^2 \leqslant (a' + b')^2 < OP(a', b'),$$

which is impossible. Similarly, $a' + b' < a + b$ is impossible; so $a + b = a' + b'$. From this and $OP(a, b) = OP(a', b')$, we get $a = a'$; and from this and

$$a + b = a' + b',$$

we get $b = b'$.

We now define β by

$$\beta(a, i) = \mu x_{x \leq a \dot{-} 1}\, \exists y_{y<a}\, \exists z_{z<a}(a = OP(y, z) \\ \& \; Div(y, 1 + (OP(x, i) + 1) \cdot z)). \tag{5}$$

In view of this explicit definition, the recursiveness of β follows if we show that Div and OP are recursive. But Div has the explicit definition

$$Div(a, b) \leftrightarrow \exists x_{x \leqslant a}(a = x \cdot b),$$

and OP has the explicit definition (3). It is also clear that $\beta(a, i) \leqslant a \dot{-} 1$.

Now let $a_0, a_1, \ldots, a_{n-1}$ be given; we shall find a as in the lemma. Let c be the largest of the $OP(a_i, i) + 1$, and let z be a non-zero number divisible by every non-zero number less than c. If $j < l < c$, then $RP(1 + jz, 1 + lz)$, as we see from (2) with $k = l - j$. It follows by a result obtained above that there is a number y such that for $j < c$, y is divisible by $1 + jz$ iff j is one of the $OP(a_i, i) + 1$. We let $a = OP(y, z)$.

We have $a_i < y < a$ and $z < a$ by the definition of OP. By (4), y and z are the only numbers satisfying $a = OP(y, z)$. Hence to prove that $\beta(a, i) = a_i$, it will suffice to show that a_i is the smallest number x such that

$$Div(y, 1 + (OP(x, i) + 1) \cdot z).$$

For this, it suffices to prove that if $x < a_i$, then $OP(x, i) < c$ and $OP(x, i)$ is not an $OP(a_j, j)$. But $OP(x, i) \leqslant OP(a_i, i) < c$ and $OP(x, i)$ is not an $OP(a_j, j)$ by (4).

We shall henceforth let β be the function defined by (5). However, the only properties of β which we use are those given by the lemma. Note that from $\beta(a, i) \leqslant a \dot{-} 1$, we get

$$\beta(0, i) = 0 \tag{6}$$

and

$$a \neq 0 \rightarrow \beta(a, i) < a. \tag{7}$$

We now assign to each n-tuple $(a_1, \ldots, a_n)$ the smallest number a such that $\beta(a, 0) = n$ and $\beta(a, i) = a_i$ for $i = 1, \ldots, n$. Such a number exists by the lemma. We call this number the *sequence number* of $(a_1, \ldots, a_n)$ and designate it by $\langle a_1, \ldots, a_n \rangle$. We allow $n = 0$; in view of (6), we have $\langle\ \rangle = 0$.

For each fixed n, $\langle a_1, \ldots, a_n\rangle$ is a recursive function of $a_1, \ldots, a_n$; for we have the explicit definition

$$\langle a_1, \ldots, a_n\rangle = \mu x\big(\beta(x, 0) = n \;\&\; \beta(x, 1) = a_1 \;\&\cdots\&\; \beta(x, n) = a_n\big).$$

Moreover, $\langle a_1, \ldots, a_n\rangle$ determines $n, a_1, \ldots, a_n$ via recursive functions. More specifically, define two recursive functions explicitly by

$$lh(a) = \beta(a, 0),$$
$$(a)_i = \beta(a, i + 1).$$

Then if a is $\langle a_0, \ldots, a_{n-1}\rangle$, we have $n = lh(a)$ and $(a)_i = a_i$ for $i < n$. We abbreviate $\big((a)_i\big)_j$ to $(a)_{i,j}$. From (7),

$$a \neq \langle\ \rangle \rightarrow lh(a) < a \;\&\; (a)_i < a. \tag{8}$$

We introduce some other recursive functions and predicates associated with sequence numbers. The set of sequence numbers is designated by *Seq*. This is recursive, since it has the explicit definition

$$Seq(a) \leftrightarrow \forall x_{x<a}\big(lh(x) \neq lh(a) \vee \exists i_{i<lh(a)}\big((x)_i \neq (a)_i\big)\big).$$

We define *In* so that

$$In(\langle a_1, \ldots, a_n\rangle, i) = \langle a_1, \ldots, a_i\rangle$$

for $i \leqslant n$:

$$In(a, i) = \mu x\big(lh(x) = i \;\&\; \forall j_{j<i}\big((x)_j = (a)_j\big)\big).$$

Finally, we define $*$ so that

$$\langle a_1, \ldots, a_n\rangle * \langle b_1, \ldots, b_m\rangle = \langle a_1, \ldots, a_n, b_1, \ldots, b_m\rangle.$$

The explicit definition of $*$ is

$$\begin{aligned} a * b = \mu x\big(lh(x) &= lh(a) + lh(b) \\ &\;\&\; \forall i_{i<lh(a)}\big((x)_i = (a)_i\big) \\ &\;\&\; \forall i_{i<lh(b)}\big((x)_{lh(a)+i} = (b)_i\big)\big). \end{aligned}$$

One use of sequence numbers is to replace n-ary functions and predicates by unary functions and predicates. If F is an n-ary function, we define a unary function $\langle F\rangle$, called the *contraction* of F, by

$$\langle F\rangle(a) = F\big((a)_0, \ldots, (a)_{n-1}\big). \tag{9}$$

We can recover F from $\langle F\rangle$ by

$$F(a_1, \ldots, a_n) = \langle F\rangle(\langle a_1, \ldots, a_n\rangle). \tag{10}$$

If P is an n-ary predicate, we define a unary predicate $\langle P\rangle$, called the *contraction* of P, by

$$\langle P\rangle(a) \leftrightarrow P\big((a)_0, \ldots, (a)_{n-1}\big). \tag{11}$$

We can recover P from $\langle P\rangle$ by

$$P(a_1, \ldots, a_n) \leftrightarrow \langle P\rangle(\langle a_1, \ldots, a_n\rangle). \tag{12}$$

We call (9) through (12) the *contraction formulas*; they imply that F is recursive iff $\langle F\rangle$ is recursive and that P is recursive iff $\langle P\rangle$ is recursive. Note also that $\langle K_P\rangle = K_{\langle P\rangle}$.

We shall now see how sequence numbers can be used to define recursive functions and predicates by induction. If F is a n-ary function with $n \neq 0$, we define a new n-ary function $\bar{F}$ by

$$\bar{F}(a, \mathfrak{a}) = \langle F(0, \mathfrak{a}), F(1, \mathfrak{a}), \ldots, F(a - 1, \mathfrak{a})\rangle. \tag{13}$$

Roughly speaking, $\bar{F}(a, \mathfrak{a})$ contains all the information supplied by values of $F(i, \mathfrak{a})$ for $i < a$.

We show that F is recursive iff $\bar{F}$ is recursive. Suppose that F is recursive. We cannot use (13) as an explicit definition, since the expression represented by the dots depends upon the value of a. However, we have the explicit definition

$$\bar{F}(a, \mathfrak{a}) = \mu x\big(lh(x) = a \ \&\ \forall i_{i<a}\big((x)_i = F(i, \mathfrak{a})\big)\big). \tag{14}$$

If $\bar{F}$ is recursive, we have the explicit definition

$$F(a, \mathfrak{a}) = \big(\bar{F}(a + 1, \mathfrak{a})\big)_a \tag{15}$$

for F.

Now suppose that G is an $(n + 1)$-ary function. The equation

$$F(a, \mathfrak{a}) = G\big(\bar{F}(a, \mathfrak{a}), a, \mathfrak{a}\big)$$

then determines the value of $F(a, \mathfrak{a})$ when the values of $F(i, \mathfrak{a})$ for $i < a$ are known. It is therefore a legitimate definition by induction of F.

R14. If G is recursive and F is defined inductively by

$$F(a, \mathfrak{a}) = G\big(\bar{F}(a, \mathfrak{a}), a, \mathfrak{a}\big),$$

then F is recursive.

Proof. Define H by

$$H(a, \mathfrak{a}) = \mu x\big(Seq(x) \ \&\ lh(x) = a \ \&\ \forall i_{i<a}\big((x)_i = G\big(In(x, i), i, \mathfrak{a}\big)\big)\big). \tag{16}$$

Then clearly H is just $\bar{F}$; so we may define F by

$$F(a, \mathfrak{a}) = G\big(H(a, \mathfrak{a}), a, \mathfrak{a}\big). \tag{17}$$

From the explicit definitions (16) and (17), we see that F is recursive.

In practical applications of R14, G is defined by an explicit definition or a definition by cases. The equation defining F then has the appearance of an explicit definition or a definition by cases of $F(a, \mathfrak{a})$, except that $\bar{F}(a, \mathfrak{a})$ may appear on the right-hand side. Thus we may define a recursive function F by

$$F(a, b) = \overline{F}(a, b) + K(b) + a,$$

where K is a previously defined recursive function. In the case of a definition by cases, we may even allow certain expressions of the form $F(\ldots, \mathfrak{a})$ to appear on the right-hand side. For example, suppose that G and H are recursive, and define F by

$$\begin{aligned} F(a, b) &= F\big(G(a), b\big) && \textit{if } \; G(a) < a, \\ &= H(a, b) && \textit{otherwise.} \end{aligned}$$

To put this in the form of R14, we note that $G(a) < a$ implies that

$$F\big(G(a), b\big) = \big(\overline{F}(a, b)\big)_{G(a)};$$

so we may replace $F\big(G(a), b\big)$ in the first line by $\big(\overline{F}(a, b)\big)_{G(a)}$. The general requirement is that if $F(\ldots, \mathfrak{a})$ appears in a case, we must be able to prove that in that case $\ldots < a$.

A frequently occurring type of inductive definition is

$$\begin{aligned} F(0, \mathfrak{a}) &= G(\mathfrak{a}), \\ F(a + 1, \mathfrak{a}) &= H\big(F(a, \mathfrak{a}), a, \mathfrak{a}\big), \end{aligned}$$

where G and H are previously defined. To see that this comes under R14, we rewrite it as

$$\begin{aligned} F(a, \mathfrak{a}) &= G(\mathfrak{a}) && \textit{if } \; a = 0, \\ &= H\big(F(a \dot{-} 1, \mathfrak{a}), a \dot{-} 1, \mathfrak{a}\big) && \textit{otherwise.} \end{aligned}$$

In an explicit definition of $P(a, \mathfrak{a})$, we may use $\overline{K_P}(a, \mathfrak{a})$ on the right. For if the definition is $P(a, \mathfrak{a}) \leftrightarrow \ldots$, we can define K_P inductively by

$$\begin{aligned} K_P(a, \mathfrak{a}) &= 0 && \textit{if } \; \ldots, \\ &= 1 && \textit{otherwise.} \end{aligned}$$

A similar remark applies to definition by cases. In such a definition we may also use $P(\ldots, \mathfrak{a})$ on the right, provided that we can show that $\ldots < a$ in the case in which it occurs. For we may replace $P(\ldots, \mathfrak{a})$ by $K_P(\ldots, \mathfrak{a}) = 0$, and then proceed as explained above.

6.5 CHURCH'S THESIS

In order to use recursiveness to discuss decision problems, we must be convinced of the truth of the following statement: every calculable function or predicate is recursive. This statement is known as *Church's thesis.*

There is an obvious difficulty in giving a proof of Church's thesis: we have not given a precise definition of *calculable*. This is not necessarily an insuperable difficulty; we proved in §6.2 that every recursive function or predicate was calculable without using such a definition. If we examine this proof, we see that we used only properties of calculable functions and predicates which were obvious even from our vague description of calculability. The question arises whether we can prove Church's thesis in the same way.

It is clear that we can prove Church's thesis for predicates if we assume Church's thesis for functions; for a predicate is calculable or recursive iff its representing function is calculable or recursive. Unfortunately, no one has given a proof of Church's thesis for functions, or even isolated the properties of calculable functions which would be needed in such a proof. Lacking such a proof, we can still hope to find evidence that Church's thesis is true. A large amount of such evidence has been collected; so much that almost all logicians have come to accept Church's thesis as correct. We shall summarize this evidence.

First of all, a great many calculable functions have been shown to be recursive. Some of these have been considered in the previous sections. The functions occurring in elementary number theory are generally defined by induction, and can be treated in the manner of the last section. For example, a^b can be defined inductively by $a^0 = 1, a^{b+1} = a^b \cdot a$. Certain calculable functions which occur in analysis can also be shown to be recursive by methods which we have discussed (see Problem 3). Still another class of calculable functions which can be shown to be recursive will be considered in the next section. Supplementing this positive evidence is some strong negative evidence: no one has produced a calculable function which cannot be shown to be recursive, or even suggested a plausible method for constructing such a function.

Further evidence along the same general lines is given by the fact that many common methods of obtaining calculable functions from calculable functions have been shown to lead from recursive functions to recursive functions. We have considered some such methods already; others will be considered later. Again there is supplementary evidence: no one has given a method which can be seen to lead from calculable functions to calculable functions but has not been shown to lead from recursive functions to recursive functions.

We get more evidence if we try to define *calculable* directly. For simplicity, consider a unary calculable function F. It is reasonable to suppose that the calculation consists of writing expressions on a sheet of paper (or that it can be reduced to this). As will become clear in the next section, there is no loss of generality in supposing that the expressions written are numbers (more precisely, expressions which designate numbers). We therefore write $a_0, a_1, \ldots, a_n$, where a_0 is a and a_n is $F(a)$. Now the decision method tells us how to derive a_i from $a_0, \ldots, a_{i-1}$ or, equivalently, from $\langle a_0, \ldots, a_{i-1}\rangle$. Hence there is a calculable function G such that $G(\langle a_0, \ldots, a_{i-1}\rangle) = a_i$. The decision method also tells us when the computation is complete; so there is a calculable predicate P such that $P(\langle a_0, \ldots, a_i\rangle)$ is false for $i < n$ and true for $i = n$.

Our attempt to define calculability thus ends in circularity, since G and P must be assumed to be calculable. However, since G describes a single step in the calculation, it must be a very simple calculable function; and the same applies to P. We can therefore expect, on the basis of other evidence for Church's thesis, that G and P will be recursive. If we assume this, we can prove that F is recursive. For if we define

$$\begin{aligned} H(i, a) &= a & &\text{if } i = 0, \\ &= G(\overline{H}(i, a)) & &\text{otherwise}, \end{aligned}$$

$$K(a) = \mu x P(\overline{H}(x+1, a)),$$

then

$$F(a) = H(K(a), a).$$

Further evidence is given by various precise definitions which have been proposed for the calculable functions. In each of these definitions, it is clear that all the functions coming under the definition are calculable, and the converse appears at least plausible. (In some cases, one can give rather convincing arguments for the converse.) These definitions are of many types. Some say that the function can be computed by a certain type of machine. Others say that the function can be computed in a suitable sort of formal system. Others are like the definition of the above paragraph, but with the possibilities for G and P specified exactly. Others are similar to the definition we have given. There are still other types of definitions; and for each type, there are several slightly different definitions.

These definitions give evidence for Church's thesis in two ways. First, all the functions coming under each of the definitions can be shown to be recursive. Thus any evidence that all calculable functions come under one of the definitions becomes evidence for Church's thesis. Second, the class of functions defined by each of these definitions is exactly equal to the class of recursive functions. This certainly suggests that this class of functions is a very natural class; and it is hard to see why this should be so, unless it is just the class of calculable functions.

We will henceforth accept Church's thesis. It will never be used in our theorems and proofs, since these will not refer to calculability. Its importance will be in showing that our theorems are solutions to problems which we have posed. Thus if we prove that a set A is not recursive, we need Church's thesis to see that we have given a negative solution of the decision problem for A.

There is another method of using Church's thesis. We can define a function and then assert that, since the function is clearly calculable, it is recursive by Church's thesis. Such uses of Church's thesis, although very convenient in some circumstances, are not really essential. The reader who does not wish to accept Church's thesis can provide a proof, based on the methods of this and the next chapter, that the function in question is recursive.

We may use Church's thesis to obtain another connection between calculability and recursiveness. We say that a predicate P is *positively calculable* if there is a method which, if applied to $\mathfrak{a}$, will give the conclusion that $P(\mathfrak{a})$ is true if this conclusion is correct and will give no conclusion if $P(\mathfrak{a})$ is false. We claim that P is positively calculable iff there is a calculable predicate Q such that

$$P(\mathfrak{a}) \leftrightarrow \exists x Q(\mathfrak{a}, x)$$

for all $\mathfrak{a}$. For if such a Q exists, we can calculate P in the above sense by calculating $Q(\mathfrak{a}, 0)$, $Q(\mathfrak{a}, 1)$, . . . until we come to one which is true and then concluding that $P(\mathfrak{a})$ is true. Conversely, suppose that P is positively calculable, and let $Q(\mathfrak{a}, x)$ mean that x steps in the calculation of $P(\mathfrak{a})$ lead to the conclusion that $P(\mathfrak{a})$ is true. Then Q is calculable, and $P(\mathfrak{a}) \leftrightarrow \exists x Q(\mathfrak{a}, x)$ for all $\mathfrak{a}$.

A predicate P is *recursively enumerable* if there is a recursive predicate Q such that $P(\mathfrak{a}) \leftrightarrow \exists x Q(\mathfrak{a}, x)$ for all $\mathfrak{a}$. From the above and the calculability of recursive predicates, we see that every recursively enumerable predicate is positively calculable. If we also assume Church's thesis, then we can conclude that a predicate is positively calculable iff it is recursively enumerable.

Every recursive predicate P is recursively enumerable; for $P(\mathfrak{a}) \leftrightarrow \exists x Q(\mathfrak{a}, x)$ where Q is the recursive predicate defined by $Q(\mathfrak{a}, x) \leftrightarrow P(\mathfrak{a})$. The converse is false, as we shall see later.

6.6 EXPRESSION NUMBERS

Before considering the decision problem for a formal system, we should know exactly what the symbols of the formal system are. In the case of a theory, this means that we must know the nonlogical symbols; for the remaining symbols are fixed. The simplest situation is when the number of nonlogical symbols is finite; we can then simply give a list of these symbols. We shall therefore suppose in the rest of the chapter that *all first-order languages and theories have only finitely many nonlogical symbols*. The reader will see that most of the results actually apply under somewhat more general conditions.

We shall now show how to connect the decision problem for theories with recursive functions. Let L be a first-order language (satisfying the above condition). We shall assign a number to each symbol of L. The number assigned to the symbol $\mathbf{u}$ is called the *symbol number* of $\mathbf{u}$, and is designated by $SN(\mathbf{u})$. If $\mathbf{z}_0, \mathbf{z}_1, \ldots$ are the variables in alphabetical order, we let $SN(\mathbf{z}_i)$ be $2i$. To the remaining symbols (of which there are only a finite number) we assign any symbol numbers, subject only to the condition that different symbol numbers shall be assigned to different symbols. We shall henceforth suppose that an assignment of symbol numbers is fixed for each first-order language which we consider.

Now we assign a number to each designator of L. The number assigned to the designator $\mathbf{u}$ is called the *expression number* of $\mathbf{u}$, and is designated by $\ulcorner \mathbf{u} \urcorner$. It is defined by induction on the length of $\mathbf{u}$. By the formation theorem, $\mathbf{u}$ is $\mathbf{v}\mathbf{v}_1 \ldots \mathbf{v}_n$, where $\mathbf{v}$ is a symbol of index n and $\mathbf{v}_1, \ldots, \mathbf{v}_n$ are designators. We then set

$$\ulcorner \mathbf{u} \urcorner = \langle SN(\mathbf{v}), \ulcorner \mathbf{v}_1 \urcorner, \ldots, \ulcorner \mathbf{v}_n \urcorner \rangle.$$

It is clear that different designators have different expression numbers. Since the functions $\langle a_1, \ldots, a_n \rangle$ are calculable, we can actually compute $\ulcorner \mathbf{u} \urcorner$ when $\mathbf{u}$ is given (provided that we have a list of the nonlogical symbols and their symbol numbers). Conversely, if we are given a number a, we can decide whether a is an expression number, and, if it is, we can find the designator of which it is the expression number. We show how to do this by induction on a. We first decide whether a is a sequence number different from $\langle\ \rangle$; we can do this because *Seq* is calculable. If it is not, then a is not an expression number. If it is, we find the numbers $a_0, a_1, \ldots, a_n$ such that $a = \langle a_0, a_1, \ldots, a_n\rangle$; we can do this because lh and $(x)_i$ are calculable. We now see whether a_0 is the symbol number of a symbol $\mathbf{v}$ of index n. If it is not, a is not an expression number. If it is, we see whether a_i is the expression number of a designator $\mathbf{v}_i$ ($i = 1, \ldots, n$). We can do this by the induction hypothesis, since $a_i < a$ by (8) of §6.4. Assuming that all this is the case, it only remains to look at the expression $\mathbf{v}\mathbf{v}_1 \ldots \mathbf{v}_n$ to see whether it is a designator.

Now suppose that T is a theory with language L, and let Thm_T be the set of expression numbers of theorems of T. We shall show that T has a decision method iff Thm_T is calculable.

Suppose that we have a decision method for T. Given a number a, we decide whether a is in Thm_T as follows. We first decide whether a is an expression number. If it is not, then a is not in Thm_T. If it is, we find the designator $\mathbf{u}$ such that $a = \ulcorner\mathbf{u}\urcorner$. Then a is in Thm_T iff $\mathbf{u}$ is a formula and is a theorem of T. Now suppose that Thm_T is calculable. Given a formula $\mathbf{A}$ of T, we decide whether or not $\mathbf{A}$ is a theorem of T by computing $\ulcorner\mathbf{A}\urcorner$ and deciding whether or not it belongs to Thm_T.

We say that T is *decidable* if Thm_T is recursive; otherwise, we say that T is *undecidable*. Combining the above discussion with Church's thesis, we see that T has a decision method iff T is decidable. (It can be shown that the decidability of T is independent of the assignment of symbol numbers; see Problem 6.)

We showed above that the set of expression numbers is calculable. The same can be shown for other important sets of expression numbers. According to Church's thesis, it follows that these sets are recursive. We shall verify this for certain of these sets. Besides giving further evidence for Church's thesis, this will be needed for some later applications.

We proceed to define some functions and predicates. For each function and predicate, we first give a formal definition (explicit, by cases, or by induction) which establishes that the function or predicate is recursive. We follow this with an explanation of the significance of the function or predicate. Sometimes this explanation is incomplete, and covers only the cases which are of interest. The symbols for all these functions and predicates should bear a subscript T to show that they relate to the theory T; but we omit this here and in other places where only one theory is being considered.

A) $Vble(a) \leftrightarrow a = \langle (a)_0\rangle \;\&\; \exists y_{y \leqslant a}((a)_0 = 2 \cdot y)$.

Vble(a) means that $a = \ulcorner\mathbf{x}\urcorner$ for some variable $\mathbf{x}$. (The bound $y \leqslant a$ is justified by (8) of §6.4, which also justifies several bounds in later definitions.)

We give definitions (B) and (C) for the special case of the theory N; but it is clear that the method is perfectly general.

$$\begin{aligned} \text{B)}\ \ Term(a) &\leftrightarrow 0 = 0 && \text{if } a = \langle SN(0)\rangle, \\ &\leftrightarrow Term\big((a)_1\big) && \text{if } a = \langle SN(S), (a)_1\rangle, \\ &\leftrightarrow Term\big((a)_1\big)\ \&\ Term\big((a)_2\big) && \text{if } a = \langle SN(+), (a)_1, (a)_2\rangle \\ & && \vee\ a = \langle SN(\cdot), (a)_1, (a)_2\rangle, \\ &\leftrightarrow Vble(a) && \textit{otherwise}. \end{aligned}$$

$Term(a)$ means that $a = \ulcorner \mathbf{a} \urcorner$ for some term $\mathbf{a}$. This is an inductive definition of the type described in §6.4.

$$\begin{aligned} \text{C)}\ \ AFor(a) \leftrightarrow a = \langle (a)_0, (a)_1, (a)_2\rangle\ \&\ \big((a)_0 = SN(=) \vee (a)_0 = SN(<)\big) \\ \&\ Term\big((a)_1\big)\ \&\ Term\big((a)_2\big). \end{aligned}$$

$AFor(a)$ means that $a = \ulcorner \mathbf{A} \urcorner$ for some atomic formula $\mathbf{A}$.

$$\begin{aligned} \text{D)}\ \ For(a) &\leftrightarrow For\big((a)_1\big) && \text{if } a = \langle SN(\neg), (a)_1\rangle, \\ &\leftrightarrow For\big((a)_1\big)\ \&\ For\big((a)_2\big) && \text{if } a = \langle SN(\vee), (a)_1, (a)_2\rangle, \\ &\leftrightarrow Vble\big((a)_1\big)\ \&\ For\big((a)_2\big) && \text{if } a = \langle SN(\exists), (a)_1, (a)_2\rangle, \\ &\leftrightarrow AFor(a) && \textit{otherwise}. \end{aligned}$$

$For(a)$ means that $a = \ulcorner \mathbf{A} \urcorner$ for some formula $\mathbf{A}$.

We give the next three definitions for a theory in which there are only unary and binary function and predicate symbols; but again the method is perfectly general.

$$\begin{aligned} \text{E)}\ \ Sub(a, b, c) &= c && \text{if } Vble(a)\ \&\ a = b, \\ &= \langle (a)_0, Sub\big((a)_1, b, c\big)\rangle && \text{if } a = \langle (a)_0, (a)_1\rangle, \\ &= \langle (a)_0, Sub\big((a)_1, b, c\big), Sub\big((a)_2, b, c\big)\rangle \\ & \qquad\qquad \text{if } a = \langle (a)_0, (a)_1, (a)_2\rangle\ \&\ (a)_0 \neq SN(\exists), \\ &= \langle (a)_0, (a)_1, Sub\big((a)_2, b, c\big)\rangle \\ & \qquad\qquad \text{if } a = \langle SN(\exists), (a)_1, (a)_2\rangle\ \&\ (a)_1 \neq b, \\ &= a && \textit{otherwise}. \end{aligned}$$

$Sub(\ulcorner \mathbf{a} \urcorner, \ulcorner \mathbf{x} \urcorner, \ulcorner \mathbf{b} \urcorner) = \ulcorner \mathbf{a_x[b]} \urcorner$; $Sub(\ulcorner \mathbf{A} \urcorner, \ulcorner \mathbf{x} \urcorner, \ulcorner \mathbf{a} \urcorner) = \ulcorner \mathbf{A_x[a]} \urcorner$.

$$\begin{aligned} \text{F)}\ \ Fr(a, b) &\leftrightarrow a = b && \text{if } Vble(a), \\ &\leftrightarrow Fr\big((a)_1, b\big) && \text{if } a = \langle (a)_0, (a)_1\rangle, \\ &\leftrightarrow Fr\big((a)_1, b\big) \vee Fr\big((a)_2, b\big) \\ & \qquad\qquad \text{if } a = \langle (a)_0, (a)_1, (a)_2\rangle\ \&\ (a)_0 \neq SN(\exists), \\ &\leftrightarrow Fr\big((a)_2, b\big)\ \&\ (a)_1 \neq b && \textit{otherwise}. \end{aligned}$$

$Fr(\ulcorner \mathbf{A} \urcorner, \ulcorner \mathbf{x} \urcorner)$ means that $\mathbf{x}$ is free in $\mathbf{A}$.

G) $Subtl(a, b, c) \leftrightarrow Subtl((a)_1, b, c) \quad \text{if } a = \langle (a)_0, (a)_1 \rangle,$

$\leftrightarrow Subtl((a)_1, b, c)\ \&\ Subtl((a)_2, b, c)$

$\text{if } a = \langle (a)_0, (a)_1, (a)_2 \rangle\ \&\ (a)_0 \neq SN(\exists),$

$\leftrightarrow Subtl((a)_2, b, c)\ \&\ (\neg Fr((a)_2, b) \vee \neg Fr(c, (a)_1))$

$\text{if } a = \langle SN(\exists), (a)_1, (a)_2 \rangle\ \&\ (a)_1 \neq b,$

$\leftrightarrow 0 = 0 \quad \textit{otherwise}.$

$Subtl(\ulcorner\mathbf{A}\urcorner, \ulcorner\mathbf{x}\urcorner, \ulcorner\mathbf{a}\urcorner)$ means that **a** is substitutible for **x** in **A**.

H) $PAx(a) \leftrightarrow \exists x_{x<a}(For(x)\ \&\ a = \langle SN(\vee), \langle SN(\neg), x \rangle, x \rangle).$

$PAx(a)$ means that a is the expression number of a propositional axiom. The next three definitions correspond similarly to substitution axioms, identity axioms, and equality axioms.

I) $SAx(a) \leftrightarrow \exists x_{x<a}\, \exists y_{y<a}\, \exists z_{z<a}(Vble(x)$
$\&\ For(y)\ \&\ Term(z)\ \&\ Subtl(y, x, z)$
$\&\ a = \langle SN(\vee), \langle SN(\neg), Sub(y, x, z) \rangle, \langle SN(\exists), x, y \rangle \rangle).$

J) $IAx(a) \leftrightarrow \exists x_{x<a}(Vble(x)\ \&\ a = \langle SN(=), x, x \rangle).$

K) $EAx(a) \leftrightarrow \ldots.$

We leave it to the reader to fill in the right-hand side of (K) for the theory N and to convince himself that the method is general.

L) $ER(a, b) \leftrightarrow b = \langle SN(\vee), (b)_1, a \rangle.$

$ER(\ulcorner\mathbf{A}\urcorner, \ulcorner\mathbf{B}\urcorner)$ means that **B** is inferrable from **A** by the expansion rule. The next four definitions correspond similarly to the contraction rule, the associative rule, the cut rule, and the $\exists$-introduction rule.

M) $CR(a, b) \leftrightarrow a = \langle SN(\vee), b, b \rangle.$

N) $AR(a, b) \leftrightarrow (a)_0 = SN(\vee)\ \&\ (a)_{2,0} = SN(\vee)$
$\&\ b = \langle SN(\vee), \langle SN(\vee), (a)_1, (a)_{2,1} \rangle, (a)_{2,2} \rangle.$

O) $TR(a, b, c) \leftrightarrow (a)_0 = SN(\vee)\ \&\ (b)_0 = SN(\vee)\ \&\ (b)_1$
$= \langle SN(\neg), (a)_1 \rangle\ \&\ c = \langle SN(\vee), (a)_2, (b)_2 \rangle.$

P) $IR(a, b) \leftrightarrow (a)_0 = SN(\vee)\ \&\ (a)_{1,0} = SN(\neg)$
$\&\ \neg Fr((a)_2, (b)_{1,1,1})$
$\&\ b = \langle SN(\vee), \langle SN(\neg), \langle SN(\exists), (b)_{1,1,1}, (a)_1 \rangle \rangle, (a)_2 \rangle.$

The set of expression numbers of nonlogical axioms of T is designated by $NLAx_T$. Since this is a completely arbitrary subset of For_T, it need not be recursive. If it is recursive, we say that T is *axiomatized.* The remaining functions and predicates which we define will be recursive only under the assumption that T is axiomatized.

Every finitely axiomatized theory is axiomatized; in particular, N is axiomatized. We can show that ACF and RCF are axiomatized by providing in each case a definition (similar to the above definitions) of $NLAx$.

Q) $Ax(a) \leftrightarrow PAx(a) \vee SAx(a) \vee IAx(a) \vee EAx(a) \vee NLAx(a)$.

Ax is the set of expression numbers of axioms.

We now assign a number to each finite sequence of expressions by assigning the number $\langle \ulcorner \mathbf{u}_1 \urcorner, \ldots, \ulcorner \mathbf{u}_n \urcorner \rangle$ to the sequence $\mathbf{u}_1, \ldots, \mathbf{u}_n$.

R) $$\begin{aligned} Prf(a) \leftrightarrow\ & Seq(a) \;\&\; lh(a) \neq 0 \;\&\; \forall i_{i<lh(a)}\big((Ax((a)_i) \\ & \vee \exists j_{j<i} \exists k_{k<i}(ER((a)_j, (a)_i) \\ & \vee CR((a)_j, (a)_i) \vee AR((a)_j, (a)_i) \\ & \vee TR((a)_j, (a)_k, (a)_i) \vee IR((a)_j, (a)_i))) \;\&\; For((a)_i)\big). \end{aligned}$$

Prf is the set of numbers of proofs.

S) $Pr(a, b) \leftrightarrow Prf(b) \;\&\; a = (b)_{lh(b) \dot{-} 1}$.

$Pr(\ulcorner \mathbf{A} \urcorner, b)$ means that b is the number of a proof of $\mathbf{A}$.

We can now define Thm by

$$Thm(a) \leftrightarrow \exists x Pr(a, x).$$

Due to the presence of the unbounded quantifier, we cannot conclude that Thm is recursive. We have, however, the following result.

Theorem. If T is an axiomatized theory, then Thm_T is recursively enumerable.

We conclude with one more definition, applicable to the theory N, which will be useful later.

T) $Num(0) = \langle SN(0) \rangle$,
$Num(a + 1) = \langle SN(S), Num(a) \rangle$.

$Num(a)$ is the expression number of the expression consisting of a S's followed by 0, i.e., the expression of N designating a.

6.7 REPRESENTABILITY

We will now show that every recursive function or predicate can, in a suitable sense, be calculated in the theory N. (In fact, the nonlogical axioms of N were chosen just for this purpose.) Throughout this section, $\vdash$ means $\vdash_N$.

The terms 0, $S0$, $SS0$, ... are called *numerals*. We use $\mathbf{k}_n$ as a name for the numeral which contains n occurrences of S. Thus the numerals are $\mathbf{k}_0, \mathbf{k}_1, \mathbf{k}_2, \ldots$.

Let F be an n-ary function; $\mathbf{A}$ a formula of N; $\mathbf{x}_1, \ldots, \mathbf{x}_n, \mathbf{y}$ distinct variables. We say that $\mathbf{A}$ *with* $\mathbf{x}_1, \ldots, \mathbf{x}_n, \mathbf{y}$ *represents* F if for every $a_1, \ldots, a_n$,

$$\vdash \mathbf{A}_{\mathbf{x}_1,\ldots,\mathbf{x}_n}[\mathbf{k}_{a_1}, \ldots, \mathbf{k}_{a_n}] \leftrightarrow \mathbf{y} = \mathbf{k}_b,$$

where $b = F(a_1, \ldots, a_n)$. We say that F is *representable* if for some $\mathbf{A}$, $\mathbf{x}_1, \ldots, \mathbf{x}_n, \mathbf{y}, \mathbf{A}$ with $\mathbf{x}_1, \ldots, \mathbf{x}_n, \mathbf{y}$ represents F.

Let P be an n-ary predicate; $\mathbf{A}$ a formula of N; $\mathbf{x}_1, \ldots, \mathbf{x}_n$ distinct variables. We say that $\mathbf{A}$ *with* $\mathbf{x}_1, \ldots, \mathbf{x}_n$ *represents* P if for every $a_1, \ldots, a_n$,

$$P(a_1, \ldots, a_n) \to \vdash \mathbf{A}_{\mathbf{x}_1,\ldots,\mathbf{x}_n}[\mathbf{k}_{a_1}, \ldots, \mathbf{k}_{a_n}]$$

and

$$\neg P(a_1, \ldots, a_n) \to \vdash \neg \mathbf{A}_{\mathbf{x}_1,\ldots,\mathbf{x}_n}[\mathbf{k}_{a_1}, \ldots, \mathbf{k}_{a_n}].$$

We say that P is *representable* if for some $\mathbf{A}$, $\mathbf{x}_1, \ldots, \mathbf{x}_n$, $\mathbf{A}$ with $\mathbf{x}_1, \ldots, \mathbf{x}_n$ represents P.

If F is representable, and $\mathbf{x}_1, \ldots, \mathbf{x}_n, \mathbf{y}$ are distinct variables, then there is a formula $\mathbf{A}$ such that $\mathbf{A}$ with $\mathbf{x}_1, \ldots, \mathbf{x}_n, \mathbf{y}$ represents F. For suppose that $\mathbf{A}'$ with $\mathbf{x}'_1, \ldots, \mathbf{x}'_n, \mathbf{y}'$ represents F. In view of the variant theorem, we may suppose that $\mathbf{x}_1, \ldots, \mathbf{x}_n, \mathbf{y}$ are not bound in $\mathbf{A}'$. Then we may take $\mathbf{A}$ to be

$$\mathbf{A}'_{\mathbf{x}'_1,\ldots,\mathbf{x}'_n,\mathbf{y}'}[\mathbf{x}_1, \ldots, \mathbf{x}_n, \mathbf{y}].$$

A similar remark applies to representable predicates.

Let F be an n-ary function; $\mathbf{a}$ a term of N; $\mathbf{x}_1, \ldots, \mathbf{x}_n$ distinct variables. We say that $\mathbf{a}$ *with* $\mathbf{x}_1, \ldots, \mathbf{x}_n$ *represents* F if for every $a_1, \ldots, a_n$,

$$\vdash \mathbf{a}_{\mathbf{x}_1,\ldots,\mathbf{x}_n}[\mathbf{k}_{a_1}, \ldots, \mathbf{k}_{a_n}] = \mathbf{k}_b,$$

where $b = F(a_1, \ldots, a_n)$. If this is the case and $\mathbf{y}$ is a new variable, then it follows from the equality theorem that $\mathbf{y} = \mathbf{a}$ with $\mathbf{x}_1, \ldots, \mathbf{x}_n, \mathbf{y}$ represents F.

We shall now consider some examples. First we show that $x = y$ with x, y represents $=$. For this we must prove

$$\vdash \mathbf{k}_m = \mathbf{k}_n \quad \textit{if} \quad m = n, \tag{1}$$

$$\vdash \mathbf{k}_m \neq \mathbf{k}_n \quad \textit{if} \quad m \neq n. \tag{2}$$

Now (1) follows from the identity axioms. In view of the symmetry theorem, it suffices to prove (2) when $m > n$. We do this by induction on n. If $n = 0$, (2) follows from N1. If $n > 0$, $\vdash \mathbf{k}_m = \mathbf{k}_n \to \mathbf{k}_{m-1} = \mathbf{k}_{n-1}$ by N2 and $\vdash \mathbf{k}_{m-1} \neq \mathbf{k}_{n-1}$ by induction hypothesis; so $\vdash \mathbf{k}_m \neq \mathbf{k}_n$ by the tautology theorem.

Next we show that $x + y$ with x, y represents $+$. We must prove

$$\vdash \mathbf{k}_m + \mathbf{k}_n = \mathbf{k}_{m+n}. \tag{3}$$

We prove this by induction on n. If $n = 0$, (3) follows from N3. Suppose that (3) holds for some n. Then by the equality theorem,

$$\vdash S(\mathbf{k}_m + \mathbf{k}_n) = \mathbf{k}_{m+n+1}.$$

From this by the equality theorem and N4,

$$\vdash \mathbf{k}_m + \mathbf{k}_{n+1} = \mathbf{k}_{m+n+1},$$

which is (3) with n replaced by $n + 1$.

A similar proof, using N5, N6, and (3), shows that

$$\vdash \mathbf{k}_m \cdot \mathbf{k}_n = \mathbf{k}_{mn}. \tag{4}$$

Hence $x \cdot y$ with x, y represents $\cdot$.

Now we show that $x < y$ with x, y represents $<$. We must prove

$$\vdash \mathbf{k}_m < \mathbf{k}_n \quad \text{if} \quad m < n, \tag{5}$$

$$\vdash \neg(\mathbf{k}_m < \mathbf{k}_n) \quad \text{if} \quad m \geqslant n. \tag{6}$$

We prove these by induction on n. If $n = 0$, (5) does not apply and (6) follows from N7. Now suppose that (5) and (6) hold for some n. By N8,

$$\vdash \mathbf{k}_m < \mathbf{k}_{n+1} \leftrightarrow (\mathbf{k}_m < \mathbf{k}_n \vee \mathbf{k}_m = \mathbf{k}_n). \tag{7}$$

Suppose $m < n + 1$. If $m < n$, then $\vdash \mathbf{k}_m < \mathbf{k}_n$ by induction hypothesis; if $m = n$, then $\vdash \mathbf{k}_m = \mathbf{k}_n$ by (1). In either case, $\vdash \mathbf{k}_m < \mathbf{k}_{n+1}$ by (7) and the tautology theorem. Now suppose $m \geqslant n + 1$. Then $\vdash \neg(\mathbf{k}_m < \mathbf{k}_n)$ by induction hypothesis and $\vdash \neg(\mathbf{k}_m = \mathbf{k}_n)$ by (2). Hence $\vdash \neg(\mathbf{k}_m < \mathbf{k}_{n+1})$ by (7) and the tautology theorem.

Lemma 1. For any predicate P, P is representable iff K_P is representable.

Proof. Suppose that $\mathbf{A}$ with $\mathbf{x}_1, \ldots, \mathbf{x}_n$ represents P. Let $\mathbf{B}$ be

$$(\mathbf{A} \ \& \ \mathbf{y} = \mathbf{k}_0) \vee (\neg\mathbf{A} \ \& \ \mathbf{y} = \mathbf{k}_1).$$

We claim that $\mathbf{B}$ with $\mathbf{x}_1, \ldots, \mathbf{x}_n, \mathbf{y}$ represents K_P. Suppose that

$$K_P(a_1, \ldots, a_n) = 0.$$

Then $P(a_1, \ldots, a_n)$; so $\vdash \mathbf{A}[\mathbf{k}_{a_1}, \ldots, \mathbf{k}_{a_n}]$. From this by the tautology theorem,

$$\vdash \mathbf{B}[\mathbf{k}_{a_1}, \ldots, \mathbf{k}_{a_n}] \leftrightarrow \mathbf{y} = \mathbf{k}_0.$$

A similar proof holds if $K_P(a_1, \ldots, a_n) = 1$.

Now suppose that $\mathbf{A}$ with $\mathbf{x}_1, \ldots, \mathbf{x}_n, \mathbf{y}$ represents K_P. We show that $\mathbf{A}_\mathbf{y}[0]$ with $\mathbf{x}_1, \ldots, \mathbf{x}_n$ represents P. If $P(a_1, \ldots, a_n)$, then $K_P(a_1, \ldots, a_n) = 0$ and hence

$$\vdash \mathbf{A}[\mathbf{k}_{a_1}, \ldots, \mathbf{k}_{a_n}] \leftrightarrow \mathbf{y} = \mathbf{k}_0.$$

Substituting $\mathbf{k}_0$ for $\mathbf{y}$ and using (1) and the tautology theorem, we have

$$\vdash \mathbf{A}[\mathbf{k}_{a_1}, \ldots, \mathbf{k}_{a_n}, \mathbf{k}_0].$$

If $\neg P(a_1, \ldots, a_n)$, we proceed similarly, using (2) in place of (1).

Representability Theorem. Every recursive function and predicate is representable.

Proof. In view of Lemma 1, it suffices to prove this for functions. We do this by induction on recursive functions; i.e., we show that the representable functions satisfy R1 through R3.

It is clear that $\mathbf{x}_i$ with $\mathbf{x}_1, \ldots, \mathbf{x}_n$ represents I_i^n. We have seen that $+$, $\cdot$, and $<$ are representable; so $K_<$ is representable by Lemma 1. This takes care of R1.

Now suppose that F is defined by

$$F(a_1, \ldots, a_n) = G(H_1(a_1, \ldots, a_n), \ldots, H_k(a_1, \ldots, a_n))$$

where $G, H_1, \ldots, H_k$ are representable. Let $\mathbf{x}_1, \ldots, \mathbf{x}_n, \mathbf{y}_1, \ldots, \mathbf{y}_k, \mathbf{z}$ be distinct variables. Choose $\mathbf{A}_i$ so that $\mathbf{A}_i$ with $\mathbf{x}_1, \ldots, \mathbf{x}_n, \mathbf{y}_i$ represents H_i, and choose $\mathbf{B}$ so that $\mathbf{B}$ with $\mathbf{y}_1, \ldots, \mathbf{y}_k, \mathbf{z}$ represents G. Let $\mathbf{C}$ be

$$\exists \mathbf{y}_1 \ldots \exists \mathbf{y}_k(\mathbf{A}_1 \ \& \cdots \& \ \mathbf{A}_k \ \& \ \mathbf{B}).$$

We show that $\mathbf{C}$ with $\mathbf{x}_1, \ldots, \mathbf{x}_n, \mathbf{z}$ represents F. Let $F(a_1, \ldots, a_n) = c$. Then $H_i(a_1, \ldots, a_n) = b_i$ and $G(b_1, \ldots, b_k) = c$. Let $\mathbf{A}_i'$ and $\mathbf{C}'$ be obtained from $\mathbf{A}_i$ and $\mathbf{C}$ by substituting $\mathbf{k}_{a_1}, \ldots, \mathbf{k}_{a_n}$ for $\mathbf{x}_1, \ldots, \mathbf{x}_n$. By the choice of $\mathbf{A}_i$,

$$\vdash \mathbf{A}_i' \leftrightarrow \mathbf{y}_i = \mathbf{k}_{b_i}.$$

Hence by the equivalence theorem

$$\vdash \mathbf{C}' \leftrightarrow \exists \mathbf{y}_1 \ldots \exists \mathbf{y}_k(\mathbf{y}_1 = \mathbf{k}_{b_1} \ \& \cdots \& \ \mathbf{y}_k = \mathbf{k}_{b_k} \ \& \ \mathbf{B}).$$

By repeated uses of the equivalence theorem and Corollary 3 to the equality theorem, we obtain

$$\vdash \mathbf{C}' \leftrightarrow \mathbf{B}_{\mathbf{y}_1, \ldots, \mathbf{y}_k}[\mathbf{k}_{b_1}, \ldots, \mathbf{k}_{b_k}].$$

Hence by the choice of $\mathbf{B}$ and the tautology theorem,

$$\vdash \mathbf{C}' \leftrightarrow \mathbf{z} = \mathbf{k}_c.$$

Before turning to R3, we need two more lemmas.

Lemma 2. $\vdash \mathbf{A}_\mathbf{x}[\mathbf{k}_0] \rightarrow \cdots \rightarrow \mathbf{A}_\mathbf{x}[\mathbf{k}_{n-1}] \rightarrow \mathbf{x} < \mathbf{k}_n \rightarrow \mathbf{A}$.

Proof. We use induction on n. For $n = 0$, the result to be proved is $\mathbf{x} < 0 \rightarrow \mathbf{A}$; this follows from N7. Now assume the lemma for some n. By N8,

$$\vdash \mathbf{x} < \mathbf{k}_{n+1} \leftrightarrow \mathbf{x} < \mathbf{k}_n \vee \mathbf{x} = \mathbf{k}_n. \tag{8}$$

From the equality theorem,

$$\vdash \mathbf{x} = \mathbf{k}_n \rightarrow (\mathbf{A} \leftrightarrow \mathbf{A}_\mathbf{x}[\mathbf{k}_n]). \tag{9}$$

From (8), (9) and the induction hypothesis by the tautology theorem,

$$\vdash \mathbf{A}_\mathbf{x}[\mathbf{k}_0] \rightarrow \cdots \rightarrow \mathbf{A}_\mathbf{x}[\mathbf{k}_{n-1}] \rightarrow \mathbf{A}_\mathbf{x}[\mathbf{k}_n] \rightarrow \mathbf{x} < \mathbf{k}_{n+1} \rightarrow \mathbf{A},$$

which is the lemma with n replaced by $n + 1$.

Lemma 3. If $\vdash \neg \mathbf{A}_\mathbf{x}[\mathbf{k}_i]$ for every $i < n$ and $\vdash \mathbf{A}_\mathbf{x}[\mathbf{k}_n]$, then

$$\vdash \mathbf{A} \ \& \ \forall \mathbf{y}(\mathbf{y} < \mathbf{x} \rightarrow \neg \mathbf{A}_\mathbf{x}[\mathbf{y}]) \leftrightarrow \mathbf{x} = \mathbf{k}_n.$$

Proof. Let **B** be the left-hand side of the equivalence to be proved. By the equality theorem

$$\vdash \mathbf{x} = \mathbf{k}_n \rightarrow (\mathbf{B} \leftrightarrow \mathbf{A_x}[\mathbf{k}_n] \ \& \ \forall \mathbf{y}(\mathbf{y} < \mathbf{k}_n \rightarrow \neg \mathbf{A_x}[\mathbf{y}])). \tag{10}$$

From Lemma 2, the detachment rule, and the generalization rule,

$$\vdash \forall \mathbf{y}(\mathbf{y} < \mathbf{k}_n \rightarrow \neg \mathbf{A_x}[\mathbf{y}]). \tag{11}$$

From (10), (11), and $\vdash \mathbf{A_x}[\mathbf{k}_n]$ by the tautology theorem,

$$\vdash \mathbf{x} = \mathbf{k}_n \rightarrow \mathbf{B}. \tag{12}$$

By the substitution theorem

$$\vdash \forall \mathbf{y}(\mathbf{y} < \mathbf{x} \rightarrow \neg \mathbf{A_x}[\mathbf{y}]) \rightarrow (\mathbf{k}_n < \mathbf{x} \rightarrow \neg \mathbf{A_x}[\mathbf{k}_n]).$$

From this and $\vdash \mathbf{A_x}[\mathbf{k}_n]$,

$$\vdash \mathbf{B} \rightarrow \neg(\mathbf{k}_n < \mathbf{x}). \tag{13}$$

By Lemma 2 and the detachment rule, $\vdash \mathbf{x} < \mathbf{k}_n \rightarrow \neg \mathbf{A}$; so

$$\vdash \mathbf{B} \rightarrow \neg(\mathbf{x} < \mathbf{k}_n). \tag{14}$$

By N9,

$$\vdash \mathbf{x} < \mathbf{k}_n \vee \mathbf{x} = \mathbf{k}_n \vee \mathbf{k}_n < \mathbf{x}. \tag{15}$$

From (12), (13), (14), and (15), we get $\vdash \mathbf{B} \leftrightarrow \mathbf{x} = \mathbf{k}_n$ by the tautology theorem.

We now show that the representable functions satisfy R3. Suppose that F is defined by

$$F(a_1, \ldots, a_n) = \mu x(G(a_1, \ldots, a_n, x) = 0),$$

where G is representable. Let **A** with $\mathbf{x}_1, \ldots, \mathbf{x}_n, \mathbf{y}, \mathbf{z}$ represent G. Let **w** be a new variable, and let **B** be

$$\mathbf{A_z}[0] \ \& \ \forall \mathbf{w}(\mathbf{w} < \mathbf{y} \rightarrow \neg \mathbf{A_{y,z}}[\mathbf{w}, 0]).$$

We show that **B** with $\mathbf{x}_1, \ldots, \mathbf{x}_n, \mathbf{y}$ represents F.

Let $F(a_1, \ldots, a_n) = b$, and set $c_i = G(a_1, \ldots, a_n, i)$. Let $\mathbf{A}'$ and $\mathbf{B}'$ be obtained from **A** and **B** by substituting $\mathbf{k}_{a_1}, \ldots, \mathbf{k}_{a_n}$ for $\mathbf{x}_1, \ldots, \mathbf{x}_n$. Then by choice of **A**,

$$\vdash \mathbf{A}'_\mathbf{y}[\mathbf{k}_i] \leftrightarrow \mathbf{z} = \mathbf{k}_{c_i}.$$

Hence

$$\vdash \mathbf{A}'_{\mathbf{y,z}}[\mathbf{k}_i, 0] \leftrightarrow \mathbf{k}_0 = \mathbf{k}_{c_i}. \tag{16}$$

If $i < b$, then $c_i \neq 0$; so by (16) and (2),

$$\vdash \neg \mathbf{A}'_{\mathbf{y,z}}[\mathbf{k}_i, 0] \qquad \text{for } i < b. \tag{17}$$

Since $c_b = 0$, we have by (16) and (1),

$$\vdash \mathbf{A}'_{\mathbf{y,z}}[\mathbf{k}_b, 0].$$

From this, (17), and Lemma 3, we get $\vdash \mathbf{B}' \leftrightarrow \mathbf{y} = \mathbf{k}_b$.

6.8 CHURCH'S THEOREM AND THE INCOMPLETENESS THEOREM

We are now in a position to apply our results to the decision problem for theories.

Let P be a binary predicate. For each number b, we define a unary predicate $P_{(b)}$ by

$$P_{(b)}(a) \leftrightarrow P(a, b).$$

The following simple lemma plays a key role.

Diagonal Lemma (*Cantor*). Let P be a binary predicate, and let Q be the unary predicate defined by $Q(a) \leftrightarrow \neg P(a, a)$. Then Q is distinct from all the $P_{(b)}$.

Proof. If Q is $P_{(b)}$, then

$$P(b, b) \leftrightarrow P_{(b)}(b) \leftrightarrow Q(b) \leftrightarrow \neg P(b, b),$$

a contradiction.

Let $\mathbf{z}$ be a fixed variable (say the first variable alphabetically). Let T be an extension of N. For each formula $\mathbf{A}$ of T, the set of n such that $\vdash_T \mathbf{A_z}[\mathbf{k}_n]$ is designated by $E(\mathbf{A})$. If T is inconsistent, each $E(\mathbf{A})$ is the set of all numbers. We shall show that if T is consistent, then every recursive set is an $E(\mathbf{A})$. Let A be a recursive set, and choose $\mathbf{A}$ by the representability theorem so that $\mathbf{A}$ with $\mathbf{z}$ represents A. If n is in A, then $\vdash_N \mathbf{A_z}[\mathbf{k}_n]$; so $\vdash_T \mathbf{A_z}[\mathbf{k}_n]$; so n is in $E(\mathbf{A})$. If n is not in A, then $\vdash_N \neg\mathbf{A_z}[\mathbf{k}_n]$; so $\vdash_T \neg\mathbf{A_z}[\mathbf{k}_n]$. By the consistency of T, n is not in $E(\mathbf{A})$.

Now define

$$P(a, b) \leftrightarrow Thm_T(Sub(b, \ulcorner\mathbf{z}\urcorner, Num(a))).$$

Then if $b = \ulcorner\mathbf{A}\urcorner$, $P_{(b)}$ is $E(\mathbf{A})$. If we define Q by $Q(a) \leftrightarrow \neg P(a, a)$, it follows by the diagonal lemma that Q is distinct from all the $E(\mathbf{A})$, and hence is not recursive. Now Q has the explicit definition

$$Q(a) \leftrightarrow \neg Thm_T(Sub(a, \ulcorner\mathbf{z}\urcorner, Num(a))),$$

and *Sub* and *Num* are recursive. It follows that Thm_T is not recursive. We have thus proved the following result.

Church's Theorem. If T is a consistent extension of N, then T is undecidable.

In particular, N is undecidable. (In view of the theorem of §6.6, this shows that Thm_N is a recursively enumerable set which is not recursive.) We shall see in the next section that Church's theorem can be used to show that many other theories are undecidable. At the moment, we are going to use Church's theorem to obtain an important result on completeness.

Negation Theorem. A predicate P is recursive iff both P and $\neg P$ are recursively enumerable.

Proof. If P is recursive, then $\neg P$ is also recursive; so P and $\neg P$ are recursively enumerable. Now suppose that P and $\neg P$ are recursively enumerable; say

$$P(\mathfrak{a}) \leftrightarrow \exists x Q(\mathfrak{a}, x) \qquad \text{and} \qquad \neg P(\mathfrak{a}) \leftrightarrow \exists x R(\mathfrak{a}, x),$$

where Q and R are recursive. For each $\mathfrak{a}$, either $P(\mathfrak{a})$ or $\neg P(\mathfrak{a})$; so there is an x such that either $Q(\mathfrak{a}, x)$ or $R(\mathfrak{a}, x)$. Hence we may define a recursive function F by

$$F(\mathfrak{a}) = \mu x(Q(\mathfrak{a}, x) \vee R(\mathfrak{a}, x)).$$

We claim that

$$P(\mathfrak{a}) \leftrightarrow Q(\mathfrak{a}, F(\mathfrak{a})). \tag{1}$$

If $Q(\mathfrak{a}, F(\mathfrak{a}))$, then $\exists x Q(\mathfrak{a}, x)$; so $P(\mathfrak{a})$. If $\neg Q(\mathfrak{a}, F(\mathfrak{a}))$, then $R(\mathfrak{a}, F(\mathfrak{a}))$; so $\exists x R(\mathfrak{a}, x)$; so $\neg P(\mathfrak{a})$. From the explicit definition (1), we see that P is recursive.

Lemma. If T is axiomatized and complete, then T is decidable.

Proof. Define recursive functions F and K by

$$\begin{aligned} F(0, a) &= a, \\ F(n+1, a) &= \langle SN(\neg), \langle SN(\exists), \langle 2n\rangle, \langle SN(\neg), F(n, a)\rangle\rangle\rangle, \\ K(a) &= F(a+1, a). \end{aligned}$$

If $a = \ulcorner \mathbf{A} \urcorner$, then $K(a) = \ulcorner \forall \mathbf{z}_0 \forall \mathbf{z}_1 \ldots \forall \mathbf{z}_a \mathbf{A} \urcorner$ where $\mathbf{z}_0, \mathbf{z}_1, \ldots$ are the variables in alphabetical order. If $\mathbf{z}_i$ occurs in $\mathbf{A}$, then $i < \ulcorner \mathbf{z}_i \urcorner < \ulcorner \mathbf{A} \urcorner = a$. Hence $\forall \mathbf{z}_0 \forall \mathbf{z}_1 \ldots \forall \mathbf{z}_a \mathbf{A}$ is closed; and by the generalization rule, it is a theorem iff $\mathbf{A}$ is a theorem. Then by the completeness of T, $\mathbf{A}$ is not a theorem iff

$$\neg \forall \mathbf{z}_0 \forall \mathbf{z}_1 \ldots \forall \mathbf{z}_a \mathbf{A}$$

is a theorem. We thus have

$$\begin{aligned} \neg Thm(a) &\leftrightarrow \neg For(a) \vee Thm(\langle SN(\neg), K(a)\rangle) \\ &\leftrightarrow \exists y(\neg For(a) \vee Pr(\langle SN(\neg), K(a)\rangle, y)). \end{aligned}$$

It follows that $\neg Thm$ is recursively enumerable. Since Thm is recursively enumerable, it follows from the negation theorem that Thm is recursive. Thus T is decidable.

This lemma gives the third application of complete theories mentioned in §5.5. With the previous results, it can be used to show that RCF and all the $ACF(n)$ are decidable.

From the lemma and Church's theorem, we obtain the following result.

Incompleteness Theorem (*Gödel-Rosser*). If T is an axiomatized extension of N, then T is not complete.

The incompleteness theorem has important implications concerning the axiomatic method. The idea of the axiomatic method is that, given certain concepts, we introduce a language for expressing facts about these concepts and then introduce an axiom system for proving facts about these concepts. The axiom system must be such that all theorems of the axiom system are true; and we hope that it will be such that all true sentences of the language will be theorems. In any case, we will certainly want the axioms and rules of the axiom system to be such that we can decide what is and what is not a proof. (Otherwise, we could achieve our object by simply adopting all true sentences as axioms.)

Now suppose that the concepts are those of $\mathfrak{N}$ and that the language selected is $L(N)$. Suppose further that we wish our axiom system to be formalizable as a theory. The requirement that we be able to recognize a proof implies that we must be able to recognize a nonlogical axiom when we see one. In view of Church's thesis, this means that our theory T must be axiomatized. If the true sentences of $\mathfrak{N}$, and only these, are to be provable in T, then T must be equivalent to $Th(\mathfrak{N})$, and hence must be a complete extension of N. But this is impossible by the incompleteness theorem.

If we are willing to argue a little more informally, we can see that the restriction to theories or to the language $L(N)$ is immaterial for the argument. Suppose that we have an axiom system in which each closed formula **A** of N is expressed by a formula **A***; and suppose that we can actually construct **A*** when **A** is given. Suppose that **A*** is provable in our axiom system iff **A** is true in $\mathfrak{N}$. Finally, suppose that our axiom system satisfies the requirement that we can recognize a proof. Then we can decide whether **A** is true or false. All we have to do is look through all sequences of formulas in our axiom system until we come to one which is either a proof of **A*** or ($\neg$**A**)*. If it is a proof of **A***, **A** is true; if it is a proof of ($\neg$**A**)*, **A** is false. Thus we have a decision method for $Th(\mathfrak{N})$, which is impossible by Church's theorem. We are thus led to the following conclusion: there is no correct axiom system in which we can prove all true facts about the natural numbers expressible in $L(N)$, much less all mathematical truths.

Although this is an important limitation, it must not be misunderstood. It does not say that there is one mathematical truth which cannot be proved in any correct axiom system. This is clearly not so, since we could obtain a new correct system by adding this truth as a new axiom. An interesting question is whether every mathematical truth (or at least every mathematical truth expressible in $L(N)$) can be proved from axioms which are evidently true. However, we cannot hope to make much progress with this question until we understand more clearly what is meant by being evidently true.

The incompleteness theorem tells us that if T is a consistent axiomatized extension of N, then some closed formula **A** of T is undecidable in T. It turns out that if we examine the details of the proof closely, we find that we can actually construct the formula **A** when T is given.

6.9 UNDECIDABILITY

We shall now develop some further methods for proving that theories are undecidable. These methods may be combined with the lemma of §6.8 to prove that theories are incomplete.

Throughout this section, we shall return to the original intuitive description of the decision problem for theories. Converting our proofs into formal proofs is merely a matter of verifying that certain functions and predicates are recursive.

Our idea throughout is to combine Church's theorem with a basic method for showing that the undecidability of one theory implies the undecidability of another. The basic method is this. Suppose that with each formula **A** of T we have associated a formula **A*** of T' such that $\vdash_T$ **A** iff $\vdash_{T'}$ **A***. Suppose also that we have a method for constructing **A*** when **A** is given. Then a decision method for T' would clearly lead to a decision method for T; so if T is undecidable, then T' is undecidable.

A *finite extension* of T is a simple extension T' of T such that there are only finitely many nonlogical axioms of T' which are not theorems of T. An interpretation I of T in T' is *faithful* if for every formula **A** of T, $\vdash_{T'} \mathbf{A}^{(I)}$ implies $\vdash_T$ **A**.

Theorem 1. If T' is a conservative extension of T, and T is undecidable, then T' is undecidable. If T is a finite extension of T', and T is undecidable, then T' is undecidable. If T is an extension by definitions of T', and T is undecidable, then T' is undecidable. If I is a faithful interpretation of T in T', and T is undecidable, then T' is undecidable.

Proof. In each case, we define an **A*** for the basic method. If T' is a conservative extension of T, **A*** is **A**. If T is a finite extension of T', **A*** is $\mathbf{B}_1 \to \cdots \to \mathbf{B}_n \to \mathbf{A}$, where $\mathbf{B}_1, \ldots, \mathbf{B}_n$ are the closures of the nonlogical axioms of T which are not theorems of T'; we then have $\vdash_T$ **A** iff $\vdash_{T'}$ **A*** by the reduction theorem. If T is an extension by definitions of T', **A*** is the translation of **A** into T'. If I is a faithful interpretation of T in T', **A*** is $\mathbf{A}^{(I)}$.

As an application, let T be the theory with the language of N and no nonlogical axioms. Then N is a finite extension of T; so T is undecidable by Theorem 1 and Church's theorem.

A structure $\mathcal{A}$ is *strongly undecidable* if every theory having $\mathcal{A}$ as a model is undecidable.

Theorem 2. The structure $\mathfrak{N}$ is strongly undecidable.

Proof. Let $\mathfrak{N}$ be a model of T. Then $\mathfrak{N}$ is a model of $T \cup N$; so $T \cup N$ is consistent. By Church's theorem, $T \cup N$ is undecidable. Since $T \cup N$ is a finite extension of T, T is undecidable by theorem 1.

If T is an extension by definitions of $Th(\mathcal{A})$, then there is a unique expansion of $\mathcal{A}$ which is a model of T. Such an expansion of $\mathcal{A}$ is called an *expansion by definitions* of $\mathcal{A}$.

Lemma. If $\mathcal{B}$ is an expansion by definitions of $\mathcal{A}$, and $\mathcal{B}$ is strongly undecidable, then $\mathcal{A}$ is strongly undecidable.

Proof. Let $\mathcal{A}$ be a model of T; we must show that T is undecidable. We know that $\mathcal{B}$ is a model of an extension by definitions U of $Th(\mathcal{A})$. Form T_1 from T by adding as new nonlogical axioms the existence and uniqueness conditions needed for the defining axioms of U. Then $\mathcal{A}$ is a model of T_1. Moreover, T_1 is a finite extension of T (since we are allowing only finitely many nonlogical symbols); so by Theorem 1, it suffices to prove that T_1 is undecidable.

Let U_1 be the extension by definitions of T_1 obtained by adding the new nonlogical symbols of U together with their defining axioms. Since $\mathcal{A}$ is a model of T_1, $Th(\mathcal{A})$ is an extension of T_1; so U is an extension of U_1. It follows that $\mathcal{B}$ is a model of U_1; so U_1 is undecidable. By Theorem 1, T_1 is undecidable.

An interpretation I of L in L' is *simple* if $\mathbf{u}_I$ is $\mathbf{u}$ for every nonlogical symbol $\mathbf{u}$ of L. Suppose that I is a simple interpretation of L in T', where $L(T') = L'$, and suppose that $\mathcal{A}$ is a model of T'. From the validity in $\mathcal{A}$ of (1) and (2) of §4.7, we see that $(U_I)_{\mathcal{A}}$ is the universe of a substructure $\mathcal{A}_I$ of $\mathcal{A}/L$.

Since an individual of $\mathcal{A}_I$ has the same name in $L(\mathcal{A}_I)$ and $L'(\mathcal{A})$, we can extend I to a simple interpretation of $L(\mathcal{A}_I)$ in $L'(\mathcal{A})$. Then $\mathcal{A}_I(\mathbf{A}) = \mathcal{A}(\mathbf{A}_I)$ for every closed formula $\mathbf{A}$ of $L(\mathcal{A}_I)$. The proof is by induction on the length of $\mathbf{A}$. If $\mathbf{A}$ is atomic, then $\mathbf{A}_I$ is $\mathbf{A}$, and $\mathcal{A}_I(\mathbf{A}) = \mathcal{A}(\mathbf{A})$. If $\mathbf{A}$ is a negation or a disjunction, the result follows immediately from the induction hypothesis. Now suppose that $\mathbf{A}$ is $\exists\mathbf{x}\mathbf{B}$. Using the induction hypothesis and recalling that

$$|\mathcal{A}_I| = (U_I)_{\mathcal{A}},$$

we get

$$\begin{aligned}
\mathcal{A}_I(\mathbf{A}) = \mathbf{T} &\leftrightarrow \mathcal{A}_I(\mathbf{B_x[i]}) = \mathbf{T} && \textit{for some } \mathbf{i} \textit{ in } L(\mathcal{A}_I)\\
&\leftrightarrow \mathcal{A}\big((\mathbf{B}_I)_\mathbf{x}[\mathbf{i}]\big) = \mathbf{T} && \textit{for some } \mathbf{i} \textit{ in } L(\mathcal{A}_I)\\
&\leftrightarrow \mathcal{A}\big(U_I\mathbf{i} \;\&\; (\mathbf{B}_I)_\mathbf{x}[\mathbf{i}]\big) = \mathbf{T} && \textit{for some } \mathbf{i} \textit{ in } L(\mathcal{A})\\
&\leftrightarrow \mathcal{A}\big(\exists\mathbf{x}(U_I\mathbf{x} \;\&\; \mathbf{B}_I)\big) = \mathbf{T}\\
&\leftrightarrow \mathcal{A}(\mathbf{A}_I) = \mathbf{T}.
\end{aligned}$$

Now let $\mathbf{A}$ be a sentence of L, and let $\mathbf{B}$ be the closure of $\mathbf{A}$. Then $\mathbf{B}_I$ is the closure of $\mathbf{A}^{(I)}$. From this and $\mathcal{A}_I(\mathbf{B}) = \mathcal{A}(\mathbf{B}_I)$, we conclude that $\mathbf{A}$ is valid in $\mathcal{A}_I$ iff $\mathbf{A}^{(I)}$ is valid in $\mathcal{A}$.

We say that a predicate in $|\mathcal{A}|$ or a function from $|\mathcal{A}|$ to $|\mathcal{A}|$ is *definable* in $\mathcal{A}$ if it occurs in some expansion by definitions of $\mathcal{A}$. It follows that a predicate p in $|\mathcal{A}|$ is definable in $\mathcal{A}$ iff there is a formula $\mathbf{D}$ such that

$$pa_1 \dots a_n \leftrightarrow \mathcal{A}(\mathbf{D}[\mathbf{i}_1, \dots, \mathbf{i}_n]) = \mathbf{T},$$

when $\mathbf{i}_1, \dots, \mathbf{i}_n$ are the names of $a_1, \dots, a_n$; and a function f from $|\mathcal{A}|$ to $|\mathcal{A}|$ is definable in $\mathcal{A}$ iff there is a formula $\mathbf{D}$ such that

$$fa_1 \dots a_n = b \leftrightarrow \mathcal{A}(\mathbf{D}[\mathbf{i}_1, \dots, \mathbf{i}_n, \mathbf{j}]) = \mathbf{T},$$

when $\mathbf{i}_1, \dots, \mathbf{i}_n, \mathbf{j}$ are the names of $a_1, \dots, a_n, b$. (It is easy to see that the last condition guarantees that the needed existence and uniqueness conditions are valid in $\mathcal{A}$ and hence are theorems of $Th(\mathcal{A})$.)

A structure $\mathcal{A}$ is *definable* in a structure $\mathcal{B}$ if the set $|\mathcal{A}|$ is a subset of $|\mathcal{B}|$ which is definable in $\mathcal{B}$, and each function or predicate of $\mathcal{A}$ is the restriction to $|\mathcal{A}|$ of a function or predicate definable in $\mathcal{B}$.

Suppose that $\mathcal{A}$ is definable in $\mathcal{B}$. We can then find an expansion by definitions $\mathcal{C}$ of $\mathcal{B}$ such that $|\mathcal{A}|$ is a predicate $\mathbf{p}_\mathcal{C}$ of $\mathcal{C}$, and such that each function or predicate of $\mathcal{A}$ is the restriction to $|\mathcal{A}|$ of a function or predicate of $\mathcal{C}$. We may suppose the notation chosen so that $\mathbf{u}_{\mathcal{A}}$ is the restriction of $\mathbf{u}_{\mathcal{C}}$ for each nonlogical symbol $\mathbf{u}$ in the language of $\mathcal{A}$. Now let I be the simple interpretation of the language of $\mathcal{A}$ in the language of $\mathcal{C}$ which has $U_I = \mathbf{p}$. Then I is an interpretation of the language of $\mathcal{A}$ in $Th(\mathcal{C})$, and $\mathcal{C}_I = \mathcal{A}$.

Theorem 3 (*Tarski*). If $\mathcal{A}$ is definable in $\mathcal{B}$ and $\mathcal{A}$ is strongly undecidable, then $\mathcal{B}$ is strongly undecidable.

Proof. Let $\mathcal{C}$ and I be as above. In view of the lemma, it will suffice to show that $\mathcal{C}$ is strongly undecidable.

Let $\mathcal{C}$ be a model of T; we must prove that T is undecidable. Obtain T_1 from T by adding an axiom $\exists \mathbf{x} U_I \mathbf{x}$ and, for each function symbol $\mathbf{f}$ of L (the language of $\mathcal{A}$), an axiom

$$U_I\mathbf{x}_1 \rightarrow \cdots \rightarrow U_I\mathbf{x}_n \rightarrow U_I\mathbf{f}\mathbf{x}_1 \ldots \mathbf{x}_n$$

with $\mathbf{x}_1, \ldots, \mathbf{x}_n$ distinct. Then T_1 is a finite extension of T; so by Theorem 1, it suffices to prove that T_1 is undecidable. Since I is an interpretation in $Th(\mathcal{C})$, the added axipms are valid in $\mathcal{C}$; so $\mathcal{C}$ is a model of T_1.

Let U be the theory with language L whose nonlogical axioms are the formulas $\mathbf{A}$ such that $\vdash_{T_1} \mathbf{A}^{(I)}$. For every such $\mathbf{A}$, $\mathbf{A}^{(I)}$ is valid in $\mathcal{C}$; so $\mathbf{A}$ is valid in $\mathcal{C}_I = \mathcal{A}$. Thus $\mathcal{A}$ is a model of U; so U is undecidable. Clearly I is a faithful interpretation of U in T_1; so T_1 is undecidable by Theorem 1.

We shall now give some applications of these results. Let $\mathcal{A}$ be the ring of integers, considered as a structure for the language of FL. We claim that $\mathfrak{N}$ is definable in $\mathcal{A}$. The most difficult part is to show that $|\mathfrak{N}|$ is definable in $\mathcal{A}$. For this, we use a theorem of Lagrange: every natural number is the sum of four squares of natural numbers. Thus if $\mathbf{A}$ is

$$\exists y \, \exists z \, \exists y' \, \exists z' (x = y \cdot y + z \cdot z + y' \cdot y' + z' \cdot z')$$

and $\mathbf{i}$ is the name of an individual a of $\mathcal{A}$, then $\mathcal{A}(\mathbf{A}_x[\mathbf{i}]) = \mathbf{T}$ iff a is a natural number. Noting that $S(a) = a + 1$ and that $a < b$ iff $a + c + 1 = b$ for some natural number c, we easily show that 0, S, $+$, $\cdot$, and $<$ are restrictions of functions and predicates definable in $\mathcal{A}$.

It follows by Theorems 2 and 3 that $\mathcal{A}$ is strongly undecidable; so any theory with the language of FL which has $\mathcal{A}$ as a model is undecidable. Among such theories are the elementary theories of rings, commutative rings, and rings of integrity.

Now let $\mathcal{A}$ be the same, and let $\mathcal{B}$ be the field of rational numbers. Then $\mathcal{A}$ is definable in $\mathcal{B}$. The only problem is to show that $|\mathcal{A}|$ is definable in $\mathcal{B}$. This has been done by Julia Robinson, using the theory of quadratic forms. (We shall not give the solution here.) It follows by Theorem 3 that $\mathcal{B}$ is strongly undecidable. From this we can obtain the undecidability of FL, $FL(0)$, and other theories.

PROBLEMS

All theories are assumed to have only finitely many nonlogical symbols.

1. The *primitive recursive functions* are defined by the generalized inductive definition:

PR1. The constant functions, the successor function S, and the I_i^n are primitive recursive.

PR2. If $G, H_1, \ldots, H_k$ are primitive recursive, and F is defined by

$$F(\mathfrak{a}) = G(H_1(\mathfrak{a}), \ldots, H_k(\mathfrak{a})),$$

then F is primitive recursive.

PR3. If G and H are primitive recursive, and F is defined by

$$F(0, \mathfrak{a}) = G(\mathfrak{a}),$$
$$F(a + 1, \mathfrak{a}) = H(F(a, \mathfrak{a}), a, \mathfrak{a}),$$

then F is primitive recursive.

A predicate is *primitive recursive* if its representing function is primitive recursive.

a) Show that R1, R2, R4, and R6 through R13 hold with *recursive* replaced by *primitive recursive.* [Prove R9 and R10 last.]

b) Show that β is primitive recursive.

c) Show that there is a primitive recursive F such that if

$$n < a, a_0 < a, \ldots, a_{n-1} < a,$$

then there is an $x < F(a)$ such that $\beta(x, i) = a_i$ for $i < n$.

d) Show that $\langle a_1, \ldots, a_n \rangle$, lh, $(a)_i$, Seq, In, and $*$ are primitive recursive. [Use (c) and R9.] Show that R14 holds with *recursive* replaced by *primitive recursive.*

e) Show that if $NLAx_T$ is primitive recursive, then the functions and predicates defined in §6.6 are primitive recursive.

2. Let H and K be recursive functions, and let F and G be defined inductively by

$$F(a, \mathfrak{a}) = H(\overline{F}(a, \mathfrak{a}), \overline{G}(a, \mathfrak{a}), a, \mathfrak{a}),$$
$$G(a, \mathfrak{a}) = K(\overline{F}(a + 1, \mathfrak{a}), \overline{G}(a, \mathfrak{a}), a, \mathfrak{a}).$$

Show that F and G are recursive. [Let $L(a, \mathfrak{a}) = \langle F(a, \mathfrak{a}), G(a, \mathfrak{a}) \rangle$, and use R14 to show that L is recursive.]

3. A real number a is *recursive* if there are recursive functions F and G such that for $n \neq 0$, we have $G(n) \neq 0$ and $\mid |a| - F(n)/G(n) \mid < 1/n$.

a) Show that every rational number is recursive.

b) Show that e and π are recursive. [Use suitable series for e and π together with estimates on the rate of convergence of these series.]

c) Show that if a and b are recursive, then $a + b$, $a - b$, and $a \cdot b$ are recursive. If also $b \neq 0$, show that a/b is recursive. [Follow the proofs of the continuity of these functions.]

d) Let $P_a(m, n)$ mean that $n \neq 0$ and $m/n < |a|$. Show that a is recursive iff P_a is recursive. [Suppose that a is recursive and irrational. Then

$$m/n < |a| \quad \text{iff} \quad m/n < F(k)/G(k) - 1/k \quad \text{for some } k,$$

and

$$m/n > |a| \quad \text{iff} \quad m/n > F(k)/G(k) + 1/k \quad \text{for some } k,$$

where F and G are as above.]

e) Show that a real number a is recursive iff there is a recursive function F such that $F(n) \leqslant 9$ for $n \neq 0$ and $|a| = \sum F(n) \cdot 10^{-n}$. [If a is recursive and irrational, then F is determined by $|a| - 10^{-k} < \sum_{n=0}^{k} F(n) \cdot 10^{-n} < |a|$.]

4. A unary function F *enumerates* a set A if the members of A are just $F(0), F(1), \ldots$.

a) Show that a nonempty set A is recursively enumerable iff it is enumerated by a recursive function.

b) Show that an infinite recursively enumerable set is enumerated by an injective recursive function.

c) Show that an infinite set A is recursive iff it is enumerated by a recursive function F such that $F(n) < F(n + 1)$ for all n.

d) Show that an infinite recursively enumerable set has an infinite recursive subset. [Use (b) and (c).]

5. A theory is *axiomatizable* if it is equivalent to an axiomatized theory.

a) Show that T is axiomatizable iff Thm_T is recursively enumerable. [If Thm_T is recursively enumerable, choose F as in 4(a). Let the nonlogical axioms of T' be the $\mathbf{A}_0 \,\&\, \mathbf{A}_1 \,\&\, \cdots \,\&\, \mathbf{A}_n$ where $\ulcorner \mathbf{A}_i \urcorner = F(i)$, and use 4(c).]

b) Give an example of an axiomatizable theory which is not axiomatized.

6. Let two assignments of symbol numbers for T be given, and let $\ulcorner \mathbf{u} \urcorner$ and '$\mathbf{u}$' be the corresponding expression numbers of $\mathbf{u}$. Show that there is a recursive function F such that $F(\ulcorner \mathbf{u} \urcorner) =$ '$\mathbf{u}$' for every designator $\mathbf{u}$. Conclude that T is decidable (axiomatized) for one assignment of symbol numbers iff it is decidable (axiomatized) for the other assignment.

7. A theory T is *numerical* if it contains the symbols 0 and S. For such a theory, we define representability as in N.

a) Show that if T is an axiomatized numerical theory, and if $\vdash_T \mathbf{k}_m = \mathbf{k}_n$ implies $m = n$ for all m and n, then every function or predicate representable in T is recursive. Conclude that a function or predicate is recursive iff it is representable in some finitely axiomatized numerical theory T such that $\vdash_T \mathbf{k}_m = \mathbf{k}_n$ implies $m = n$ for all m and n.

b) If T is an extension of N, show that the functions representable in T satisfy R1 through R3. Conclude that if T is consistent, then Thm_T is not representable in T. [Use the proof of Church's theorem.]

c) A formula $\mathbf{A}$ in a numerical theory T is a *truth definition* if for every closed formula $\mathbf{B}$ of T, $\vdash_T \mathbf{B} \leftrightarrow \mathbf{A}_\mathbf{x}[\mathbf{k}_{\ulcorner \mathbf{B} \urcorner}]$. Show that if T is a consistent extension of N, then there is no truth definition in T. [Use (b) and Lindenbaum's theorem.]

8. Let T be a numerical theory. A formula $\mathbf{A}$ with the distinct variables $\mathbf{x}_1, \ldots, \mathbf{x}_n$ *weakly represents P* if for every $a_1, \ldots, a_n$,

$$\vdash_T \mathbf{A}_{\mathbf{x}_1,\ldots,\mathbf{x}_n}[\mathbf{k}_{a_1}, \ldots, \mathbf{k}_{a_n}] \quad \text{iff} \quad P(a_1, \ldots, a_n).$$

The predicate P is *weakly representable* in T if it is weakly represented in T by some sentence with some sequence of variables.

a) Show that if T is consistent, then every predicate which is representable in T is weakly representable in T.

b) Let P and Q be n-ary recursively enumerable predicates. Show that there is a formula $\mathbf{A}$ of N and distinct variables $\mathbf{x}_1, \ldots, \mathbf{x}_n$ such that

$$P(a_1, \ldots, a_n) \;\&\; \neg Q(a_1, \ldots, a_n) \rightarrow \vdash_N \mathbf{A}[\mathbf{k}_{a_1}, \ldots, \mathbf{k}_{a_n}],$$
$$Q(a_1, \ldots, a_n) \;\&\; \neg P(a_1, \ldots, a_n) \rightarrow \vdash_N \neg\mathbf{A}[\mathbf{k}_{a_1}, \ldots, \mathbf{k}_{a_n}].$$

[Let $\mathbf{A}$ be $\exists\mathbf{x}(\mathbf{B} \;\&\; \forall\mathbf{y}(\mathbf{y} < \mathbf{x} \rightarrow \neg\mathbf{C}))$, where $\mathbf{B}$ represents a predicate P_1 such that $P(\mathfrak{a}) \leftrightarrow \exists x P_1(\mathfrak{a}, x)$ and $\mathbf{C}$ represents a predicate Q_1 such that $Q(\mathfrak{a}) \leftrightarrow \exists x Q_1(\mathfrak{a}, x)$.]

c) Show that if T is a consistent axiomatized extension of N, then every recursively enumerable predicate P is weakly representable in T. [Choose a recursively enumerable Q such that

$$Q(a_1, \ldots, a_n, \ulcorner\mathbf{A}\urcorner) \leftrightarrow \vdash_T \mathbf{A}_{\mathbf{x}_1,\ldots,\mathbf{x}_n,\mathbf{y}}[\mathbf{k}_{a_1}, \ldots, \mathbf{k}_{a_n}, \mathbf{k}_{\ulcorner\mathbf{A}\urcorner}]$$

for all $\mathbf{A}$. Take $\mathbf{A}$ as in (b) and let $\mathbf{B}$ be $\mathbf{A}_{\mathbf{y}}[\mathbf{k}_{\ulcorner\mathbf{A}\urcorner}]$.]

d) Show that if T is axiomatized, then every predicate which is weakly representable in T is recursively enumerable. Conclude that a predicate is recursively enumerable iff it is weakly representable in some finitely axiomatized numerical theory. [Use (c).]

9. Let $\mathbf{C}_n$ be a formula asserting that there are exactly n individuals.

a) Let T be finitely axiomatized. Show that there is a decision method for formulas of the form $\mathbf{C}_n \rightarrow \mathbf{A}$. [Use 13(b) of Chapter 4.]

b) Show that if T is finitely axiomatized and $\mathfrak{m}$-categorical for some infinite cardinal $\mathfrak{m}$, then T is decidable. [Let T' be obtained from T by adding all the $\neg\mathbf{C}_n$ as axioms. Use the Łoś-Vaught theorem and the lemma of §6.8 to show that T' is decidable. Show that, given a closed theorem $\mathbf{A}$ of T', we can find an n such that $\vdash_T \neg\mathbf{C}_1 \rightarrow \cdots \rightarrow \neg\mathbf{C}_n \rightarrow \mathbf{A}$. Then use (a).]

c) Show that (b) becomes false if *finitely axiomatized* is replaced by *axiomatized.* [Let A be a recursively enumerable set which is not recursive. Let T be the theory with no nonlogical symbols whose nonlogical axioms are the $\neg\mathbf{C}_n$ for n in A, and use 5(a).]

10. A structure $\mathfrak{A}$ is *undecidable* if $Th(\mathfrak{A})$ is undecidable.

a) If $\mathfrak{A}$ is undecidable, show that every expansion of $\mathfrak{A}$ is undecidable.

b) Show that if $\mathfrak{B}$ is an expansion by definitions of $\mathfrak{A}$, and $\mathfrak{B}$ is undecidable, then $\mathfrak{A}$ is undecidable. [Show that an extension by definitions of a complete theory is complete. Use Lemma 1 of §5.5 to show that $Th(\mathfrak{B})$ is equivalent to an extension by definitions of $Th(\mathfrak{A})$.]

c) Show that if $\mathfrak{A}$ is definable in $\mathfrak{B}$ and $\mathfrak{A}$ is undecidable, then $\mathfrak{B}$ is undecidable. [Show that $Th(\mathfrak{A})$ has a faithful interpretation in $Th(\mathfrak{B})$.]

11. A consistent theory T is *essentially undecidable* (*strongly undecidable*) if every model of T is undecidable (strongly undecidable).

a) Show that if T is consistent and decidable, then T has a complete decidable simple extension. [Use the proof of Lindenbaum's theorem described in Problem 1 of Chapter 4.]

b) Show that a consistent theory T is essentially undecidable iff every consistent simple extension of T is undecidable. [Use (a) and Lemma 1 of §5.5.]

c) A theory T' is *compatible* with a theory T if $L(T') = L(T)$ and $T \cup T'$ is consistent. Show that a consistent theory T is strongly undecidable iff every theory compatible with T is undecidable.

d) Show that if T is finitely axiomatized and essentially undecidable, then T is strongly undecidable. [Use the proof of Theorem 2 of §6.9.] Conclude that N is strongly undecidable. [Use (b).]

e) Show that if T is essentially undecidable (strongly undecidable), then every consistent extension of T is essentially undecidable (strongly undecidable). [Show that an expansion of a strongly undecidable structure is strongly undecidable, and use this and 10(a).]

f) Show that if T' is an extension by definitions of T, and T' is essentially undecidable (strongly undecidable), then T is essentially undecidable (strongly undecidable). [Use 10(b) and the lemma of §6.9.]

g) Show that if T is interpretable in the consistent theory T', and T is essentially undecidable (strongly undecidable), then T' is essentially undecidable (strongly undecidable). [By (f), we may assume that there is an interpretation of T in T'. Show that if $\mathfrak{B}$ is a model of T', then some model of T is definable in $\mathfrak{B}$. Then use 10(c) and Theorem 3 of §6.9.]

12. A nonempty set Γ of closed formulas of T is a *base* of T if for every two models $\mathfrak{A}$ and $\mathfrak{B}$ of T, if $\mathfrak{A}(\mathbf{A}) = \mathfrak{B}(\mathbf{A})$ for every $\mathbf{A}$ in Γ, then $\mathfrak{A}$ and $\mathfrak{B}$ are elementarily equivalent. We write $\mathfrak{D}(\Gamma)$ for the set of disjunctions of formulas which are either formulas in Γ or negations of formulas in Γ, and $\mathfrak{B}(\Gamma)$ for the set of conjunctions of formulas in $\mathfrak{D}(\Gamma)$.

a) Show that if Γ is a base for T, then every closed formula in T is equivalent in T to a formula in $\mathfrak{B}(\Gamma)$. [Use the proof of Lemma 4 of §5.5.]

b) Suppose that T is axiomatized, that Γ is a base of T, and that we have a method of deciding whether a given formula is in Γ. Show that, given a closed formula $\mathbf{A}$ of T, we may find a formula in $\mathfrak{B}(\Gamma)$ which is equivalent to $\mathbf{A}$ in T.

c) Let T be axiomatized, and let Γ be a base of T. Suppose that we have a method of deciding whether a given formula is in Γ, and a method of deciding whether a given sentence of $\mathfrak{D}(\Gamma)$ is a theorem of T. Show that T is decidable. [Use (b).]

d) Show that EQ is decidable. [Use (c) and 21 of Chapter 5.]

e) Show that if T is finitely axiomatized and all the nonlogical symbols of T are unary predicate symbols, then T is decidable. [Show that we may assume that T has no nonlogical axioms, and use (c) and 22 of Chapter 5.]

13. A set R *separates* the sets P and Q if P is a subset of R, and Q and R are disjoint. The disjoint sets P and Q are *recursively inseparable* if no recursive set separates P and Q. The theory T is recursively inseparable if the set Thm_T is recursively inseparable from the set of expression numbers of negations of theorems of T.

a) Let P and Q be disjoint sets, and let T be a numerical theory such that for every recursive set A, there is a formula $\mathbf{A}$ in T such that for each n, $P(\ulcorner\mathbf{A}[\mathbf{k}_n]\urcorner)$ if n is in A and $Q(\ulcorner\mathbf{A}[\mathbf{k}_n]\urcorner)$ if n is not in A. Show that P and Q are recursively inseparable. [If R separates P and Q, let $E'(\mathbf{A})$ be the set of n such that $R(\ulcorner\mathbf{A}[\mathbf{k}_n]\urcorner)$, and proceed as in the proof of Church's theorem.]

b) Show that if every recursive set is weakly representable in T, then T is undecidable. [Use (a).]

c) Show that if T is consistent and if every recursive set is representable in T, then T is recursively inseparable. [Use (a).] Conclude that there exist recursively inseparable recursively enumerable sets.

d) Show that if T is recursively inseparable, then T is essentially undecidable. [Use 11(b).]

14. A theory T is *hereditarily undecidable* if every theory having T as a simple extension is undecidable.

a) Show that $\mathfrak{A}$ is strongly undecidable iff $Th(\mathfrak{A})$ is hereditarily undecidable.

b) Show that if T has a strongly undecidable model, then T is hereditarily undecidable. Conclude that every strongly undecidable theory is hereditarily undecidable.

c) Show that the elementary theory of rings is hereditarily undecidable but not essentially undecidable. [Use (b) and 11(b).]

d) Let A and B be recursively inseparable recursively enumerable sets [see 13(c)]. Let $\mathbf{A}_{n,k}$ be as in 21(a) of Chapter 5. Let T be formed from EQ by adding $\mathbf{A}_{1,k}$ as an axiom for each k in A and adding $\neg\mathbf{A}_{1,k}$ as an axiom for each k in B. Show that T is axiomatizable and essentially undecidable. [Use 5(a) and 13(d).] Show that every finite part of T is decidable, and conclude that T is not hereditarily undecidable. [Use 12(d) and Theorem 1 of §6.9.]

15. a) Show that if T is axiomatizable and has only finitely many inequivalent complete simple extensions, then T is decidable. [Show that if $\mathbf{A}$ is a closed formula and if $T[\mathbf{A}]$ and $T[\neg\mathbf{A}]$ are decidable, then T is decidable. Then use induction on the number of inequivalent complete simple extensions.]

b) Show that there is an axiomatizable undecidable theory which has only countably many inequivalent complete simple extensions, all of which are decidable. [Use the example of 9(c).]

16. a) Show that there is a strongly undecidable structure for a language whose only nonlogical symbol is a 4-ary predicate symbol R. [Let $|\mathfrak{A}|$ be the set of natural numbers, and let $R_{\mathfrak{A}}$ be the set of all 4-tuples $(1, m, n, m + n)$ and $(0, m, n, m \cdot n)$. Note that $R(m, m, m, m)$ iff $m = 0$. Show that $\mathfrak{N}$ is definable in $\mathfrak{A}$.]

b) An expansion $\mathfrak{B}$ of $\mathfrak{A}$ is *inessential* if the only symbols of the language of $\mathfrak{B}$ which are not symbols of the language of $\mathfrak{A}$ are constants. Show that if $\mathfrak{B}$ is an inessential expansion of $\mathfrak{A}$ and $\mathfrak{B}$ is strongly undecidable, then $\mathfrak{A}$ is strongly undecidable. [Let $\mathfrak{A}$ be a model of T. Obtain T' from T by adding the new constants of $\mathfrak{B}$, and apply the theorem on constants.]

c) Show that there is a strongly undecidable structure for the language whose only nonlogical symbol is a binary predicate symbol P. [Let $\mathfrak{A}$ be as in (a). Let $|\mathfrak{B}|$ consist of the elements of $|\mathfrak{A}|$, the ordered pairs of elements of $|\mathfrak{A}|$, and a new element u. Let $P_{\mathfrak{B}}$ consist of the following ordered pairs of elements of $|\mathfrak{B}|$ (where a, b, c, d designate elements of $|\mathfrak{A}|$): all $((a, b), (c, d))$ such that $R_{\mathfrak{A}}(a, b, c, d)$; all $(a, (a, b))$; all $((a, b), b)$; all (u, a); and all $((a, b), u)$. Show that $\mathfrak{A}$ is definable in an inessential expansion of $\mathfrak{B}$, and use (b).]

17. a) Let T be the theory whose only nonlogical symbol is the binary predicate symbol P, and whose nonlogical axioms are $\neg P(x, x)$ and $P(x, y) \rightarrow P(y, x)$. Show that T has a strongly undecidable model. [Let $\mathfrak{A}$ be as in 16(c). For each element a of $|\mathfrak{A}|$, let $|\mathfrak{B}|$ contain three elements a_1, a_2, a_3; and, in addition, let $|\mathfrak{B}|$ contain two elements u and v. Let $P_{\mathfrak{B}}$ consist of all pairs (a_1, a_2), (a_2, a_3), (u, a_1), (v, a_2); all pairs (a_1, b_3) such that (a, b) is in $P_{\mathfrak{A}}$; and the inverse pairs of these. Show that $\mathfrak{A}$ is definable in an inessential expansion of $\mathfrak{B}$.]

b) Let PO be the theory whose only nonlogical symbol is $<$, and whose axioms are $\neg(x < x)$ and $x < y \rightarrow y < z \rightarrow x < z$. Then the models of PO are the partially ordered sets. Let LT be the extension of PO obtained by adding axioms asserting that every two elements have a least upper bound and a greatest lower bound. Show that LT has a strongly undecidable model. [Let $\mathfrak{A}$ be as in (a). Let E consist of the elements of $|\mathfrak{A}|$ and the ordered pairs in $P_{\mathfrak{A}}$. Let $|\mathfrak{B}|$ be the class of all subsets F of E such that whenever (a, b) is in F, then a and b are in F. Show that the union and intersection of two members of $|\mathfrak{B}|$ is in $|\mathfrak{B}|$. Let $F <_{\mathfrak{B}} F'$ if F is a proper subset of F'. Show that a structure isomorphic to $\mathfrak{A}$ is definable in $\mathfrak{B}$.]

c) Show that there is a strongly undecidable structure whose language has two unary function symbols as its only nonlogical symbols. [Let $\mathfrak{A}$ be as in (a). Let $|\mathfrak{B}|$ consist of the individuals of $\mathfrak{A}$ and the pairs in $P_{\mathfrak{A}}$. Let

$$\mathbf{f}_{\mathfrak{B}}((a, b)) = a, \qquad \mathbf{g}_{\mathfrak{B}}((a, b)) = b, \qquad \mathbf{f}_{\mathfrak{B}}(a) = \mathbf{g}_{\mathfrak{B}}(a) = a$$

for a, b in $|\mathfrak{A}|$. Show that $\mathfrak{A}$ is definable in $\mathfrak{B}$.]

18. Let $\mathfrak{A}$ be the set of bijective mappings from the set of integers to itself. We consider $\mathfrak{A}$ as a group (and hence a model of G) with composition of functions as the group operation. Let S be the element of $\mathfrak{A}$ defined by $S(i) = i + 1$.

a) Show that the mapping from i to S^i is a bijective mapping from the set of integers to the set of individuals of $\mathfrak{A}$ commuting with S. [If F commutes with S, then $F(S^i(0)) = S^i(F(0))$.]

b) Show that for any integers i and j, j is divisible by i iff S^j commutes with every individual of $\mathfrak{A}$ with which S^i commutes. [If $i \neq 0$ and the condition holds, apply it to the F defined by $F(k) = k + i$ if k is divisible by i, $F(k) = k$ otherwise.]

c) Let $\mathfrak{B}$ be the following structure: $|\mathfrak{B}|$ is the set of integers, $1_{\mathfrak{B}}$ is the integer 1, $+_{\mathfrak{B}}$ is the addition function, and $D_{\mathfrak{B}}$ is the divisibility predicate (that is, $D_{\mathfrak{B}}(i, j)$ iff i is divisible by j). Show that $\mathfrak{B}$ is strongly undecidable. [Show that the ring of integers is definable in $\mathfrak{B}$. For this, note that i^2 is the unique integer j such that $j + i$ is a least common multiple of i and $i + 1$ and $j - i$ is a least common multiple of i and $i - 1$; and that $i \cdot j$ is determined by $(i + j)^2 = i^2 + i \cdot j + i \cdot j + j^2$.]

d) Show that $\mathfrak{A}$ is strongly undecidable. [Use (a) and (b) to define a model isomorphic to $\mathfrak{B}$ in an inessential expansion of $\mathfrak{A}$.]

19. Let J_0 be the theory obtained from N by replacing N2 and N8 by the three axioms $S0 \neq 0$,

$$Sx \neq x \rightarrow Sy \neq y \rightarrow Sx = Sy \rightarrow x = y,$$
$$Sx \neq x \rightarrow (y < Sx \leftrightarrow y < x \vee y = x).$$

Let J be obtained from J_0 by adding all sentences $\mathbf{k}_{n+1} \neq \mathbf{k}_n$ as new axioms.

a) Show that if $m \neq n$,

$$\vdash_{J_0} \mathbf{k}_{m+1} \neq \mathbf{k}_m \rightarrow \mathbf{k}_{n+1} \neq \mathbf{k}_n \rightarrow \mathbf{k}_m \neq \mathbf{k}_n.$$

b) Show that if F is recursive, then there is a formula $\mathbf{A}$ and distinct variables $\mathbf{x}_1, \ldots, \mathbf{x}_n, \mathbf{y}$ such that if $F(a_1, \ldots, a_n) = b$, then

$$\vdash_{J_0} \mathbf{A}[\mathbf{k}_{a_1}, \ldots, \mathbf{k}_{a_n}, \mathbf{k}_b],$$
$$\vdash_J \mathbf{A}[\mathbf{k}_{a_1}, \ldots, \mathbf{k}_{a_n}] \leftrightarrow \mathbf{y} = \mathbf{k}_b.$$

[Follow the proof of the representability theorem. If F is $K_<$, let $\mathbf{A}$ be

$$S\mathbf{x}_1 \neq \mathbf{x}_1 \rightarrow S\mathbf{x}_2 \neq \mathbf{x}_2 \rightarrow (\mathbf{x}_1 < \mathbf{x}_2 \;\&\; \mathbf{y} = \mathbf{k}_0) \vee (\neg(\mathbf{x}_1 < \mathbf{x}_2) \;\&\; \mathbf{y} = \mathbf{k}_1).$$

In treating R3, note that G may be taken so that $G(\mathfrak{a}, b) \leqslant 1$, since we may replace $G(\mathfrak{a}, b)$ by $K_<(G(\mathfrak{a}, b), 1)$.]

c) Show that if P is recursive, then there is a formula $\mathbf{A}$ and distinct variables $\mathbf{x}_1, \ldots, \mathbf{x}_n$ such that

$$P(a_1, \ldots, a_n) \rightarrow \vdash_{J_0} \mathbf{A}[\mathbf{k}_{a_1}, \ldots, \mathbf{k}_{a_n}],$$
$$\neg P(a_1, \ldots, a_n) \rightarrow \vdash_J \neg\mathbf{A}[\mathbf{k}_{a_1}, \ldots, \mathbf{k}_{a_n}].$$

[Use (b).]

d) Show that J is strongly undecidable. [Let T be compatible with J. Let P and $\mathbf{A}$ be as in (c), and let $\mathbf{B}$ be the conjunction of the closures of the nonlogical axioms of J_0. Show that $\mathbf{B} \rightarrow \mathbf{A}$ weakly represents P in T, and use 13(b) and 11(c).]

e) Show that every finitely axiomatized part of J has a finite model. [For each natural number n, consider the structure $\mathfrak{A}$ with individuals $0, 1, \ldots, n$ and

$$S_{\mathfrak{A}}(a) = min(a + 1, n), \qquad a +_{\mathfrak{A}} b = min(a + b, n),$$
$$a \cdot_{\mathfrak{A}} b = min(a \cdot b, n), \qquad a <_{\mathfrak{A}} b \leftrightarrow a < b.$$

f) Let T be the theory with language $L(N)$ whose nonlogical axioms are all sentences which are valid in every finite structure for this language. Show that T is undecidable. [Use (d), (e), 11(c), and the compactness theorem.]

CHAPTER 7

RECURSION THEORY

7.1 PARTIAL FUNCTIONS

A mapping F from A to B is completely determined by any decision method for F: the value $F(a)$ must be the element of B obtained when the method is applied to a. However, a method for obtaining an element of B from an element of A may not be a decision method for a mapping from A to B, since the method may not lead to an answer for certain a in A. For example, the method may require us to solve a set of equations, and the equations may have no solution for some a. Again, there may be an a for which the calculations which the method directs us to make never end. This will happen, for example, if we try to calculate $\mu x R(a, x)$ by the usual method and a is such that $\forall x \neg R(a, x)$.

This leads us to a new definition. A *partial mapping* F from A to B is a mapping from a subset of A, called the *domain* of F, to B. Thus a mapping from A to B is a partial mapping from A to B whose domain is A. When we wish to contrast mappings with partial mappings, we say *total mapping* for *mapping*.

A *decision method* for a partial mapping F from A to B is a method which, if applied to an element a of A, will give the value $F(a)$ if a is in the domain of F and will give no result if a is not in the domain of F. We say that F is *calculable* if it has a decision method. These definitions agree with our earlier ones when F is total.

It may seem unnatural to require that the method give no result when applied to an element of A not in the domain of F. However, this ensures that a decision method for a partial mapping F completely determines F. For the domain of F is the set of a such that the method gives a result when applied to a; and for such a, $F(a)$ is the value obtained by applying the method to a. From this, we see that every method for obtaining an element of B when an element of A is given is a decision method for a uniquely determined partial mapping from A to B. If a method gives the result $F(a)$ for every a in the domain of F, then the partial mapping associated with the method is an extension of F; so we have not really lost anything by our restriction.

A *partial subset* A of E consists of a subset of E, called the *domain* of A, and a determination for each element of the domain that it is *in A* or *out of A*. Thus a partial subset of E can be specified by specifying the domain and the subset of the domain consisting of the elements which are in the partial subset. A subset of E is simply a partial subset of E whose domain is E. Again we say *total subset* for *subset* when we wish to emphasize the contrast with partial subsets.

A *decision method* for a partial subset A of E is a method which, applied to an element in the domain of A, will tell us whether this element is in A or out of A; and which, applied to an element of E not in the domain of A, will lead to no result. A partial subset is *calculable* if it has a decision method.

These notions extend to functions and predicates. Thus an *n-ary partial function* from A to B is a partial mapping from the set of n-tuples in A to B; and an *n-ary partial predicate* in A is a partial subset of the set of n-tuples in A. Again we say *total function* and *total predicate* for *function* and *predicate*.

In this chapter, we adopt the conventions stated at the beginning of §6.2 and extend them as follows: *partial function* means *partial function from the set of natural numbers to the set of natural numbers*, and *partial predicate* means *partial predicate in the set of natural numbers*.

The letters formerly used to designate functions and predicates will now be used to designate partial functions and partial predicates. A consequence of this is that certain expressions now have no meaning, or, as we shall say, are *undefined*, for certain values of the variables. Thus $F(a)$ is undefined if a is not in the domain of F. Likewise, $P(a)$ is undefined if a is not in the domain of P. (If a is in the domain of P, then $P(a)$ is true if a is in P and false if a is out of P.)

We agree that a complicated expression is defined only if all its parts are defined. Thus $F(G(a))$ is defined iff a is in the domain of G and $G(a)$ is in the domain of F. Again, $P(a) \vee Q(a)$ is defined iff a is in both the domain of P and the domain of Q.

If we attempt to define a partial function F explicitly by $F(a) = G(H(a))$, where G and H are partial, we run into difficulty. If $G(H(a))$ is undefined, then $F(a) = G(H(a))$ is undefined, and hence can tell us nothing about $F(a)$. We avoid this difficulty by introducing a new symbol. If ___ and ... are expressions (designating numbers) which may be undefined, then ___ $\simeq$ ... means that either ___ and ... are both defined, and have the same value, or ___ and ... are both undefined. (This is an exception to the rule stated above; ___ $\simeq$ ... is always defined, even if one or both of ___ and ... are undefined.) We may now define a partial function F explicitly by $F(a) \simeq G(H(a))$. This tells us the domain of F consists of those a for which $G(H(a))$ is defined, and that for such a, $F(a) = G(H(a))$.

We introduce another symbol to be used in explicit definitions of partial predicates. If ___ and ... are sentences which may be undefined, then ___ $\overset{\sim}{\leftrightarrow}$ ... means that ___ and ... are both defined and true, or both defined and false, or both undefined. Thus we might define a partial predicate P explicitly by

$$P(a) \overset{\sim}{\leftrightarrow} Q(F(a))$$

(where Q and F are partial).

We shall generally avoid using bound variables in expressions which may be undefined. However, we do wish to give a meaning to $\mu x(\ldots x \ldots)$, where $\ldots x \ldots$ is a sentence which is undefined for some values of x. We wish to do this in such a way that if R is calculable and F is defined by

$$F(\mathfrak{a}) \simeq \mu x R(\mathfrak{a}, x),$$

then F is calculable. Suppose that we calculate $F(\mathfrak{a})$ in the same manner that we would if R were total; i.e., we calculate $R(\mathfrak{a}, 0)$, $R(\mathfrak{a}, 1), \ldots$ until we come to an a such that $R(\mathfrak{a}, a)$ is true, and then conclude that $F(\mathfrak{a}) = a$. We will then obtain a as a value for $F(\mathfrak{a})$ iff a is the smallest x such that $R(\mathfrak{a}, x)$, *and* $R(\mathfrak{a}, x)$ is defined for all x less than a. We therefore define $\mu x(\ldots x \ldots)$ to be the smallest number a such that $\ldots a \ldots$ is defined and true, provided that $\ldots x \ldots$ is defined for all x less than a. If this last condition is not satisfied, or if there is no a such that $\ldots a \ldots$ is defined and true, then $\mu x(\ldots x \ldots)$ is undefined. Note that this agrees with our previous definition if $\ldots x \ldots$ is defined for all x.

The *representing partial function* K_P of a partial predicate P is the partial function with the same domain as P which is defined for arguments $\mathfrak{a}$ in that domain as follows: $K_P(\mathfrak{a}) = 0$ if $P(\mathfrak{a})$ and $K_P(\mathfrak{a}) = 1$ if $\neg P(\mathfrak{a})$. Clearly P is calculable iff K_P is calculable.

If F is an n-ary partial function, the *graph* of F, designated by $\mathfrak{G}_F$, is the $(n + 1)$-ary total predicate defined by

$$\mathfrak{G}_F(\mathfrak{a}, a) \leftrightarrow F(\mathfrak{a}) \simeq a.$$

We then have

$$F(\mathfrak{a}) \simeq \mu x \mathfrak{G}_F(\mathfrak{a}, x);$$

so F may be recovered from $\mathfrak{G}_F$.

We shall show that F is calculable iff $\mathfrak{G}_F$ is positively calculable. Suppose that F is calculable. We calculate $\mathfrak{G}_F(\mathfrak{a}, a)$ as follows: given $\mathfrak{a}$ and a, we calculate $F(\mathfrak{a})$; if we obtain a value and that value is a we conclude that $\mathfrak{G}_F(\mathfrak{a}, a)$ is true. This clearly gives the correct answer if $\mathfrak{G}_F(\mathfrak{a}, a)$ and no answer if $\neg\mathfrak{G}_F(\mathfrak{a}, a)$; so $\mathfrak{G}_F$ is positively calculable.

For the converse, we observe first that

$$\forall y(\mathfrak{G}_F(\mathfrak{a}, y) \leftrightarrow \exists x R(\mathfrak{a}, y, x)) \rightarrow F(\mathfrak{a}) \simeq (\mu z R(\mathfrak{a}, (z)_0, (z)_1))_0. \tag{1}$$

For if $F(\mathfrak{a})$ is defined, let $y = F(\mathfrak{a})$ and choose x so that $R(\mathfrak{a}, y, x)$. Setting $z = \langle y, x \rangle$, we have $R(\mathfrak{a}, (z)_0, (z)_1)$; so

$$(\mu z R(\mathfrak{a}, (z)_0, (z)_1))_0$$

is defined. Now suppose that

$$(\mu z R(\mathfrak{a}, (z)_0, (z)_1))_0$$

is defined, and let

$$u = \mu z R(\mathfrak{a}, (z)_0, (z)_1).$$

Then $R(\mathfrak{a}, (u)_0, (u)_1)$; so $\exists x R(\mathfrak{a}, (u)_0, x)$; so $\mathfrak{G}_F(\mathfrak{a}, (u)_0)$; so $F(\mathfrak{a}) = (u)_0$. This proves (1).

Now suppose that $\mathfrak{G}_F$ is positively calculable. Then there is a calculable predicate R such that

$$\mathfrak{G}_F(\mathfrak{a}, y) \leftrightarrow \exists x R(\mathfrak{a}, y, x)$$

for all $\mathfrak{a}$ and y. Hence by (1),

$$F(\mathfrak{a}) \simeq (\mu z R(\mathfrak{a}, (z)_0, (z)_1))_0.$$

In view of the calculability of $(a)_i$, it follows that F is calculable.

A partial function F is *recursive* if its graph is recursively enumerable. A partial predicate P is *recursive* if its representing partial function is recursive. By the above and Church's thesis, a partial function or predicate is recursive iff it is calculable.

Since a total function is a partial function, we now have two definitions of *recursive* for total functions. To see that they are the same, we must prove the following: a total function F is recursive (in the old sense) iff its graph is recursively enumerable. Now for F total, $\mathfrak{G}_F$ has the explicit definition

$$\mathfrak{G}_F(\mathfrak{a}, a) \leftrightarrow F(\mathfrak{a}) = a.$$

Hence if F is recursive, then $\mathfrak{G}_F$ is recursive and hence recursively enumerable. Now suppose that $\mathfrak{G}_F$ is recursively enumerable; say

$$\mathfrak{G}_F(\mathfrak{a}, y) \leftrightarrow \exists x R(\mathfrak{a}, y, x)$$

for all $\mathfrak{a}$ and y. Then by (1),

$$F(\mathfrak{a}) = (\mu z R(\mathfrak{a}, (z)_0, (z)_1))_0,$$

and hence F is recursive. It follows at once that the two definitions of *recursive* for total predicates coincide.

Before considering properties of recursive partial functions and predicates, we shall introduce a further generalization of recursiveness.

7.2 FUNCTIONALS AND RELATIONS

We have considered the decision problem for a mapping from A to B only in the case in which the elements of A and B were concrete objects. We shall now discuss the case in which the elements of A and B are the simplest type of abstract objects, viz., functions.

An n-ary function F may be thought of as made up of an infinite number of parts, each of which consists of the value of F for one n-tuple of numbers. Each part may be replaced by a concrete object, e.g., an equation giving the value of F for the n-tuple. But we have no way of replacing F by a single concrete object.

Suppose that Φ is a mapping from a set A of concrete objects to a set B of n-ary functions. Given an element a of A, we obviously cannot find the infinitely many parts of $\Phi(a)$ in a finite calculation. The nearest that we can come to this is to obtain a decision method for $\Phi(a)$ as a result of the calculation. Hence we define a decision method for Φ to be a method by which, given an element a of A, we can find a decision method for $\Phi(a)$ in a finite number of steps. (Of course, such a method can exist only if all of the values of Φ are calculable functions.)

Suppose that we have a decision method for Φ. Given an element a of A and an n-tuple of numbers $x_1, \ldots, x_n$, we can clearly calculate the value of $\Phi(a)$ at $(x_1, \ldots, x_n)$. If we designate this value by $\Phi'(a, x_1, \ldots, x_n)$, this means that we have a decision method for Φ'. It is also clear that a decision method for Φ' provides us with a decision method for Φ.

We can use this to define *decision method* for a mapping Φ from an *arbitrary* set A to a set B of n-ary functions. We define Φ' as above, and then define a decision method for Φ to be a decision method for Φ'. Thus we can always reduce to the case of mappings whose values are natural numbers.

Now we consider a mapping Φ from a set A of n-ary functions to the set of natural numbers. To see how information about F is utilized in the calculation of $\Phi(F)$, we consider a specific example. Let $n = 1$, and let

$$\Phi(F) = 2 \cdot F(F(0) + 2).$$

To evaluate $\Phi(F)$, we must first know the value $F(0)$. Suppose that this is 3. We then add 2 to this, obtaining 5. To proceed further, we must know the value $F(5)$. Suppose that this is 4. We then multiply this by 2, obtaining 8. We conclude that $\Phi(F) = 8$. Note that for the calculation, it is immaterial how we obtain the values $F(0)$ and $F(5)$; so long as $F(0) = 3$ and $F(5) = 4$, we will have $\Phi(F) = 8$.

We thus see that F is utilized in the computation of $\Phi(F)$ by utilizing the values of F for certain n-tuples $(a_1, \ldots, a_n)$. In some cases, $a_1, \ldots, a_n$ are given by the directions for calculating $\Phi(F)$; in other cases, $a_1, \ldots, a_n$ are computed in the course of the calculation, possibly making use of other values of F. In our example, the argument 0 was given by the directions for calculation, while the argument 5 was computed, using the value $F(0) = 3$.

As we have remarked, it is immaterial for the description of the decision method how the values of F are obtained. As an aid to the imagination, we suppose that at the beginning of the calculation of $\Phi(F)$, we are furnished with an object which, when supplied with an n-tuple of numbers, will give the value of F at this n-tuple. Such an object cannot be a machine, since F may not be calculable. Following Turing, we call such an object an *oracle* for F (because it supplies correct answers without any apparent method of doing so). We can then say that a decision method for Φ is a method by which, given an oracle for F, we can calculate $\Phi(F)$ in a finite number of steps.

An oracle for F can be replaced by an oracle for $\langle F \rangle$ and vice versa. For we can find $F(\mathfrak{a})$ by computing $\langle \mathfrak{a} \rangle$ and asking the oracle for $\langle F \rangle$ for $\langle F \rangle(\langle \mathfrak{a} \rangle)$; and we can find $\langle F \rangle(a)$ by computing $(a)_0, \ldots, (a)_{n-1}$ and asking the oracle for F for $F((a)_0, \ldots, (a)_{n-1})$. Hence there will be no loss of generality if we restrict the functions in A to be unary.

We are thus led to the following definitions. Let $N_{m,n}$ be the set of $(m + n)$-tuples $(\alpha_1, \ldots, \alpha_m, a_1, \ldots, a_n)$, where $\alpha_1, \ldots, \alpha_m$ are unary total functions and

$a_1, \ldots, a_n$ are numbers. An (m, n)-ary *partial* (*total*) *functional* is a partial (total) mapping from $N_{m,n}$ to the set of numbers. A total functional is called simply a *functional*. The $(0, n)$-ary partial (total) functionals are clearly just the n-ary partial (total) functions.

An (m, n)-ary *partial* (*total*) *relation* is a partial (total) subset of $N_{m,n}$. A total relation is called simply a *relation*. The $(0, n)$-ary partial (total) relations are just the n-ary partial (total) predicates.

The letters formerly used to represent partial functions and partial predicates will now be used to represent partial functionals and partial relations respectively. We use small Greek letters to represent total unary functions. When used as variables, these letters will be called *function variables*; small Latin letters used as variables will be called *number variables*. We shall use capital German letters to stand for finite sequences of distinct Greek and Latin letters in which the Greek letters (if any) precede all the Latin letters (if any). If two distinct German letters appear in the same context, it is understood that the two sequences abbreviated by them have no letter in common. As with small German letters, we suppose that the number of letters in the sequence is chosen to suit the context. Thus if F is (m, n)-ary and we write $F(\mathfrak{A})$, we assume that $\mathfrak{A}$ contains m Greek letters and n Latin letters.

We have restricted the arguments to be $(m + n)$-tuples in which the m functions precede the n numbers. There is clearly no real loss of generality in this restriction. For notational convenience, we may sometimes write some of the symbols for functions after some of the symbols for numbers. When this happens, it is understood that the symbols for functions are to be moved to the front of the symbols for numbers without otherwise changing the order of the symbols. For example, if we write $F(\mathfrak{A}, \mathfrak{B})$, it is understood that the function variables in $\mathfrak{B}$ must be moved to the front of the number variables in $\mathfrak{A}$.

The *representing partial functional* K_P of a partial relation P is the partial functional with the same domain as P defined for arguments $\mathfrak{A}$ in that domain as follows: $K_P(\mathfrak{A}) = 0$ if $P(\mathfrak{A})$ and $K_P(\mathfrak{A}) = 1$ if $\neg P(\mathfrak{A})$.

We say that we are *given* an $(m + n)$-tuple $\mathfrak{A}$ if we are given the n numbers in $\mathfrak{A}$ and oracles for the m functions in $\mathfrak{A}$. This enables us to use the definitions of the last section to define *decision method* and *calculable* for partial functionals and partial relations. It is clear that a partial relation is calculable iff its representing partial functional is calculable.

A total relation P is *positively calculable* if there is a method which when applied to $\mathfrak{A}$ will give the conclusion that $P(\mathfrak{A})$ is true if this conclusion is correct and will give no conclusion if $P(\mathfrak{A})$ is false. We shall show that this can be reduced to the notion of a calculable total predicate. We first introduce the following notation: if $\mathfrak{a}$ is $\alpha_1, \ldots, \alpha_m, a_1, \ldots, a_n$, then $\overline{\mathfrak{a}}(x)$ is $\overline{\alpha_1}(x), \ldots, \overline{\alpha_m}(x), a_1, \ldots, a_n$. We then claim that an (m, n)-ary total relation P is positively calculable iff there is a calculable $(m + n + 1)$-ary total predicate Q such that

$$P(\mathfrak{A}) \leftrightarrow \exists x Q(\overline{\mathfrak{A}}(x), x)$$

for all $\mathfrak{A}$.

To save some notation, suppose that $\mathfrak{A}$ is α, a. Suppose that P is positively calculable. We define a decision method for a 3-ary total predicate Q. To calculate $Q(b, a, x)$, we do x steps in the calculation of $P(\alpha, a)$. However, we do not use an oracle for α. Instead, when a value $\alpha(i)$ is required, we supply the value $(b)_i$ if $i < x$ and stop the calculation if $i \geqslant x$. If this calculation leads to the conclusion that $P(\alpha, a)$ is true, then $Q(b, a, x)$ is true; otherwise, $Q(b, a, x)$ is false. Now the calculation of $Q(\bar{\alpha}(x), a, x)$ is simply a part of the correct calculation of $P(\alpha, a)$; so if $\exists x Q(\bar{\alpha}(x), a, x)$, then $P(\alpha, a)$. Conversely, suppose that $P(\alpha, a)$. Choose x so large that the calculation of $P(\alpha, a)$ requires less than x steps and the arguments for which values of α are required in the calculation are all less than x. Then $Q(\bar{\alpha}(x), a, x)$.

Now suppose that there is a calculable Q such that $P(\alpha, a) \leftrightarrow \exists x Q(\bar{\alpha}(x), a, x)$ for all α and a. For the xth step in the calculation of $P(\alpha, a)$, we use the oracle for α to calculate $\bar{\alpha}(x)$ and then calculate $Q(\bar{\alpha}(x), a, x)$. If $Q(\bar{\alpha}(x), a, x)$ is true, we conclude that $P(\alpha, a)$ is true; otherwise, we proceed to the next step. Clearly this gives the correct answer if $P(\alpha, a)$ and no answer if $\neg P(\alpha, a)$.

An (m, n)-ary total relation P is *recursively enumerable* if there is a recursive $(m + n + 1)$-ary total predicate Q such that

$$P(\mathfrak{A}) \leftrightarrow \exists x Q(\bar{\mathfrak{A}}(x), x)$$

for all $\mathfrak{A}$. By the above and Church's thesis, a relation is recursively enumerable iff it is positively calculable. Note that our definition agrees with our previous definition of recursive enumerability for predicates.

If F is an (m, n)-ary partial functional, we define an $(m, n + 1)$-ary total relation $\mathfrak{G}_F$, called the *graph* of F, by

$$\mathfrak{G}_F(\mathfrak{A}, a) \leftrightarrow F(\mathfrak{A}) \simeq a.$$

We can extend (1) of §7.1 to this case:

$$\forall y(\mathfrak{G}_F(\mathfrak{A}, y) \leftrightarrow \exists x R(\mathfrak{A}, y, x)) \rightarrow F(\mathfrak{A}) \simeq (\mu z R(\mathfrak{A}, (z)_0, (z)_1))_0. \tag{1}$$

We can then prove as in the case of partial functions that a partial functional is calculable iff its graph is positively calculable.

A partial functional is *recursive* if its graph is recursively enumerable. A partial relation is *recursive* if its representing partial functional is recursive. Then by the above and Church's thesis, a partial functional or relation is recursive iff it is calculable.

7.3 PROPERTIES OF RECURSIVE FUNCTIONALS

We begin with some rules for obtaining recursively enumerable relations.

RE1. If Q is recursively enumerable, and P is defined by

$$P(\alpha_1, \ldots, \alpha_m, \mathfrak{a}) \leftrightarrow Q(\alpha_{i_1}, \ldots \alpha_{i_r}, F_1(\mathfrak{a}), \ldots, F_k(\mathfrak{a})),$$

where $i_1, \ldots, i_r$ are among $1, \ldots, m$ and $F_1, \ldots, F_k$ are recursive total functions, then P is recursively enumerable.

Proof. For some recursive predicate R,

$$Q(\mathfrak{A}) \leftrightarrow \exists x R(\overline{\mathfrak{A}}(x), x).$$

Hence

$$P(\alpha_1, \ldots, \alpha_m, \mathfrak{a}) \leftrightarrow \exists x R(\bar{\alpha}_{i_1}(x), \ldots, \bar{\alpha}_{i_r}(x), F_1(\mathfrak{a}), \ldots, F_k(\mathfrak{a}), x)$$
$$\leftrightarrow \exists x R'(\bar{\alpha}_1(x), \ldots, \bar{\alpha}_m(x), \mathfrak{a}, x),$$

where R' is the recursive predicate defined by

$$R'(b_1, \ldots, b_m, \mathfrak{a}, x) \leftrightarrow R(b_{i_1}, \ldots, b_{i_r}, F_1(\mathfrak{a}), \ldots, F_k(\mathfrak{a}), x).$$

Thus P is recursively enumerable.

Suppose that the relation P has an explicit definition using only variables and symbols for recursively enumerable relations and recursive functions. Suppose that the function variables occur only as arguments to the recursively enumerable relations, and that the symbols for recursive functions do not occur as arguments to these relations. Then P is recursively enumerable. For the definition must have the form

$$P(\alpha_1, \ldots, \alpha_m, \mathfrak{a}) \leftrightarrow Q(\alpha_{i_1}, \ldots, \alpha_{i_r}, A_1, \ldots, A_k),$$

where A_i consists of number variables and symbols for recursive functions. We then define $F_i(\mathfrak{a}) = A_i$ and apply RE1.

As a particular case, every recursive relation P is recursively enumerable; for $P(\mathfrak{A}) \leftrightarrow \mathfrak{G}_{K_P}(\mathfrak{A}, 0)$, and $\mathfrak{G}_{K_P}$ is recursively enumerable.

If $\mathfrak{A}$ is $\alpha_1, \ldots, \alpha_m, a_1, \ldots, a_n$, we write $In(\overline{\mathfrak{A}}(x), y)$ for

$$In(\overline{\alpha_1}(x), y), \ldots, In(\overline{\alpha_m}(x), y), a_1, \ldots, a_n.$$

Then if $y \leq x$, $In(\overline{\mathfrak{A}}(x), y)$ is just $\overline{\mathfrak{A}}(y)$.

RE2. If Q is recursively enumerable and P is defined by

$$P(\mathfrak{A}) \leftrightarrow \exists x Q(\mathfrak{A}, x),$$

then P is recursively enumerable.

Proof. We have $Q(\mathfrak{A}, x) \leftrightarrow \exists y R(\overline{\mathfrak{A}}(y), x, y)$ with R recursive. Then

$$P(\mathfrak{A}) \leftrightarrow \exists x \exists y R(\overline{\mathfrak{A}}(y), x, y). \tag{1}$$

Now as z runs through all numbers, $((z)_0, (z)_1)$ runs through all pairs (x, y) of numbers. Hence we may rewrite (1) as

$$P(\mathfrak{A}) \leftrightarrow \exists z R(\overline{\mathfrak{A}}((z)_1), (z)_0, (z)_1). \tag{2}$$

Since $(z)_1 \leqslant z$ by (8) of §6.4,

$$\overline{\mathfrak{A}}((z)_1) = In(\overline{\mathfrak{A}}(z), (z)_1).$$

Hence
$$P(\mathfrak{A}) \leftrightarrow \exists z R\big(In(\overline{\mathfrak{A}}(z), (z)_1), (z)_0, (z)_1\big)$$
$$\leftrightarrow \exists z R'\big(\overline{\mathfrak{A}}(z), z\big),$$

where R' is the recursive predicate defined by

$$R'(\mathfrak{a}, z) \leftrightarrow R\big(In(\mathfrak{a}, (z)_1), (z)_0, (z)_1\big).$$

Thus P is recursively enumerable.

The step from (1) to (2) is called *contraction of quantifiers.* It may also be used to contract two adjacent universal quantifiers to a single universal quantifier.

It follows from RE2 that we may use existential quantifiers on number variables in explicit definitions of recursively enumerable relations.

RE3. If Q and R are recursively enumerable, then $Q \vee R$ and $Q \;\&\; R$ are recursively enumerable.

Proof. Let P be $Q \vee R$. Let $Q(\mathfrak{A}) \leftrightarrow \exists x Q_1\big(\overline{\mathfrak{A}}(x), x\big)$, $R(\mathfrak{A}) \leftrightarrow \exists y R_1\big(\overline{\mathfrak{A}}(y), y\big)$ with Q_1 and R_1 recursive. Then by the prenex operations and contraction of quantifiers,

$$P(\mathfrak{A}) \leftrightarrow \exists x Q_1\big(\overline{\mathfrak{A}}(x), x\big) \vee \exists y R_1\big(\overline{\mathfrak{A}}(y), y\big)$$
$$\leftrightarrow \exists x\, \exists y\big(Q_1\big(\overline{\mathfrak{A}}(x), x\big) \vee R_1\big(\overline{\mathfrak{A}}(y), y\big)\big)$$
$$\leftrightarrow \exists z\big(Q_1\big(\overline{\mathfrak{A}}((z)_0), (z)_0\big) \vee R_1\big(\overline{\mathfrak{A}}((z)_1), (z)_1\big)\big)$$
$$\leftrightarrow \exists z\big(Q_1\big(In(\overline{\mathfrak{A}}(z), (z)_0), (z)_0\big) \vee R_1\big(In(\overline{\mathfrak{A}}(z), (z)_1), (z)_1\big)\big).$$

It follows that P is recursively enumerable. We treat $Q \;\&\; R$ similarly.

It follows from RE3 that we may use $\vee$ and & in explicit definitions of recursively enumerable relations.

We note that

$$\forall x_{x<a}\, \exists y P(x, y, a) \leftrightarrow \exists z\, \forall x_{x<a} P\big(x, (z)_x, a\big). \tag{3}$$

For both sides say that there are numbers $y_0, y_1, \ldots, y_{a-1}$ such that $P(x, y_x, a)$ for all $x < a$.

RE4. If Q is recursively enumerable and P is defined by

$$P(a, \mathfrak{A}) \leftrightarrow \forall x_{x<a} Q(\mathfrak{A}, x),$$

then P is recursively enumerable.

Proof. Let $Q(\mathfrak{A}, x) \leftrightarrow \exists y R\big(\overline{\mathfrak{A}}(y), x, y\big)$ with R recursive. Then, using (3),

$$P(a, \mathfrak{A}) \leftrightarrow \forall x_{x<a}\, \exists y R\big(\overline{\mathfrak{A}}(y), x, y\big)$$
$$\leftrightarrow \exists z\, \forall x_{x<a} R\big(\overline{\mathfrak{A}}((z)_x), x, (z)_x\big)$$
$$\leftrightarrow \exists z\, \forall x_{x<a} R\big(In(\overline{\mathfrak{A}}(z), (z)_x), x, (z)_x\big).$$

It follows that P is recursively enumerable.

It follows from RE4 that we may use bounded universal quantifiers in explicit definitions of recursively enumerable relations.

We shall now generalize R1 through R14 to partial functionals and relations. Since the generalizations include the previous results, we shall continue to call them R1 through R14. When the generalization is straightforward, we omit the proof, and sometimes the statement, of the rule.

If $1 \leqslant i \leqslant n$, we define an (m, n)-ary functional $I_i^{m,n}$ by

$$I_i^{m,n}(\alpha_1, \ldots, \alpha_m, a_1, \ldots, a_n) = a_i.$$

We also define a $(1, 1)$-ary functional Ap by

$$Ap(\alpha, a) = \alpha(a).$$

R1. The functionals $I_i^{m,n}$, $+$, $\cdot$, $K_<$, and Ap are recursive.

Proof. We need only consider $I_i^{m,n}$ and Ap. Writing I for $I_i^{m,n}$, we have

$$\mathfrak{G}_I(\alpha_1, \ldots, \alpha_m, a_1, \ldots, a_n, b) \leftrightarrow a_i = b.$$

Thus $\mathfrak{G}_I$ is recursively enumerable. Also

$$\begin{aligned}\mathfrak{G}_{Ap}(\alpha, a, b) &\leftrightarrow \alpha(a) = b \\ &\leftrightarrow \exists x\big(x > a \ \&\ (\bar{\alpha}(x))_a = b\big).\end{aligned}$$

Hence $\mathfrak{G}_{Ap}$ is recursively enumerable.

R2. If $G, H_1, \ldots, H_k$ are recursive, and F is defined by

$$\begin{aligned}&F(\alpha_1, \ldots, \alpha_m, \mathfrak{a}) \\ &\quad \simeq G\big(\alpha_{i_1}, \ldots, \alpha_{i_r}, H_1(\alpha_1, \ldots, \alpha_m, \mathfrak{a}), \ldots, H_k(\alpha_1, \ldots, \alpha_m, \mathfrak{a})\big),\end{aligned}$$

where $i_1, \ldots, i_r$ are among $1, \ldots, m$, then F is recursive.

Proof. We have

$$\begin{aligned}\mathfrak{G}_F(\alpha_1, \ldots, \alpha_m, \mathfrak{a}, a) \leftrightarrow \exists z_1 \ldots \exists z_k\big(&\mathfrak{G}_{H_1}(\alpha_1, \ldots, \alpha_m, \mathfrak{a}, z_1) \ \&\ \cdots \\ &\&\ \mathfrak{G}_{H_k}(\alpha_1, \ldots, \alpha_m, \mathfrak{a}, z_k) \ \&\ \mathfrak{G}_G(\alpha_{i_1}, \ldots, \alpha_{i_r}, z_1, \ldots, z_k, a)\big).\end{aligned}$$

Hence by the results obtained above, $\mathfrak{G}_F$ is recursively enumerable.

R3. If G is recursive and F is defined by

$$F(\mathfrak{A}) \simeq \mu x\big(G(\mathfrak{A}, x) = 0\big),$$

then F is recursive.

Proof. We have

$$\mathfrak{G}_F(\mathfrak{A}, a) \leftrightarrow \mathfrak{G}_G(\mathfrak{A}, a, 0) \ \&\ \forall x_{x<a} \exists y\big(y > 0 \ \&\ \mathfrak{G}_G(\mathfrak{A}, x, y)\big).$$

Hence by the results obtained above, $\mathfrak{G}_F$ is recursively enumerable.

R4. If $Q, H_1, \ldots, H_k$ are recursive and P is defined by

$$\begin{aligned}&P(\alpha_1, \ldots, \alpha_m, \mathfrak{a}) \\ &\quad \rightleftarrows Q\big(\alpha_{i_1}, \ldots, \alpha_{i_r}, H_1(\alpha_1, \ldots, \alpha_m, \mathfrak{a}), \ldots, H_k(\alpha_1, \ldots, \alpha_m, \mathfrak{a})\big),\end{aligned}$$

where $i_1, \ldots, i_r$ are among $1, \ldots, m$, then P is recursive.

R5. If P is recursive and F is defined by

$$F(\mathfrak{A}) \simeq \mu x P(\mathfrak{A}, x),$$

then F is recursive.

We can use R1 through R5 much as before to show that partial functionals and relations defined explicitly are recursive. For example, suppose that F is defined by

$$F(\alpha, \beta, x) \simeq \beta(G(\alpha, x) + \alpha(x)),$$

where G is recursive. We can then define successively the recursive partial functionals

$$\begin{aligned}
F_1(\alpha, \beta, x) &\simeq G(\alpha, x) \simeq G(\alpha, I_1^{2,1}(\alpha, \beta, x)),\\
F_2(\alpha, \beta, x) &\simeq \alpha(x) \simeq Ap(\alpha, x),\\
F_3(\alpha, \beta, x) &\simeq G(\alpha, x) + \alpha(x) \simeq F_1(\alpha, \beta, x) + F_2(\alpha, \beta, x),\\
F(\alpha, \beta, x) &\simeq \beta(G(\alpha, x) + \alpha(x)) \simeq Ap(\beta, F_3(\alpha, \beta, x)).
\end{aligned}$$

In extending R7, it is understood that $P \vee Q$ is defined by

$$(P \vee Q)(\mathfrak{A}) \rightleftarrows P(\mathfrak{A}) \vee Q(\mathfrak{A}),$$

and hence is defined at $\mathfrak{A}$ iff both $P(\mathfrak{A})$ and $Q(\mathfrak{A})$ are defined. We then have

$$K_{P \vee Q}(\mathfrak{A}) \simeq K_P(\mathfrak{A}) \cdot K_Q(\mathfrak{A}).$$

Similar remarks apply to $\neg P$, $P \rightarrow Q$, $P \ \& \ Q$, and $P \leftrightarrow Q$.

We can treat bounded μ-operators and bounded quantifiers as before. However, a bounded μ-operator or bounded quantifier on x has its usual meaning only when the sentence following is defined for all x.

R12. Let $G_1, \ldots, G_k$ be recursive. Let $R_1, \ldots, R_k$ be either recursive partial relations or recursively enumerable relations such that for each $\mathfrak{A}$, at most one of $R_1(\mathfrak{A}), \ldots, R_k(\mathfrak{A})$ is defined and true. If F is defined by

$$\begin{aligned}
F(\mathfrak{A}) &\simeq G_1(\mathfrak{A}) \quad \textit{if} \quad R_1(\mathfrak{A}),\\
&\vdots\\
&\simeq G_k(\mathfrak{A}) \quad \textit{if} \quad R_k(\mathfrak{A})
\end{aligned}$$

(where it is understood that $F(\mathfrak{A})$ is undefined if none of $R_1(\mathfrak{A}), \ldots, R_k(\mathfrak{A})$ is defined and true), then F is recursive.

Proof. If R_i is a recursive partial relation, and R'_i is defined by

$$R'_i(\mathfrak{A}) \leftrightarrow \mathfrak{G}_{K_{R_i}}(\mathfrak{A}, 0),$$

then R'_i is recursively enumerable and $R'_i(\mathfrak{A})$ is true iff $R_i(\mathfrak{A})$ is defined and true. Hence we may as well suppose that the R_i are recursively enumerable relations. Then

$$\mathfrak{G}_F(\mathfrak{A}, a) \leftrightarrow (\mathfrak{G}_{G_1}(\mathfrak{A}, a) \ \& \ R_1(\mathfrak{A})) \vee \cdots \vee (\mathfrak{G}_{G_k}(\mathfrak{A}, a) \ \& \ R_k(\mathfrak{A}));$$

so $\mathfrak{G}_F$ is recursively enumerable.

R13. Let $Q_1, \ldots, Q_k$ be recursive. Let $R_1, \ldots, R_k$ be either recursive partial relations or recursively enumerable relations such that for each $\mathfrak{A}$, at most one of $R_1(\mathfrak{A}), \ldots, R_k(\mathfrak{A})$ is defined and true. If P is defined by

$$\begin{aligned} P(\mathfrak{A}) &\simeq Q_1(\mathfrak{A}) \quad \text{if } R_1(\mathfrak{A}), \\ &\vdots \\ &\simeq Q_k(\mathfrak{A}) \quad \text{if } R_k(\mathfrak{A}) \end{aligned}$$

(where it is understood that $P(\mathfrak{A})$ is undefined if none of $R_1(\mathfrak{A}), \ldots, R_k(\mathfrak{A})$ is defined and true), then P is recursive.

We define $\bar{F}$ by

$$\bar{F}(b, \mathfrak{A}) \simeq \langle F(0, \mathfrak{A}), \ldots, F(b-1, \mathfrak{A}) \rangle.$$

The explicit definition (14) of §6.4 is still correct; so if F is recursive, then $\bar{F}$ is recursive. The definition (15) of §6.4 does not hold when F is partial. We can extend R14 in the obvious manner to recursive partial functionals.

We note that

$$\overline{Ap}(\alpha, a) = \bar{\alpha}(a).$$

Hence by using $\overline{Ap}$ in the way that we formerly used Ap, we see that barred function variables may be used in explicit definitions of recursive partial functionals and relations.

A consequence of our results on explicit definitions is that if $G(\mathfrak{A}, x)$ and $H(\mathfrak{A})$ are recursive, and we substitute $H(\mathfrak{A})$ for x in $G(\mathfrak{A}, x)$, then we obtain a recursive partial functional of $\mathfrak{A}$. We are going to prove a similar substitution rule for function variables. Clearly we cannot substitute $H(\mathfrak{A})$ for α, since $H(\mathfrak{A})$ is a number. We therefore first introduce some notation.

Let $..x..$ be an expression which, when defined, represents a number. We then use $\lambda x..x..$ to designate the unary partial function F defined by $F(x) \simeq ..x..$. Note that x is bound in $\lambda x..x..$. For example, $\lambda x(x+y)$ is the function F defined by $F(x) = x + y$; what function this is depends upon the value of y but not upon the value of x. The partial function $\lambda x..x..$ is total if $..x..$ is defined for all x.

Substitution Theorem. If G and H are recursive partial functionals, then there is a recursive partial functional F such that

$$F(\mathfrak{A}) \simeq G(\lambda x H(x, \mathfrak{A}), \mathfrak{A})$$

for all $\mathfrak{A}$ for which $\lambda x H(x, \mathfrak{A})$ is a total function.

Proof. Define F' by

$$F'(\mathfrak{A}) \simeq G(\lambda x H(x, \mathfrak{A}), \mathfrak{A}).$$

Let R be a recursive predicate such that

$$\mathfrak{G}_G(\alpha, \mathfrak{A}, y) \leftrightarrow \exists x R(\bar{\alpha}(x), \bar{\mathfrak{A}}(x), y, x).$$

Then for $\mathfrak{A}$ such that $\lambda x H(x, \mathfrak{A})$ is total,

$$\mathfrak{G}_{F'}(\mathfrak{A}, y) \leftrightarrow \exists x R(\overline{H}(x, \mathfrak{A}), \overline{\mathfrak{A}}(x), y, x);$$

so by (1) of §7.2,

$$F'(\mathfrak{A}) \simeq (\mu z R(\overline{H}((z)_1, \mathfrak{A}), \overline{\mathfrak{A}}((z)_1), (z)_0, (z)_1))_0.$$

We may therefore define F explicitly by setting $F(\mathfrak{A})$ equal to the right-hand side of the last equation.

Remark. If $\mathfrak{A}$ is such that $\lambda x H(x, \mathfrak{A})$ is not total, then $F'(\mathfrak{A})$ is certainly undefined; but $F(\mathfrak{A})$ may be defined.

Now consider an explicit definition

$$F(\mathfrak{A}) \simeq \_\_\_ \lambda x(\ldots x \ldots) \_\_\_ \tag{4}$$

involving λ, and suppose that all the other symbols appearing have been seen to be allowed in explicit definitions of recursive partial functionals. We may then define recursive partial functionals G and H by

$$G(\alpha, \mathfrak{A}) \simeq \_\_\_ \alpha \_\_\_,$$
$$H(x, \mathfrak{A}) \simeq \ldots x \ldots .$$

Then (4) becomes

$$F(\mathfrak{A}) \simeq G(\lambda x H(x, \mathfrak{A}), \mathfrak{A}).$$

We cannot conclude that F is recursive; but by the substitution theorem, there is a recursive F such that (4) holds whenever $\ldots x \ldots$ is defined for all x. If $\ldots x \ldots$ is defined for all values of x and $\mathfrak{A}$, we can conclude that the F defined by (4) is recursive. Entirely similar remarks apply to the use of λ in explicit definitions of recursive partial relations.

We write $(\alpha)_i$ for the function $\lambda x \alpha(\langle i, x \rangle)$. In view of the above, $(\alpha)_i$ may be used in explicit definitions of recursive partial functionals and relations. If $\alpha_0, \alpha_1, \ldots$ is an infinite sequence of functions, then there is an α such that $(\alpha)_i = \alpha_i$ for all i; for we can define α by $\alpha(x) = \alpha_{(x)_0}((x)_1)$.

7.4 INDICES

We are now going to assign a number to each recursive partial functional in such a way that the partial functional may be recovered from the number.

Let $\mathbf{z}_1, \mathbf{z}_2, \ldots$ be the variables of N in alphabetical order. If $\mathbf{A}$ is a formula of N, let $E_n(\mathbf{A})$ be the set of n-tuples $(a_1, \ldots, a_n)$ such that

$$\vdash_N \mathbf{A}_{\mathbf{z}_1, \ldots, \mathbf{z}_n}[\mathbf{k}_{a_1}, \ldots, \mathbf{k}_{a_n}].$$

Exactly as in §6.8, we can show that every recursive n-ary total predicate is $E_n(\mathbf{A})$ for some $\mathbf{A}$.

We shall now obtain an explicit definition of $E_n(\mathbf{A})$. Define a recursive function S_n by

$$S_n(e, a) = Sub(e, \langle 2n \rangle, Num(a)).$$

Then $$S_n(\ulcorner \mathbf{A} \urcorner, a) = \ulcorner \mathbf{A}_{\mathbf{z}_{n+1}}[\mathbf{k}_a] \urcorner.$$

Now we shall define a recursive function Sb_n such that

$$Sb_n(\ulcorner \mathbf{A} \urcorner, a_1, \ldots, a_n) = \ulcorner \mathbf{A}_{\mathbf{z}_1, \ldots, \mathbf{z}_n}[\mathbf{k}_{a_1}, \ldots, \mathbf{k}_{a_n}] \urcorner.$$

Proceeding by induction on n, we define

$$Sb_0(e) = e$$

and

$$Sb_{n+1}(e, a_1, \ldots, a_{n+1}) = Sb_n\big(S_n(e, a_{n+1}), a_1, \ldots, a_n\big). \tag{1}$$

We then have

$$(a_1, \ldots, a_n) \in E_n(\mathbf{A}) \leftrightarrow \exists v\, Pr_N\big(Sb_n(\ulcorner \mathbf{A} \urcorner, a_1, \ldots, a_n), v\big). \tag{2}$$

Now let F be a recursive (m, n)-ary partial functional. We can then choose a recursive predicate R such that

$$\mathfrak{G}_F(\mathfrak{A}, y) \leftrightarrow \exists w R\big(y, w, \overline{\mathfrak{A}}(w)\big).$$

Choose $\mathbf{A}$ so that R is $E_{m+n+2}(\mathbf{A})$, and let $f = \ulcorner \mathbf{A} \urcorner$. Then by (2),

$$\begin{aligned}\mathfrak{G}_F(\mathfrak{A}, y) &\leftrightarrow \exists w\, \exists v\, Pr_N\big(Sb_{m+n+2}\big(f, y, w, \overline{\mathfrak{A}}(w)\big), v\big)\\ &\leftrightarrow \exists x\, Pr_N\big(Sb_{m+n+2}\big(f, y, (x)_0, \overline{\mathfrak{A}}\big((x)_0\big)\big), (x)_1\big)\end{aligned}$$

by contraction of quantifiers. Hence by (1) of §7.2,

$$F(\mathfrak{A}) \simeq \big(\mu z Pr_N\big(Sb_{m+n+2}\big(f, (z)_0, (z)_{1,0}, \overline{\mathfrak{A}}\big((z)_{1,0}\big)\big), (z)_{1,1}\big)\big)_0.$$

We define a recursive relation $T_{m,n}$ and a recursive function U by

$$T_{m,n}(f, \mathfrak{A}, z) \leftrightarrow Pr_N\big(Sb_{m+n+2}\big(f, (z)_0, (z)_{1,0}, \overline{\mathfrak{A}}\big((z)_{1,0}\big)\big), (z)_{1,1}\big),$$
$$U(z) = (z)_0.$$

(We write T_n for $T_{0,n}$.) We then have

$$F(\mathfrak{A}) \simeq U\big(\mu z T_{m,n}(f, \mathfrak{A}, z)\big). \tag{3}$$

A number f is an *index* of an (m, n)-ary partial functional F if (3) holds for all $\mathfrak{A}$. We have just proved that every recursive partial functional has an index. Conversely, if f is an index of F, then F has the explicit definition (3) and hence is recursive. We thus have the following result.

Normal Form Theorem (*Kleene*). A partial functional is recursive iff it has an index.

Every number f is an index of a unique (m, n)-ary partial functional, viz., the F defined by (3). We designate this partial functional by $\{f\}^{m,n}$, omitting the superscripts when no confusion results. Thus

$$\{f\}^{m,n}(\mathfrak{A}) \simeq U\big(\mu z T_{m,n}(f, \mathfrak{A}, z)\big) \tag{4}$$

and hence

$$\{f\}^{m,n}(\mathfrak{A}) \textit{ is defined} \leftrightarrow \exists z T_{m,n}(f, \mathfrak{A}, z). \tag{5}$$

It follows from (4) that $\{f\}(\mathfrak{A})$ is a recursive partial functional of the arguments $f, \mathfrak{A}$. This shows that we may use $\{f\}(\mathfrak{A})$ in explicit definitions of recursive partial functionals and relations. It also shows (in view of the definition of a recursive partial functional) that $\{f\}(\mathfrak{A}) \simeq b$ is a recursively enumerable relation of the arguments $f, \mathfrak{A}, b$.

Now let P be a recursively enumerable (m, n)-ary relation; say

$$P(\mathfrak{A}) \leftrightarrow \exists x R(\overline{\mathfrak{A}}(x), x)$$

with R recursive. Define a recursive partial functional F by $F(\mathfrak{A}) \simeq \mu x R(\overline{\mathfrak{A}}(x), x)$, and let p be an index of F. Then $\{p\}(\mathfrak{A})$ is defined iff $P(\mathfrak{A})$. From this and (5),

$$P(\mathfrak{A}) \leftrightarrow \exists z T_{m,n}(p, \mathfrak{A}, z). \tag{6}$$

A number p is an *RE-index* of an (m, n)-ary relation P if (6) holds for all $\mathfrak{A}$. We have seen that every recursively enumerable relation has an *RE*-index. A relation P having an *RE*-index p has the explicit definition (6) and hence is recursively enumerable. Thus:

Enumeration Theorem (*Kleene*). A relation is recursively enumerable iff it has an *RE*-index.

Each number p is an *RE*-index of a unique (m, n)-ary relation P, which is defined by (6). We designate this relation by $W_p^{m,n}$, omitting the superscripts when no confusion results.

If F is an $(m, n + k)$-ary partial functional, we define for each k-tuple $\mathfrak{a}$ an (m, n)-ary partial functional $F_{(\mathfrak{a})}$ by

$$F_{(\mathfrak{a})}(\mathfrak{A}) \simeq F(\mathfrak{A}, \mathfrak{a}).$$

If P is an $(m, n + k)$-ary relation, we define for each k-tuple $\mathfrak{a}$ an (m, n)-ary relation $P_{(\mathfrak{a})}$ by

$$P_{(\mathfrak{a})}(\mathfrak{A}) \leftrightarrow P(\mathfrak{A}, \mathfrak{a}).$$

(We could give a similar definition for partial P, but we shall not have occasion to make use of this.) We say that an $(m, n + 1)$-ary relation P *enumerates* the class of (m, n)-ary relations consisting of $P_{(0)}, P_{(1)}, \ldots$.

We now show that there is a recursive function $S_{m,n,k}$ such that

$$T_{m,n+k}(f, \mathfrak{A}, \mathfrak{a}, z) \leftrightarrow T_{m,n}(S_{m,n,k}(f, \mathfrak{a}), \mathfrak{A}, z). \tag{7}$$

We can take $S_{m,n,0}(f) = f$. Comparing (1) and the definition of $T_{m,n}$, we see that we may define

$$S_{m,n,1}(f, a) = S_{m+n+2}(f, a).$$

Then, proceeding by induction on k, we define

$$S_{m,n,k+1}(f, \mathfrak{a}, a) = S_{m,n,k}(S_{m,n+k,1}(f, a), \mathfrak{a}).$$

We sometimes omit the subscripts on $S_{m,n,k}$.

From (7) we have

$$\{f\}^{m,n+k}(\mathfrak{A}, \mathfrak{a}) \simeq \{S_{m,n,k}(f, \mathfrak{a})\}(\mathfrak{A})$$

and

$$W_p^{m,n+k}(\mathfrak{A}, \mathfrak{a}) \leftrightarrow W_{S_{m,n,k}(p,\mathfrak{a})}(\mathfrak{A}). \tag{8}$$

It follows that if f is an index of F, then $S(f, \mathfrak{a})$ is an index of $F_{(\mathfrak{a})}$; if p is an *RE*-index of P, then $S(p, \mathfrak{a})$ is an *RE*-index of $P_{(\mathfrak{a})}$.

We shall now use indices to obtain another method of defining recursive partial functionals. In an *implicit* definition of a partial functional F, the value of $F(\mathfrak{A})$ is given in terms of F as well as $\mathfrak{A}$. Thus we set

$$F(\mathfrak{A}) \simeq \Phi(F, \mathfrak{A}), \tag{9}$$

where Φ is a mapping whose values are numbers. Of course, this does not really define F; it merely indicates that we are looking for an F such that the equation holds for all $\mathfrak{A}$. There may be no solution, a unique solution, or many solutions.

We want to show that under suitable hypotheses on Φ, there is at least one recursive solution. We cannot adopt the hypothesis that Φ is recursive, since Φ is not a functional. We shall therefore replace F on the right-hand side of (9) by an index of F. In other words, we define a partial functional Φ' by

$$\Phi'(f, \mathfrak{A}) \simeq \Phi(\{f\}, \mathfrak{A}).$$

We can then hypothesize that Φ' is recursive. The conclusion we want now is that there is a recursive F with an index f such that $F(\mathfrak{A}) \simeq \Phi'(f, \mathfrak{A})$ for all $\mathfrak{A}$.

Recursion Theorem (*Kleene*). If G is a recursive $(m, n + 1)$-ary partial functional, then there is a recursive (m, n)-ary partial functional F with an index f such that

$$F(\mathfrak{A}) \simeq G(\mathfrak{A}, f)$$

for all $\mathfrak{A}$.

Proof. Define a recursive H by

$$H(\mathfrak{A}, e) \simeq G(\mathfrak{A}, S_{m,n,1}(e, e)),$$

and let h be an index of H. Let F be $H_{(h)}$ and let $f = S_{m,n,1}(h, h)$. Then f is an index of F, and

$$\begin{aligned} F(\mathfrak{A}) &\simeq H(\mathfrak{A}, h) \\ &\simeq G(\mathfrak{A}, S_{m,n,1}(h, h)) \\ &\simeq G(\mathfrak{A}, f). \end{aligned}$$

The recursion theorem is a valuable tool for showing that partial functionals are recursive. Given an F, we try to construct Φ so that (9) has F as its unique solution and so that the corresponding Φ' is recursive. We can then conclude that F is recursive. For an example of this technique, see Problem 1.

In using the recursion theorem, we shall generally say: define a recursive partial functional F with an index f by $F(\mathfrak{A}) \simeq G(\mathfrak{A}, f)$. Of course F and f are not unique in general; we mean that an F and an f having these properties should be selected. Usually G will be given by an explicit definition or a definition by cases. Hence $F(\mathfrak{A}) \simeq G(\mathfrak{A}, f)$ will have the form of an explicit definition or a definition by cases of F, except that the index f of F can appear on the right-hand side. We shall also allow expressions $F(\ldots)$ to appear on the right-hand side; such an expression is to be regarded as an abbreviation of $\{f\}(\ldots)$. Thus we might define

$$\begin{aligned} F(\mathfrak{A}) &\simeq H(F(\mathfrak{A})) && \textit{if } P(\mathfrak{A}), \\ &\simeq K(\mathfrak{A}, f) && \textit{otherwise}, \end{aligned}$$

where H and K are recursive partial functionals and P is a recursive relation.

We may also use the recursion theorem to define a recursive partial relation P by $P(\mathfrak{A}) \rightleftarrows Q(\mathfrak{A}, p)$ where p is an index of K_P; for this definition is equivalent to $K_P(\mathfrak{A}) \simeq K_Q(\mathfrak{A}, p)$. Again, Q will generally be given by an explicit definition or a definition by cases. We may then allow the right-hand side to contain expressions $P(\ldots)$; such an expression is regarded as an abbreviation of $\{p\}(\ldots) = 0$.

7.5 THE ARITHMETICAL HIERARCHY

We have already seen that the use of quantifiers in explicit definitions leads outside of the class of recursive relations. We are going to study the relations which may be obtained by applying quantifiers to recursive relations. In this section, we consider only quantifiers on number variables; quantifiers on function variables will be considered later.

A relation P is *arithmetical* if it has an explicit definition

$$P(\mathfrak{A}) \leftrightarrow Qx_1 \ldots Qx_n R(\mathfrak{A}, x_1, \ldots, x_n), \tag{1}$$

where R is a recursive relation and each Qx_i is either $\exists x_i$ or $\forall x_i$.

Two quantifiers are *of the same kind* if they are both existential or both universal. If we are given a definition (1) in which there are two adjacent quantifiers of the same kind, we can use contraction of quantifiers to replace them by a single quantifier. Thus if the original definition is

$$P(\mathfrak{A}) \leftrightarrow \exists x\, \forall y\, \forall z\, \exists w R(\mathfrak{A}, x, y, z, w),$$

we have a new definition

$$P(\mathfrak{A}) \leftrightarrow \exists x\, \forall v\, \exists w R'(\mathfrak{A}, x, v, w)$$

where R' is the recursive relation defined by

$$R'(\mathfrak{A}, x, v, w) \leftrightarrow R(\mathfrak{A}, x, (v)_0, (v)_1, w).$$

If $n \geqslant 1$, a relation is Σ_n^0 (Π_n^0) if it has an explicit definition (1) with R recursive in which no two adjacent quantifiers are of the same kind and in which the first quantifier is existential (universal). We have thus seen that every arithmetical relation is either recursive or Σ_n^0 or Π_n^0 for some $n \geqslant 1$. This classification of arithmetical relations is called the *arithmetical hierarchy*.

We now give some rules for defining Σ_n^0 and Π_n^0 relations.

A1. If P is recursive, then P is Σ_n^0 and Π_n^0 for all n. If P is Σ_m^0 or Π_m^0, then P is Σ_n^0 and Π_n^0 for all $n > m$.

Proof. By adding superfluous quantifiers. For example, suppose that P is Σ_2^0 with a definition

$$P(\mathfrak{A}) \leftrightarrow \exists x\, \forall y R(\mathfrak{A}, x, y).$$

To show that P is Σ_3^0 and Π_3^0,

$$\begin{aligned} P(\mathfrak{A}) &\leftrightarrow \exists x\, \forall y\, \exists z R(\mathfrak{A}, x, y) \\ &\leftrightarrow \forall z\, \exists x\, \forall y R(\mathfrak{A}, x, y). \end{aligned}$$

A2. If Q is Σ_n^0 (Π_n^0) and $F_1, \ldots, F_r, G_1, \ldots, G_k$ are recursive functionals, then the relation P defined by

$$P(\mathfrak{A}) \leftrightarrow Q\big(\lambda x F_1(\mathfrak{A}, x), \ldots, \lambda x F_r(\mathfrak{A}, x), G_1(\mathfrak{A}), \ldots, G_k(\mathfrak{A})\big)$$

is Σ_n^0 (Π_n^0).

Proof. Suppose that Q is Σ_2^0 with a definition

$$Q(\mathfrak{A}) \leftrightarrow \exists y\, \forall z R(\mathfrak{A}, y, z).$$

Then $P(\mathfrak{A}) \leftrightarrow \exists y\, \forall z R'(\mathfrak{A}, y, z)$, where R' is the recursive relation defined by

$$R'(\mathfrak{A}, y, z) \leftrightarrow R\big(\lambda x F_1(\mathfrak{A}, x), \ldots, \lambda x F_r(\mathfrak{A}, x), G_1(\mathfrak{A}), \ldots, G_k(\mathfrak{A}), y, z\big).$$

A3. If Q is Σ_n^0 and P is defined by $P(\mathfrak{A}) \leftrightarrow \exists x Q(\mathfrak{A}, x)$, then P is Σ_n^0. If Q is Π_n^0 and P is defined by $P(\mathfrak{A}) \leftrightarrow \forall x Q(\mathfrak{A}, x)$, then P is Π_n^0.

Proof. By contraction of quantifiers. Thus if Q is Σ_2^0 and is defined by

$$Q(\mathfrak{A}, x) \leftrightarrow \exists y\, \forall z R(\mathfrak{A}, x, y, z),$$

then

$$P(\mathfrak{A}) \leftrightarrow \exists x\, \exists y\, \forall z R(\mathfrak{A}, x, y, z),$$

and hence P is Σ_2^0 by contraction of quantifiers.

A4. If P and Q are Σ_n^0 (Π_n^0), then $P \vee Q$ and $P \,\&\, Q$ are Σ_n^0 (Π_n^0).

Proof. Suppose that P and Q are Σ_2^0; say

$$P(\mathfrak{A}) \leftrightarrow \exists x\, \forall y P_1(\mathfrak{A}, x, y)$$

and

$$Q(\mathfrak{A}) \leftrightarrow \exists z\, \forall w Q_1(\mathfrak{A}, z, w).$$

Then by the prenex operations,

$$\begin{aligned} (P \vee Q)(\mathfrak{A}) &\leftrightarrow \exists x\, \forall y P_1(\mathfrak{A}, x, y) \vee \exists z\, \forall w Q_1(\mathfrak{A}, z, w) \\ &\leftrightarrow \exists x\, \exists z\, \forall y\, \forall w\big(P_1(\mathfrak{A}, x, y) \vee Q_1(\mathfrak{A}, z, w)\big). \end{aligned}$$

Then by contraction of quantifiers $P \vee Q$ is Σ_2^0.

A5. If P is Σ_n^0 (Π_n^0), then $\neg P$ is Π_n^0 (Σ_n^0).

Proof. By the prenex operations. For example, if P is Σ_2^0 with a definition

$$P(\mathfrak{A}) \leftrightarrow \exists x \, \forall y R(\mathfrak{A}, x, y),$$

then

$$\begin{aligned} \neg P(\mathfrak{A}) &\leftrightarrow \neg \exists x \, \forall y R(\mathfrak{A}, x, y) \\ &\leftrightarrow \forall x \, \exists y \, \neg R(\mathfrak{A}, x, y). \end{aligned}$$

To treat bounded quantifiers, we need the equivalences

$$\exists x_{x<a} \, \exists y P(x, y, a) \leftrightarrow \exists y \, \exists x_{x<a} P(x, y, a), \tag{2}$$

$$\forall x_{x<a} \, \forall y P(x, y, a) \leftrightarrow \forall y \, \forall x_{x<a} P(x, y, a), \tag{3}$$

$$\forall x_{x<a} \, \exists y P(x, y, a) \leftrightarrow \exists y \, \forall x_{x<a} P\big(x, (y)_x, a\big), \tag{4}$$

$$\exists x_{x<a} \, \forall y P(x, y, a) \leftrightarrow \forall y \, \exists x_{x<a} P\big(x, (y)_x, a\big). \tag{5}$$

The first two are obvious, and (4) is (3) of §7.3. To obtain (5), we put $\neg P$ for P in (4) and bring the negation signs to the front by the prenex operations. We find that the negations of the two sides of (5) are equivalent; so (5) follows.

A6. If P is Σ_n^0 (Π_n^0) and Q and R are defined by

$$Q(a, \mathfrak{A}) \leftrightarrow \exists x_{x<a} P(\mathfrak{A}, x), \qquad R(a, \mathfrak{A}) \leftrightarrow \forall x_{x<a} P(\mathfrak{A}, x),$$

then Q and R are Σ_n^0 (Π_n^0).

Proof. Suppose that P is Σ_2^0; say

$$P(\mathfrak{A}, x) \leftrightarrow \exists y \, \forall z R(\mathfrak{A}, x, y, z).$$

Then by (2) and (5),

$$\begin{aligned} Q(a, \mathfrak{A}) &\leftrightarrow \exists x_{x<a} \, \exists y \, \forall z R(\mathfrak{A}, x, y, z) \\ &\leftrightarrow \exists y \, \exists x_{x<a} \, \forall z R(\mathfrak{A}, x, y, z) \\ &\leftrightarrow \exists y \, \forall z \, \exists x_{x<a} R\big(\mathfrak{A}, x, y, (z)_x\big). \end{aligned}$$

These results can be used to show that various explicitly defined relations are Σ_n^0 or Π_n^0; the technique is the same as that used for recursive partial relations and recursively enumerable relations.

The Σ_1^0 relations are just the recursively enumerable relations. For a Σ_1^0 relation is recursively enumerable by the results of §7.3, and the converse is evident.

If P is a Σ_1^0 relation having at least one function argument, then P has a definition

$$P(\alpha, \mathfrak{A}) \leftrightarrow \exists x R\big(\bar{\alpha}(x), \mathfrak{A}\big) \tag{6}$$

with R recursive. For, since P is recursively enumerable,

$$P(\alpha, \mathfrak{A}) \leftrightarrow \exists x Q\big(\bar{\alpha}(x), \bar{\mathfrak{A}}(x), x\big)$$

with Q recursive; and we may then define R by

$$R(a, \mathfrak{A}) \leftrightarrow Q\big(a, \bar{\mathfrak{A}}(lh(a)), lh(a)\big).$$

From this and A5, we see that if P is a Π_1^0 relation having at least one function argument, then

$$P(\alpha, \mathfrak{A}) \leftrightarrow \neg \exists x Q(\bar{\alpha}(x), \mathfrak{A})$$

with Q recursive. Using prenex operation on the right-hand side, we find that P has a definition

$$P(\alpha, \mathfrak{A}) \leftrightarrow \forall x R(\bar{\alpha}(x), \mathfrak{A}) \tag{7}$$

with R recursive.

Arithmetical Enumeration Theorem (*Kleene*). For each m and n and each $k \geqslant 1$, there is a Σ_k^0 (Π_k^0) $(m, n + 1)$-ary relation which enumerates the set of Σ_k^0 (Π_k^0) (m, n)-ary relations.

Proof. Suppose, e.g., that $k = 3$. If P is Σ_3^0, then

$$P(\mathfrak{A}) \leftrightarrow \exists x \, \forall y \, \exists z R(x, y, z, \mathfrak{A})$$

with R recursive. By the enumeration theorem, there is a p such that

$$\exists z R(x, y, z, \mathfrak{A}) \leftrightarrow \exists z T_{m,n+2}(p, x, y, \mathfrak{A}, z);$$

so

$$P(\mathfrak{A}) \leftrightarrow \exists x \, \forall y \, \exists z T_{m,n+2}(p, x, y, \mathfrak{A}, z).$$

Conversely, the P defined by this equivalence is Σ_3^0. Thus the enumerating relation for Σ_3^0 relations is defined by

$$Q(\mathfrak{A}, p) \leftrightarrow \exists x \, \forall y \, \exists z T_{m,n+2}(p, x, y, \mathfrak{A}, z).$$

By A5, the Π_3^0 relations are just the negations of the Σ_3^0 relations. It follows that $\neg Q$ is the enumerating relation for the Π_3^0 relations.

We can now show that the relations between the Σ_n^0 and Π_n^0 relations given by A1 are the only ones of their type.

Arithmetical Hierarchy Theorem (*Kleene*). For each $n \geqslant 1$, there is a Σ_n^0 unary predicate P which is not Π_n^0 and hence not Σ_m^0 or Π_m^0 for any $m < n$. Then $\neg P$ is Π_n^0 but not Σ_n^0 and hence not Σ_m^0 or Π_m^0 for any $m < n$.

Proof. Let Q be a binary Σ_n^0 predicate which enumerates the set of unary Σ_n^0 predicates, and define P by $P(a) \leftrightarrow Q(a, a)$. Then P is Σ_n^0 by A2, while $\neg P$ is not Σ_n^0 by the diagonal lemma. The required conclusions then follow from A5 and A1.

A relation P is Δ_n^0 if it is both Σ_n^0 and Π_n^0. By A5, P is Δ_n^0 iff both P and $\neg P$ are Σ_n^0, and iff both P and $\neg P$ are Π_n^0. In particular, P is Δ_1^0 iff both P and $\neg P$ are recursively enumerable. Hence by the negation theorem, the Δ_1^0 predicates are just the recursive predicates. We can easily use the proof of the negation theorem to show that the Δ_1^0 relations are just the recursive relations; we shall also refer to this result as the negation theorem. We will obtain a similar characterization of Δ_n^0 predicates for $n > 1$ in the next section.

7.6 RELATIVE RECURSIVENESS

Let Φ be a set of total functions. We define the functions which are *recursive in* Φ (or *recursive relative to* Φ) by a generalized inductive definition consisting of four rules $R0_\Phi$, $R1_\Phi$, $R2_\Phi$, and $R3_\Phi$. The last three rules are obtained from R1, R2, and R3 by replacing *recursive* by *recursive in* Φ. The first rule is:

$R0_\Phi$. Every function in Φ is recursive in Φ.

Now let Φ be a set of total functions and predicates, and let Φ' be the set consisting of the functions in Φ and the representing functions of the predicates in Φ. We say that a function is *recursive in* Φ if it is recursive in Φ'. If Φ consists of $F_1, \ldots, F_m, R_1, \ldots, R_n$, then we say recursive in $F_1, \ldots, F_m, R_1, \ldots, R_n$ for recursive in Φ.

Since all the notions of recursion theory are defined in terms of the notion of a recursive function, we can obtain from each of these notions a notion relative to Φ by replacing *recursive* by *recursive in* Φ in the definition of the notion. This gives the following definitions.

A predicate is *recursive in* Φ if its representing function is recursive in Φ. (This obviously implies that every predicate in Φ is recursive in Φ.) A relation P is *recursively enumerable in* Φ if there is a predicate R recursive in Φ such that

$$P(\mathfrak{A}) \leftrightarrow \exists x R(\overline{\mathfrak{A}}(x), x)$$

for all $\mathfrak{A}$. A partial functional is *recursive in* Φ if its graph is recursively enumerable in Φ. A partial relation is *recursive in* Φ if its representing partial functional is recursive in Φ. We leave to the reader the definitions of arithmetical, Σ_n^0, Π_n^0, and Δ_n^0 in Φ.

Many of the results which we have proved depend only on the fact that the recursive functions satisfy R1 through R3. These results clearly also hold in the relative case. Thus R1 through R14 (in both versions) and RE1 through RE4 extend, as does the substitution theorem. The result that every arithmetical relation is recursive, Σ_n^0, or Π_n^0 and the result that the Σ_1^0 relations are just the recursively enumerable relations extend, as do A1 through A6. The negation theorem also extends. Of course, results which are proved by induction on recursive functions (like the representability theorem) cannot be extended so simply.

If Φ is empty, the functions recursive in Φ are just the recursive functions. Thus relative recursion theory includes ordinary recursion theory as a special case.

Let Ψ be another set of total functions and predicates. We say that Φ is *recursive in* Ψ if every member of Φ is recursive in Ψ.

Transitivity Lemma. If Φ is recursive in Ψ, then every function recursive in Φ is recursive in Ψ.

Proof. By induction on functions recursive in Φ.

Corollary 1. A function recursive in Φ is recursive in every set including Φ. In particular, every recursive function is recursive in Φ for all Φ.

Corollary 2. If a function F is recursive in a set whose members are all either recursive or in Φ, then F is recursive in Φ.

The transitivity lemma extends to recursive partial functionals and relations, recursively enumerable and arithmetical relations, and Σ_n^0, Π_n^0, and Δ_n^0 relations. This means, for example, that if Φ is recursive in Ψ, then every relation arithmetical in Φ is arithmetical in Ψ. The corollaries also extend to these cases.

Finiteness Lemma. If a function is recursive in Φ, it is recursive in some finite subset of Φ.

Proof. By induction on functions recursive in Φ.

The finiteness lemma also extends to all of the notions mentioned above.

We shall now investigate the case in which Φ is finite. It is useful here to suppose that the members of Φ are arranged in some order, so that Φ is a finite sequence rather than a finite set. We then define a unary function Φ^*, called the *contraction* of Φ, as follows. Let $F_0, \ldots, F_{k-1}$ be the sequence obtained from Φ by replacing each function by its contraction and each predicate by the contraction of its representing function, and define

$$\Phi^*(a) = \langle F_0(a), \ldots, F_{k-1}(a) \rangle.$$

Then

$$F_i(x) = (\Phi^*(x))_i.$$

These equations show that Φ^* is recursive in $F_0, \ldots, F_{k-1}$ and that each F_i is recursive in Φ^*. From the contraction formulas, each member of Φ is recursive in $F_0, \ldots, F_{k-1}$ and each F_i is recursive in Φ. When these facts are combined with the transitivity lemma, it follows that Φ^* is recursive in Φ and vice versa. Then by the transitivity lemma, the functions recursive in Φ are just the functions recursive in Φ^*; and similarly for recursive partial functionals, recursively enumerable relations, etc.

Replacement Lemma. A partial functional F is recursive in a finite sequence Φ iff there is a recursive partial functional F' such that $F(\mathfrak{A}) \simeq F'(\Phi^*, \mathfrak{A})$ for all $\mathfrak{A}$. A relation P is recursively enumerable in a finite sequence Φ iff there is a recursively enumerable relation P' such that $P(\mathfrak{A}) \leftrightarrow P'(\Phi^*, \mathfrak{A})$ for all $\mathfrak{A}$.

Proof. If such an F' exists, then the explicit definition

$$F(\mathfrak{A}) \simeq F'(\lambda x \Phi^*(x), \mathfrak{A})$$

shows that F is recursive in Φ^* and hence in Φ. The "if" part of the second half of the lemma is proved similarly.

We now prove the "only if" part when F is a total function. Since F is recursive in Φ^*, we may use induction on functions recursive in Φ^*. To treat $R0_{\Phi^*}$, we note that

$$\Phi^*(a) = Ap(\Phi^*, a).$$

The case of $R1_{\Phi^*}$ is trivial. Since $R2_{\Phi^*}$ and $R3_{\Phi^*}$ are treated similarly, we consider only the latter. Suppose that F is defined by $F(\mathfrak{a}) = \mu x(G(\mathfrak{a}, x) = 0)$. By induction hypothesis,

$$G(\mathfrak{a}, x) = G'(\Phi^*, \mathfrak{a}, x)$$

for a recursive partial functional G'. Defining F' by

$$F'(\alpha, \mathfrak{a}) \simeq \mu x(G'(\alpha, \mathfrak{a}, x) = 0),$$

we have $F(\mathfrak{a}) = F'(\Phi^*, \mathfrak{a})$.

Now suppose that P is recursively enumerable in Φ; say

$$P(\mathfrak{A}) \leftrightarrow \exists x R(\overline{\mathfrak{A}}(x), x)$$

with R recursive in Φ. Choose F' recursive by the above so that

$$K_R(\mathfrak{a}, x) = F'(\Phi^*, \mathfrak{a}, x).$$

Then

$$\begin{aligned} P(\mathfrak{A}) &\leftrightarrow \exists x(F'(\Phi^*, \overline{\mathfrak{A}}(x), x) = 0) \\ &\leftrightarrow \exists x \mathfrak{G}_{F'}(\Phi^*, \overline{\mathfrak{A}}(x), x, 0). \end{aligned}$$

Defining P' by

$$P'(\alpha, \mathfrak{A}) \leftrightarrow \exists x \mathfrak{G}_{F'}(\alpha, \overline{\mathfrak{A}}(x), x, 0),$$

we find that P' is recursively enumerable and $P(\mathfrak{A}) \leftrightarrow P'(\Phi^*, \mathfrak{A})$.

Finally, let F be a partial functional recursive in Φ. Using the result just proved and the fact that recursively enumerable relations are Σ_1^0, we have

$$\mathfrak{G}_F(\mathfrak{A}, a) \leftrightarrow \exists x R'(\Phi^*, \mathfrak{A}, a, x)$$

with R' a recursive relation. Then by (1) of §7.2,

$$F(\mathfrak{A}) \simeq (\mu z R'(\Phi^*, \mathfrak{A}, (z)_0, (z)_1))_0;$$

so we may define F' by

$$F'(\alpha, \mathfrak{A}) \simeq (\mu z R'(\alpha, \mathfrak{A}, (z)_0, (z)_1))_0.$$

Corollary 1. A partial relation P is recursive in a finite sequence Φ iff there is a recursive partial relation P' such that $P(\mathfrak{A}) \rightleftarrows P'(\Phi^*, \mathfrak{A})$ for all $\mathfrak{A}$.

Corollary 2. A relation P is arithmetical (Σ_n^0) (Π_n^0) in a finite sequence Φ iff there is an arithmetical (Σ_n^0) (Π_n^0) relation P' such that $P(\mathfrak{A}) \leftrightarrow P'(\Phi^*, \mathfrak{A})$ for all $\mathfrak{A}$.

Proof. For the Σ_1^0 case, this follows from the theorem (since Σ_1^0 is the same as recursively enumerable). We then get the Π_1^0 case by taking negations. The remaining cases follow from these two cases by adding quantifiers in front.

Remark. The corresponding result does not hold for recursive total functionals, recursive total relations, or Δ_n^0 relations; see Problems 11(d) and 21(b).

We shall now use relative recursiveness to characterize the Δ_n^0 predicates for $n > 1$.

Lemma. A predicate P is Σ_{n+1}^0 iff it is recursively enumerable in the set of Π_n^0 predicates.

Proof. If P is Σ_{n+1}^0, then $P(\mathfrak{a}) \leftrightarrow \exists x Q(\mathfrak{a}, x)$, where Q is Π_n^0; so P is recursively enumerable in the set of Π_n^0 predicates. Now suppose that P is recursively enumerable in this set. The contraction formulas show that the contraction of a Π_n^0 predicate is Π_n^0 and that a predicate is recursive in its contraction. Thus P is recursively enumerable in the set of unary Π_n^0 predicates, and hence, by the finiteness lemma, in a finite sequence Φ of such predicates. Let the predicates in Φ be $R_1, \ldots, R_k$, and let Q_i be the graph of K_{R_i}. Then

$$Q_i(a, b) \leftrightarrow (R_i(a) \;\&\; b = 0) \vee (\neg R_i(a) \;\&\; b = 1);$$

so Q_i is Σ_{n+1}^0 by A1 through A6. Now

$$\mathfrak{G}_{\Phi^*}(a, b) \leftrightarrow b = \langle (b)_0, \ldots, (b)_{k-1} \rangle \;\&\; Q_1(a, (b)_0) \;\&\; \cdots \;\&\; Q_k(a, (b)_{k-1});$$

so $\mathfrak{G}_{\Phi^*}$ is Σ_{n+1}^0.

By the replacement lemma, there is a recursive predicate M such that

$$\begin{aligned} P(\mathfrak{a}) &\leftrightarrow \exists x M(\overline{\Phi^*}(x), \mathfrak{a}, x) \\ &\leftrightarrow \exists x \exists y (y = \overline{\Phi^*}(x) \;\&\; M(y, \mathfrak{a}, x)) \\ &\leftrightarrow \exists x \exists y (Seq(y) \;\&\; lh(y) = x \;\&\; \forall z_{z<x} \mathfrak{G}_{\Phi^*}(z, (y)_z) \;\&\; M(y, \mathfrak{a}, x)). \end{aligned}$$

It follows by A1 through A6 that P is Σ_{n+1}^0.

By combining the lemma with the relativized negation theorem, we get the following result.

Post's Theorem. A predicate is Δ_{n+1}^0 iff it is recursive in the set of Π_n^0 predicates.

Using A5 and the fact that P and $\neg P$ are recursive in each other, we see that we could replace Π_n^0 by Σ_n^0 in Post's theorem.

Let Φ be a finite sequence. A number f is an *index from* Φ of an (m, n)-ary partial functional F if

$$F(\mathfrak{A}) \simeq \{f\}^{m+1,n}(\Phi^*, \mathfrak{A}) \tag{1}$$

for all $\mathfrak{A}$. By the normal form theorem and the replacement lemma, a partial functional is recursive in Φ iff it has an index from Φ. Each f is an index from Φ of a unique (m, n)-ary F, viz., the F defined by (1). We designate this F by $\{f\}^\Phi$.

Again let Φ be a finite sequence. A number p is an *RE-index from* Φ of an (m, n)-ary relation P if

$$P(\mathfrak{A}) \leftrightarrow W_p^{m+1,n}(\Phi^*, \mathfrak{A})$$

for all $\mathfrak{A}$. Then a relation is recursively enumerable in Φ iff it has an *RE*-index from Φ. Again each p is an *RE*-index from Φ of a unique relation; we designate this relation by W_p^Φ.

We have

$$\{f\}^{\Phi}(\mathfrak{A}, \mathfrak{a}) \simeq \{S_{m+1,n,k}(f, \mathfrak{a})\}^{\Phi}(\mathfrak{A}),$$
$$W_p^{\Phi}(\mathfrak{A}, \mathfrak{a}) \leftrightarrow W^{\Phi}_{S_{m+1,n,k}(p,\mathfrak{a})}(\mathfrak{A}).$$

We can then obtain a relativized version of the recursion theorem.

A partial functional is *functionally recursive* if it is recursive in the set of all total functions. We define *functionally recursive* partial relation, *functionally recursively enumerable*, and *functionally arithmetical* similarly. A relation is $\mathbf{\Sigma}_n^0$ if it is Σ_n^0 in the set of all functions; we define $\mathbf{\Pi}_n^0$ and $\mathbf{\Delta}_n^0$ similarly. Clearly every function or predicate is functionally recursive.

Suppose that F is a functionally recursive (m, n)-ary partial functional. By the finiteness lemma, F is recursive in some finite sequence Φ. Let f be an index of F from Φ, and let α be the function defined by $\alpha(0) = f$, $\alpha(n + 1) = \Phi^*(n)$. Then

$$F(\mathfrak{A}) \simeq \{\alpha(0)\}^{m+1,n}(\lambda x\alpha(x + 1), \mathfrak{A}). \tag{2}$$

By a *functional index* of F, we mean a unary function α such that (2) holds for all $\mathfrak{A}$. Since the F defined by (2) is recursive in α and hence functionally recursive, we see that a partial functional is functionally recursive iff it has a functional index. Each unary function α is an index of a unique (m, n)-ary partial functional F, viz., the F defined by (2). We designate this F by $\{\alpha\}^{m,n}$ or simply $\{\alpha\}$.

From (2),

$$\{\alpha\}^{m,n}(\mathfrak{A}) \simeq U(\mu z T_{m+1,n}(\alpha(0), \lambda x\alpha(x + 1), \mathfrak{A}, z)).$$

We define a recursive functional $\mathbf{T}_{m,n}$ by

$$\mathbf{T}_{m,n}(\alpha, \mathfrak{A}, z) \leftrightarrow T_{m+1,n}(\alpha(0), \lambda x\alpha(x + 1), \mathfrak{A}, z).$$

Then

$$\{\alpha\}^{m,n}(\mathfrak{A}) \simeq U(\mu z \mathbf{T}_{m,n}(\alpha, \mathfrak{A}, z)). \tag{3}$$

By a *functional RE-index* of a relation P, we mean a function α such that

$$P(\mathfrak{A}) \leftrightarrow \exists z \mathbf{T}_{m,n}(\alpha, \mathfrak{A}, z)$$

for all $\mathfrak{A}$. Using the proof of the enumeration theorem, we show that a relation is functionally recursively enumerable iff it has a functional *RE*-index.

If P is an $(m + 1, n)$-ary relation, we define for each α an (m, n)-ary relation $P_{(\alpha)}$ by

$$P_{(\alpha)}(\mathfrak{A}) \leftrightarrow P(\mathfrak{A}, \alpha).$$

We say that P *functionally enumerates* the class of relations consisting of the $P_{(\alpha)}$.

We can now obtain an analogue of the arithmetical enumeration theorem for the functionally arithmetic case. From this, we get an analogue of the arithmetical hierarchy theorem for (1, 0)-ary relations. (Of course, we cannot get such an analogue for unary predicates, since every predicate is functionally recursive.)

Projection Lemma. If $F_{(a)}$ is functionally recursive for each a, then F is functionally recursive.

Proof. For each a, let α_a be a functional index of $F_{(a)}$, and choose α so that $(\alpha)_a = \alpha_a$ for all a. Define a recursive G by

$$G(\beta, \mathfrak{A}, a) \simeq \{(\beta)_a\}(\mathfrak{A}).$$

Then

$$\begin{aligned} F(\mathfrak{A}, a) &\simeq F_{(a)}(\mathfrak{A}) \\ &\simeq \{\alpha_a\}(\mathfrak{A}) \\ &\simeq G(\alpha, \mathfrak{A}, a). \end{aligned}$$

Thus F is recursive in α, and hence functionally recursive, by the replacement lemma.

The projection lemma extends to functionally recursive partial relations, since $(K_P)_{(a)}$ is the representing partial functional of $P_{(a)}$. It also extends to $\mathbf{\Sigma}_n^0$ relations (in particular, functionally recursively enumerable relations), $\mathbf{\Pi}_n^0$ relations, and $\mathbf{\Delta}_n^0$ relations, and hence to functionally arithmetical relations. For example, suppose that $P_{(a)}$ is $\mathbf{\Sigma}_2^0$ for all a. Then

$$P_{(a)}(\mathfrak{A}) \leftrightarrow \exists x \, \forall y R_a(\mathfrak{A}, x, y)$$

with R_a functionally recursive. Define

$$R(\mathfrak{A}, x, y, a) \leftrightarrow R_a(\mathfrak{A}, x, y).$$

Then $R_{(a)} = R_a$ is functionally recursive for each a; so R is functionally recursive. Also

$$P(\mathfrak{A}, a) \leftrightarrow \exists x \, \forall y R(\mathfrak{A}, x, y, a);$$

so P is $\mathbf{\Sigma}_2^0$.

7.7 DEGREES

Let Φ be the set of all total functions and predicates. We say two members of Φ are *equivalent* if each is recursive in the other. Using the transitivity lemma, we easily verify that this is an equivalence relation. The equivalence classes are called *degrees of recursive unsolvability*, or simply *degrees*. We use small boldface letters to designate degrees.

Roughly speaking, two functions or predicates are equivalent if they are equally difficult to calculate. Thus the degree of a function or predicate is a measure of the difficulty of calculating it.

Clearly P is equivalent to K_P and F is equivalent to $\mathfrak{G}_F$. Hence every degree contains a function and a predicate. The contraction formulas show that every function or predicate is equivalent to its contraction; so every degree contains a unary function and a set. For this reason, we often deal only with sets.

If $\mathbf{a}$ and $\mathbf{b}$ are degrees, we write $\mathbf{a} \leqslant \mathbf{b}$ to mean that there is a set A in $\mathbf{a}$ and a set B in $\mathbf{b}$ such that A is recursive in B. If this is the case, then any function or predicate in $\mathbf{a}$ is recursive in any function or predicate in $\mathbf{b}$ by the transitivity lemma.

Using this and the transitivity lemma, we easily verify that $\leqslant$ has the basic properties of a partial ordering:

$$\mathbf{a} \leqslant \mathbf{a},$$
$$\mathbf{a} \leqslant \mathbf{b} \rightarrow \mathbf{b} \leqslant \mathbf{a} \rightarrow \mathbf{a} = \mathbf{b},$$
$$\mathbf{a} \leqslant \mathbf{b} \rightarrow \mathbf{b} \leqslant \mathbf{c} \rightarrow \mathbf{a} \leqslant \mathbf{c}.$$

We write $\mathbf{a} < \mathbf{b}$ to mean that $\mathbf{a} \leqslant \mathbf{b}$ and $\mathbf{a} \neq \mathbf{b}$.

The set of all recursive functions and predicates is clearly a degree; we designate this degree by $\mathbf{0}$. We have $\mathbf{0} \leqslant \mathbf{a}$ for all $\mathbf{a}$; that is, $\mathbf{0}$ is the smallest degree.

We say that $\mathbf{a}$ is *recursively enumerable in* $\mathbf{b}$ if there is a set A in $\mathbf{a}$ and a set B in $\mathbf{b}$ such that A is recursively enumerable in B. Then by the transitivity lemma A is recursively enumerable in every set in $\mathbf{b}$. A degree $\mathbf{a}$ is *recursively enumerable* if it is recursively enumerable in $\mathbf{0}$, that is, if it contains a recursively enumerable set. If $\mathbf{a}$ is recursively enumerable in $\mathbf{b}$ and $\mathbf{b} \leqslant \mathbf{c}$, then $\mathbf{a}$ is recursively enumerable in $\mathbf{c}$ by the transitivity lemma. In particular, a recursively enumerable degree is recursively enumerable in every degree.

Among the degrees recursively enumerable in $\mathbf{a}$, there is a largest. For let A be a set in $\mathbf{a}$, and define

$$B(a) \leftrightarrow \exists x T_{1,1}((a)_0, K_A, (a)_1, x).$$

Then the degree $\mathbf{b}$ of B is recursively enumerable in $\mathbf{a}$. We show that it is the largest such degree. Let $\mathbf{c}$ be recursively enumerable in $\mathbf{a}$. Then $\mathbf{c}$ contains a set C which is recursively enumerable in A. By the relativized enumeration theorem, there is an e such that

$$\begin{aligned} C(a) &\leftrightarrow \exists x T_{1,1}(e, K_A, a, x) \\ &\leftrightarrow B(\langle e, a \rangle) \end{aligned}$$

for all a. Thus C is recursive in B; so $\mathbf{c} \leqslant \mathbf{b}$.

The largest degree which is recursively enumerable in $\mathbf{a}$ is called the *jump* of $\mathbf{a}$, and is designated by $\mathbf{a}'$. Thus $\mathbf{0}'$ is the largest recursively enumerable degree. If $\mathbf{a} \leqslant \mathbf{b}$, then $\mathbf{a}'$ is recursively enumerable in $\mathbf{a}$ and hence in $\mathbf{b}$; so $\mathbf{a}' \leqslant \mathbf{b}'$. Since every set recursive in A is recursively enumerable in A but not conversely (by the relativized arithmetical hierarchy theorem), we have $\mathbf{a} < \mathbf{a}'$. It follows that there is no largest degree.

Two questions are suggested by the above. The first is: Is the set of degrees linearly ordered? The second, known as Post's problem, is: Are there any recursively enumerable degrees other than $\mathbf{0}$ and $\mathbf{0}'$? We shall now prove a theorem which answers both questions.

Friedberg-Muchnik Theorem. There are recursively enumerable sets A and B such that A is not recursive in B and B is not recursive in A.

Assuming this theorem, let $\mathbf{a}$ and $\mathbf{b}$ be the degrees of A and B respectively. Then $\mathbf{a} \leqslant \mathbf{b}$ and $\mathbf{b} \leqslant \mathbf{a}$ are both false; so the set of degrees is not linearly ordered. Since $\mathbf{0} \leqslant \mathbf{c} \leqslant \mathbf{0}'$ for every recursively enumerable degree $\mathbf{c}$, $\mathbf{a}$ and $\mathbf{b}$ are recursively enumerable degrees different from $\mathbf{0}$ and $\mathbf{0}'$; this solves Post's problem.

We first give a general description of the proof. We shall construct A and B in stages. At each stage, we shall either do nothing or put exactly one number in A or in B. Given n, it will be possible to find out exactly what is done at the nth stage. It follows from this that A and B are positively calculable and hence recursively enumerable.

To ensure that B is not recursive in A, we must ensure that K_B is different from all the $\{e\}^A$. Our idea is this. We continually try to compute $\{e\}^A(x_{2e})$ for a certain number x_{2e}. If we succeed and the value is 1, we put x_{2e} in B. Thus $K_B(x_{2e}) = 0$ iff $\{e\}^A(x_{2e}) \simeq 1$; so K_B is not $\{e\}^A$. Similarly, we guarantee that K_A is not $\{e\}^B$ by putting a certain number x_{2e+1} in A iff $\{e\}^B(x_{2e+1}) \simeq 1$.

The difficulty with this idea is that at the nth stage, we do not know A, but only the finite set A_n of numbers which have been put in A before the nth stage. We therefore try to compute $\{e\}^{A_n}(x_{2e})$; if we succeed and the value is 1, then we put x_{2e} in B. The trouble with this is that we may later put more numbers in A and thus change $\{e\}^A(x_{2e})$ to 0; we will then have $\{e\}^A(x_{2e}) = K_B(x_{2e})$.

We can remedy this by allowing x_e to change. In computing that

$$\{e\}^{A_n}(x_{2e}) = 1,$$

we used only a finite number of values of K_{A_n}. Let y be bigger than all of these arguments. If we never again put a number less than y in A, all will be well. Now the only numbers put in A are the numbers x_{2f+1}. Hence we simply change all the numbers x_{2f+1} to be greater than y.

There is still one difficulty. An x_e may change its value infinitely often, so that we never settle on a value of x_e to use. We avoid this difficulty by modifying our procedure; when we put x_{2e} in B, we change x_{2f+1} only if $2f + 1 > 2e$. Thus an x_{2f+1} can change value only because of the finite number of x_{2e} with $2e < 2f + 1$; so it changes values only finitely often. But what of the x_{2f+1} with $2f + 1 < 2e$ which we do not change? If x_{2f+1} is never put in A, then there is no problem. If it is put in A, we will simultaneously change x_{2e}, since $2e > 2f + 1$. Thus the fact that we may have $\{e\}^A(x_{2e}) = K_B(x_{2e})$ for the old value of x_{2e} is immaterial.

We now give the proof. If $F(\alpha, e, a) \simeq \{e\}^\alpha(a)$, then F is recursive; so $\mathfrak{G}_F$ is recursively enumerable and hence Σ_1^0. By (6) of §7.5, there is a recursive predicate R such that

$$\{e\}^\alpha(a) \simeq b \leftrightarrow \exists y R(\bar{\alpha}(y), e, a, b).$$

We define sets A_n and B_n and functions F_n by induction on n. ($F_n(e)$ will be the value of x_e before the nth stage.) We let A_0 and B_0 be empty, and set $F_0(e) = 2^e$. Now suppose that A_n, B_n, and F_n are chosen. Suppose first that $(n)_0$ is even; say $(n)_0 = 2e$. We search for a number y such that

$$y < n \;\&\; R(\bar{K}_{A_n}(y), e, F_n(2e), 1) \;\&\; F_n(2e) \notin B_n. \tag{1}$$

If there is no such y, then $A_{n+1} = A_n$, $B_{n+1} = B_n$, $F_{n+1} = F_n$. Otherwise, we pick the smallest such y. We put $F_n(2e)$ in B_{n+1} and set

$$F_{n+1}(2f + 1) = 3^y \cdot F_n(2f + 1)$$

for $2f + 1 > 2e$. Otherwise, A_{n+1}, B_{n+1}, and F_{n+1} are the same as A_n, B_n, and F_n. Now suppose that $(n)_0$ is odd; say $(n)_0 = 2f + 1$. Then we proceed as above, interchanging A and B, e and f, and $2e$ and $2f + 1$.

We claim that $K_{A_n}(x)$, $K_{B_n}(x)$, and $F_n(x)$ are recursive functions of n and x. This is clear from Church's thesis. The simplest way to give a rigorous proof is to set $G(x) = \langle K_{A_n}(x), K_{B_n}(x), F(x)\rangle$ and define G by the recursion theorem. We leave this to the reader.

Now we show that for each e, $F_n(e)$ changes only finitely often. The proof is by induction on e. If $F_{n+1}(e) \neq F_n(e)$, we have $F_n(f) \in A_{n+1}$ and $F_n(f) \notin A_n$ for some $f < e$. (This is for e even; for e odd, we must interchange A and B.) By induction hypothesis, the total number of $F_n(f)$ with $f < e$ is finite. Hence $F_n(e)$ can change only finitely often.

Let A be the union of the A_n and let B be the union of the B_n. Since

$$A(x) \leftrightarrow \exists n\big(K_{A_n}(x) = 0\big)$$

and similarly for B, A and B are recursively enumerable. Let x_e be the final value of $F_n(e)$. We shall show that $x_{2e} \in B$ iff $\{e\}^A(x_{2e}) \simeq 1$. This shows that B is not recursive in A; and a symmetric argument shows that A is not recursive in B.

Suppose that $\{e\}^A(x_{2e}) \simeq 1$. Then there is a y such that

$$R\big(\overline{K_A}(y), e, x_{2e}, 1\big).$$

Now there are infinitely many n with $(n)_0 = 2e$. Choose one such n such that $n > y$, $F_n(2e) = x_{2e}$, and $\overline{K_{A_n}}(y) = \overline{K_A}(y)$. Then either x_{2e} is in B_n or it is put in B_{n+1}. Hence $x_{2e} \in B$.

Now suppose that $x_{2e} \in B$. Choose n so that $x_{2e} \in B_{n+1}$ and $x_{2e} \notin B_n$. We then have $(n)_0 = 2e$ & $F_n(e) = x_{2e}$, and (1) holds for some y. (We are using the fact that $F_n(e)$ always has the form 2^e3^z, and hence is different for different e.) If $\overline{K_{A_n}}(y) = \overline{K_A}(y)$, then (1) implies that $\{e\}^A(x_{2e}) \simeq 1$. Otherwise, there is a number z such that $z < y$ and $K_{A_n}(z) \neq K_A(z)$. This means that there is an $m > n$ such that $z \in A_{m+1}$, $z \notin A_m$. We then have

$$(m)_0 = 2f + 1 \quad \text{and} \quad z = F_m(2f + 1).$$

Now if $2f + 1 > 2e$, we have

$$z = F_m(2f + 1) \geqslant F_{n+1}(2f + 1) = 3^y \cdot F_n(2f + 1) > y,$$

a contradiction. Hence $2f + 1 < 2e$; so

$$F_{m+1}(2e) > F_m(2e) \geqslant F_n(2e) = x_{2e}.$$

This is impossible by the definition of x_{2e}.

Many other results on degrees can be proved by the techniques developed in the above proof; but we shall not pursue the matter here.

7.8 THE ANALYTICAL HIERARCHY

We are now going to study the result of prefixing quantifiers on function variables to recursive relations.

A relation P is *analytical* if it has an explicit definition of the form

$$P(\mathfrak{A}) \leftrightarrow Q_1 \ldots Q_k R(\mathfrak{A}, \mathfrak{B}), \tag{1}$$

where R is recursive and $Q_1, \ldots, Q_k$ are quantifiers, one on each variable in $\mathfrak{B}$.

We say that two quantifiers are *of the same kind* if they are both existential or both universal. We say that two quantifiers are *of the same type* if they are both on function variables or both on number variables.

We now suppose that we have a definition of the form (1), and list some simplifications which can be made in the string of quantifiers $Q_1 \ldots Q_k$. In each case, it is understood that the simplification requires replacing R by a new recursive relation (as was the case for contraction of quantifiers in the arithmetical case).

i) A number quantifier may be replaced by a function quantifier of the same kind.

This follows from the equivalences

$$\exists x P(x) \leftrightarrow \exists \alpha P(\alpha(0)),$$
$$\forall x P(x) \leftrightarrow \forall \alpha P(\alpha(0)).$$

ii) Two adjacent quantifiers of the same kind and type may be replaced by a single quantifier of the same kind and type.

For number quantifiers, this is just contraction of quantifiers. For function quantifiers we use the same method, replacing $(a)_i$ by $(\alpha)_i$.

iii) If a function quantifier immediately follows a number quantifier, it may be brought to the front of that number quantifier.

This follows from the equivalences

$$\exists x\, \exists \alpha P(\alpha, x) \leftrightarrow \exists \alpha\, \exists x P(\alpha, x),$$
$$\forall x\, \forall \alpha P(\alpha, x) \leftrightarrow \forall \alpha\, \forall x P(\alpha, x),$$
$$\forall x\, \exists \alpha P(\alpha, x) \leftrightarrow \exists \alpha\, \forall x P((\alpha)_x, x),$$
$$\exists x\, \forall \alpha P(\alpha, x) \leftrightarrow \forall \alpha\, \exists x P((\alpha)_x, x).$$

The first two are evident; the next two are proved in the same manner as (4) and (5) of §7.5.

The prefix $Q_1 \ldots Q_k$ in (1) is *normalized* if $k \geqslant 2$; $Q_1, \ldots, Q_{k-1}$ are function quantifiers and Q_k is a number quantifier; and the quantifiers alternate in kind. We shall show that if we start with a definition (1) in which at least one function quantifier occurs, then we may simplify the prefix to a normalized prefix. We first bring all function quantifiers to the front by (iii). If there are now number quantifiers of the same kind as the last function quantifier, we eliminate each of these number quantifiers, beginning with the first, as follows: we change the quantifier to a function quantifier by (i), bring it to the front of all preceding number quantifiers by (iii), and contract it with the last function quantifier by (ii). The number quantifiers are now all at the end and all opposite in kind to the last function quantifier. If there are no such number quantifiers, we add at the end a superfluous number quantifier opposite in kind to the last function quantifier. If we now perform all possible contractions under (ii), we obtain a normalized prefix. Note that we would have obtained the same final result if we had proceeded as follows: strike out all number quantifiers, perform all possible contractions under (ii), and add at the end one number quantifier opposite in kind to the last function quantifier.

A relation is Σ_n^1 (Π_n^1) for $n \geqslant 1$ if it has a definition (1) in which the prefix is normalized, the number of function quantifiers is n, and the first quantifier is existential (universal). We have then seen that every analytical relation is either arithmetical or Σ_n^1 or Π_n^1 for some n. This classification of analytical relations is called the *analytical hierarchy*.

We shall now give analogues of A1 through A6. When no proof is given, the proof is the same as in the arithmetical case.

Y1. If P is arithmetical, then P is Σ_n^1 and Π_n^1 for every n. If P is Σ_m^1 or Π_m^1, then P is Σ_n^1 and Π_n^1 for all $n > m$.

Proof. By adding superfluous quantifiers and simplifying as above. Thus if P is Σ_2^1, then

$$\begin{aligned} P(\mathfrak{A}) &\leftrightarrow \exists\alpha\, \forall\beta\, \exists x R(\mathfrak{A}, \alpha, \beta, x) \\ &\leftrightarrow \forall\gamma\, \exists\alpha\, \forall\beta\, \exists x R(\mathfrak{A}, \alpha, \beta, x) \\ &\leftrightarrow \exists\alpha\, \forall\beta\, \exists\gamma\, \exists x R(\mathfrak{A}, \alpha, \beta, x). \end{aligned}$$

After simplification, the second line shows that P is Π_3^1 and the third line shows that P is Σ_3^1.

Y2. If Q is Σ_n^1 (Π_n^1) and $F_1, \ldots, F_r, G_1, \ldots, G_k$ are recursive functionals, then the relation P defined by

$$P(\mathfrak{A}) \leftrightarrow Q\big(\lambda x F_1(\mathfrak{A}, x), \ldots, \lambda x F_r(\mathfrak{A}, x), G_1(\mathfrak{A}), \ldots, G_k(\mathfrak{A})\big)$$

is Σ_n^1 (Π_n^1).

Y3. If Q is Σ_n^1 and P is defined by $P(\mathfrak{A}) \leftrightarrow \exists\alpha Q(\mathfrak{A}, \alpha)$, then P is Σ_n^1. If Q is Π_n^1 and P is defined by $P(\mathfrak{A}) \leftrightarrow \forall\alpha Q(\mathfrak{A}, \alpha)$, then P is Π_n^1.

Y4. If P and Q are Σ_n^1 (Π_n^1), then $P \vee Q$ and $P \;\&\; Q$ are Σ_n^1 (Π_n^1).

Y5. If P is Σ_n^1 (Π_n^1), then $\neg P$ is Π_n^1 (Σ_n^1).

Y6. If P is Σ_n^1 (Π_n^1) and Q and R are defined by $Q(\mathfrak{A}) \leftrightarrow \exists x P(\mathfrak{A}, x)$ and $R(\mathfrak{A}) \leftrightarrow \forall x P(\mathfrak{A}, x)$, then Q and R are Σ_n^1 (Π_n^1).

Proof. By simplification as described above.

Note that Y6 implies the corresponding result for bounded quantifiers, since we may rewrite $\exists x_{x<b}P(\mathfrak{A}, x)$ and $\forall x_{x<b}P(\mathfrak{A}, x)$ as $\exists x(x < b \mathbin{\&} P(\mathfrak{A}, x))$ and $\forall x(x < b \rightarrow P(\mathfrak{A}, x))$.

If P is Π_1^1, then $P(\mathfrak{A}) \leftrightarrow \forall\alpha Q(\mathfrak{A}, \alpha)$, where Q is Σ_1^0. Hence by (6) of §7.5, P has a definition

$$P(\mathfrak{A}) \leftrightarrow \forall\alpha\, \exists x R(\bar{\alpha}(x), \mathfrak{A}) \tag{2}$$

with R recursive. Similarly, using (7) of §7.5, we see that a Σ_1^1 relation P has a definition

$$P(\mathfrak{A}) \leftrightarrow \exists\alpha\, \forall x R(\bar{\alpha}(x), \mathfrak{A}) \tag{3}$$

with R recursive.

We can prove an analytical enumeration theorem and an analytical hierarchy theorem by the same methods used in the arithmetical case. We leave these to the reader.

A relation is Δ_n^1 if it is both Σ_n^1 and Π_n^1. By Y5, P is Δ_n^1 iff both P and $\neg P$ are Σ_n^1, and iff both P and $\neg P$ are Π_n^1.

We define *analytical in* Φ, Σ_n^1 *in* Φ, Π_n^1 *in* Φ, and Δ_n^1 *in* Φ by the usual method of defining relative notions. For example, P is analytical in Φ if it has a definition of the form (1) with R recursive in Φ. The fact that every analytical relation is either arithmetical or Σ_n^1 or Π_n^1 for some n extends to the relative case, as do Y1 through Y6. The analytical enumeration theorem and the analytical hierarchy theorem extend when Φ is a finite sequence. The transitivity lemma and the finiteness lemma hold for these notions. The replacement lemma holds for analytical in Φ, Σ_n^1 in Φ, and Π_n^1 in Φ (but not for Δ_n^1 in Φ).

A relation P is *projective* ($\boldsymbol{\Sigma}_n^1$) ($\boldsymbol{\Pi}_n^1$) ($\boldsymbol{\Delta}_n^1$) if it is analytical (Σ_n^1) (Π_n^1) (Δ_n^1) in the set of all functions. We can extend the analytical enumeration theorem and the analytical hierarchy theorem to this case by using functional enumeration, just as in the arithmetical case. The projection lemma extends to all of these notions. A consequence is that if we define P by

$$P(\mathfrak{A}) \leftrightarrow \exists x Q_x(\mathfrak{A}),$$

where each Q_x is $\boldsymbol{\Sigma}_n^1$ ($\boldsymbol{\Pi}_n^1$), then P is $\boldsymbol{\Sigma}_n^1$ ($\boldsymbol{\Pi}_n^1$). For if we define Q by

$$Q(\mathfrak{A}, x) \leftrightarrow Q_x(\mathfrak{A}),$$

then Q is $\boldsymbol{\Sigma}_n^1$ ($\boldsymbol{\Pi}_n^1$) by the projection lemma; and we can then apply Y6.

7.9 HYPERARITHMETICAL RELATIONS

The negation theorem tells us that the Δ_1^0 relations, which apparently require quantifiers in their definitions, are actually recursive, and hence can be defined without quantifiers. This suggests that we look for a characterization of Δ_1^1 relations which shows that these relations, which apparently require function quantifiers in their definitions, can actually be defined without function quantifiers. A first conjecture might be that the Δ_1^1 relations are just the arithmetical relations;

but it turns out that the arithmetical relations are a proper subclass of the Δ_1^1 relations. To obtain the proper characterization of Δ_1^1 relations, we must extend the arithmetical hierarchy.

Since a relation is a subset of $N_{m,n}$, we may perform set-theoretical operations on relations. Thus A4 and A5 may be viewed as telling us how arithmetical relations behave under taking unions, intersections, and complements. Moreover, every Σ_{n+1}^0 (Π_{n+1}^0) relation is a countable union (intersection) of Π_n^0 (Σ_n^0) relations. For example, if P is Σ_{n+1}^0, then there is a Π_n^0 relation Q such that

$$\begin{aligned} P(\mathfrak{A}) &\leftrightarrow \exists x Q(\mathfrak{A}, x) \\ &\leftrightarrow \exists x Q_{(x)}(\mathfrak{A}). \end{aligned}$$

Thus P is the union of the $Q_{(x)}$; and $Q_{(x)}$ is Π_n^0, since $Q_{(x)}(\mathfrak{A}) \leftrightarrow Q(\mathfrak{A}, x)$. However, only particularly simple countable unions and intersections of arithmetical relations are arithmetical. For example, every set is the union of countably many finite, and hence recursive, sets.

We shall obtain the hyperarithmetical relations by starting with the recursively enumerable relations and repeatedly taking complements and certain countable unions. To describe these unions, we assign an index to each hyperarithmetical relation. Then if A is a recursively enumerable set of indices, we take the union of the relations whose indices are in A.

The following three rules constitute a generalized inductive definition of an *H-index*:

I1. For each e, $\langle 0, e\rangle$ is an H-index.

I2. If e is an H-index, then $\langle 1, e\rangle$ is an H-index.

I3. If every number in $W_e^{0,1}$ is an H-index, then $\langle 2, e\rangle$ is an H-index.

For each H-index i, we define an (m, n)-ary relation $J_i^{m,n}$ as follows. If $i = \langle 0, e\rangle$, then $J_i^{m,n}$ is $W_e^{m,n}$. If $i = \langle 1, e\rangle$ where e is an H-index, then $J_i^{m,n}$ is $\neg J_e^{m,n}$. If $i = \langle 2, e\rangle$ where every member of $W_e^{0,1}$ is an H-index, then $J_i^{m,n}$ is the union of the $J_x^{m,n}$ for x in $W_e^{0,1}$; that is

$$J_i^{m,n}(\mathfrak{A}) \leftrightarrow \exists x\big(W_e^{0,1}(x) \ \&\ J_x^{m,n}(\mathfrak{A})\big). \tag{1}$$

We must show that this is a legitimate method of defining the $J_i^{m,n}$. To define these, it is obviously sufficient to define for each m and n the set $\Phi_{m,n}$ of pairs $(i, J_i^{m,n})$. Now the three parts of the above definition may be regarded as three rules in a generalized inductive definition of $\Phi_{m,n}$. For example, the rule corresponding to the case $i = \langle 2, e\rangle$ is: if $i = \langle 2, e\rangle$ is an H-index, and if (x, P_x) is in $\Phi_{m,n}$ for each x in $W_e^{0,1}$, and if P is the union of the P_x, then (i, P) is in $\Phi_{m,n}$.

An (m, n)-ary relation P is *hyperarithmetical* if $P = J_i^{m,n}$ for some H-index i; any such i is then called an H-index *of P*.

We now derive some rules for obtaining hyperarithmetical relations and H-indices of such relations. We omit the superscripts on W_e and J_i when no confusion results.

H1. If R is a recursively enumerable predicate, then there is a recursive function F such that

$$J_{F(\mathfrak{a})}(\mathfrak{A}) \leftrightarrow \exists x\big(R(x, \mathfrak{a}) \ \& \ J_x(\mathfrak{A})\big)$$

whenever $\mathfrak{a}$ is such that for each x, $R(x, \mathfrak{a})$ implies that x is an H-index.

Proof. Choose e so that R is W_e. We have

$$W_{S(e,\mathfrak{a})}(x) \leftrightarrow W_e(x, \mathfrak{a}).$$

Setting $F(\mathfrak{a}) = \langle 2, S(e, \mathfrak{a})\rangle$, we have for $\mathfrak{a}$ as described in H1

$$\begin{aligned} J_{F(\mathfrak{a})}(\mathfrak{A}) &\leftrightarrow \exists x\big(W_{S(e,\mathfrak{a})}(x) \ \& \ J_x(\mathfrak{A})\big) \\ &\leftrightarrow \exists x\big(R(x, \mathfrak{a}) \ \& \ J_x(\mathfrak{A})\big). \end{aligned}$$

H2. Let $F_1, \ldots, F_r, G_1, \ldots, G_k$ be recursive functionals. Then there is a recursive function H such that

$$J_{H(i,\mathfrak{a})}(\mathfrak{A}) \leftrightarrow J_i\big(\lambda x F_1(\mathfrak{A}, \mathfrak{a}, x), \ldots, \lambda x F_r(\mathfrak{A}, \mathfrak{a}, x), G_1(\mathfrak{A}, \mathfrak{a}), \ldots, G_k(\mathfrak{A}, \mathfrak{a})\big)$$

for every H-index i.

Proof. We first prove the equivalence, noting what properties of H are needed; we then define an H having these properties.

Our proof of the equivalence is by induction on H-indices. We write ... for the set of arguments to J_i on the right of the equivalence. Now $W_e(\ldots)$ is a recursively enumerable relation of the arguments $\mathfrak{A}, \mathfrak{a}, e$ by A1. Hence by (8) of §7.4, there is a recursive function L such that

$$W_{L(\mathfrak{a},e)}(\mathfrak{A}) \leftrightarrow W_e(\ldots).$$

Then if $i = \langle 0, e\rangle$,

$$\begin{aligned} J_i(\ldots) &\leftrightarrow W_{L(\mathfrak{a},e)}(\mathfrak{A}) \\ &\leftrightarrow J_{\langle 0, L(\mathfrak{a},e)\rangle}(\mathfrak{A}). \end{aligned}$$

Hence to obtain the equivalence $J_{H(i,\mathfrak{a})}(\mathfrak{A}) \leftrightarrow J_i(\ldots)$ in this case it suffices to have

$$H(i, \mathfrak{a}) = \langle 0, L(\mathfrak{a}, (i)_1)\rangle \quad \textit{if} \quad (i)_0 = 0. \tag{2}$$

Now let $i = \langle 1, e\rangle$. Then, using the induction hypothesis, we have

$$\begin{aligned} J_i(\ldots) &\leftrightarrow \neg J_e(\ldots) \\ &\leftrightarrow \neg J_{H(e,\mathfrak{a})}(\mathfrak{A}) \\ &\leftrightarrow J_{\langle 1, H(e,\mathfrak{a})\rangle}(\mathfrak{A}). \end{aligned}$$

Hence for this case it suffices to have

$$H(i, \mathfrak{a}) = \langle 1, H((i)_1, \mathfrak{a})\rangle \quad \textit{if} \quad (i)_0 = 1. \tag{3}$$

Now let $i = \langle 2, e\rangle$. Then, using the induction hypothesis, we obtain

$$\begin{aligned} J_i(\ldots) &\leftrightarrow \exists x\big(W_e(x) \ \& \ J_x(\ldots)\big) \\ &\leftrightarrow \exists x\big(W_e(x) \ \& \ J_{H(x,\mathfrak{a})}(\mathfrak{A})\big) \\ &\leftrightarrow \exists y\big(\exists x\big(W_e(x) \ \& \ H(x, \mathfrak{a}) = y\big) \ \& \ J_y(\mathfrak{A})\big). \end{aligned}$$

Let h be an index of H. Then the above gives

$$J_i(\ldots) \leftrightarrow \exists y(\exists x(W_e(x) \;\&\; \{h\}(x, \mathfrak{a}) = y) \;\&\; J_y(\mathfrak{A})).$$

Now $\exists x(W_e(x) \;\&\; \{h\}(x, \mathfrak{a}) = y)$ is a recursively enumerable predicate of the arguments y, e, h, $\mathfrak{a}$. Hence by H1, there is a recursive function K (independent of H and h) such that

$$J_i(\ldots) \leftrightarrow J_{K(e,h,\mathfrak{a})}(\mathfrak{A}).$$

Hence for this case it suffices to have

$$H(i, \mathfrak{a}) = K((i)_1, h, \mathfrak{a}) \quad \textit{if} \quad (i)_0 > 1. \tag{4}$$

It remains to define a recursive function H with an index h such that (2), (3), and (4) hold. By the recursion theorem, we may define a recursive partial function H with an index h so that (2), (3), and (4) hold with $=$ replaced by $\simeq$. Now in (3), we have $(i)_1 < i$ by (8) of §6.4. With this in mind, it is easy to prove by induction on i that $H(i, \mathfrak{a})$ is defined for all i and $\mathfrak{a}$. This completes the proof.

H3. There is a recursive function F such that $J_{F(i)}$ is $\neg J_i$ for all H-indices i. There is a recursive function G such that $J_{G(i,j)}$ is $J_i \vee J_j$ for all H-indices i and j. There are similar recursive functions for $\rightarrow$, &, and $\leftrightarrow$.

Proof. We can take $F(i) = \langle 1, i \rangle$. Now

$$(J_i \vee J_j)(\mathfrak{A}) \leftrightarrow \exists x((x = i \vee x = j) \;\&\; J_x(\mathfrak{A})).$$

From this and H1 we obtain G. The remaining functions are obtained by using the definitions of $\rightarrow$, &, and $\leftrightarrow$ in terms of $\neg$ and $\vee$. For example, the function H for $\rightarrow$ is defined by $H(i, j) = G(F(i), j)$.

H4. There are recursive functions H and K such that

$$J_{H(i)}(\mathfrak{A}) \leftrightarrow \exists x J_i(\mathfrak{A}, x),$$
$$J_{K(i)}(\mathfrak{A}) \leftrightarrow \forall x J_i(\mathfrak{A}, x)$$

for every H-index i.

Proof. By H2, there is a recursive G such that $J_{G(i,x)}(\mathfrak{A}) \leftrightarrow J_i(\mathfrak{A}, x)$. Then

$$\begin{aligned}\exists x J_i(\mathfrak{A}, x) &\leftrightarrow \exists x J_{G(i,x)}(\mathfrak{A}) \\ &\leftrightarrow \exists y(\exists x(y = G(i, x)) \;\&\; J_y(\mathfrak{A})).\end{aligned}$$

We then obtain the desired H by H1. To obtain K, set $K(i) = F(H(F(i)))$, where F is as in H3.

H5. Every arithmetical relation is hyperarithmetical.

Proof. We show by induction on n that every Σ_n^0 or Π_n^0 relation is hyperarithmetical. A Σ_1^0 relation is recursively enumerable and hence hyperarithmetical; so a Π_1^0 relation is hyperarithmetical by H3. The step from n to $n + 1$ follows from H4.

Remark. The converse of H5 is false; see Problem 22.

Since the definition of hyperarithmetical involves *RE*-indices, we cannot directly relativize it to an arbitrary Φ. We thus first consider a finite sequence Φ. We define *H-index from* Φ and the J_i^{Φ} for i an H-index from Φ as before, except that W_e is replaced by W_e^{Φ}. We say that P is *hyperarithmetical in* Φ if there is an i such that P is J_i^{Φ}; any such i is then an H-index from Φ *of P*. We can then relativize H1 through H5.

Now let Φ be a class of total functions and relations. We say that P is *hyperarithmetical in* Φ if it is hyperarithmetical in a finite sequence of elements of Φ. A relation is *Borel* if it is hyperarithmetical in the class of all functions.

7.10 THE CHARACTERIZATION THEOREM

We can now proceed to our characterization of Δ_1^1 relations.

Let $J(i, \mathfrak{A})$ mean that i is an H-index and that $J_i(\mathfrak{A})$. (We are omitting superscripts.) We shall obtain an explicit definition of J.

Let $Q(\mathfrak{A}, \alpha, \beta, e)$ be the conjunction of the statements

$$\begin{aligned}
&\alpha(\langle 0, e\rangle) = 0,\\
&\beta(\langle 0, e\rangle) = 0 \leftrightarrow W_e(\mathfrak{A}),\\
&\alpha(e) = 0 \rightarrow \alpha(\langle 1, e\rangle) = 0,\\
&\alpha(\langle 1, e\rangle) = 0 \rightarrow \big(\beta(\langle 1, e\rangle) = 0 \leftrightarrow \beta(e) \neq 0\big),\\
&\forall x\big(W_e(x) \rightarrow \alpha(x) = 0\big) \rightarrow \alpha(\langle 2, e\rangle) = 0,\\
&\alpha(\langle 2, e\rangle) = 0 \rightarrow \big(\beta(\langle 2, e\rangle) = 0 \leftrightarrow \exists x\big(W_e(x) \,\&\, \beta(x) = 0\big)\big).
\end{aligned}$$

We show that

$$J(i, \mathfrak{A}) \leftrightarrow \forall\alpha\, \forall\beta\big(\forall e Q(\mathfrak{A}, \alpha, \beta, e) \rightarrow \alpha(i) = 0 \,\&\, \beta(i) = 0\big). \tag{1}$$

Suppose that $J(i, \mathfrak{A})$. Let α and β be such that $\forall e Q(\mathfrak{A}, \alpha, \beta, e)$. Using induction on H-indices, we find that for each H-index j, $\alpha(j) = 0$ and $\beta(j) = 0 \leftrightarrow J_j(\mathfrak{A})$. Taking $j = i$, we find that $\alpha(i) = \beta(i) = 0$. Now suppose that the right-hand side of (1) holds. Let α be the representing function of the set of H-indices, and let β be the representing function of the set of H-indices j such that $J_j(\mathfrak{A})$. Clearly $Q(\mathfrak{A}, \alpha, \beta, e)$ for all e; so by hypothesis, $\alpha(i) = \beta(i) = 0$. It follows that $J(i, \mathfrak{A})$.

Since Q is clearly arithmetical, it follows from (1) that J is Π_1^1. If i is an H-index, then $J_i(\mathfrak{A}) \leftrightarrow J(i, \mathfrak{A})$ for all $\mathfrak{A}$, and hence J_i is Π_1^1. Thus every hyperarithmetical relation is Π_1^1. If P is hyperarithmetical, then $\neg P$ is hyperarithmetical; so P and $\neg P$ are Π_1^1. Thus:

Lemma 1. Every hyperarithmetical relation is Δ_1^1.

A sequence number $\langle a_1, \ldots, a_n\rangle$ is an *extension* of a sequence number $\langle b_1, \ldots, b_m\rangle$ if $n \geq m$ and $a_i = b_i$ for $i = 1, \ldots, m$. If also $n > m$, we say that $\langle a_1, \ldots, a_n\rangle$ is a *proper* extension of $\langle b_1, \ldots, b_m\rangle$. We shall write $a <^* b$ to mean that a is a proper extension of b. Clearly

$$a <^* b \rightarrow b <^* c \rightarrow a <^* c. \tag{2}$$

The predicate $<^*$ is recursive, since it has the explicit definition

$$a <^* b \leftrightarrow Seq(a) \ \& \ \exists x_{x<lh(a)}(b = In(a, x)).$$

A *descending sequence for* $<^*$ is an infinite sequence $a_0, a_1, \ldots$ such that $a_{n+1} <^* a_n$ for all n. A set A of sequence numbers is a *tree* if there is no descending sequence for $<^*$ whose members are all in A. It is clear that every subset of a tree is a tree.

We designate by SS the class of representing functions of sets of sequence numbers, and by Tr the class of representing functions of trees. Both are (1, 0)-ary relations. We have the explicit definitions

$$SS(\alpha) \leftrightarrow \forall x(\alpha(x) \leqslant 1) \ \& \ \forall x(\alpha(x) = 0 \rightarrow Seq(x)),$$
$$Tr(\alpha) \leftrightarrow SS(\alpha) \ \& \ \neg\exists\beta \, \forall x(\alpha(\beta(x)) = 0 \ \& \ \beta(x+1) <^* \beta(x)).$$

It follows that SS is Π^0_1 and Tr is Π^1_1.

Tree Theorem. If P is a Π^1_1 relation, then there is a recursive functional F such that for all $\mathfrak{A}$, $SS(\lambda x F(\mathfrak{A}, x))$ and

$$P(\mathfrak{A}) \leftrightarrow Tr(\lambda x F(\mathfrak{A}, x)).$$

Proof. By (2) of §7.8, there is a recursive relation R such that

$$P(\mathfrak{A}) \leftrightarrow \forall\alpha \, \exists x R(\bar{\alpha}(x), \mathfrak{A}). \tag{3}$$

Let $U(\mathfrak{A})$ be the set of sequence numbers $\langle x_0, \ldots, x_{n-1}\rangle$ such that

$$\neg R(\langle x_0, \ldots, x_{i-1}\rangle, \mathfrak{A}) \qquad \text{for all } i \leqslant n.$$

We claim that

$$\exists\alpha \, \forall x \neg R(\bar{\alpha}(x), \mathfrak{A}) \leftrightarrow U(\mathfrak{A}) \textit{ is not a tree.} \tag{4}$$

Suppose that the left-hand side holds, and choose α such that $\neg R(\bar{\alpha}(x), \mathfrak{A})$ for all x. Then $\bar{\alpha}(0), \bar{\alpha}(1), \ldots$ is a descending sequence in $U(\mathfrak{A})$. Now suppose that $a_0, a_1, \ldots$ is a descending sequence in $U(\mathfrak{A})$. By the definition of $<^*$, there is a function α such that each $\bar{\alpha}(x)$ has some a_i as an extension. Since a_i is in $U(\mathfrak{A})$, we have $\neg R(\bar{\alpha}(x), \mathfrak{A})$.

If we bring the negation sign in (4) to the front by prenex operations, drop the negation from both sides, and use (3), we get

$$P(\mathfrak{A}) \leftrightarrow U(\mathfrak{A}) \textit{ is a tree.}$$

It therefore suffices to choose a recursive F so that $\lambda x F(\mathfrak{A}, x)$ is the representing function of $U(\mathfrak{A})$. Such an F is defined by

$$\begin{aligned} F(\mathfrak{A}, x) &= 0 \quad \textit{if} \ \ Seq(x) \ \& \ \forall i_{i \leqslant lh(x)} \neg R(In(x, i), \mathfrak{A}), \\ &= 1 \quad \textit{otherwise}. \end{aligned}$$

If A is a set of sequence numbers and a is in A, then $A_{[a]}$ is the set of b in A such that $b <^* a$. From (2),

$$b \in A_{[a]} \rightarrow A_{[a][b]} = A_{[b]}. \tag{5}$$

We also define

$$\begin{aligned} \alpha_{[a]}(b) &= \alpha(b) \qquad \textit{if} \quad b <^* a \;\&\; \alpha(a) = 0, \\ &= 1 \qquad \textit{otherwise.} \end{aligned}$$

Then if α is the representing function of A and a is in A, $\alpha_{[a]}$ is the representing function of $A_{[a]}$. We note that $\alpha_{[a]}$ is $\lambda x F(\alpha, a, x)$ for a certain recursive functional F, and hence can be used in explicit definitions of recursive partial functionals and relations.

We shall now assume that the reader is familiar with the elementary properties of ordinals. (All necessary material will be found in Chapter 9.) We use σ, τ, and ρ to designate ordinals.

We shall define a class $TO(\sigma)$ of sets of sequence numbers for each ordinal σ by transfinite induction in σ. A set of sequence numbers A is in $TO(\sigma)$ if for every a in A, $A_{[a]}$ is in $TO(\tau)$ for some $\tau < \sigma$. It is clear that

$$\sigma \leqslant \tau \rightarrow TO(\sigma) \subset TO(\tau). \tag{6}$$

We now show that a set belongs to some $TO(\sigma)$ iff it is a tree. First we prove by transfinite induction on σ that every set in $TO(\sigma)$ is a tree. Suppose that some set A in $TO(\sigma)$ has a descending sequence $a_0, a_1, \ldots$. Then $A_{[a_0]}$ is in $TO(\tau)$ for some $\tau < \sigma$ and $a_1, a_2, \ldots$ is a descending sequence in $A_{[a_0]}$ by (2). This contradicts the induction hypothesis.

Now suppose that A is in no $TO(\sigma)$. We show that A is not a tree by defining inductively a descending sequence $a_0, a_1, \ldots$ in A such that each $A_{[a_n]}$ is in no $TO(\sigma)$. Suppose a_i chosen for $i < n$. Let $B = A$ if $n = 0$ and let $B = A_{[a_{n-1}]}$ otherwise. In view of (5), it suffices to choose a_n in B so that $B_{[a_n]}$ is in no $TO(\sigma)$. If this is impossible, then for each a in B there is a σ_a such that $B_{[a]} \in TO(\sigma_a)$. Choosing σ larger than all the σ_a, we have $B \in TO(\sigma)$, contradicting the fact that B is in no $TO(\sigma)$.

If A is a tree, the smallest ordinal σ such that $A \in TO(\sigma)$ is called the ordinal of A, and is designated by $\|A\|$. We understand $\|A\| \leqslant \sigma$ to mean A *is a tree and* $\|A\| \leqslant \sigma$, and similarly for $\|A\| < \sigma$. Then

$$A \in TO(\sigma) \leftrightarrow \|A\| \leqslant \sigma. \tag{7}$$

The implication from left to right is clear. If A is a tree, then $A \in TO(\|A\|)$; so if $\|A\| \leqslant \sigma$, then $A \in TO(\sigma)$ by (6). From (7) and the definition of TO, we obtain

$$\|A\| \leqslant \sigma \leftrightarrow \forall x\big(A(x) \rightarrow \|A_{[x]}\| < \sigma\big) \tag{8}$$

for every set A of sequence numbers. It follows that if A is a tree, then

$$A(x) \rightarrow \|A_{[x]}\| < \|A\|. \tag{9}$$

We also have a converse to (9): if A is a tree then

$$\sigma < \|A\| \rightarrow \exists x\big(A(x) \;\&\; \sigma = \|A_{[x]}\|\big). \tag{10}$$

We prove this by transfinite induction on $\|A\|$. If $\sigma < \|A\|$, then by (8) there is an x in A such that $\sigma \leqslant \|A_{[x]}\|$. If $\sigma = \|A_{[x]}\|$, then we are through. Otherwise,

$\sigma < \|A_{[x]}\| < \|A\|$ by (9); so by induction hypothesis and (5), there is a y in $A_{[x]}$ such that $\sigma = \|A_{[x][y]}\| = \|A_{[y]}\|$.

Let A and B be sets of sequence numbers. A mapping F from A to B is *monotone* if

$$a <^* a' \rightarrow F(a) <^* F(a')$$

for all a and a' in A.

Lemma 2. Let A be a set of sequence numbers, and let B be a tree. Then $\|A\| \leqslant \|B\|$ iff there is a monotone mapping from A to B.

Proof. We use transfinite induction on $\|B\|$. Suppose that F is a monotone mapping from A to B. If a is in A, then some restriction of F is a monotone mapping from $A_{[a]}$ to $B_{[F(a)]}$; so $\|A_{[a]}\| \leqslant \|B_{[F(a)]}\|$ by (9) and the induction hypothesis. From this, (9), and (8), we get $\|A\| \leqslant \|B\|$.

Now suppose that $\|A\| \leqslant \|B\|$. Fixing a in A, we have $\|A_{[a]}\| < \|A\| \leqslant \|B\|$ by (9). By (10), $\|A_{[a]}\| = \|B_{[b]}\|$ for some b in B. By (9) and the induction hypothesis, there is a monotone mapping F_a from $A_{[a]}$ to $B_{[b]}$ and hence to B. If we set $F_a(a) = b$, then F_a becomes a monotone mapping from A_a to B, where A_a is the set of extensions of a which belong to A.

An element of A is *maximal* if it is not a proper extension of any element of A. Clearly every element of A is in A_a for a unique maximal element a of A. Hence there is a mapping F from A to B such that for each maximal element a, the restriction of F to A_a is F_a. If a and a' are distinct maximal elements, then no element of A_a can be an extension of an element of $A_{a'}$. It follows that F is monotone.

If α is the representing function of a tree A, we write $\|\alpha\|$ for $\|A\|$. We understand $\|\alpha\| \leqslant \sigma$ or $\|\alpha\| < \sigma$ to imply that α is the representing function of a tree.

Corollary. There are Σ^1_1 relations $T_{\leqslant}$ and $T_{<}$ such that if $Tr(\beta)$, then

$$T_{\leqslant}(\alpha, \beta) \leftrightarrow \|\alpha\| \leqslant \|\beta\|$$

and

$$T_{<}(\alpha, \beta) \leftrightarrow \|\alpha\| < \|\beta\|.$$

Proof. By the lemma, we may set

$$T_{\leqslant}(\alpha, \beta) \leftrightarrow SS(\alpha) \;\&\; \exists\gamma\big(\forall x(\alpha(x) = 0 \rightarrow \beta(\gamma(x)) = 0\big) \;\&\; \forall x\, \forall y(\alpha(x) = 0 \rightarrow \alpha(y) = 0 \rightarrow x <^* y \rightarrow \gamma(x) <^* \gamma(y))\big).$$

By (10), we may set

$$T_{<}(\alpha, \beta) \leftrightarrow \exists x(\beta(x) = 0 \;\&\; T_{\leqslant}(\alpha, \beta_{[x]})).$$

Both are Σ^1_1 by Y1 through Y6.

An ordinal σ is *recursive* if $\sigma = \|A\|$ for some recursive tree A. Every ordinal less than a recursive ordinal is recursive. For let $\tau < \sigma = \|A\|$, where A is a recursive tree. By (10), $\tau = \|A_{[a]}\|$ for some a in A. But $A_{[a]}$ is a recursive tree, since

$$A_{[a]}(x) \leftrightarrow A(x) \;\&\; x <^* a.$$

The first ordinal which is not recursive is designated by κ. By the result just proved, an ordinal is recursive iff it is less than κ.

Remark. We have $\omega < \kappa$. To prove this, let A be the set of $\langle a_1, \ldots, a_n \rangle$ where $a_1 > a_2 > \cdots > a_n$. It is easy to see that A is a tree; and it is recursive, since

$$A(a) \leftrightarrow Seq(a) \;\&\; \forall i_{i<lh(a)} \forall j_{j<i}((a)_i < (a)_j).$$

Fixing n, let $a_i = \langle n, n-1, \ldots, i \rangle$ for $i \leqslant n$. Then a_i is in A, and

$$a_0 <^* a_1 <^* \cdots <^* a_n.$$

Then by (9) and (5),

$$\|A_{[a_0]}\| < \|A_{[a_1]}\| < \cdots < \|A_{[a_n]}\| < \|A\|.$$

It follows that $n < \|A\|$. Since this holds for all n, it follows that $\omega \leqslant \|A\| < \kappa$.

For each σ, we define

$$Tr_\sigma(\alpha) \leftrightarrow \|\alpha\| \leqslant \sigma,$$
$$Tr'_\sigma(\alpha) \leftrightarrow \|\alpha\| < \sigma.$$

By (8),

$$Tr_\sigma(\alpha) \leftrightarrow SS(\alpha) \;\&\; \forall x(\alpha(x) = 0 \rightarrow Tr'_\sigma(\alpha_{[x]})).$$

It follows by H1 through H5 that there is a recursive function L such that if Tr'_σ is J_i, then Tr_σ is $J_{L(i)}$.

Lemma 3. If σ is recursive, then Tr_σ and Tr'_σ are hyperarithmetical.

Proof. By the result just proved, we need only consider Tr'_σ. Let A be a recursive tree such that $\sigma = \|A\|$; and for x in A, let $\sigma_x = \|A_{[x]}\|$. By (9) and (10),

$$Tr'_\sigma(\alpha) \leftrightarrow \exists x(A(x) \;\&\; Tr_{\sigma_x}(\alpha)).$$

We shall show that there is a recurisve function F such that for each x in A, Tr_{σ_x} is $J_{F(x)}$. It will follow that

$$Tr'_\sigma(\alpha) \leftrightarrow \exists x(A(x) \;\&\; J_{F(x)}(\alpha))$$
$$\leftrightarrow \exists y(\exists x(A(x) \;\&\; F(x) = y) \;\&\; J_y(\alpha)).$$

Hence Tr'_σ is hyperarithmetical by H1.

We first prove that Tr_{σ_x} is $J_{F(x)}$ by transfinite induction on σ_x, noting what properties of F are needed; we then define an F with these properties. By (9), (10), and (5), the ordinals less than σ_x are the σ_y for y in A and $y <^* x$. Hence

$$Tr'_{\sigma_x}(\alpha) \leftrightarrow \exists y(A(y) \;\&\; y <^* x \;\&\; J_{F(y)}(\alpha))$$

by induction hypothesis. If f is an index of F, this may be written

$$Tr'_{\sigma_x}(\alpha) \leftrightarrow \exists z(\exists y(A(y) \;\&\; y <^* x \;\&\; \{f\}(y) = z) \;\&\; J_z(\alpha)).$$

It follows by H1 that there is a recursive function M (independent of F and f) such that

$$Tr'_{\sigma_x}(\alpha) \leftrightarrow J_{M(f,x)}(\alpha).$$

If L is as above, then

$$Tr_{\sigma_x}(\alpha) \leftrightarrow J_{L(M(f,x))}(\alpha).$$

Hence to conclude that Tr_{σ_x} is $J_{F(x)}$, it suffices to have $F(x) = L(M(f, x))$. Now we can define a recursive partial function F with an index f such that

$$F(x) \simeq L(M(f, x))$$

for all x by the recursion theorem. Since L and M are total, F is also, and $F(x) = L(M(f, x))$.

Boundedness Theorem. If P is a Σ_1^1 subclass of Tr, then there is a recursive σ such that P is a subset of Tr_σ.

Proof. We suppose that there is no such σ, and prove that every Π_1^1 unary predicate Q is Σ_1^1; this will contradict the analytical hierarchy theorem. By the tree theorem, there is a recursive function F such that

$$Q(a) \leftrightarrow Tr(\lambda x F(a, x)). \tag{11}$$

We show that

$$Q(a) \leftrightarrow \exists\alpha(P(\alpha) \;\&\; T_{\leqslant}(\lambda x F(a, x), \alpha));$$

this will imply that Q is Σ_1^1.

Suppose that $Q(a)$. Then $Tr(\lambda x F(a, x))$ by (11). Let $\sigma = \|\lambda x F(a, x)\|$. Since F is recursive, σ is recursive; so there is an α in P such that $\neg Tr_\sigma(\alpha)$. We then have $Tr(\alpha)$ and $\sigma \leqslant \|\alpha\|$; so $T_{\leqslant}(\lambda x F(a, x), \alpha)$. Now suppose that there is an α in P such that $T_{\leqslant}(\lambda x F(a, x), \alpha)$. Then $Tr(\alpha)$ and hence $Tr(\lambda x F(a, x))$. Hence $Q(a)$ by (11).

Let P, Q, and R be (m, n)-ary relations. We say that P and Q are *disjoint* if $\neg(P(\mathfrak{A}) \;\&\; Q(\mathfrak{A}))$ for all $\mathfrak{A}$. We say that R *separates* P and Q if $P(\mathfrak{A}) \rightarrow R(\mathfrak{A})$ for all $\mathfrak{A}$ and R and Q are disjoint; this clearly implies that P and Q are disjoint.

Separation Theorem (*Lusin-Addison*). If P and Q are disjoint Σ_1^1 relations, then there is a hyperarithmetical relation R which separates P and Q.

Proof. By the tree theorem, there is a recursive functional F such that

$$\neg Q(\mathfrak{A}) \leftrightarrow Tr(\lambda x F(\mathfrak{A}, x)).$$

Let S be the class of all $\lambda x F(\mathfrak{A}, x)$ for $\mathfrak{A}$ such that $P(\mathfrak{A})$. Since P and Q are disjoint, S is a subset of Tr. Since

$$S(\alpha) \leftrightarrow \exists\mathfrak{A}(P(\mathfrak{A}) \;\&\; \forall x(\alpha(x) = F(\mathfrak{A}, x))),$$

S is Σ_1^1. It follows by the boundedness theorem that there is a recursive ordinal σ such that S is a subclass of Tr_σ. We define R by

$$R(\mathfrak{A}) \leftrightarrow Tr_\sigma(\lambda x F(\mathfrak{A}, x)).$$

Since Tr_σ is hyperarithmetical by Lemma 3, R is hyperarithmetical. If $P(\mathfrak{A})$, then $\lambda x F(\mathfrak{A}, x)$ is in S and hence in Tr_σ; so $R(\mathfrak{A})$. If $Q(\mathfrak{A})$, then $\lambda x F(\mathfrak{A}, x)$ is not in Tr and hence not in Tr_σ; so $\neg R(\mathfrak{A})$. Thus R separates P and Q.

Characterization Theorem (*Souslin-Kleene*). A relation is Δ_1^1 iff it is hyperarithmetical.

Proof. If P is hyperarithmetical, it is Δ_1^1 by Lemma 1. If P is Δ_1^1, then P and $\neg P$ are disjoint Σ_1^1 sets. By the separation theorem, there is a hyperarithmetical R which separates P and $\neg P$. This implies that $R = P$; so P is hyperarithmetical.

We can relativize the entire argument of this section to a finite sequence Φ. The relativized characterization theorem together with the finiteness lemma implies that a relation is $\mathbf{\Delta}_1^1$ iff it is Borel.

Having obtained this analogue of the negation theorem, it is tempting to conjecture the following analogue of Post's theorem: a predicate is Δ_{n+1}^1 iff it is hyperarithmetical in the class of Π_n^1 predicates. This is false, however; the predicates hyperarithmetical in the class of Π_n^1 predicates form a proper subclass of the class of Δ_{n+1}^1 predicates (Problem 18). Some useful characterizations of Δ_2^1 predicates are known (see Problem 27); but no such characterization of Δ_n^1 is known for any $n > 2$.

7.11 BASIS THEOREMS

A function is Σ_n^i (Π_n^i) (Δ_n^i) if its graph is Σ_n^i (Π_n^i) (Δ_n^i) (where $i = 0$ or $i = 1$). Now for a total function F

$$\neg \mathfrak{G}_F(\mathfrak{a}, a) \leftrightarrow \exists x(x \neq a \;\&\; \mathfrak{G}_F(\mathfrak{a}, x)).$$

Hence if F is Σ_n^0, then $\neg\mathfrak{G}_F$ is Σ_n^0 and hence F is Δ_n^0; while if F is Σ_n^1 or Π_n^1, then $\neg\mathfrak{G}_F$ is Σ_n^1 or Π_n^1 respectively, and F is Δ_n^1. For this reason, we shall only discuss Δ_n^i functions.

A function F is Δ_1^0 iff $\mathfrak{G}_F$ is recursive (by the negation theorem) and hence iff F is recursive. A function F is Δ_1^1 iff $\mathfrak{G}_F$ is hyperarithmetical (by the characterization theorem); in this case, we say that F is *hyperarithmetical.*

We are interested in the following problem. Suppose that P is a class of unary functions, i.e., a (1, 0)-ary relation. Suppose that we know something about the classification of P in one of the hierarchies. What can be said about the classification of the functions in P? We cannot hope to say anything about all the functions in P; e.g., if P is the set of all functions, P is recursive, but some members of P have no classification. The best that we can hope to prove is that if P has a simple classification, then some member of P has a simple classification.

We are thus led to the following definition. A class B of unary functions is a *basis* for a collection Φ of classes of unary functions if for every P in Φ,

$$\exists \alpha P(\alpha) \rightarrow \exists \alpha(B(\alpha) \;\&\; P(\alpha)).$$

Example. The class of functions which have the value 0 for all but a finite number of arguments is a basis for the collection of Σ_1^0 classes of functions. For suppose that P is Σ_1^0 and nonempty. For some R, $P(\alpha) \leftrightarrow \exists x R(\bar{\alpha}(x))$. Since P is nonempty, there is a sequence number s such that $R(s)$. Setting $\alpha(x) = (s)_x$ for $x < lh(s)$ and $\alpha(x) = 0$ for $x \geqslant lh(s)$, we obtain an α in P.

We note that if B is a basis for the collection of Π_1^0 classes of functions, then the class of functions $(\gamma)_0$ with γ in B is a basis for the collection of Σ_1^1 classes of functions. For let P be Σ_1^1 and nonempty. Then $P(\alpha) \leftrightarrow \exists\beta Q(\alpha, \beta)$ where Q is Π_1^0. Define a Π_1^0 relation R by $R(\gamma) \leftrightarrow Q((\gamma)_0, (\gamma)_1)$. Since P is nonempty, Q is nonempty; so R is nonempty. Thus R contains a function γ in B; and $(\gamma)_0$ is a function in P. In a similar way, we see that if B is a basis for the collection of Π_n^1 classes of functions, then the class of functions $(\gamma)_0$ with γ in B is a basis for the collection of Σ_{n+1}^1 classes of functions.

For P a class of functions, we let I_P be the set of numbers $\bar{\alpha}(x)$ with α in P and x arbitrary. This has the explicit definition

$$I_P(a) \leftrightarrow \exists\alpha(P(\alpha) \ \& \ \bar{\alpha}(lh(a)) = a). \tag{1}$$

Lemma. If P is a nonempty Π_1^0 class of functions, then P contains a function recursive in I_P.

Proof. Define F inductively by

$$F(n) \simeq \mu z I_P(\bar{F}(n) * \langle z \rangle).$$

By R14, F is recursive in I_P. We show by induction on n that $\bar{F}(n)$ is defined and in I_P. Since P is nonempty, $\bar{F}(0) = \langle\ \rangle$ is in I_P. Now suppose that $\bar{F}(n)$ is defined and in I_P. Then $\bar{F}(n) = \bar{\alpha}(n)$ for some α in P. Since

$$I_P(\bar{\alpha}(n+1)) \quad \text{and} \quad \bar{\alpha}(n+1) = \bar{\alpha}(n) * \langle \alpha(n) \rangle,$$

$F(n)$ is defined and $I_P(\bar{F}(n) * \langle F(n) \rangle)$. Hence $\bar{F}(n+1)$ is defined and in I_P.

It remains to show that F is in P. We have $P(\alpha) \leftrightarrow \forall x R(\bar{\alpha}(x))$ for some predicate R. Clearly I_P is a subset of R; so $\forall x R(\bar{F}(x))$, and hence F is in P.

Kleene Basis Theorem. The class of functions which are recursive in the class of Σ_1^1 predicates is a basis for the collection of Π_1^0 classes of functions and hence for the collection of Σ_1^1 classes of functions. The class of hyperarithmetical functions is not a basis for the collection of Π_1^0 classes of functions.

Proof. If P is Π_1^0, then I_P is Σ_1^1 by (1); so the first result follows from the lemma. For the second result, it suffices to prove that the class H of hyperarithmetical functions is not a basis for the collection of Σ_1^1 classes of functions. Since $\neg H$ is not empty and contains no member of H, it will suffice to show that $\neg H$ is Σ_1^1, or, equivalently, that H is Π_1^1. For this we have the explicit definition

$$H(\alpha) \leftrightarrow \exists i \, \forall x \, \forall y((\alpha(x) = y \rightarrow J(i, x, y)) \ \& \ (\alpha(x) \neq y \rightarrow J(\langle 1, i \rangle, x, y))). \tag{2}$$

(Note that the right-hand side implies that i is an H-index, since it implies that $J(i, x, y)$ for some x and y.)

We may also apply our considerations to classes of sets. In order not to have to introduce new terminology, we consider instead classes of representing functions of sets, i.e., unary functions having only 0 and 1 as values. We use small Greek letters with asterisks as variables which vary through such functions.

Infinity Lemma (*Brouwer-König*). For any predicate P,

$$\exists\alpha^* \, \forall x P(\overline{\alpha^*}(x)) \leftrightarrow \forall n \, \exists\alpha^* \, \forall x_{x \leqslant n} P(\overline{\alpha^*}(x)).$$

Proof. The implication from left to right is obvious. Assume that the right-hand side holds. Let A be the set of sequence numbers $\langle a_0, \ldots, a_n\rangle$ such that $a_i \leqslant 1$ and $P(\langle a_0, \ldots, a_i\rangle)$ for all $i \leqslant n$. We shall define $\alpha^*(x)$ by induction on x so that for all x, $\overline{\alpha^*}(x)$ has infinitely many extensions in A; this will imply that $\forall x P(\overline{\alpha^*}(x))$. The right-hand side of the equivalence implies that $\overline{\alpha^*}(0) = \langle\ \rangle$ has infinitely many extensions in A. Now suppose that $\overline{\alpha^*}(x)$ is defined and has infinitely many extensions in A. Every proper extension of $\overline{\alpha^*}(x)$ in A is an extension of either $\overline{\alpha^*}(x) * \langle 0\rangle$ or $\overline{\alpha^*}(x) * \langle 1\rangle$. It follows that for a suitable choice of $\alpha^*(x)$,

$$\overline{\alpha^*}(x + 1) = \overline{\alpha^*}(x) * \langle\alpha(x)\rangle$$

has infinitely many extensions in A.

Corollary. If P is Π^0_1 and Q is defined by

$$Q(\mathfrak{A}) \leftrightarrow \exists\alpha^* P(\mathfrak{A}, \alpha^*),$$

then Q is Π^0_1.

Proof. We have $P(\mathfrak{A}, \alpha) \leftrightarrow \forall x R(\bar{\alpha}(x), \mathfrak{A})$ with R recursive. Hence by the lemma

$$Q(\mathfrak{A}) \leftrightarrow \forall n \, \exists\alpha^* \, \forall x_{x<n} R(\overline{\alpha^*}(x), \mathfrak{A}).$$

It will therefore suffice to prove that the part following $\forall n$ is a recursive relation of $\mathfrak{A}, n$.

Define a recursive function F by

$$F(0) = \langle\ \rangle,$$
$$F(n + 1) = \mu z \, \forall y_{y \leqslant F(n)}(y * \langle 0\rangle \leqslant z \;\&\; y * \langle 1\rangle \leqslant z).$$

Then if $a_i \leqslant 1$ for $i < n$, we have $\langle a_0, \ldots, a_{n-1}\rangle \leqslant F(n)$. Hence

$$\exists\alpha^* \, \forall x_{x \leqslant n} R(\overline{\alpha^*}(x), \mathfrak{A})$$
$$\leftrightarrow \exists s_{s \leqslant F(n)}(Seq(s) \;\&\; lh(s) = n \;\&\; \forall i_{i<n}((s)_i \leqslant 1) \;\&\; \forall i_{i \leqslant n} R(In(s, i), \mathfrak{A})).$$

This gives the desired result.

Kreisel Basis Theorem. The class of Δ^0_2 functions is a basis for the collection of Π^0_1 classes of representing functions.

Proof. Let P be a nonempty Π_1^0 class of representing functions. By (1) and the corollary to the infinity lemma, I_P is Π_1^0. By the lemma and Post's theorem, P contains a Δ_2^0 function.

If P is the class having the unary function F as its only member, we say that P *defines* F *implicitly*. Then F is Δ_n^1 iff P is Δ_n^1. For assume that P is Δ_n^1 and hence Σ_n^1. Since

$$\mathfrak{G}_F(a, b) \leftrightarrow \exists\alpha(P(\alpha) \,\&\, \alpha(a) = b)$$
$$\leftrightarrow \forall\alpha(P(\alpha) \to \alpha(a) = b),$$

F is Δ_n^1. Now suppose that F is Δ_n^1. Then $\mathfrak{G}_F$ and $\neg\mathfrak{G}_F$ are Δ_n^1. Since

$$P(\alpha) \leftrightarrow \forall x\, \forall y(\alpha(x) = y \leftrightarrow \mathfrak{G}_F(x, y))$$
$$\leftrightarrow \forall x\, \forall y((\alpha(x) = y \,\&\, \mathfrak{G}_F(x, y)) \vee (\alpha(x) \neq y \,\&\, \neg\mathfrak{G}_F(x, y))),$$

P is Δ_n^1.

It follows that in order to prove that the class of Δ_n^1 functions is a basis for Φ, it is sufficient to prove that every nonempty class in Φ contains a function defined implicitly by a Δ_n^1 class; equivalently, that every nonempty class in Φ has a Δ_n^1 subclass which contains exactly one function. We shall apply this method to show that the class of Δ_2^1 functions is a basis for the collection of Π_1^1 classes of functions.

Uniformization Theorem (*Novikoff-Kondo-Addison*). If P is a Π_1^1 relation, then there is a Π_1^1 relation Q such that for all $\mathfrak{A}$, α, and β:

a) $Q(\alpha, \mathfrak{A}) \to P(\alpha, \mathfrak{A})$;

b) $Q(\alpha, \mathfrak{A}) \,\&\, Q(\beta, \mathfrak{A}) \to \alpha = \beta$;

c) $\exists\alpha P(\alpha, \mathfrak{A}) \to \exists\alpha Q(\alpha, \mathfrak{A})$.

Proof. Since $\mathfrak{A}$ remains fixed, we shall omit it to simplify the notation. By the tree theorem, there is a recursive functional F such that $P(\alpha) \leftrightarrow Tr(\lambda x F(\alpha, x))$. We write F_α for $\lambda x F(\alpha, x)$ and $F_{\alpha,n}$ for $(F_\alpha)_{[n]}$.

If P is empty, we take Q to be empty. Now suppose that P is not empty. We shall define a nonempty subclass P_n of P for each n by induction on n. Let σ be the smallest of the ordinals $\|F_\alpha\|$ for α in P, and let P_0 be the class of α in P such that $\|F_\alpha\| = \sigma$. Now suppose that P_n has been defined. We let s_n be the smallest of the numbers $\bar{\alpha}(n)$ for α in P_n, and let σ_n be the smallest of the ordinals $\|F_{\alpha,n}\|$ for α in P_n and $\bar{\alpha}(n) = s_n$. We then let P_{n+1} be the set of α in P_n such that $\bar{\alpha}(n) = s_n$ and $\|F_{\alpha,n}\| = \sigma_n$. We then have

$$\alpha \in P_n \,\&\, \bar{\alpha}(n) \leqslant s_n \,\&\, \|F_{\alpha,n}\| \leqslant \sigma_n \to \alpha \in P_{n+1}. \tag{3}$$

We let Q be the intersection of the P_n. Property (a) is obvious. If α is in Q, then $\bar{\alpha}(n) = s_n$ for all n; this proves (b). To prove (c), we must show that Q is not empty. Taking α in P_{n+2}, we have $\alpha \in P_{n+1}$; so $\bar{\alpha}(n + 1) = s_{n+1}$ and $\bar{\alpha}(n) = s_n$. Hence s_{n+1} is an extension of s_n. Since also $lh(s_n) = n$, there is a unique function γ such that $\bar{\gamma}(n) = s_n$ for all n. We shall show that γ is in Q.

We first show that

$$F_\gamma(m) = F_\gamma(n) = 0 \;\&\; m <^* n \rightarrow \sigma_m < \sigma_n. \tag{4}$$

Since F is recursive, we have $F(\alpha, a) \simeq b \leftrightarrow \exists z R(\bar{\alpha}(z), a, b)$. From this and $F_\gamma(m) = F_\gamma(n) = 0$, we see that there is a number k such that if $\bar{\alpha}(k) = \bar{\gamma}(k)$, then $F_\alpha(m) = F_\alpha(n) = 0$. We may also suppose that $k > m, n$. Choose α in P_{k+1}. Then $\bar{\alpha}(k) = s_k = \bar{\gamma}(k)$, so $F_\alpha(m) = F_\alpha(n) = 0$. From this and $m <^* n$ we get $\|F_{\alpha,m}\| = \|(F_{\alpha,n})_{[m]}\| < \|F_{\alpha,n}\|$. Since α is in P_{m+1} and P_{n+1}, this inequality becomes $\sigma_m < \sigma_n$. A similar (but somewhat simpler) proof shows that

$$F_\gamma(n) = 0 \rightarrow \sigma_n < \sigma. \tag{5}$$

Next we show that γ is in P. Otherwise, there would be a descending sequence $m_0, m_1, \ldots$ such that $F_\gamma(m_i) = 0$ for all i. Then by (4), $\sigma_{m_0} > \sigma_{m_1} > \cdots$, which is impossible.

We now prove

$$F_\gamma(n) = 0 \rightarrow \|F_{\gamma,n}\| \leqslant \sigma_n \tag{6}$$

by transfinite induction on σ_n. If $F_{\gamma,n}(m) = 0$, then $F_\gamma(m) = 0$ and $m <^* n$; so $\sigma_m < \sigma_n$ by (4). Hence by the induction hypothesis,

$$\|(F_{\gamma,n})_{[m]}\| = \|F_{\gamma,m}\| \leqslant \sigma_m < \sigma_n.$$

Using (8) of §7.10, it follows that $\|F_{\gamma,n}\| \leqslant \sigma_n$.

From (5) and (6), $\|F_\gamma\| \leqslant \sigma$; so γ is in P_0. If γ is in P_n, it follows from (3) and (6) that γ is in P_{n+1}. Thus by induction, γ is in all P_n and hence in Q.

It remains to show that Q is Π^1_1. In view of (3), we have

$$P_n(\beta) \leftrightarrow \|F_\beta\| \leqslant \sigma \;\&\; \forall m_{m<n}(\bar{\beta}(m) \leqslant s_m \;\&\; \|F_{\beta,m}\| \leqslant \sigma_m).$$

Hence if α is in P_n, then

$$P_n(\beta) \leftrightarrow T_{\leqslant}(F_\beta, F_\alpha) \;\&\; \forall m_{m<n}(\bar{\beta}(m) \leqslant \bar{\alpha}(m) \;\&\; T_{\leqslant}(F_{\beta,m}, F_{\alpha,m})).$$

The right-hand side of this equivalence can be written as $R(\alpha, \beta, n)$ where R is Σ^1_1. Then for α in P_n,

$$\begin{aligned} \neg P_{n+1}(\alpha) &\leftrightarrow \exists\beta(P_n(\beta) \;\&\; [\bar{\beta}(n) < \bar{\alpha}(n) \vee (\bar{\beta}(n) = \bar{\alpha}(n) \;\&\; \|F_{\beta,n}\| < \|F_{\alpha,n}\|)]) \\ &\leftrightarrow \exists\beta(R(\alpha, \beta, n) \;\&\; [\bar{\beta}(n) < \bar{\alpha}(n) \vee (\bar{\beta}(n) = \bar{\alpha}(n) \;\&\; T_{<}(F_{\beta,n}, F_{\alpha,n}))]) \\ &\leftrightarrow R'(\alpha, n) \end{aligned}$$

where R' is Σ^1_1. From this,

$$\begin{aligned} Q(\alpha) &\leftrightarrow P_0(\alpha) \;\&\; \forall n \,\neg R'(\alpha, n) \\ &\leftrightarrow P(\alpha) \;\&\; \forall\beta \,\neg T_{<}(F_\beta, F_\alpha) \;\&\; \forall n \,\neg R'(\alpha, n). \end{aligned}$$

Thus Q is Π^1_1. If P and Q are empty, we can use the same definition for Q.

Corollary 1. The class of Δ^1_2 functions is a basis for the collection of Π^1_1 classes of functions.

Corollary 2. Then class of Δ_2^1 functions is a basis for the collection of Σ_2^1 classes of functions.

The next problem would be to find a basis for the collection of Π_2^1 classes of functions. However, Lévy has shown that with the usual axioms for set theory, it cannot be proved that the class of functions definable in set theory is a basis for this collection. Since the definable functions include, e.g., all functions recursive in the set of analytical predicates, we see that the problem cannot be solved satisfactorily without new axioms. On the other hand, Addison has shown that it is consistent with the present axioms to assume that the class of Δ_3^1 functions is a basis for the collection of Π_2^1 classes of functions.

Since we are now obviously far removed from the decision problems which first led us to recursive functions, it is natural to ask what we hope to accomplish by the study of hierarchies. One answer is that we hope to help clarify the notion of a set. From this point of view, we think of the hierarchies as classifying certain important sets according to the complexity of their definition. It is therefore not surprising that, as just indicated, many of the unsolved problems in hierarchy theory are connected with problems in axiomatic set theory.

PROBLEMS

1. a) Let F be defined inductively by

$$F(0, a) = G(a),$$
$$F(b + 1, a) = H(F(b, K(F(b, a), b, a)), b, a)$$

where G, H, and K are recursive functions. Show that F is recursive. [Define by the recursion theorem a recursive partial function F' satisfying the equations with $=$ replaced by $\simeq$, and then prove $F'(b, a) \simeq F(b, a)$ by induction on b.]

b) Show that there is a binary recursive function F such that for each n-ary primitive recursive function G, there is a number g such that $F_{(g)} = \langle G \rangle$. [Assign an index g to each primitive recursive function G. Set $F(a, g) = \langle G \rangle(a)$ if g is an index of G, and $F(a, g) = 0$ if g is not an index. Use the method of (a) to show that F is recursive.] Conclude that F is not primitive recursive. [Show that the H defined by $H(a) = F(a, a) + 1$ is not primitive recursive.]

2. a) Show that the domain of a recursive partial functional is recursively enumerable.

b) Let F be a recursive partial functional, and let A be a subset of the domain of F. Show that the restriction of F to A is recursive iff A is recursively enumerable. [Use (a) and R12.]

c) Let P be a recursive predicate such that $\forall x P(x, a)$ is not a recursively enumerable predicate of a. Let $H(x, a) \simeq \mu z P(x, a)$, $G(\alpha, a) = 0$, $F(a) \simeq G(\lambda x H(a, x), a)$. Show that F is not recursive. [Use (a).]

3. A partial functional F is a *selector* for a relation P if for all $\mathfrak{A}$ such that $\exists x P(x, \mathfrak{A})$, $F(\mathfrak{A})$ is defined and $P(F(\mathfrak{A}), \mathfrak{A})$.

a) Show that if P is recursively enumerable, then there is a recursive selector for P. [If $P(x, \mathfrak{A}) \leftrightarrow \exists y R(x, \mathfrak{A}, y)$, let $F(\mathfrak{A}) \simeq (\mu z R((z)_0, \mathfrak{A}, (z)_1))_0$.]

b) Show that for a suitable recursively enumerable P, the selector $\mu x P(x, a)$ for P is not recursive. [Let $P(x, a) \leftrightarrow R(a) \vee x = 1$ where R is recursively enumerable but not recursive.]

c) Suppose that P is recursive. Show that there is a recursive total functional which is a selector for P iff $\exists x P(x, \mathfrak{A})$ is a recursive relation of $\mathfrak{A}$.

d) Let $F_1, \ldots, F_n$ be recursive partial functionals. Show that there is a recursive partial functional F such that for each $\mathfrak{A}$, $F(\mathfrak{A})$ is defined iff at least one of the $F_i(\mathfrak{A})$ is defined, and, in this case, $F(\mathfrak{A}) = F_i(\mathfrak{A})$ for some i. [Use (a).]

4. A recursive partial function F is a *creating function* for a set A if for each e such that A and W_e are disjoint, $F(e)$ is defined and is in neither A nor W_e. A recursively enumerable set A is *creative* if it has a creating function. A recursively enumerable set A is *simple* if its complement is infinite but includes no infinite recursively enumerable set.

a) Show that a creative set exists. [Let $A(e) \leftrightarrow W_e(e)$.]

b) Show that a creative set is not recursive. [Use the negation theorem.]

c) Show that a creative set A has a total creating function. [Define a recursive G whose domain is the set of e such that W_e and A are not disjoint, and use 3(d).]

d) Show that a simple set exists. [Let $P(x, e) \leftrightarrow W_e(x) \mathbin{\&} x > 2e$. Let F be a recursive selector for P and let A be the set of values of F. Show that for each k, there are at most k numbers in A which are $\leqslant 2k$.]

e) Show that a simple set is not recursive. [Use the negation theorem.]

f) Show that a simple set is not creative.

5. A set A is *many-one reducible* to a set B if there is a recursive function F such that $A(a) \leftrightarrow B(F(a))$ for all a. If, in addition, F may be chosen injective (bijective), then A is *one-one reducible* to B (A is *recursively isomorphic* to B).

a) Show that if A and B are recursively isomorphic, then each is one-one reducible to the other.

b) Let A be one-one reducible to B. Let $R(a, b)$ mean that $a = \langle a_1, \ldots, a_n \rangle$, $b = \langle b_1, \ldots, b_n \rangle$, and $a_i = a_j \leftrightarrow b_i = b_j$ and $A(a_i) \leftrightarrow B(b_i)$ for all i, j. Show that there is a recursive function G such that

$$R(a, b) \rightarrow R(a * \langle x \rangle, b * \langle G(a, b, x) \rangle).$$

[Let $A(a) \leftrightarrow B(F(a))$ with F injective. If $x = a_i$, let $G(a, b, x) = b_i$. Otherwise, if $F(x)$ is not a b_i, let $G(a, b, x) = F(x)$; if $F(x) = b_{i_1}$ and $F(a_{i_1})$ is not a b_i, let $G(a, b, x) = F(a_{i_1})$; etc.]

c) Show that if each of A and B is one-one reducible to the other, then A is recursively isomorphic to B. [Use (b) and the technique of the proof of Ryll-Nardjewski's theorem.]

d) Show that there are recursively enumerable sets A and B such that A is one-one reducible to B, B is many-one reducible to A, and B is not one-one reducible to A. [Let A be simple, and obtain B from A by omitting an infinite recursive subset.]

6. a) Show that if P is recursively enumerable, then there is a recursive function H such that $W_{H(x)}(a) \leftrightarrow P(a, x, H(x))$ for all a and x. [Define a recursive G with an index g such that $G(a, x) = 0$ if $P(a, x, S(g, x))$ and $G(a, x)$ is undefined otherwise, and set $H(x) = S(g, x)$.]

b) Show that if A is creative and P is recursively enumerable, then there is a recursive function F such that $P(x, F(x)) \leftrightarrow A(F(x))$ for all x. [Let G be a total creating function for A. Choose H by (a) so that

$$W_{H(x)}(a) \leftrightarrow a = G(H(x)) \;\&\; P(x, G(H(x)))$$

and let $F(x) = G(H(x))$.]

c) Show that if A is creative, then there is a total creating function F for A such that if A and W_e are not disjoint, then $A(F(e))$. [Let $P(e, y) \leftrightarrow W_e(y) \vee \exists x(W_e(x) \;\&\; A(x))$ and apply (b).]

d) Let A be creative. Show that there is a recursive function F such that if $A(x)$, then $W_{F(x)} \subset A$, and if $\neg A(x)$, then $W_{F(x)}$ is infinite and disjoint from A. [Use (c).]

e) Show that if A is creative, then there is a recursive function F such that for all x and y, $F(x, y) \geqslant y$ and $A(x) \leftrightarrow A(F(x, y))$. [Use (d) and 3(a).]

f) Show that if A is creative and B is many-one reducible to A, then B is one-one reducible to A. [If F is recursive, use (e) to define inductively an injective recursive function F' such that $A(F(a)) \leftrightarrow A(F'(a))$.]

g) Show that for a recursively enumerable set A, the following are equivalent:

i) A is creative;

ii) every recursively enumerable set is many-one reducible to A;

iii) every recursively enumerable set is one-one reducible to A.

[Use (b), (f), and 4(a).]

7. Two disjoint sets A and B are *effectively recursively inseparable* if there is a recursive partial function F such that if $A \subset W_e$ and $B \subset W_f$, then $F(e, f)$ is defined and belongs to neither W_e nor W_f.

a) Show that effectively recursively inseparable sets are recursively inseparable. [Use the negation theorem.]

b) Show that effectively recursively inseparable recursively enumerable sets exist. [Let

$$A(x) \leftrightarrow \exists y(T_1((x)_1, x, y) \;\&\; \forall z_{z \leqslant y} \neg T_1((x)_0, x, z)),$$
$$B(x) \leftrightarrow \exists y(T_1((x)_0, x, y) \;\&\; \forall z_{z < y} \neg T_1((x)_1, x, z)),$$

and $F(e, f) = \langle e, f \rangle$.]

c) Show that if A and B are effectively recursively inseparable recursively enumerable sets, then A, B, and $A \vee B$ are creative.

d) Show that there are recursively inseparable recursively enumerable sets A and B which are not effectively recursively inseparable. [Construct A and B as follows. At the nth stage, let $e = (n)_0$, $x = (n)_1$, $y = (n)_2$. Do nothing unless $T_1(e, x, y) \;\&\; x > 3e$. In this case, put x in A if it has not been put in B and no number has been put in A at a stage $m < n$ with $(m)_0 = e$. Otherwise, put x in B if it has not been put in A and no number has been put in B at a stage $m < n$ with $(m)_0 = e$. Show that $A \vee B$ is simple, and that if W_e & $\neg(A \vee B)$ is infinite, then each of A and B contains a member of W_e. Use (c) and 4(f).]

8. Let Ux mean *for infinitely many x*. A predicate P is U_n (V_n) if it has a definition

$$P(\mathfrak{a}) \leftrightarrow Ux_1 \ldots Ux_n R(\mathfrak{a}, x_1, \ldots, x_n)$$

where R is recursive (Π_1^0).

a) Show that a U_n predicate is Π_{2n}^0 and that a V_n predicate is Π_{2n+1}^0.

b) If P is Π_{n+1}^0, show that $P(a) \leftrightarrow UxQ(a, x)$ for a Σ_n^0 predicate Q. [Use the equivalence $\forall z R(z) \leftrightarrow Uw \forall z_{z<w} R(w)$.]

c) If P is Π_{n+2}^0 (Π_2^0), show that

$$P(\mathfrak{a}) \leftrightarrow Ux(P_1(\mathfrak{a}, x) \;\&\; P_2(\mathfrak{a}, x))$$

where P_1 is Π_n^0 (recursive) and P_2 is Σ_n^0 (recursive). [Use (b) and the equivalence

$$Ux\, \exists y Q(x, y) \leftrightarrow Uz\big(Q((z)_0, (z)_1) \;\&\; \forall w_{w<(z)_1} \neg Q((z)_0, w)\big).]$$

d) If P is Π_{n+2}^0, show that $P(\mathfrak{a}) \leftrightarrow UxP_1(\mathfrak{a}, x)$ where P_1 is Π_n^0. [Use (c) and the equivalence

$$Ux(\exists y Q(x, y) \;\&\; \forall v R(x, v)) \leftrightarrow Uz\big(Q((z)_0, (z)_1) \;\&\; \forall w_{w<(z)_1} \neg Q((z)_0, w) \;\&\; \forall v R((z)_0, v)\big).]$$

e) Let P be Π_n^0. Show that if $n = 2m$, then P is U_m, and if $n = 2m + 1$, then P is V_m. [Use (c), (d), and induction on m.]

9. A *complete* Σ_n^0 (Π_n^0) set is a Σ_n^0 (Π_n^0) set A such that every Σ_n^0 (Π_n^0) set is many-one reducible to A.

a) Let P be a binary Σ_n^0 (Π_n^0) predicate which enumerates the class of unary Σ_n^0 (Π_n^0) predicates, and define $A(a) \leftrightarrow P((a)_0, (a)_1)$. Show that A is a complete Σ_n^0 (Π_n^0) set.

b) Show that the set of e such that W_e is infinite is a complete Π_2^0 set. [Use 8(e).]

c) Show that the set of e such that $\neg W_e$ is infinite is a complete Π_3^0 set. [Use 8(a) and 8(e).]

d) Show that the set of e such that W_e is recursive is a complete Σ_3^0 set. [Use the negation theorem to show that this set is Σ_3^0. Let A be a recursively enumerable set which is not recursive. Define a recursive function F such that

$$W_{F(e)}(x) \leftrightarrow \big(W_e((x)_0) \;\&\; (x)_1 = 0\big) \vee \big((x)_0 < (x)_1 \;\&\; A((x)_0)\big) \vee \big((x)_1 > 0 \;\&\; W_e((x)_1)\big).$$

Show that $W_{F(e)}$ is recursive iff $\neg W_e$ is finite, and use (c).]

10. A set A is *truth-table reducible* to a set B if there is a recursive function F and a recursive predicate P such that $A(x) \leftrightarrow P(\overline{K_B}(F(x)))$ for all x.

a) Show that if A is many-one reducible to B, then A is truth-table reducible to B.

b) Show that for every recursively enumerable set A, there is a simple set B such that A is truth-table reducible to B. [We may suppose that $\neg A$ is infinite. Let C be a simple set such that for each n, there are at most n numbers $<2n + 2$ in C. Let D_n be the set of x such that $2^n \leqslant x < 2^{n+1}$, and let B be the union of C and the D_n for n in A.] Conclude that the converse of (a) is false, even for recursively enumerable sets. [Use 4(f).]

c) Show that A is truth-table reducible to B iff there is a recursive total functional F such that $K_A(x) = F(K_B, x)$ for all x. [Suppose that such an F exists. Choose a recursive R such that $\mathfrak{G}_F(\alpha, x, a) \leftrightarrow \exists z R(\bar{\alpha}(z), x, a)$. Note that for each x

$$\forall \alpha^* \exists z R(\overline{\alpha^*}((z)_0), x, (z)_1),$$

and apply the infinity lemma.]

11. A recursively enumerable set A is *hypersimple* if the complement of A is infinite and there is no recursive function F such that $\forall x \exists y(\neg A(y) \,\&\, x < y < F(x))$.

a) Show that if A is hypersimple, then A is simple and hence not recursive.

b) Suppose that the creative set A is truth-table reducible to the recursively enumerable set B. Show that B is not hypersimple. [Let $A(a) \leftrightarrow P(\overline{K_B}(F(x)))$ with P and F recursive. Choose a recursive G by 6(b) so that $A(G(e)) \leftrightarrow W_e(G(e))$. Let $J_n(x) = (n)_x$ if $x < lh(n)$, $J_n(x) = 0$ otherwise. Choose a recursive H such that

$$W_{H(n)}(a) \leftrightarrow \neg P(\overline{J_n}(F(a))).$$

Let $L(n) = F(G(H(n)))$. Show that

$$\neg P(\overline{J_n}(L(n))) \leftrightarrow P(\overline{K_B}(L(n))),$$

and conclude that $J_n(i) \neq K_B(i)$ for some $i < L(n)$. Use L to define a recursive function M such that $\forall x \exists y(\neg B(y) \,\&\, x < y < M(x))$.]

c) Show that for every nonrecursive recursively enumerable set A, there is a hypersimple set B such that B is truth-table reducible to A and A is recursive in B. [Let F be an injective recursive function enumerating A, and define B by

$$B(a) \leftrightarrow \exists x(a < x \,\&\, F(x) < F(a)).$$

Show that the complement of B is infinite, and that

$$\neg B(w) \,\&\, F(w) > x \rightarrow (A(x) \leftrightarrow \exists z_{z<w}(x = F(z))).$$

Assume that G is a recursive function such that $\forall x \exists y(\neg B(y) \,\&\, x < y < G(x))$. Show that we can determine if a is in A by searching for an x such that

$$a < min(F(x), F(x + 1), \ldots, F(G(x))).]$$

d) Show that there are recursively enumerable sets A and B such that A is recursive in B but not truth-table reducible to B. [Use (b) and (c).] Conclude that the replacement lemma does not hold for recursive total functions and predicates. [Use 10(c).]

12. Let $\mathfrak{F}$ be the class of recursive unary partial functions. A subclass $\mathfrak{F}'$ of $\mathfrak{F}$ is *completely recursively enumerable* (*completely recursive*) if the set of indices of partial functions in $\mathfrak{F}'$ is recursively enumerable (recursive). A mapping Φ from a subclass $\mathfrak{F}'$ of $\mathfrak{F}$ to the set of numbers is an *effective operation* if there is a recursive partial function F such that $F(e) \simeq \Phi(\{e\})$ whenever $\{e\}$ is in $\mathfrak{F}'$.

a) Show that if $\mathfrak{F}'$ is completely recursively enumerable, then every extension of a partial function in $\mathfrak{F}'$ which is in $\mathfrak{F}$ is in $\mathfrak{F}'$. [Suppose that G is in $\mathfrak{F}'$ and its extension H is not. Let A be recursively enumerable but not recursive, and define F_1, F_2 by

$$F_1(x, y) \simeq H(x) \quad \text{if} \quad A(y),$$
$$F_2(x, y) \simeq G(x).$$

Define F as in 3(d), and choose a recursive function K so that $\{K(y)\}(x) \simeq F(x, y)$. Show that $\{K(y)\}$ is in $\mathfrak{F}'$ iff $\neg A(y)$.]

b) Show that if $\mathfrak{F}'$ is completely recursively enumerable and G is in $\mathfrak{F}'$, then G is an extension of a partial function G' in $\mathfrak{F}'$ which has a finite domain. [Suppose that there is no such G'. Choose e so that W_e is not recursive. Define F by

$$F(y, x) \simeq G(x) \quad \text{if} \quad \neg \exists z_{z<x} T_1(e, y, z)$$

and choose a recursive function K such that $\{K(y)\}(x) \simeq F(x, y)$. Show that $\{K(y)\}$ is in $\mathfrak{F}'$ iff $\neg W_e(y)$.]

c) Let $\mathfrak{F}_{s,t,n}$ be the set of G in $\mathfrak{F}$ such that $G((s)_i) \simeq (t)_i$ for $i < n$. Show that a subset $\mathfrak{F}'$ of $\mathfrak{F}$ is completely recursively enumerable iff there is a recursively enumerable predicate P such that $\mathfrak{F}'$ is the union of the $\mathfrak{F}_{s,t,n}$ for s, t, n such that $P(s, t, n)$. [Use (a) and (b).]

d) Show that the only completely recursive subclasses of $\mathfrak{F}$ are $\mathfrak{F}$ and the empty class. [Use (a) and the negation theorem.]

e) Show that a mapping Φ from a completely recursively enumerable class $\mathfrak{F}'$ to the set of numbers is an effective operation iff there is a recursive partial function H such that for F in $\mathfrak{F}'$,

$$\Phi(F) \simeq a \leftrightarrow \exists s \exists t \exists n (F \in \mathfrak{F}_{s,t,n} \;\&\; H(s, t, n) \simeq a).$$

[If Φ is an effective operation, the set of F such that $\Phi(F) \simeq a$ is completely recursively enumerable. Using this, imitate the proof of (c).]

f) Show that the requirement in (e) that $\mathfrak{F}'$ be completely recursively enumerable cannot be replaced by the requirement that every partial function in $\mathfrak{F}'$ be total. [Let $\Phi(F) = 0$ if $F(x) = 0$ for all x, and $\Phi(F) = 1$ if F is a recursive total function and $\exists x(x \leqslant e \;\&\; F(x) \neq 0)$ for every index e of F. Show that for each k, there is an F such that $F(x) = 0$ for $x \leqslant k$ and $\Phi(F) = 1$.]

13. Let $\mathfrak{R}$ be the class of recursive unary functions. A subclass $\mathfrak{R}'$ of $\mathfrak{R}$ is *totally recursively enumerable* (*totally recursive*) if there is a recursively enumerable (recursive) set A such that $\{e\} \in \mathfrak{R}' \leftrightarrow A(e)$ whenever $\{e\}$ is total.

a) Let $\mathfrak{R}'$ be totally recursively enumerable. Show that for every G in $\mathfrak{R}'$ and every z, there is a G' in $\mathfrak{R}'$ such that $G'(x) = G(x)$ for $x < z$ and $G'(x) = 0$ for all but a finite number of x. [Like 12(b).]

b) Let Φ be a mapping from a totally recursively enumerable class $\mathfrak{R}'$ to the set of numbers. Show that Φ is an effective operation iff it has an extension which is a recursive partial functional. [Suppose that Φ is an effective operation; say $F(e) \simeq \Phi(\{e\})$ for $\{e\}$ in $\mathfrak{R}'$. Let K be a recursive function such that $\{K(n)\}(x) = (n)_x$ if $Seq(n) \;\&\; x < lh(n)$, $\{K(n)\}(x) = 0$ otherwise. Use 3(a) to define a recursive partial function M such that

$M(f, y)$ is a number n such that $\{K(n)\}$ is in $\mathfrak{R}'$, $\overline{\{f\}}(y)$ has n as an extension, and $F(f) \neq F(K(n))$, provided that such an n exists. Choose a recursive function L such that

$$\{L(e, f)\}(x) \simeq \{f\}(x) \quad \text{if} \quad \forall y_{y<x} \neg T_1(e, e, y),$$
$$\simeq (M(f, \mu y T_1(e, e, y)))_x \quad \text{otherwise.}$$

If $\neg W_e(e)$, then $\{L(e,f)\} = \{f\}$; if $y = \mu y T_1(e, e, y)$ and $M(f, y)$ is defined, then

$$\{L(e, f)\} = \{K(M(f, y))\}.$$

Choose a recursive function N such that $W_{N(f)}$ is the set of e such that $F(L(e, f)) \simeq F(f)$. Show that if $\{f\} \in \mathfrak{R}'$, then $N(f) \in W_{N(f)}$. If $F(f)$ is defined and

$$y = \mu y T_1(N(f), N(f), y)$$

and G is in $\mathfrak{R}'$ and $\overline{G}(y) = \overline{\{f\}}(y)$, then $\Phi(G) = F(f)$; for otherwise $M(f, y)$ is defined by (a), so $F(L(N(f)), f) = F(K(M(f, y))) \neq F(f)$ and hence $N(f) \notin W_{N(f)}$. Let $Q(f, \alpha)$ mean that there is a y such that $F(f)$ is defined and $y = \mu y T_1(N(f), N(f), y)$ and $\bar{\alpha}(y) = \overline{\{f\}}(y)$. Let H be a recursive selector for Q. Show that $\Phi(\alpha) = F(H(\alpha))$ for α in $\mathfrak{R}'$.]

c) Show that a subclass $\mathfrak{R}'$ of $\mathfrak{R}$ is totally recursive iff there are disjoint recursively enumerable sets A and B such that for $\alpha \in \mathfrak{R}$,

$$\alpha \in \mathfrak{R}' \leftrightarrow \exists x A(\bar{\alpha}(x)) \quad \text{and} \quad \alpha \notin \mathfrak{R}' \leftrightarrow \exists x B(\bar{\alpha}(x)).$$

[If $\mathfrak{R}'$ is totally recursive, use (b) to find a recursive H such that $H(\alpha) = 0$ if α is in $\mathfrak{R}'$ and $H(\alpha) = 1$ for α recursive and not in $\mathfrak{R}'$. Express $\mathfrak{G}_H$ in terms of a recursive predicate.]

d) Show that if A is recursively enumerable, then the set of α in $\mathfrak{R}$ such that $\exists x A(\bar{\alpha}(x))$ is totally recursively enumerable. Show that not every totally recursively enumerable class can be obtained in this way. [Let an F in $\mathfrak{R}$ be in $\mathfrak{R}'$ if either $\forall x(F(x) = 0)$ or, setting $z = \mu x(F(x) \neq 0)$, $\exists e_{e<z} \forall x_{x\leqslant z}(\{e\}(x) = F(x))$. Show that if $\{f\}$ is in $\mathfrak{R}$ and $\{f\}(x) = 0$ for $x \leqslant f$, then $\{f\}$ is in $\mathfrak{R}'$, and conclude that $\mathfrak{R}'$ is totally recursively enumerable. Show that for every k there is an F in $\mathfrak{R}$ but not in $\mathfrak{R}'$ such that $F(x) = 0$ for all $x \leqslant k$.]

14. a) Show that if **a** and **b** are degrees, then the set consisting of **a** and **b** has a least upper bound (for the partial ordering $\leqslant$ of degrees). [Let A and B be sets in **a** and **b** respectively, and let **c** be the degree of the set C defined by

$$C(x) \leftrightarrow A((x)_0) \quad \text{if} \quad (x)_1 = 0,$$
$$C(x) \leftrightarrow B((x)_0) \quad \text{otherwise.}]$$

We designate this degree by $\mathbf{a} \cup \mathbf{b}$.

b) Let A and B be disjoint recursively enumerable sets having degrees **a** and **b** respectively. Show that $A \cup B$ has degree $\mathbf{a} \cup \mathbf{b}$.

c) Show that if $\mathbf{0}' \leqslant \mathbf{a}$, then there is a degree **b** such that $\mathbf{a} = \mathbf{b}' = \mathbf{b} \cup \mathbf{0}'$. [Let A be a set in **a**. Choose a recursive R such that

$$\exists x T_{1,1}((a)_0, \alpha, (a)_1, x) \leftrightarrow \exists x R(\bar{\alpha}(x), a)$$

and set

$$Q(s, a) \leftrightarrow \exists t(t <^* s \;\&\; R(t, a)).$$

Set $G(0) = \langle\ \rangle$ and

$$G(a+1) = \mu t(t <^* G(a) * \langle K_A(a)\rangle \ \&\ R(t,a)) \quad \textit{if} \quad Q(G(a) * \langle K_A(a)\rangle, a)$$
$$= G(a) * \langle K_A(a)\rangle \quad \textit{otherwise}.$$

Using the fact that Q is recursive in A, show that G and A are equivalent. Let

$$F(a) = (G(a+1))_a$$

and let $\mathbf{b}$ be the degree of F. Show that $\exists x R(\overline{F}(x), a)$ is recursive in G, and conclude that $\mathbf{b}' \leqslant \mathbf{a}$. Show that G is recursive in F and Q, and conclude that $\mathbf{a} \leqslant \mathbf{b} \cup \mathbf{0}'$.]

15. a) Let A be a recursively enumerable set which is not recursive. Show that there are disjoint recursively enumerable sets A_0 and A_1 whose union is A such that A is recursive in neither A_0 nor A_1. [Let F be an injective recursive function enumerating A, and let R be as in the proof of the Friedberg-Muchnik theorem. Define A_0 and A_1 in stages, simultaneously defining sets B_k. Let $A_{i,n}$ be the set of numbers put in A_i before the nth stage, and let D_n be the set of $F(x)$ for $x < n$. At the nth stage, for $j < n$ and $i = 0, 1$, let $x_{j,i}$ be the largest number less than n such that for all $x < x_{j,i}$,

$$\exists y_{y<n} R(\overline{K_{A_{i,n}}}(y), j, x, K_{D_n}(x)).$$

If $x < x_{j,i}$, put every number less than

$$x + \mu y R(\overline{K_{A_{i,n}}}(y), j, x, K_{D_n}(x))$$

in B_{2j+i}. Now pick $2j + i$ minimal such that $F(n)$ is in B_{2j+i} and put $F(n)$ in A_{1-i}; or, if $F(n)$ is in no B_k, put $F(n)$ in A_0. Show that if $\{j\}^{A_i} \neq K_A$, then B_{2j+i} is finite. Then prove $\{j\}^{A_i} \neq K_A$ by induction on $2j + i$ as follows. Show that there is a stage n_0 after which no number already in B_{2j+i} is put in A_i. Assume $\{j\}^{A_i} = K_A$, and show that $K_A(x)$ may be computed by looking for a stage $n > n_0$ at which $x < x_{i,j}$ and putting $K_A(x) = K_{D_n}(x)$.]

b) Show that every recursively enumerable nonrecursive set is the union of two disjoint recursively enumerable nonrecursive sets. [Use (a) and 14(b).]

c) Show that if $\mathbf{a}$ is a recursively enumerable degree different from $\mathbf{0}$, then there is a recursively enumerable degree $\mathbf{b}$ such that $\mathbf{0} < \mathbf{b} < \mathbf{a}$. [Use (a) and 14(b).]

16. Give each $N_{m,n}$ a topology as follows. Give $N_{0,1}$ the discrete topology. Consider $N_{1,0}$ as the product of countably many copies of $N_{0,1}$ and give it the product topology. Consider $N_{m,n}$ as the product of m copies of $N_{1,0}$ and n copies of $N_{0,1}$ and give it the product topology.

a) Let B_s be the class of all $\mathfrak{A}$ in $N_{m,n}$ such that $\overline{\mathfrak{A}}((s)_0) = ((s)_1, \ldots, (s)_{m+n})$. Show that the B_s form a base for $N_{m,n}$.

b) Show that a subset P of $N_{m,n}$ is open iff it is a Σ_1^0 relation. [If P is open, P is the union of the $B_{\alpha(n)}$ for some α by (a). If P is Σ_1^0, $P(\mathfrak{A}) \leftrightarrow \exists x R(\bar{\alpha}(x), \overline{\mathfrak{A}}(x))$ for some R and α.]

c) Show that a mapping Φ from $N_{m,n}$ to $N_{1,0}$ is continuous iff the functional F defined by $F(\mathfrak{A}, b) = (\Phi(\mathfrak{A}))(b)$ is functionally recursive. [If F is functionally recursive, use (b) to prove that the inverse under Φ of an open set is open. If Φ is continuous, show that the graph of F is open and use (b).]

d) Show that the class of (m, n)-ary Borel relations is the smallest class of subclasses of $N_{m,n}$ which contains the open sets and is closed under taking complements and countable unions. [Use (b).]

17. a) Show that if P is Σ_1^1, then there is a recursive relation R such that

$$P(\mathfrak{A}) \leftrightarrow \exists\alpha^* \,\forall x\, \exists y R(\mathfrak{A}, \alpha^*, x, y).$$

[Let $P(\mathfrak{A}) \leftrightarrow \exists\alpha\, \forall x Q(\bar{\mathfrak{A}}(x), \bar{\alpha}(x))$ with Q recursive. Then $P(\mathfrak{A})$ iff there is an α^* which is the representing function of some $\langle \mathfrak{S}_{\bar{\beta}} \rangle$ such that $\forall x\, \exists y (\alpha^*(\langle x, y\rangle) = 0\ \&\ Q(\bar{\mathfrak{A}}(x), y)$.]

b) Show that if P is defined by $P(\mathfrak{A}) \leftrightarrow \exists\alpha R(\mathfrak{A}, \alpha)$ with R recursive, then P is recursively enumerable. [Write the right-hand side as $\exists\alpha\, \exists x Q(\mathfrak{A}, \bar{\alpha}(x))$ with Q recursive.]

18. a) Show that if Φ is a finite sequence of Δ_n^1 predicates, then Φ^* is Δ_n^1.

b) Suppose that P is Σ_k^1 (Π_k^1) (Δ_k^1) in the set of Δ_n^1 predicates. Show that if $k \geqslant n$, then P is Σ_k^1 (Π_k^1) (Δ_k^1), and that if $k < n$, then P is Δ_n^1. [Note that if $P(\mathfrak{A}) \leftrightarrow Q(\Phi^*, \mathfrak{A})$ and R is the class whose only member is Φ^*, then

$$\begin{aligned} P(\mathfrak{A}) &\leftrightarrow \exists\alpha(R(\alpha)\ \&\ Q(\alpha, \mathfrak{A})) \\ &\leftrightarrow \forall\alpha(R(\alpha) \rightarrow Q(\alpha, \mathfrak{A})). \end{aligned}$$

Use this, the finiteness lemma, and (a).]

c) Show that every relation which is hyperarithmetical in the set of hyperarithmetical predicates is hyperarithmetical. [Use (b) and the characterization theorem.]

d) Show that the predicates which are hyperarithmetical in the set of Π_n^1 predicates form a proper subclass of the class of Δ_{n+1}^1 predicates. [Show that there is a Π_n^1 predicate P in which every Π_n^1 predicate is recursive, and choose a predicate which is Π_1^1 in P but not Σ_1^1 in P. Apply (b) and the relativized characterization theorem to this predicate.]

19. A *relational set* consists of a set A and a binary predicate $<_A$. A *descending sequence* in such a relational set is a sequence $a_0, a_1, \ldots$ of elements of A such that $a_{n+1} <_A a_n$ for all n. A relational set is *well founded* if it has no descending sequences. Thus a tree is a well-founded relational set with the predicate $<^*$. A relational set is *well-ordered* if it is well-founded and $<_A$ linearly orders A. If x is in A, then $A_{[x]}$ is the set of y in A such that $y <_A x$, considered as a relational set with the predicate $<_A$.

a) Show that ordinals may be assigned to well-founded relational sets in the same manner as to trees. The ordinal assigned to the well-founded relational set A is designated by $|A|$.

b) If A is a relational set, $DS(A)$ is the set of all numbers $\langle a_1, \ldots, a_n\rangle$ such that $a_n <_A a_{n-1} <_A \cdots <_A a_1$. Show that A is well-founded iff $DS(A)$ is a tree, and that in this case, $|A| = \|DS(A)\|$. [If A is well-founded and $a = \langle a_1, \ldots, a_n\rangle$ is in $DS(A)$, show that $|A_{[a_n]}| = \|DS(A)_{[a]}\|$ by transfinite induction on $|A_{[a_n]}|$.]

c) Show that for every countable ordinal σ, there is a tree which has ordinal σ. [Use (b).]

d) A relational set A is *recursive* (*hyperarithmetical*) if both the set A and the predicate $<_A$ are recursive (hyperarithmetical). Show that if A is a well-founded hyperarithmetical relational set, then $|A|$ is recursive. [Show that $DS(A)$ is hyperarithmetical, and apply the boundedness theorem to the class defining $K_{DS(A)}$ implicitly.]

e) If a and b are sequence numbers, $a <^\circ b$ means that either $a <^* b$, or there is a j such that $j < lh(a), j < lh(b), (a)_j < (b)_j$, and $\forall i_{i<j}((a)_i = (b)_i)$. Show that if $a_0, a_1, \ldots$ is a sequence such that $a_{n+1} <^\circ a_n$ for all n, then there is a descending sequence $b_0, b_1, \ldots$ for $<^*$ such that for each n, we have $a_m <^* b_n$ for all sufficiently large m. [Define b_n by induction on n.] Show that $<^\circ$ linearly orders *Seq*.

f) Let A be a relational set, and consider $DS(A)$ as a relational set with the predicate $<^\circ$. Show that A is well-founded iff $DS(A)$ is well-ordered, and that in this case, $|A| \leqslant |DS(A)|$. [Use (e) and the method of (b).] Conclude that the recursive ordinals are just the ordinals of the recursive well-ordered relational sets.

g) Show that if σ and τ are recursive, then $\sigma + \tau$ and $\sigma \cdot \tau$ are recursive. [Use (f).] Conclude that κ is a limit number.

20. Two predicates are *H-equivalent* if each is hyperarithmetical in the other.

a) Show that H-equivalence is an equivalence relation. [Use a relativized version of 18(c).] The equivalence classes are called *hyperdegrees*. Define $\leqslant$ and $<$ for hyperdegrees as for degrees. Show that the class of hyperarithmetical predicates is a hyperdegree $\mathbf{0}$, and that $\mathbf{0} \leqslant \mathbf{a}$ for every hyperdegree $\mathbf{a}$.

b) A hyperdegree $\mathbf{a}$ is Π_1^1 in a hyperdegree $\mathbf{b}$ if there is a set A in $\mathbf{a}$ and a set B in $\mathbf{b}$ such that A is Π_1^1 in B. Show that this implies that A is Π_1^1 in every set in $\mathbf{b}$. Show that among the hyperdegrees Π_1^1 in $\mathbf{a}$ there is a largest. It is called the *hyperjump* of $\mathbf{a}$ and is designated by $\mathbf{a}'$. Show that $\mathbf{a} \leqslant \mathbf{b} \rightarrow \mathbf{a}' \leqslant \mathbf{b}'$ and that $\mathbf{a} < \mathbf{a}'$.

c) For Φ a finite sequence, let κ^Φ be the smallest ordinal not recursive in Φ. Show that if $\mathbf{a}$ is a hyperdegree, then κ^A is the same for all sets A in $\mathbf{a}$. [Use a relativized version of 19(d).] This ordinal is designated by $\kappa^{\mathbf{a}}$. Show that $\mathbf{a} \leqslant \mathbf{b} \rightarrow \kappa^{\mathbf{a}} \leqslant \kappa^{\mathbf{b}}$.

d) Show that if $\mathbf{a}$ is Π_1^1 in $\mathbf{0}$ and $\kappa^{\mathbf{a}} = \kappa$, then $\mathbf{a} = \mathbf{0}$. [Let A be a Π_1^1 set in $\mathbf{a}$. Then $A(a) \leftrightarrow Tr(\lambda x F(a, x))$ with F recursive. As in the proof of the separation theorem, show that there is a σ recursive in A such that $A(a) \leftrightarrow Tr_\sigma(\lambda x F(a, x))$. Note that σ is recursive, and show that A is hyperarithmetical.]

e) Show that $\kappa^{\mathbf{a}} \neq \kappa \leftrightarrow \mathbf{0}' \leqslant \mathbf{a}$. [If $\kappa < \kappa^{\mathbf{a}}$, then Tr_κ is hyperarithmetical in every set in $\mathbf{a}$. Use the tree theorem to show that $\mathbf{0}' \leqslant \mathbf{a}$. For the converse, note that $\kappa < \kappa^{\mathbf{0}'}$ by (d), and use (c).]

f) Show that the only degrees Π_1^1 in $\mathbf{0}$ are $\mathbf{0}$ and $\mathbf{0}'$. [Use (d) and (e).]

21. a) Show that if σ is recursive, then there is a hyperarithmetical set A such that for no recursive function F do we have $A(a) \leftrightarrow Tr_\sigma(\lambda x F(a, x))$ for all a. [Let $Q(e, a)$ mean that $\{e\}^{0,2}$ is total and that $Tr_\sigma(\lambda x \{e\}(a, x))$. Show that Q is hyperarithmetical, and apply the diagonal lemma.]

b) Show that the replacement lemma does not hold for hyperarithmetical relations. [Use 19(c) to choose Φ so that $\kappa^\Phi > \kappa$. Suppose that $P(a) \leftrightarrow P'(\Phi^*, a)$ with P' hyperarithmetical. Choose a recursive F such that $P'(\alpha, a) \leftrightarrow Tr(\lambda x F(\alpha, a, x))$. As in the proof of the separation theorem, show that $P'(\alpha, a) \leftrightarrow Tr_\kappa(\lambda x F(\alpha, a, x))$ and hence $P(a) \leftrightarrow Tr_\kappa(\lambda x F(\Phi^*, a, x))$, and use a relativized version of (a).]

22. Define L_n inductively by

$$L_0(a) \leftrightarrow a = a, \qquad L_{n+1}(a) \leftrightarrow \exists z T_{1,1}((a)_0, K_{L_n}, (a)_1, z).$$

Define $L(n, a) \leftrightarrow L_n(a)$.

a) Show that if a set P is recursively enumerable in L_n, then P is recursive in L_{n+1}. [Use the relativized enumeration theorem.]

b) Show that every arithmetical predicate is recursive in L. [Show by induction on n that every Σ_n^0 predicate is recursive in L_n, using (*a*).] Conclude that L is not arithmetical. [Use the arithmetical hierarchy theorem and Post's theorem.]

c) Show that there is a recursive function F such that for each n, $F(n)$ is an H-index of L_n. Conclude that L is hyperarithmetical. [Use H1.]

23. Let Q be a (1, 1)-ary relation such that for all sets A and B,

$$A \subset B \rightarrow \forall x(Q(K_A, x) \rightarrow Q(K_B, x)).$$

A set A is *Q-closed* if $\forall x(Q(K_A, x) \rightarrow A(x))$. Let A be the set of numbers which are in every Q-closed set.

a) Show that A is Q-closed.

b) Show that

$$A(a) \leftrightarrow \forall \alpha(\forall x(Q(\alpha, x) \rightarrow \alpha(x) = 0) \rightarrow \alpha(a) = 0).$$

Conclude that if Q is Π_1^1, then A is Π_1^1.

c) Show that in the equivalence of (b), we may replace α by α^*. Conclude that if Q is recursively enumerable, then A is recursively enumerable. [Use the corollary to the infinity lemma.]

d) Using transfinite induction on σ, define for each σ a set A_σ by

$$A_\sigma(a) \leftrightarrow \exists \tau(\tau < \sigma \;\&\; Q(K_{A_\tau}, a)).$$

Show that $\sigma \leqslant \tau \rightarrow A_\sigma \subset A_\tau$, and conclude that there is a countable ordinal σ such that $A_\sigma = A_{\sigma+1}$. Let ρ be the smallest such ordinal. Show that $A_\sigma = A$ for $\sigma \geqslant \rho$.

e) Show that A_0 is the empty set, that $A_{\sigma+1}$ is the set of a such that $Q(K_{A_\sigma}, a)$, and that if σ is a limit number, then A_σ is the union of the A_τ for $\tau < \sigma$.

f) Show that if Q is recursively enumerable, then $\rho \leqslant \omega$. [Note that

$$Q(K_A, a) \leftrightarrow \exists x R(\overline{K_A}(x), a)$$

for some R, and use (e) to show that $A_{\omega+1} \subset A_\omega$.]

g) Given a relation P_σ for each σ, we say that P_σ is Π_1^1 *uniformly in* σ if there is a Π_1^1 relation P' such that $P_{\|B\|}(\mathfrak{A}) \leftrightarrow P'(K_B, \mathfrak{A})$ for every tree B. Show that if Q is Π_1^1, then A_σ is Π_1^1 uniformly in σ. [For B a tree, define P_B by

$$P_B(0, a) \leftrightarrow A_{\|B\|}(a),$$
$$P_B(b + 1, a) \leftrightarrow B(b) \;\&\; A_{\|B_{[b]}\|}(a).$$

Show that for a suitable Π_1^1 relation Q'_B, $\langle P_B \rangle$ is the set of numbers belonging to every Q'_B-closed set, and use the equivalence of (b).]

h) If P_σ is Π_1^1 uniformly in σ and

$$\forall \mathfrak{A}\, \exists \sigma(\sigma < \kappa \;\&\; P_\sigma(\mathfrak{A})),$$

then there is a recursive τ such that

$$\forall \mathfrak{A}\, \exists \sigma(\sigma < \tau \;\&\; P_\sigma(\mathfrak{A})).$$

[Let $V(\alpha) \leftrightarrow \exists \mathfrak{A}\, \forall \sigma(\sigma < \kappa \;\&\; P_\sigma(\mathfrak{A}) \rightarrow \|\alpha\| \leqslant \sigma)$. Apply the boundedness theorem to V.]

i) Let

$$R_\sigma(\alpha, a) \leftrightarrow (\forall x(A_\sigma(x) \rightarrow \alpha(x) = 0) \rightarrow Q(\alpha, a)).$$

Show that $\sigma \leqslant \tau \rightarrow R_\sigma(\alpha, a) \rightarrow R_\tau(\alpha, a)$. Show that

$$A_{\sigma+1}(a) \leftrightarrow \forall \alpha R_\sigma(\alpha, a)$$

and that for σ a limit number,

$$R_\sigma(\alpha, a) \rightarrow \exists \tau(\tau < \sigma \;\&\; R_\tau(\alpha, a)).$$

[Use (e).] Show that if Q is Π_1^1, then R_σ is Π_1^1 uniformly in σ. [Use (g).]

j) Show that if Q is Π_1^1, then $\rho \leqslant \kappa$. [Use (i), (h), and 19(g) to show that $A_{\kappa+1} \subset A_\kappa$.] Conclude that if Q is Π_1^1, then A_σ is Π_1^1 for all σ. [Use (g), (d), and (b).]

24. a) Show that if P is Π_1^1, then there is a Π_1^1 relation Q such that

$$Q(x, \mathfrak{A}) \rightarrow P(x, \mathfrak{A}),$$

$$Q(x, \mathfrak{A}) \;\&\; Q(y, \mathfrak{A}) \rightarrow x = y,$$

and

$$\exists x P(x, \mathfrak{A}) \rightarrow \exists x Q(x, \mathfrak{A}).$$

[Use the uniformization theorem.]

b) Show that if P is Π_1^1 and $\forall x\, \exists y P(y, x)$, then there is a hyperarithmetical α such that $\forall x P(\alpha(x), x)$. [Using (a), show that there is an α such that $\forall x P(\alpha(x), x)$ and $\mathfrak{G}_\alpha$ is Π_1^1.]

c) Show that if P is Π_1^1 and if for every x there is a hyperarithmetical α such that $P(\alpha, x)$, then there is a hyperarithmetical α such that $\forall x P((\alpha)_x, x)$. [Let $Q(i, x)$ mean that i is an H-index of the graph of an α such that $P(\alpha, x)$. Show that Q is Π_1^1. Choose a hyperarithmetical β by (b) so that $\forall x Q(\beta(x), x)$, and use β to construct an α such that $\forall x P((\alpha)_x, x)$ and $\mathfrak{G}_\alpha$ is Π_1^1.]

25. Let $\mathfrak{P}$ be a collection of subclasses of some space, and let $\mathfrak{P}_c$ be the collection of complements of classes in $\mathfrak{P}$. We say that $\mathfrak{P}$ satisfies the *reduction principle* if for every pair A and B of classes in $\mathfrak{P}$, there are disjoint classes A_1 and B_1 in $\mathfrak{P}$ such that $A_1 \subset A$, $B_1 \subset B$, and $A_1 \cup B_1 = A \cup B$. We say that $\mathfrak{P}$ satisfies the *separation principle* if every pair of disjoint classes in $\mathfrak{P}$ can be separated by a class which is in both $\mathfrak{P}$ and $\mathfrak{P}_c$.

a) Show that if $\mathfrak{P}$ satisfies the reduction principle, then $\mathfrak{P}_c$ satisfies the separation principle.

b) Show that the collection of Σ_k^0 subclasses of $N_{m,n}$ satisfies the reduction principle. [Let $A(\mathfrak{A}) \leftrightarrow \exists x P(\mathfrak{A}, x)$, $B(\mathfrak{A}) \leftrightarrow \exists x Q(\mathfrak{A}, x)$. Define

$$A_1(\mathfrak{A}) \leftrightarrow \exists x(P(\mathfrak{A}, x) \;\&\; \forall y_{y<x} \neg Q(\mathfrak{A}, y)),$$

and define B_1 similarly but with $<$ replaced by $\leqslant$.]

c) Show that the collection of Π^1_1 subclasses of $N_{m,n}$ satisfies the reduction principle. [Let $A(\mathfrak{A}) \leftrightarrow Tr(\lambda x F(\mathfrak{A}, x))$, $B(\mathfrak{A}) \leftrightarrow Tr(\lambda x G(\mathfrak{A}, x))$. Define

$$A_1(\mathfrak{A}) \leftrightarrow Tr(\lambda x F(\mathfrak{A}, x)) \;\&\; \neg T_<(\lambda x G(\mathfrak{A}, x), \lambda x F(\mathfrak{A}, x)),$$

and define B_1 similarly but with $<$ replaced by $\leqslant$.]

d) Show that the collection of Σ^1_2 subclasses of $N_{m,n}$ satisfies the reduction principle. [Let $A(\mathfrak{A}) \leftrightarrow \exists\beta Tr(\lambda x F(\mathfrak{A}, \beta, x))$, $B(\mathfrak{A}) \leftrightarrow \exists\beta Tr(\lambda x G(\mathfrak{A}, \beta, x))$. Set

$$A_1(\mathfrak{A}) \leftrightarrow \exists\beta[Tr(\lambda x F(\mathfrak{A}, \beta, x)) \;\&\; \neg\exists\alpha T_<(\lambda x G(\mathfrak{A}, \alpha, x), \lambda x F(\mathfrak{A}, \beta, x))].$$

e) Show that the collection of Σ^0_k (Π^1_1) (Σ^1_2) subclasses of $N_{0,2}$ does not satisfy the separation principle. [Let P be a binary Σ^0_k predicate which enumerates the class of unary Σ^0_k predicates. Let $A(a, b) \leftrightarrow P(a, (b)_0)$, $B(a, b) \leftrightarrow P(a, (b)_1)$. Take A_1 and B_1 as in the reduction principle. If a Δ^0_k predicate separated A_1 and B_1 it would enumerate the class of unary Δ^0_k predicates. This is impossible by the diagonal lemma.]

26. a) Show that the class of functions whose degree is less than $\mathbf{0}'$ is a basis for the collection of Π^0_1 classes of representing functions. [Let P be a nonempty Π^0_1 class of representing functions. Let Q be the class of α^* such that if $\beta = (\alpha^*)_0$ and $\gamma = (\alpha^*)_1$, then $P(\beta)$ and $\gamma(e) \neq \{e\}(\beta, e)$ whenever $\{e\}(\beta, e)$ is defined. Apply the Kreisel basis theorem to Q.]

b) Show that the class of functions whose hyperdegree is less than $\mathbf{0}'$ is a basis for the collection of Π^0_1 classes of functions. [Let P be a nonempty Π^0_1 class of functions. Let Q be the class of α such that if $\beta = (\alpha)_0$ and $\gamma = (\alpha)_1$, then $P(\beta)$ and $\gamma(e) \neq K_R(e)$ if e is an H-index of R from β. Use a relativized version of J to show that Q is Σ^1_1, and apply the Kleene basis theorem.]

c) Show that there is a hyperdegree $\mathbf{a}$ such that $\mathbf{0} < \mathbf{a} < \mathbf{0}'$. [Use (b) and the Kleene basis theorem.]

d) Show that the class of representing functions of recursively enumerable sets is not a basis for the collection of Π^0_1 classes of representing functions. [Show that if P and Q are recursively enumerable, then the class of representing functions of sets separating P and Q is Π^0_1, and use 7(b).]

27. For i an H-index, define A_i by

$$\begin{aligned} A_i(x) &\leftrightarrow x = i \vee A_{(i)_1}(x) && \text{if } (i)_0 = 1,\\ &\leftrightarrow x = i \vee \exists y(y \in W_{(i)_1} \;\&\; A_y(x)) && \text{if } (i)_0 = 2,\\ &\leftrightarrow x = i && \text{otherwise.} \end{aligned}$$

a) Show that there is a recursive function F such that $A_i = W_{F(i)}$ for every H-index i. [Use the recursion theorem.]

b) For i an H-index, let

$$J_i(j, a) \leftrightarrow A_i(j) \;\&\; J(j, x).$$

Show that there is an arithmetical relation Q such that for each H-index i, $Q_{(i)}$ defines $\langle K_{J_i}\rangle$ implicitly. [Use (a).]

c) Show that every hyperarithmetical predicate is recursive in a function defined implicitly by an arithmetical class. [Use (b).]

d) Show that if F is defined implicitly by a Π^0_{2n+1} class, then F is recursive in a function G defined implicitly by a Π^0_{2n-1} class. [Let $\alpha = F \leftrightarrow \forall x \exists y\, R(\alpha, x, y)$ and define $G(x) = \langle F(x), \mu y R(F, x, y)\rangle$.]

e) Show that a predicate is hyperarithmetical iff it is recursive in a function defined implicitly by a Π^0_1 class. [Use (c) and (d).]

f) Show that a predicate is Δ^1_2 iff it is hyperarithmetical in a function defined implicitly by a Π^1_1 class. [For the "if" part use 18(b). Suppose that

$$P(a) \leftrightarrow \exists \alpha R(\alpha, a) \quad \text{and} \quad \neg P(a) \leftrightarrow \exists \alpha Q(\alpha, a)$$

where R and Q are Π^1_1. By the uniformization theorem, we may suppose that for each a there is a unique α_a such that $R(\alpha_a, a) \vee Q(\alpha_a, a)$. Choose F so that $(F)_a = \alpha_a$ for all a, and show that P is Δ^1_1 in F.]

28. Let $J(i, a)$ be as in §7.10, and let H be the set of H-indices.

a) Show that J is not hyperarithmetical. [Use the diagonal lemma.]

b) Show that H is Π^1_1 but not hyperarithmetical. [Use (a) and the equivalences

$$H(i) \leftrightarrow \exists x(J(i, x) \vee J(\langle 1, i\rangle, x)),$$
$$J(i, a) \leftrightarrow H(i) \;\&\; \neg J(\langle 1, i\rangle, a).]$$

CHAPTER 8

THE NATURAL NUMBERS

8.1 PEANO ARITHMETIC

So far we have studied the general theory of axiom systems. In this and the next chapter, we shall study axiom systems for two fundamental notions of mathematics: *natural number* and *set*.

The theory N is not a satisfactory axiom system for studying natural numbers. It contains only a few of the basic functions and predicates of number theory; and even for these, it contains only some of the evident axioms. In particular, the induction axiom, which is essential in most proofs in number theory, is missing. We shall show that by adding a suitable form of the induction axiom, we obtain a more satisfactory theory.

One statement of the induction axiom is that if a set contains 0 and contains the successor of every natural number in the set, then it contains every natural number. We cannot express this form in $L(N)$, since we have no variables which vary through sets of natural numbers. Another form of the induction axiom states that if 0 has some property, and if the successor of every natural number having this property also has this property, then every natural number has this property. We cannot fully express this either, since we have no variables which vary through properties of natural numbers. However, we can express it for each property of natural numbers which can be expressed in $L(N)$.

If $\mathbf{A}$ is a formula of $L(N)$, then the formula

$$\mathbf{A_x}[0] \ \& \ \forall\mathbf{x}(\mathbf{A} \rightarrow \mathbf{A_x}[S\mathbf{x}]) \rightarrow \mathbf{A} \qquad (1)$$

is called an *induction axiom*. The theory obtained from N by omitting N9 and adding all the induction axioms as new nonlogical axioms is called *Peano arithmetic*, and is designated by P.

Our first observation is that many of the usual proofs using induction can be formalized in P. As an example, we prove N9 in P, thus proving that P is an extension of N. Let $\mathbf{A}$ be $0 = \mathbf{y} \vee 0 < \mathbf{y}$. Then $\vdash\mathbf{A_y}[0]$ by the identity axioms. By N8, $\vdash\mathbf{A} \leftrightarrow 0 < S\mathbf{y}$; so $\vdash\mathbf{A} \rightarrow \mathbf{A_y}[S\mathbf{y}]$. Hence $\vdash\mathbf{A}$ by the induction axioms. Now let $\mathbf{B}$ be $\mathbf{x} < \mathbf{y} \rightarrow S\mathbf{x} < S\mathbf{y}$. Then $\vdash\mathbf{B_y}[0]$ by N7. By N8

$$\vdash\mathbf{B_y}[S\mathbf{y}] \leftrightarrow (\mathbf{x} < \mathbf{y} \vee \mathbf{x} = \mathbf{y} \rightarrow S\mathbf{x} < S\mathbf{y} \vee S\mathbf{x} = S\mathbf{y}).$$

From this and the equality theorem we get $\vdash\mathbf{B} \rightarrow \mathbf{B_y}[S\mathbf{y}]$. Thus $\vdash\mathbf{B}$ by the induction axioms. Now let $\mathbf{C}$ be $\mathbf{x} < \mathbf{y} \vee \mathbf{x} = \mathbf{y} \vee \mathbf{y} < \mathbf{x}$. From $\vdash\mathbf{A}$ we get $\vdash\mathbf{C_x}[0]$.

From $\vdash\mathbf{B}$ and N8,

$$\vdash\mathbf{x} < \mathbf{y} \to S\mathbf{x} < \mathbf{y} \vee S\mathbf{x} = \mathbf{y},$$
$$\vdash\mathbf{y} < \mathbf{x} \vee \mathbf{y} = \mathbf{x} \to \mathbf{y} < S\mathbf{x}.$$

From these, $\vdash\mathbf{C} \to \mathbf{C}_\mathbf{x}[S\mathbf{x}]$. Hence $\vdash\mathbf{C}$ by the induction axioms.

Now we observe that some common consequences of the induction axiom can be proved in P. Thus we have the *principle of complete induction*:

$$\vdash\forall\mathbf{x}(\forall\mathbf{y}(\mathbf{y} < \mathbf{x} \to \mathbf{A}_\mathbf{x}[\mathbf{y}]) \to \mathbf{A}) \to \forall\mathbf{x}\mathbf{A} \tag{2}$$

provided that $\mathbf{y}$ is different from $\mathbf{x}$ and does not occur in $\mathbf{A}$. To prove this, let $\mathbf{B}$ be $\forall\mathbf{y}(\mathbf{y} < \mathbf{x} \to \mathbf{A}_\mathbf{x}[\mathbf{y}])$. Then $\vdash\mathbf{B}_\mathbf{x}[0]$ by N7. From N8 by some elementary transformations

$$\vdash\mathbf{B}_\mathbf{x}[S\mathbf{x}] \leftrightarrow \mathbf{B}\ \&\ \mathbf{A}. \tag{3}$$

Now if $\mathbf{C}$ is the left side of (2), then

$$\vdash\mathbf{C} \to \mathbf{B} \to \mathbf{A} \tag{4}$$

by the substitution theorem. From (3), (4), and the $\forall$-introduction rule,

$$\vdash\mathbf{C} \to \forall\mathbf{x}(\mathbf{B} \to \mathbf{B}_\mathbf{x}[S\mathbf{x}]).$$

From this, $\vdash\mathbf{B}_\mathbf{x}[0]$, and the induction axioms, $\vdash\mathbf{C} \to \mathbf{B}$. Using this, (4), and the $\forall$-introduction rule, we get (2).

We can now obtain the least number principle:

$$\exists\mathbf{x}\mathbf{A} \to \exists\mathbf{x}(\mathbf{A}\ \&\ \forall\mathbf{y}(\mathbf{y} < \mathbf{x} \to \neg\mathbf{A}_\mathbf{x}[\mathbf{y}])), \tag{5}$$

where $\mathbf{y}$ is different from $\mathbf{x}$ and does not occur in $\mathbf{A}$. To obtain this, we put $\neg\mathbf{A}$ for $\mathbf{A}$ in (2) and make some elementary transformations.

We remark that (1), (2), and (5) also hold in every extension by definitions of P. This is because the translation into P of a formula of the form (1) is again of the form (1), and similarly for (2) and (5).

Now consider the problem of introducing new functions and predicates. Suppose that in an informal development of number theory from the Peano axioms, we have a definition of a new function or predicate u. If this definition can be expressed in P, then we can introduce an extension by definitions P' of P in which some new nonlogical symbol $\mathbf{u}$ designates u; the defining axiom of $\mathbf{u}$ will express formally the given informal definition of u. We shall then say that we have *introduced* u in P'. Informal proofs of properties of u can then be replaced by formal proofs in P', provided that the techniques of the proof do not go beyond what is available in P'. (It is of course essential to formalize the given informal definition of u, or at least formalize some informal definition which can be proved in P' to be equivalent to the given one.)

We shall show that many fundamental functions and predicates of number theory can be introduced in a special type of extension by definitions of P. Before describing this type of extension, we introduce a formal analogue of the μ-operator.

Suppose that P' is an extension by definitions of P, and that $\mathbf{A}$ is a formula of P' in which no variable other than $\mathbf{x}_1, \ldots, \mathbf{x}_n, \mathbf{y}$ is free. Suppose also that $\vdash_{P'} \exists\mathbf{y}\mathbf{A}$. We can then form an extension by definitions P'' of P' by introducing an n-ary symbol $\mathbf{f}$ and a defining axiom

$$\mathbf{A}_\mathbf{y}[\mathbf{f}\mathbf{x}_1 \ldots \mathbf{x}_n] \,\&\, \forall\mathbf{y}(\mathbf{y} < \mathbf{f}\mathbf{x}_1 \ldots \mathbf{x}_n \rightarrow \neg\mathbf{A}). \tag{6}$$

To see this, we must prove the existence and uniqueness conditions for $\mathbf{f}$ in P'. The existence condition follows from $\vdash_{P'} \exists\mathbf{y}\mathbf{A}$ and (5); the uniqueness condition is an easy consequence of N9. We shall usually abbreviate (6) to

$$\mathbf{f}\mathbf{x}_1 \ldots \mathbf{x}_n = \mu\mathbf{y}\mathbf{A} \tag{7}$$

and call $\exists\mathbf{y}\mathbf{A}$ the *existence condition* for $\mathbf{f}$.

By a *recursive extension* of P, we mean an extension by definitions of P in which the defining axioms for predicate symbols are open and the defining axioms for function symbols are of the form (7) with $\mathbf{A}$ open. It is an easy consequence of the results of §6.3 that the functions and predicates introduced in recursive extensions of P are recursive. We shall now show that a great many important recursive functions and predicates can be introduced in such extensions.

We note first that the existence condition for a definition

$$\mathbf{f}\mathbf{x}_1 \ldots \mathbf{x}_n = \mu\mathbf{x}(\mathbf{x} = \mathbf{a})$$

is $\exists\mathbf{x}(\mathbf{x} = \mathbf{a})$, which is provable by the identity axioms and the substitution theorem. Moreover, this axiom implies

$$\mathbf{f}\mathbf{x}_1 \ldots \mathbf{x}_n = \mathbf{a}. \tag{8}$$

We shall generally simply say that we are introducing $\mathbf{f}$ by the axiom (8).

We shall now examine R1 through R14. In doing so, we shall allow ourselves to use all Latin letters (possibly with subscripts) as if they were variables of P. If a function or predicate has been given a name informally, we use that name as a nonlogical symbol which designates that function or predicate.

We observe first that if R has been introduced, then K_R can be introduced by the axiom

$$K_R(a_1, \ldots, a_n) = \mu x\big((R(a_1, \ldots, a_n) \,\&\, x = 0) \vee (\neg R(a_1, \ldots, a_n) \,\&\, x = 1)\big)$$

(where 1 abbreviates $S0$). The existence condition for K_R is easily proved. From the axiom for K_R we can prove

$$R(a_1, \ldots, a_n) \rightarrow K_R(a_1, \ldots, a_n) = 0,$$
$$\neg R(a_1, \ldots, a_n) \rightarrow K_R(a_1, \ldots, a_n) = 1.$$

If K_R has been introduced, we can introduce R by the axiom

$$R(a_1, \ldots, a_n) \leftrightarrow K_R(a_1, \ldots, a_n) = 0.$$

We can introduce I_i^n by the axiom

$$I_i^n(a_1, \ldots, a_n) = a_i.$$

We already have $+$, $\cdot$, and $<$; so we may introduce $K_<$. If F is defined by

$$F(a_1, \ldots, a_n) = G(H_1(a_1, \ldots, a_n), \ldots, H_k(a_1, \ldots, a_n)) \qquad (9)$$

where $G, H_1, \ldots, H_k$ have already been introduced, then we can use (9) as an axiom to introduce F. We treat R4 similarly. If F is defined by

$$F(a_1, \ldots, a_n) = \mu x(G(a_1, \ldots, a_n, x) = 0) \qquad (10)$$

where G has already been introduced, and if the formula

$$\exists x(G(a_1, \ldots, a_n, x) = 0)$$

(which asserts that F is well-defined) is provable, then we may use (10) as an axiom to introduce F. We treat R5 similarly.

Obviously every constant function can be introduced by a definition

$$F(a_1, \ldots, a_n) = \mathbf{k}_m.$$

It is also clear that if P and Q have been introduced, then $\neg P$, $P \vee Q$, $P \rightarrow Q$, $P \mathbin{\&} Q$, and $P \leftrightarrow Q$ may be introduced. We already have $<$ and $=$; the remaining predicates of R8 can be introduced by the definitions given in the proof of R8.

If F is defined by

$$F(b, a_1, \ldots, a_n) = \mu x_{x<b} R(a_1, \ldots, a_n, x)$$

where R has already been introduced, then we can introduce F by

$$F(b, a_1, \ldots, a_n) = \mu x(R(a_1, \ldots, a_n, x) \vee x = b). \qquad (11)$$

The existence condition for F is then provable from the identity axioms and the substitution theorem.

If P is defined by

$$P(b, a_1, \ldots, a_n) \leftrightarrow \exists x_{x<b} R(a_1, \ldots, a_n, x)$$

where R has been introduced, then we introduce F by (11) and then introduce P by

$$P(b, a_1, \ldots, a_n) \leftrightarrow F(b, a_1, \ldots, a_n) < b.$$

We can then prove

$$P(b, a_1, \ldots, a_n) \leftrightarrow \exists x(x < b \mathbin{\&} R(a_1, \ldots, a_n, x)).$$

We introduce bounded universal quantifiers by defining them in terms of bounded existential quantifiers, as in §6.3.

We introduce $\dot{-}$ by

$$x \dot{-} y = \mu z(y + z = x \vee x < y);$$

the existence condition $\exists z(y + z = x \vee x < y)$ can be proved in P by induction on y.

Now suppose that F is defined by

$$\begin{aligned} F(a_1, \ldots, a_n) &= G_1(a_1, \ldots, a_n) \quad &&\textit{if} \;\; P_1(a_1, \ldots, a_n), \\ &\;\;\vdots && \;\;\;\;\vdots \\ &= G_k(a_1, \ldots, a_n) \quad &&\textit{if} \;\; P_k(a_1, \ldots, a_n), \end{aligned}$$

where $G_1, \ldots, G_k, P_1, \ldots, P_k$ have already been introduced. Suppose also that

$$P_1(a_1, \ldots, a_n) \vee \cdots \vee P_k(a_1, \ldots, a_n) \tag{12}$$

and

$$\neg(P_i(a_1, \ldots, a_n) \;\&\; P_j(a_1, \ldots, a_n)), \qquad 1 \leqslant i < j \leqslant k, \tag{13}$$

are provable. We then introduce F by

$$\begin{aligned} F(a_1, \ldots, a_n) = \mu x((P_1(a_1, \ldots, a_n) \;\&\; x = G_1(a_1, \ldots, a_n)) \vee \cdots \\ \vee (P_k(a_1, \ldots, a_n) \;\&\; x = G_k(a_1, \ldots, a_n))). \end{aligned}$$

The existence condition for F follows from (12); and using (13), we can prove

$$P_i(a_1, \ldots, a_n) \rightarrow F(a_1, \ldots, a_n) = G_i(a_1, \ldots, a_n).$$

If R is defined by

$$\begin{aligned} R(a_1, \ldots, a_n) &\leftrightarrow Q_1(a_1, \ldots, a_n) \quad &&\textit{if} \;\; P_1(a_1, \ldots, a_n), \\ &\;\;\vdots && \;\;\;\;\vdots \\ &\leftrightarrow Q_k(a_1, \ldots, a_n) \quad &&\textit{if} \;\; P_k(a_1, \ldots, a_n), \end{aligned}$$

where $Q_1, \ldots, Q_k, P_1, \ldots, P_k$ have been introduced and (12) and (13) are provable, then we introduce R by

$$\begin{aligned} R(a_1, \ldots, a_n) \leftrightarrow ((P_1(a_1, \ldots, a_n) \;\&\; Q_1(a_1, \ldots, a_n)) \vee \cdots \\ \vee (P_k(a_1, \ldots, a_n) \;\&\; Q_k(a_1, \ldots, a_n))). \end{aligned}$$

We can introduce OP, Div, and β by the explicit definitions given in §6.4. The basic property of β is expressed by

$$\exists \mathbf{x}\, \forall \mathbf{y}(\mathbf{y} < \mathbf{z} \rightarrow \beta(\mathbf{x}, \mathbf{y}) = \mathbf{a}), \tag{14}$$

where **x**, **y**, and **z** are distinct and **x** and **z** do not occur in **a**. This can be proved by formalizing the proof given in §6.4. This gives the existence condition for introducing the functions $\langle a_1, \ldots, a_n \rangle$. The remaining functions and predicates concerned with sequence numbers can be introduced without difficulty.

If F has been introduced, we introduce $\overline{F}$ by using the explicit definition (14) of §6.4. Again (14) gives the existence condition. Now suppose that F is defined by

$$F(b, a_1, \ldots, a_n) = G(\overline{F}(b, a_1, \ldots, a_n), b, a_1, \ldots, a_n), \tag{15}$$

where G has already been introduced. We then introduce F by using the definitions (16) and (17) of §6.4. Using elementary properties of the functions involved, we can then prove (15). We shall simply say that we have introduced F by the axiom (15).

If F is defined by

$$F(0, a_1, \ldots, a_n) = G(a_1, \ldots, a_n),$$
$$F(b + 1, a_1, \ldots, a_n) = H(F(b, a_1, \ldots, a_n), b, a_1, \ldots, a_n), \tag{16}$$

where G and H have been introduced, then we may introduce F by a definition of the form (15), as explained in §6.4; we may then prove (16).

With these tools, we can introduce the functions of ordinary number theory (exponentials, factorials, etc.) and formalize the more elementary proofs given in textbooks on number theory. Actually, many advanced proofs can be formalized by modifying them a little. Thus many proofs in number theory make use of contour integrals in the complex plane. Such proofs can often be formalized by replacing the integrals by suitable approximating Riemann sums. We shall not enter into any of the details of this, but merely note that such methods can be used to increase the scope of the results proved in the remainder of this chapter.

8.2 THE THEOREM ON CONSISTENCY PROOFS

We have seen that a considerable portion of number theory can be formalized in P. We shall now investigate the properties of P.

It is clear that $\mathfrak{N}$ is a model of P. We call it the *standard* model of P, and say that a formula of P is *true* if it is valid in $\mathfrak{N}$.

We now apply the results of Chapter 6. Since P has $\mathfrak{N}$ as a model, it is a consistent extension of N; so by Church's theorem, P is undecidable. It is easy to give an explicit definition of the set of expression numbers of induction axioms which shows that this set is recursive; and from this, it follows that P is axiomatized. Hence by the incompleteness theorem, P is incomplete. These results extend immediately to recursive extensions of P.

Let P' be a recursive extension of P, and let $\mathfrak{N}'$ be the expansion of $\mathfrak{N}$ to a model of P'. We say that a formula of P' is *true* if it is valid in $\mathfrak{N}'$. Thus every theorem of P' is true. If $\mathbf{A}$ is a closed formula of P', then either $\mathbf{A}$ or $\neg\mathbf{A}$ is true. Hence by the incompleteness of P', there is a true formula of P' which is not a theorem of P'. We shall investigate this situation more closely.

Let P' be a recursive extension of P. We define the *R-formulas* of P' by the generalized inductive definition.

i) Every formula $\mathbf{fx}_1 \ldots \mathbf{x}_n = \mathbf{y}$ or $\mathbf{px}_1 \ldots \mathbf{x}_n$ or $\neg\mathbf{px}_1 \ldots \mathbf{x}_n$ is an R-formula.

ii) If $\mathbf{A}$ and $\mathbf{B}$ are R-formulas, then $\mathbf{A} \vee \mathbf{B}$ and $\mathbf{A} \;\&\; \mathbf{B}$ are R-formulas.

iii) If $\mathbf{A}$ is an R-formula and $\mathbf{x}$ and $\mathbf{y}$ are distinct, then $\forall\mathbf{x}(\mathbf{x} < \mathbf{y} \to \mathbf{A})$ is an R-formula.

iv) If $\mathbf{A}$ is an R-formula, then $\exists\mathbf{x}\mathbf{A}$ is an R-formula.

Lemma 1. If P' is a recursive extension of P, then every existential formula of P' is equivalent in P' to an R-formula.

Proof. We first prove this for formulas of the form $\mathbf{x} = \mathbf{a}$, using induction on the length of $\mathbf{a}$. If $\mathbf{a}$ is a variable, then $\mathbf{x} = \mathbf{a}$ is an R-formula. Otherwise, $\mathbf{a}$ is $\mathbf{fa}_1 \ldots \mathbf{a}_n$. Then by Corollary 3 to the equality theorem,

$$\vdash \mathbf{x} = \mathbf{a} \leftrightarrow \exists \mathbf{y}_1 \ldots \exists \mathbf{y}_n(\mathbf{y}_1 = \mathbf{a}_1 \,\&\, \cdots \,\&\, \mathbf{y}_n = \mathbf{a}_n \,\&\, \mathbf{x} = \mathbf{fy}_1 \ldots \mathbf{y}_n).$$

By induction hypothesis, $\mathbf{y}_i = \mathbf{a}_i$ is equivalent to an R-formula; and by the symmetry theorem, $\mathbf{x} = \mathbf{fy}_1 \ldots \mathbf{y}_n$ is equivalent to an R-formula. Hence $\mathbf{x} = \mathbf{a}$ is equivalent to an R-formula by the equivalence theorem.

It is clearly sufficient to prove that every open formula $\mathbf{A}$ is equivalent to an R-formula. We prove this by induction on the length of $\mathbf{A}$. If $\mathbf{A}$ is an atomic formula $\mathbf{pa}_1 \ldots \mathbf{a}_n$, then

$$\vdash \mathbf{A} \leftrightarrow \exists \mathbf{y}_1 \ldots \exists \mathbf{y}_n(\mathbf{y}_1 = \mathbf{a}_1 \,\&\, \cdots \,\&\, \mathbf{y}_n = \mathbf{a}_n \,\&\, \mathbf{py}_1 \ldots \mathbf{y}_n).$$

Since $\mathbf{y}_i = \mathbf{a}_i$ is equivalent to an R-formula by the above, $\mathbf{A}$ is equivalent to an R-formula. A similar proof holds if $\mathbf{A}$ is the negation of an atomic formula. It remains to consider the cases in which $\mathbf{A}$ is $\neg\neg\mathbf{B}$, $\neg(\mathbf{B} \vee \mathbf{C})$, and $\mathbf{B} \vee \mathbf{C}$. If $\mathbf{A}$ is $\neg\neg\mathbf{B}$, then $\mathbf{B}$ is equivalent to an R-formula by induction hypothesis; since $\mathbf{A}$ is equivalent to $\mathbf{B}$, it is equivalent to an R-formula. If $\mathbf{A}$ is $\neg(\mathbf{B} \vee \mathbf{C})$, then $\mathbf{A}$ is equivalent to $\neg\mathbf{B} \,\&\, \neg\mathbf{C}$. By induction hypothesis, $\neg\mathbf{B}$ and $\neg\mathbf{C}$ are equivalent to R-formulas; so by the equivalence theorem, $\mathbf{A}$ is equivalent to an R-formula. A similar proof holds if $\mathbf{A}$ is $\mathbf{B} \vee \mathbf{C}$.

Lemma 2. If P' is recursive extension of P, then every R-formula in P' is equivalent in P' to an R-formula in P.

Proof. It is clearly sufficient to show that if P' is obtained from P'' by adding one new nonlogical symbol, then every R-formula of P' is equivalent to an R-formula of P''. It is also sufficient to consider only R-formulas of the form $\mathbf{fx}_1 \ldots \mathbf{x}_n = \mathbf{y}$ or $\mathbf{px}_1 \ldots \mathbf{x}_n$ or $\neg\mathbf{px}_1 \ldots \mathbf{x}_n$ where $\mathbf{f}$ or $\mathbf{p}$ is the new nonlogical symbol. From the form of the defining axiom of $\mathbf{p}$ and the substitution rule and the equivalence theorem, we see that $\mathbf{px}_1 \ldots \mathbf{x}_n$ and $\neg\mathbf{px}_1 \ldots \mathbf{x}_n$ are equivalent to open formulas of P''. Hence by Lemma 1, they are equivalent to R-formulas of P''. Similarly, $\mathbf{fx}_1 \ldots \mathbf{x}_n = \mathbf{y}$ is equivalent to a formula of P'' of the form $\mathbf{A} \,\&\, \forall \mathbf{z}(\mathbf{z} < \mathbf{y} \rightarrow \mathbf{A}')$ where $\mathbf{A}$ and $\mathbf{A}'$ are open. Hence by Lemma 1 and the equivalence theorem, it is equivalent to an R-formula of P''.

A *numerical instance* of a formula $\mathbf{A}$ is a closed formula of the form

$$\mathbf{A}[\mathbf{k}_{a_1}, \ldots, \mathbf{k}_{a_n}].$$

Lemma 3. If $\mathbf{A}$ is an R-formula of P, then every true numerical instance of $\mathbf{A}$ is a theorem of P.

Proof. We use induction on the length of $\mathbf{A}$. If $\mathbf{A}$ is $0 = \mathbf{y}$, the only true numerical instance is $0 = 0$, which is provable. If $\mathbf{A}$ is $S\mathbf{x} = \mathbf{y}$, every true numerical instance has the form $\mathbf{k}_{n+1} = \mathbf{k}_{n+1}$ and hence is provable. If $\mathbf{A}$ is $\mathbf{x} + \mathbf{x}' = \mathbf{y}$ or $\mathbf{x} \cdot \mathbf{x}' = \mathbf{y}$ or $\mathbf{x} = \mathbf{y}$ or $\mathbf{x} \neq \mathbf{y}$ or $\mathbf{x} < \mathbf{y}$ or $\neg(\mathbf{x} < \mathbf{y})$, then the result follows from (1) through

(6) of §6.7. If **A** is **B** ∨ **C** or **B** & **C**, the result follows from the induction hypothesis and the tautology theorem. Suppose that **A** is $\forall\mathbf{x}(\mathbf{x} < \mathbf{y} \rightarrow \mathbf{B})$. A true numerical instance of **A** has the form $\forall\mathbf{x}(\mathbf{x} < \mathbf{k}_n \rightarrow \mathbf{B}')$. For each $i < n$, $\mathbf{B}'[\mathbf{k}_i]$ is a true numerical instance of **B** and hence is provable by the induction hypothesis. Hence $\vdash\forall\mathbf{x}(\mathbf{x} < \mathbf{k}_n \rightarrow \mathbf{B})$ by Lemma 2 of §6.7, the ∀-introduction rule, and the detachment rule. Finally, suppose that **A** is $\exists\mathbf{x}\mathbf{B}$. A true numerical instance of **A** has the form $\exists\mathbf{x}\mathbf{B}'$. Since $\exists\mathbf{x}\mathbf{B}'$ is true, $\mathbf{B}'_{\mathbf{x}}[\mathbf{k}_i]$ is true for some i. Since it is a numerical instance of **B**, it is provable by the induction hypothesis. Hence $\vdash\exists\mathbf{x}\mathbf{B}'$ by the substitution theorem.

Combining the three lemmas, we obtain the following result.

Theorem. If P' is a recursive extension of P, then every true closed existential formula of P' is a theorem of P'.

We shall show that this does not extend to universal formulas. Our procedure for producing a true but unprovable formula is based on the proof of Church's theorem. In that proof we constructed a predicate Q different from all of the $E(\mathbf{A})$. We shall produce a formula **A** such that $\mathbf{A}[\mathbf{k}_n]$ is true iff $Q(n)$. Since $Q \neq E(\mathbf{A})$, there is an n such that $\vdash_P \mathbf{A}[\mathbf{k}_n]$ is not equivalent to $Q(n)$. Since provable formulas are true, it must be that $\mathbf{A}[\mathbf{k}_n]$ is true and unprovable.

We recall that Q was defined by

$$Q(a) \leftrightarrow \neg Thm_P(Sub(a, \ulcorner\mathbf{z}\urcorner, Num(a))).$$

Then if $a = \ulcorner\mathbf{A}\urcorner$, $Q(a)$ is true iff $\mathbf{A}[\mathbf{k}_a]$ is not a theorem of P.

Now let P' be a recursive extension of P in which all the functions and predicates of §6.6 (for the theory P) have been introduced. Let **B** be the formula

$$\neg\exists\mathbf{y}\, Pr_P(Sub(\mathbf{z}, \mathbf{k}_{\ulcorner\mathbf{z}\urcorner}, Num(\mathbf{z})), \mathbf{y})$$

(where **y** is distinct from **z**), and let **A** be the translation of **B** into P. Then $\mathbf{B}[\mathbf{k}_n]$ and $\mathbf{A}[\mathbf{k}_n]$ are true iff $Q(n)$. If $a = \ulcorner\mathbf{A}\urcorner$, then $\mathbf{A}[\mathbf{k}_a]$ is true iff $Q(a)$, hence, by the above, iff $\mathbf{A}[\mathbf{k}_a]$ is not a theorem of P. Since all theorems of P are true, it follows that $\mathbf{A}[\mathbf{k}_a]$ is true and not a theorem of P. Hence $\mathbf{B}[\mathbf{k}_a]$ is true and not a theorem of P'. The prenex form of $\mathbf{B}[\mathbf{k}_a]$ is then a true universal formula of P' which is not a theorem of P'.

Our proof that $\mathbf{A}[\mathbf{k}_a]$ is true depends on the fact that the theorems of P are true. Using our lemmas, we can instead give a proof based on the consistency of P. By Lemmas 1 and 2, there is an R-formula **C** of P which is equivalent in P' to $\neg\mathbf{B}[\mathbf{k}_a]$. Then **C** is equivalent to $\neg\mathbf{A}[\mathbf{k}_a]$ in P' and hence in P. Thus if $\mathbf{A}[\mathbf{k}_a]$ is false, then **C** is true; so $\vdash_P \mathbf{C}$ by Lemma 3; so $\vdash_P \neg\mathbf{A}[\mathbf{k}_a]$. But if $\mathbf{A}[\mathbf{k}_a]$ is false, then $Q(a)$ is false and hence $\vdash_P \mathbf{A}[\mathbf{k}_a]$. From these two results and the consistency of P, we conclude that $\mathbf{A}[\mathbf{k}_a]$ is true.

Let us formalize these steps in P'. Let $Thm_P(\mathbf{a})$ abbreviate $\exists\mathbf{y}\, Pr_P(\mathbf{a}, \mathbf{y})$. Let $c = \ulcorner\mathbf{C}\urcorner$, $d = \ulcorner\mathbf{A}[\mathbf{k}_a]\urcorner$. The statement *if* $\mathbf{A}[\mathbf{k}_a]$ *is false, then* **C** *is true* becomes

$$\neg\mathbf{A}[\mathbf{k}_a] \rightarrow \mathbf{C}. \tag{1}$$

The statement *if* $\mathbf{C}$ *is true, then* $\vdash_P \mathbf{C}$ becomes

$$\mathbf{C} \to Thm_P(\mathbf{k}_c). \tag{2}$$

The statement *if* $\vdash_P \mathbf{C}$ *then* $\vdash_P \neg\mathbf{A}[\mathbf{k}_a]$ becomes

$$Thm_P(\mathbf{k}_c) \to Thm_P(Neg(\mathbf{k}_d)), \tag{3}$$

where $Neg(\mathbf{a})$ abbreviates $\langle \mathbf{k}_{SN(\neg)}, \mathbf{a}\rangle$. The statement *if* $\mathbf{A}[\mathbf{k}_a]$ *is false, then* $\vdash_P \mathbf{A}[\mathbf{k}_a]$ becomes

$$\neg\mathbf{A}[\mathbf{k}_a] \to Thm_P(\mathbf{k}_d). \tag{4}$$

The (tacitly used) statement *if* $\vdash_P \mathbf{A}[\mathbf{k}_a]$ *and* $\vdash_P \neg\mathbf{A}[\mathbf{k}_a]$, *then P is inconsistent* becomes

$$Thm_P(\mathbf{k}_d) \to Thm_P(Neg(\mathbf{k}_d)) \to \neg Con_P, \tag{5}$$

where Con_P abbreviates $\neg\forall\mathbf{x}(For_P(\mathbf{x}) \to Thm_P(\mathbf{x}))$. The final conclusion that *if P is consistent then* $\mathbf{A}[\mathbf{k}_a]$ *is true* becomes

$$Con_P \to \mathbf{A}[\mathbf{k}_a]. \tag{6}$$

We shall show that all of these are provable in P'. For (1), this follows from the fact that $\neg\mathbf{A}[\mathbf{k}_a]$ is equivalent to $\mathbf{C}$. We postpone (2) for a moment. For (3), we note that, since $\vdash_P \mathbf{C} \to \neg\mathbf{A}[\mathbf{k}_a]$,

$$Thm_P(\langle \mathbf{k}_{SN(\vee)}, Neg(\mathbf{k}_c), Neg(\mathbf{k}_d)\rangle) \tag{7}$$

is a true closed existential formula of P' and hence a theorem of P'. If we also formalize the proof of the detachment rule in P', we can obtain (3) from (7). To prove (4), note that

$$\mathbf{k}_d = Sub(\mathbf{k}_a, \mathbf{k}_{\ulcorner \mathbf{z} \urcorner}, Num(\mathbf{k}_a))$$

is a true variable-free formula of P' and hence a theorem of P'. We then obtain (4) from the equality theorem and the choice of $\mathbf{A}$. The proof of (5) in P' is just the formalization of the proof of a very simple syntactical lemma. Finally, (6) is a tautological consequence of (1) through (5).

Now we examine the proof of (2). We let $S(\mathbf{a}, \mathbf{b}_1)$ be an abbreviation of

$$Sub(\mathbf{a}, \mathbf{k}_{\ulcorner \mathbf{x}_1 \urcorner}, Num(\mathbf{b}_1));$$

$S(\mathbf{a}, \mathbf{b}_1, \mathbf{b}_2)$ bc an abbreviation of

$$Sub(S(\mathbf{a}, \mathbf{b}_1), \mathbf{k}_{\ulcorner \mathbf{x}_2 \urcorner}, Num(\mathbf{b}_2));$$

and so on. (Of course, S depends on the choice of $\mathbf{x}_1, \mathbf{x}_2, \ldots$; but we shall not find it necessary to indicate this by the notation.) We shall prove that if $\mathbf{D}$ is an R-formula of P, and $\mathbf{x}_1, \ldots, \mathbf{x}_n$ are the variables free in $\mathbf{D}$, then

$$\vdash_{P'} \mathbf{D} \to Thm_P(S(\mathbf{k}_{\ulcorner \mathbf{D} \urcorner}, \mathbf{x}_1, \ldots, \mathbf{x}_n)); \tag{8}$$

(2) will then follow as a special case. The proof is by induction on the length of $\mathbf{D}$. Since it is merely a formalization of the proof of Lemma 3, we shall only consider a few cases briefly.

Suppose that $\mathbf{D}$ is $0 = \mathbf{x}$. From properties of *Sub*, we can prove

$$S(\mathbf{k}_{\ulcorner \mathbf{D} \urcorner}, \mathbf{x}) = \langle \mathbf{k}_{SN(=)}, Num(0), Num(\mathbf{x}) \rangle.$$

Hence we need only prove

$$0 = \mathbf{x} \rightarrow Thm_P(\langle \mathbf{k}_{SN(=)}, Num(0), Num(\mathbf{x}) \rangle).$$

In view of the equality theorem, it suffices to prove

$$Thm_P(\langle \mathbf{k}_{SN(=)}, Num(0), Num(0) \rangle).$$

This is a true closed existential formula and hence provable.

Suppose that $\mathbf{D}$ is $\mathbf{x} + \mathbf{y} = \mathbf{z}$. As above, it suffices to prove

$$\mathbf{x} + \mathbf{y} = \mathbf{z} \rightarrow Thm_P(\langle \mathbf{k}_{SN(=)}, \langle \mathbf{k}_{SN(+)}, Num(\mathbf{x}), Num(\mathbf{y}) \rangle, Num(\mathbf{z}) \rangle);$$

and for this, it suffices to prove

$$Thm_P(\langle \mathbf{k}_{SN(=)}, \langle \mathbf{k}_{SN(+)}, Num(\mathbf{x}), Num(\mathbf{y}) \rangle, Num(\mathbf{x} + \mathbf{y}) \rangle).$$

This is a formalized version of (3) of §6.7, and is proved by induction on $\mathbf{y}$.

Suppose that $\mathbf{D}$ is $\exists \mathbf{z} \mathbf{D}'$. If $\mathbf{a}$ is $S(\mathbf{k}_{\ulcorner \mathbf{D}' \urcorner}, \mathbf{x}_1, \ldots, \mathbf{x}_n)$, then by properties of *Sub*,

$$\vdash S(\mathbf{k}_{\ulcorner \mathbf{D} \urcorner}, \mathbf{x}_1, \ldots, \mathbf{x}_n) = \langle \mathbf{k}_{SN(\exists)}, \mathbf{k}_{\ulcorner \mathbf{z} \urcorner}, \mathbf{a} \rangle. \tag{9}$$

By induction hypothesis

$$\vdash \mathbf{D}' \rightarrow Thm_P(Sub(\mathbf{a}, \mathbf{k}_{\ulcorner \mathbf{z} \urcorner}, Num(\mathbf{z})));$$

so by the distribution rule

$$\vdash \mathbf{D} \rightarrow \exists \mathbf{z} Thm_P(Sub(\mathbf{a}, \mathbf{k}_{\ulcorner \mathbf{z} \urcorner}, Num(\mathbf{z}))). \tag{10}$$

But also

$$\vdash \exists \mathbf{z} Thm_P(Sub(\mathbf{a}, \mathbf{k}_{\ulcorner \mathbf{z} \urcorner}, Num(\mathbf{z}))) \rightarrow Thm_P(\langle \mathbf{k}_{SN(\exists)}, \mathbf{k}_{\ulcorner \mathbf{z} \urcorner}, \mathbf{a} \rangle); \tag{11}$$

for this is merely a formal statement that if $\vdash \mathbf{D}'_{\mathbf{z}}[\mathbf{k}_n]$ for some n, then $\vdash \exists \mathbf{z} \mathbf{D}'$. From (10), (11), and (9) we obtain (8). The remaining cases are treated similarly, using a formalization of Lemma 2 of §6.7 when $\mathbf{D}$ is $\forall \mathbf{z}(\mathbf{z} < \mathbf{x}_i \rightarrow \mathbf{D}')$.

We have now shown that $Con_P \rightarrow \mathbf{A}[\mathbf{k}_a]$ is a theorem of P'. Since $\mathbf{A}[\mathbf{k}_a]$ is not a theorem of P', it follows that Con_P is not a theorem of P'. If we refer to the translation of Con_P into P as *the formula of P which states that P is consistent*, then we have the following result.

Theorem on Consistency Proofs (*Gödel*). The formula of P which states that P is consistent is not a theorem of P.

As with Church's theorem and the incompleteness theorem, this result can be extended to more general theories; and we can argue informally that it extends to more general axiom systems. The general conclusion is that if an axiom system contains as much number theory as P, then we cannot prove the consistency of that axiom system from the axioms of that system.

The theorem on consistency proofs is a limitation on the type of consistency proof which we can give for P. For this to be of any significance, we must know that some types of consistency proofs can be formalized in P. Now it is reasonable to suggest that every finitary consistency proof can be formalized in P (or, equivalently, in a recursive extension of P). First, a finitary proof deals only with concrete objects, and these may be replaced by natural numbers by assigning such a number to each object (as we have done for expressions). Second, the proof deals with these objects in a constructive fashion; so we can expect the functions and predicates which arise to be introducible in recursive extensions of P.

An examination of specific finitary consistency proofs confirms this suggestion. For example, the consistency proof for N given in Chapter 4 can be formalized in P. It is a tedious but elementary exercise to formalize the proof of the consistency theorem. We then have to check that the set of expression numbers of true variable-free formulas of N can be introduced in a recursive extension of P; and this is also straightforward.

We cannot, of course, state with assurance that every finitary consistency proof can be formalized in P, since we have not specified exactly what methods are finitary. One might try to prove the result from axioms about finitary methods which are evident even from our imprecise definition; but not much progress has been made in this direction. However, investigations by Kreisel have shown that a consistency proof which could not be formalized in P would have to use some quite different principles from those used in known finitary proofs.

We conclude that it is reasonable to give up hope of finding a finitary consistency proof for P. This does not mean, however, that we should be satisfied with the consistency proof by means of the standard model. The main trouble with this proof is not that it is nonfinitary, but that it is so uninformative. The consistency proof for N in Chapter 4 was accompanied by side results which eventually led to a solution of the characterization problem. The consistency proof for P by means of the standard model has no such side results. It does not even increase our understanding of P, since nothing goes into it which we did not put into P in the first place.

There is a second reason for continuing to look for a consistency proof for P. A finitary proof has two features: it deals with concrete objects, and it does so in a constructive fashion. Now we can hope to find a consistency proof which deals with abstract objects, but is still constructive. The first such proof was found by Gentzen; and several others have been discovered since. We shall present one such proof, due to Gödel.

8.3 THE CONSISTENCY PROOF

We now introduce a formal system P' which is really just a variation of P. The symbols of P' are the same as the symbols of P, except that $\exists$ is replaced by $\forall$. *Terms* and *formulas* and *free* and *bound* occurrences of variables are defined as in P, except that again $\exists$ is replaced by $\forall$.

The axioms and rules of P' are obtained from those of P by three changes. First, the substitution axioms of P are replaced by the *substitution axioms of* P', which are the formulas of the form $\forall \mathbf{x}\mathbf{A} \to \mathbf{A}_{\mathbf{x}}[\mathbf{a}]$. Second, the $\exists$-introduction rule is replaced by the following $\forall$-*introduction rule*: infer $\forall \mathbf{x}\mathbf{A} \lor \mathbf{B}$ from $\mathbf{A} \lor \mathbf{B}$, provided that $\mathbf{x}$ is not free in $\mathbf{B}$. Third, the induction axioms are replaced by the *induction rule*: infer $\mathbf{A}$ from $\mathbf{A}_{\mathbf{x}}[0]$ and $\mathbf{A} \to \mathbf{A}_{\mathbf{x}}[S\mathbf{x}]$.

It is clear that the tautology theorem holds in P', since the modified axioms and rules of P were not used in its proof.

We introduce $\exists$ as an abbreviation in P' by letting $\exists \mathbf{x}\mathbf{A}$ be an abbreviation for $\neg\forall \mathbf{x} \neg \mathbf{A}$. Then every formula of P is a defined formula of P'. We show that if $\vdash_P \mathbf{A}$, then $\vdash_{P'} \mathbf{A}$. For this it obviously suffices to show that the substitution axioms of P and the induction axioms are provable in P', and that the $\exists$-introduction rule holds in P'.

By the substitution axioms of P', $\vdash_{P'} \forall \mathbf{x} \neg \mathbf{A} \to \neg \mathbf{A}_{\mathbf{x}}[\mathbf{a}]$. Then by the tautology theorem, $\vdash_{P'} \mathbf{A}_{\mathbf{x}}[\mathbf{a}] \to \neg\forall \mathbf{x} \neg \mathbf{A}$, that is, $\vdash_{P'} \mathbf{A}_{\mathbf{x}}[\mathbf{a}] \to \exists \mathbf{x}\mathbf{A}$.

Suppose that $\vdash_{P'} \mathbf{A} \to \mathbf{B}$ and that $\mathbf{x}$ is not free in $\mathbf{B}$. By the definition of $\to$ and the $\forall$-introduction rule, $\vdash_{P'} \forall \mathbf{x} \neg \mathbf{A} \lor \mathbf{B}$. By the tautology theorem, $\vdash_{P'} \neg\forall \mathbf{x} \neg \mathbf{A} \to \mathbf{B}$, that is, $\vdash_{P'} \exists \mathbf{x}\mathbf{A} \to \mathbf{B}$.

Suppose that $\mathbf{A}$ is an induction axiom

$$\mathbf{B}_{\mathbf{x}}[0] \;\&\; \forall \mathbf{x}(\mathbf{B} \to \mathbf{B}_{\mathbf{x}}[S\mathbf{x}]) \to \mathbf{B}.$$

It is easy to see that $\mathbf{A}_{\mathbf{x}}[0]$ and $\mathbf{A} \to \mathbf{A}_{\mathbf{x}}[S\mathbf{x}]$ are provable in P without use of induction axioms. Hence by what has already been proved, they are theorems of P'. By the induction rule, $\vdash_{P'} \mathbf{A}$.

The theorem on consistency proofs shows that every consistency proof for P will have to use something which is not formalizable in P. Gentzen's proof used transfinite induction. We shall instead use functionals of higher types, which we now describe.

We define the *type symbols* by the generalized inductive definition:

a) o is a type symbol;

b) if r and s are type symbols, then $(r \to s)$ is a type symbol.

We now define the *functionals of type* r for each type symbol r by induction on the length of r. A functional of type o is a natural number. A functional of type $(r \to s)$ is a mapping from the set of functionals of type r to the set of functionals of type s. Henceforth *functional* will mean functional of some type.

If F is a functional of type $(r \to (s \to t))$ and x is a functional of type r, then $F(x)$ is a functional of type $(s \to t)$; so if y is a functional of type s, then $F(x)(y)$ is a functional of type t. More generally, if F is a functional of type

$$(r_1 \to (r_2 \to \cdots \to (r_n \to s) \ldots)) \tag{1}$$

and $x_1, \ldots, x_n$ are functionals of types $r_1, \ldots, r_n$ respectively, then $F(x_1) \ldots (x_n)$ is a functional of type s. If we set

$$F'(x_1, \ldots, x_n) = F(x_1) \ldots (x_n),$$

we get a mapping F' from the set A of n-tuples $(x_1, \ldots, x_n)$ with x_i of type r_i for $i = 1, \ldots, n$ to the set B of functionals of type s. Moreover, every mapping from A to B may be obtained in this way from a unique functional F of type (1). We shall therefore identify F and F'. Notationally, this means that we can write $F(x_1, \ldots, x_n)$ in place of $F(x_1) \ldots (x_n)$. To emphasize this identification, we shall sometimes write the type (1) as $(r_1, \ldots, r_n \to s)$.

It is clear that an (m, n)-ary functional in the sense of the preceding chapter is a functional of type

$$(r_1, \ldots, r_m, s_1, \ldots, s_n \to o),$$

where each r_i is $(o \to o)$ and each s_i is o.

We shall now introduce a language Y for discussing functionals. The symbols of Y are the following.

i) For each type symbol r, the *variables of type* r:

$$x_r, y_r, z_r, w_r, x'_r, \ldots.$$

ii) For each type symbol r, the constants of type r. These will be described below.

iii) The symbols Ap, $=$, $\neg$, and $\vee$.

The variables of type r will vary through the functionals of type r. Each constant of type r will designate a particular functional of type r, as explained below. We intend Ap to designate the application of a function to its argument; so we abbreviate $Ap\mathbf{uv}$ to $\mathbf{u}(\mathbf{v})$ and $\mathbf{u}(\mathbf{v}_1) \ldots (\mathbf{v}_n)$ to $\mathbf{u}(\mathbf{v}_1, \ldots, \mathbf{v}_n)$. The symbols $=$, $\neg$, and $\vee$ have their usual meaning.

We use $\mathbf{x}$, $\mathbf{y}$, and $\mathbf{z}$ as syntactical variables which vary through the variables of Y.

The *constants* and *terms* of Y are defined by the following generalized inductive definition.

a) The symbol 0 is a constant of type o; the symbol S is a constant of type $(o \to o)$.

b) Every variable or constant of type r is a term of type r.

c) If $\mathbf{u}$ is a term of type $(r \to s)$ and $\mathbf{v}$ is a term of type r, then $\mathbf{u}(\mathbf{v})$ is a term of type s.

d) Suppose that $\mathbf{x}_1, \ldots, \mathbf{x}_n$ are distinct variables of types $r_1, \ldots, r_n$ respectively, and that $\mathbf{u}$ is a term of type s in which no variable other than $\mathbf{x}_1, \ldots, \mathbf{x}_n$ occurs. We introduce a new constant $\mathbf{v}$ of type $(r_1, \ldots, r_n \to s)$ with the *defining equation* $\mathbf{v}\mathbf{x}_1 \ldots \mathbf{x}_n = \mathbf{u}$.

e) Suppose that $\mathbf{x}$ is a variable of type o, $\mathbf{u}$ is a constant of type r, and $\mathbf{u}'$ is a constant of type $(r, o \to r)$. We introduce a new constant $\mathbf{v}$ of type $(o \to r)$ with the *defining equations* $\mathbf{v}(0) = \mathbf{u}$ and $\mathbf{v}(S(\mathbf{x})) = \mathbf{u}'(\mathbf{v}(\mathbf{x}), \mathbf{x})$.

The constant 0 designates the natural number zero and the constant S designates the successor function. The meanings of the remaining constants are given by their defining equations. (This is the only function of the defining equations. They are not definitions in the sense of §1.2, since the constants are not defined symbols.)

We use $\mathbf{f}$, $\mathbf{g}$, and $\mathbf{h}$ as syntactical variables which vary through constants, and $\mathbf{a}$, $\mathbf{b}$, and $\mathbf{c}$ as syntactical variables which vary through terms.

An *atomic formula* is an expression $\mathbf{a} = \mathbf{b}$ where $\mathbf{a}$ and $\mathbf{b}$ are of type o. The formulas are defined by the generalized inductive definition.

i) An atomic formula is a formula.

ii) If $\mathbf{u}$ is a formula, then $\neg\mathbf{u}$ is a formula.

iii) If $\mathbf{u}$ and $\mathbf{v}$ are formulas, then $\vee\mathbf{uv}$ is a formula.

(Note that defining equations are usually not formulas, since they are equations between terms not necessarily of type o.)

Since we have explained the meaning of all the symbols of Y, it will be clear what is meant by saying that a formula of Y is *true* for certain values of its variables. We use $\vdash_Y \mathbf{A}$ to mean that $\mathbf{A}$ is true for all values of its variables, sometimes omitting the subscript.

The abbreviations used in theories are, when appropriate, assumed introduced in Y. To avoid constant references to type, we adopt the following convention: the types of all variables and constants are to be such that all expressions which occur are formed according to the rules for forming terms, formulas, and defining equations. Thus if we refer to the formula $\mathbf{f}(\mathbf{x}) = \mathbf{a}$, it is understood that for some r, $\mathbf{f}$ is of type $(r \to o)$, $\mathbf{x}$ is of type r, and $\mathbf{a}$ is of type o.

We let $\mathbf{u_x}[\mathbf{a}]$ be the expression obtained from $\mathbf{u}$ by replacing all occurrences of $\mathbf{x}$ by $\mathbf{a}$. When this symbol occurs, it is understood that $\mathbf{x}$ and $\mathbf{a}$ have the same type. We use $\mathbf{u}_{\mathbf{x}_1,\ldots,\mathbf{x}_n}[\mathbf{a}_1, \ldots, \mathbf{a}_n]$ similarly.

We remark that there is at least one constant of each type. For 0 is a constant of type o; and if $\mathbf{f}$ is of type s and $\mathbf{x}$ is a variable of type r, we can introduce a $\mathbf{g}$ of type $(r \to s)$ by the defining equation $\mathbf{g}(\mathbf{x}) = \mathbf{f}$.

Let $\mathbf{x}_1, \ldots, \mathbf{x}_n, \mathbf{y}, \mathbf{z}$ be distinct variables with $\mathbf{y}$ of type o. Let $\mathbf{a}$ and $\mathbf{b}$ be terms of type o such that $\mathbf{a}$ contains no variable other than $\mathbf{x}_1, \ldots, \mathbf{x}_n$ and $\mathbf{b}$ contains no variable other than $\mathbf{x}_1, \ldots, \mathbf{x}_n, \mathbf{y}, \mathbf{z}$. Then we can find a constant $\mathbf{f}$ such that

$$\begin{aligned} &\vdash_Y \mathbf{f}(0, \mathbf{x}_1, \ldots, \mathbf{x}_n) = \mathbf{a}, \\ &\vdash_Y \mathbf{f}\big(S(\mathbf{y}), \mathbf{x}_1, \ldots, \mathbf{x}_n\big) = \mathbf{b_z}[\mathbf{f}(\mathbf{y}, \mathbf{x}_1, \ldots, \mathbf{x}_n)]. \end{aligned} \tag{2}$$

For this we introduce $\mathbf{f}$ by the defining equations $\mathbf{f}(0) = \mathbf{g}$, $\mathbf{f}\big(S(\mathbf{y})\big) = \mathbf{h}\big(\mathbf{f}(\mathbf{y}), \mathbf{y}\big)$, where $\mathbf{g}$ and $\mathbf{h}$ have the defining equations

$$\begin{aligned} \mathbf{g}(\mathbf{x}_1, \ldots, \mathbf{x}_n) &= \mathbf{a}, \\ \mathbf{h}(\mathbf{w}, \mathbf{y}, \mathbf{x}_1, \ldots, \mathbf{x}_n) &= \mathbf{b_z}[\mathbf{w}(\mathbf{x}_1, \ldots, \mathbf{x}_n)]. \end{aligned}$$

When we choose an $\mathbf{f}$ to satisfy (2), we say that we are *introducing* $\mathbf{f}$ by (2). In particular, we introduce $+$ by

$$\begin{aligned} 0 + x &= x, \\ S(y) + x &= S(x + y), \end{aligned}$$

and then introduce $\cdot$ by

$$0 \cdot x = 0,$$
$$S(y) \cdot x = (y \cdot x) + x.$$

We also introduce P by

$$P(0) = 0,$$
$$P(S(y)) = y,$$

and then introduce L by

$$L(0, x) = x,$$
$$L(S(y), x) = P(L(y, x)).$$

We then abbreviate $L(\mathbf{a}, \mathbf{b}) \neq 0$ to $\mathbf{a} < \mathbf{b}$. Clearly this gives $<$ its usual meaning.

We identify the variables of P' with the variables of type o of Y. Then every term of P' is a term of type o of Y, and every open formula of P' is a (possibly defined) formula of Y. Moreover, such a formula has the same meaning in P' as in Y.

Next we observe that for each formula $\mathbf{A}$ of Y, there is a term $\mathbf{a}$ of type o containing no variable not in $\mathbf{A}$ such that

$$\vdash_Y \mathbf{A} \leftrightarrow \mathbf{a} = 0.$$

We can easily obtain $\mathbf{a}$ by induction on the length of $\mathbf{A}$ if we have constants E, N, and D such that

$$\mathbf{a} = \mathbf{b} \leftrightarrow E(\mathbf{a}, \mathbf{b}) = 0,$$
$$\mathbf{a} \neq 0 \leftrightarrow N(\mathbf{a}) = 0,$$
$$\mathbf{a} = 0 \vee \mathbf{b} = 0 \leftrightarrow D(\mathbf{a}, \mathbf{b}) = 0.$$

These can be introduced by

$$E(x, y) = L(x, y) + L(y, x),$$
$$N(0) = 1, N(S(y)) = 0,$$
$$D(x, y) = x \cdot y.$$

Given distinct variables $\mathbf{x}_1, \ldots, \mathbf{x}_n$ which include all the variables occurring in $\mathbf{A}$, $\mathbf{a}$, and $\mathbf{b}$, we can find a constant $\mathbf{f}$ such that

$$\vdash \mathbf{A} \rightarrow \mathbf{f}(\mathbf{x}_1, \ldots, \mathbf{x}_n) = \mathbf{a},$$
$$\vdash \neg\mathbf{A} \rightarrow \mathbf{f}(\mathbf{x}_1, \ldots, \mathbf{x}_n) = \mathbf{b}. \tag{3}$$

For we choose $\mathbf{c}$ by the above so that $\vdash \mathbf{A} \leftrightarrow \mathbf{c} = 0$ and introduce $\mathbf{f}$ by the defining equation

$$\mathbf{f}(\mathbf{x}_1, \ldots, \mathbf{x}_n) = \mathbf{a} \cdot N(\mathbf{c}) + \mathbf{b} \cdot N(N(\mathbf{c})).$$

When we choose $\mathbf{f}$ to satisfy (3), we say that we are *introducing* $\mathbf{f}$ by (3).

A *generalized formula* is an expression $\forall \mathbf{x}_1 \ldots \forall \mathbf{x}_m \exists \mathbf{y}_1 \ldots \exists \mathbf{y}_n \mathbf{A}$, where $\mathbf{x}_1, \ldots, \mathbf{x}_m, \mathbf{y}_1, \ldots, \mathbf{y}_n$ are distinct and $\mathbf{A}$ is a formula of Y. Of course this is not even an expression of Y (unless $m = n = 0$); but it is clear what it is intended to mean. We regard two generalized formulas as being the same if they differ only in the choice of the quantified variables (i.e., if they are variants).

It is convenient to modify our use of the syntactical variables for the rest of this section. We shall use $\mathbf{x}$, $\mathbf{y}$, and $\mathbf{z}$ to represent sequences (possibly empty) of distinct variables. (We continue to let $\mathbf{w}$ vary through variables of P'.) If $\mathbf{x}$ and $\mathbf{y}$ appear in the same context, the variables in $\mathbf{x}$ are to be distinct from the variables in $\mathbf{y}$, and similarly for other pairs of syntactical variables. If $\mathbf{x}$ is the sequence $\mathbf{x}_1, \ldots, \mathbf{x}_n$, then $\exists\mathbf{x}$ is $\exists\mathbf{x}_1 \ldots \exists\mathbf{x}_n$ and $\forall\mathbf{x}$ is $\forall\mathbf{x}_1 \ldots \forall\mathbf{x}_n$. We use $\mathbf{a}$, $\mathbf{b}$, and $\mathbf{c}$ to represent sequences of (not necessarily distinct) terms and $\mathbf{f}$, $\mathbf{g}$, and $\mathbf{h}$ to represent sequences of constants. (We continue to let $\mathbf{d}$ vary through terms of P'.) If $\mathbf{a}$ is the sequence $\mathbf{a}_1, \ldots, \mathbf{a}_n$ and $\mathbf{b}$ is the sequence $\mathbf{b}_1, \ldots, \mathbf{b}_m$, then $\mathbf{a}(\mathbf{b})$ is the sequence

$$\mathbf{a}_1(\mathbf{b}_1, \ldots, \mathbf{b}_m), \ldots, \mathbf{a}_n(\mathbf{b}_1, \ldots, \mathbf{b}_m).$$

A similar meaning is given to $\mathbf{f}(\mathbf{a})$, $\mathbf{b}(\mathbf{x})$, etc. If $\mathbf{a}$ is the sequence $\mathbf{a}_1, \ldots, \mathbf{a}_n$ and $\mathbf{b}$ is the sequence $\mathbf{b}_1, \ldots, \mathbf{b}_n$, then $\mathbf{a} = \mathbf{b}$ stands for the sequence of equalities $\mathbf{a}_1 = \mathbf{b}_1, \ldots, \mathbf{a}_n = \mathbf{b}_n$. Thus we might introduce a sequence of constants $\mathbf{f}$ by the sequence of defining equations $\mathbf{f}(\mathbf{x}) = \mathbf{a}$.

In order to avoid subscript notation, we write a generalized formula $\forall\mathbf{x}\,\exists\mathbf{y}\mathbf{A}$ as $\forall\mathbf{x}\,\exists\mathbf{y}\mathbf{A}[\mathbf{x}, \mathbf{y}, \mathbf{z}]$, where $\mathbf{z}$ consists of the variables free in $\mathbf{A}$, and, possibly, variables not appearing in $\mathbf{A}$. We then write $\mathbf{A}[\mathbf{a}, \mathbf{b}, \mathbf{c}]$ for $\mathbf{A}_{\mathbf{x},\mathbf{y},\mathbf{z}}[\mathbf{a}, \mathbf{b}, \mathbf{c}]$. Since we allow variables not in $\mathbf{A}$ to appear in $\mathbf{z}$ and since we may change quantified variables without changing a generalized formula, we see that any two generalized formulas can be written as $\forall\mathbf{x}\,\exists\mathbf{y}\mathbf{A}[\mathbf{x}, \mathbf{y}, \mathbf{z}]$ and $\forall\mathbf{x}'\,\exists\mathbf{y}'\mathbf{B}[\mathbf{x}', \mathbf{y}', \mathbf{z}]$ with the usual convention that the variables in $\mathbf{x}$, $\mathbf{y}$, $\mathbf{x}'$, $\mathbf{y}'$, and $\mathbf{z}$ are all distinct.

We shall now assign a generalized formula $\mathbf{A}^*$ to each formula $\mathbf{A}$ of P'. The definition of $\mathbf{A}^*$ is by induction on the length of $\mathbf{A}$.

i) If $\mathbf{A}$ is atomic, then $\mathbf{A}^*$ is $\mathbf{A}$.

ii) If $\mathbf{A}$ is $\neg\mathbf{B}$ and $\mathbf{B}^*$ is $\forall\mathbf{x}\,\exists\mathbf{y}\mathbf{B}'[\mathbf{x}, \mathbf{y}, \mathbf{z}]$, then $\mathbf{A}^*$ is $\forall\mathbf{y}'\,\exists\mathbf{x}\neg\mathbf{B}'[\mathbf{x}, \mathbf{y}'(\mathbf{x}), \mathbf{z}]$, where $\mathbf{y}'$ is a sequence of new variables of the appropriate types.

iii) If $\mathbf{A}$ is $\mathbf{B} \vee \mathbf{C}$, $\mathbf{B}^*$ is $\forall\mathbf{x}\,\exists\mathbf{y}\mathbf{B}'[\mathbf{x}, \mathbf{y}, \mathbf{z}]$, and $\mathbf{C}^*$ is $\forall\mathbf{x}'\,\exists\mathbf{y}'\mathbf{C}'[\mathbf{x}', \mathbf{y}', \mathbf{z}]$, then $\mathbf{A}^*$ is $\forall\mathbf{x}\,\forall\mathbf{x}'\,\exists\mathbf{y}\,\exists\mathbf{y}'(\mathbf{B}'[\mathbf{x}, \mathbf{y}, \mathbf{z}] \vee \mathbf{C}'[\mathbf{x}', \mathbf{y}', \mathbf{z}])$.

iv) If $\mathbf{A}$ is $\forall\mathbf{w}\mathbf{B}$ and $\mathbf{B}^*$ is $\forall\mathbf{x}\,\exists\mathbf{y}\mathbf{B}'[\mathbf{x}, \mathbf{y}, \mathbf{z}]$, then $\mathbf{A}^*$ is $\forall\mathbf{w}\,\forall\mathbf{x}\,\exists\mathbf{y}\mathbf{B}'[\mathbf{x}, \mathbf{y}, \mathbf{z}]$.

Although we shall not use the fact in our proof, it is worth noting that $\mathbf{A}$ and $\mathbf{A}^*$ have the same meaning. This is obvious except, perhaps, in case (ii). There we observe that $\mathbf{B}^*$ has the same meaning as $\exists\mathbf{y}'\,\forall\mathbf{x}\mathbf{B}'[\mathbf{x}, \mathbf{y}'(\mathbf{x}), \mathbf{z}]$, so that $\mathbf{A}$ has the same meaning as $\neg\exists\mathbf{y}'\,\forall\mathbf{x}\mathbf{B}[\mathbf{x}, \mathbf{y}'(\mathbf{x}), \mathbf{z}]$; and then we apply the prenex operations.

If $\mathbf{A}$ is open, then $\mathbf{A}^*$ is $\mathbf{A}$. It is easy to verify by induction on the length of $\mathbf{A}$ that $\mathbf{A}_\mathbf{w}[\mathbf{d}]^*$ is $\mathbf{A}^*_\mathbf{w}[\mathbf{d}]$.

A generalized formula $\forall\mathbf{x}\,\exists\mathbf{y}\mathbf{A}[\mathbf{x}, \mathbf{y}, \mathbf{z}]$ is *valid* if there is a sequence of terms $\mathbf{a}$ such that $\vdash_Y \mathbf{A}[\mathbf{x}, \mathbf{a}, \mathbf{z}]$. (This of course implies that the generalized formula is true for all values of its variables.)

Theorem. If $\vdash_{P'} \mathbf{A}$, then $\mathbf{A}^*$ is valid.

Proof. We use induction on theorems of P'. First suppose that $\mathbf{A}$ is a propositional axiom $\neg\mathbf{B} \vee \mathbf{B}$. If $\mathbf{B}^*$ is $\forall\mathbf{x}\,\exists\mathbf{y}\mathbf{B}'[\mathbf{x}, \mathbf{y}, \mathbf{z}]$, then $(\neg\mathbf{B})^*$ is (after a change of variable) $\forall\mathbf{y}'\,\exists\mathbf{x}'\,\neg\mathbf{B}'[\mathbf{x}', \mathbf{y}'(\mathbf{x}'), \mathbf{z}]$; so $\mathbf{A}^*$ is

$$\forall\mathbf{y}'\,\forall\mathbf{x}\,\exists\mathbf{x}'\,\exists\mathbf{y}(\neg\mathbf{B}'[\mathbf{x}', \mathbf{y}'(\mathbf{x}'), \mathbf{z}] \vee \mathbf{B}'[\mathbf{x}, \mathbf{y}, \mathbf{z}]).$$

Hence we must find sequences of terms $\mathbf{a}$ and $\mathbf{b}$ such that

$$\vdash \neg\mathbf{B}'[\mathbf{a}, \mathbf{y}'(\mathbf{a}), \mathbf{z}] \vee \mathbf{B}'[\mathbf{x}, \mathbf{b}, \mathbf{z}].$$

We take $\mathbf{a}$ to be $\mathbf{x}$ and $\mathbf{b}$ to be $\mathbf{y}'(\mathbf{x})$.

Now suppose that $\mathbf{A}$ is a substitution axiom $\forall\mathbf{w}\mathbf{B} \rightarrow \mathbf{B}_{\mathbf{w}}[\mathbf{d}]$. If $\mathbf{B}^*$ is

$$\forall\mathbf{x}\,\exists\mathbf{y}\mathbf{B}'[\mathbf{x}, \mathbf{y}, \mathbf{z}, \mathbf{w}],$$

then $(\neg\forall\mathbf{w}\mathbf{B})^*$ is

$$\forall\mathbf{y}'\,\exists\mathbf{w}\,\exists\mathbf{x}'\,\neg\mathbf{B}'[\mathbf{x}', \mathbf{y}'(\mathbf{w}, \mathbf{x}'), \mathbf{z}, \mathbf{w}]$$

and $\mathbf{B}_{\mathbf{w}}[\mathbf{d}]^*$ is

$$\forall\mathbf{x}\,\exists\mathbf{y}\mathbf{B}'[\mathbf{x}, \mathbf{y}, \mathbf{z}, \mathbf{d}].$$

Hence $\mathbf{A}^*$ is

$$\forall\mathbf{y}'\,\forall\mathbf{x}\,\exists\mathbf{w}\,\exists\mathbf{x}'\,\exists\mathbf{y}\big(\mathbf{B}'[\mathbf{x}', \mathbf{y}'(\mathbf{w}, \mathbf{x}'), \mathbf{z}, \mathbf{w}] \rightarrow \mathbf{B}'[\mathbf{x}, \mathbf{y}, \mathbf{z}, \mathbf{d}]\big).$$

We must find a term $\mathbf{a}$ and sequences of terms $\mathbf{b}$ and $\mathbf{c}$ such that

$$\vdash \mathbf{B}'[\mathbf{b}, \mathbf{y}'(\mathbf{a}, \mathbf{b}), \mathbf{z}, \mathbf{a}] \rightarrow \mathbf{B}'[\mathbf{x}, \mathbf{c}, \mathbf{z}, \mathbf{d}].$$

We take $\mathbf{a}$ to be $\mathbf{d}$, $\mathbf{b}$ to be $\mathbf{x}$, and $\mathbf{c}$ to be $\mathbf{y}'(\mathbf{a}, \mathbf{b})$.

If $\mathbf{A}$ is an identity axiom or an equality axiom or one of N1 through N8, then $\mathbf{A}$ is open and hence $\mathbf{A}^*$ is $\mathbf{A}$. Clearly $\vdash_Y \mathbf{A}$.

Suppose that $\mathbf{A}$ is inferred from $\mathbf{B}$ by the expansion rule; say $\mathbf{A}$ is $\mathbf{C} \vee \mathbf{B}$. If $\mathbf{B}^*$ is $\forall\mathbf{x}\,\exists\mathbf{y}\mathbf{B}'[\mathbf{x}, \mathbf{y}, \mathbf{z}]$ and $\mathbf{C}^*$ is $\forall\mathbf{x}'\,\exists\mathbf{y}'\mathbf{C}'[\mathbf{x}', \mathbf{y}', \mathbf{z}]$, then $\mathbf{A}^*$ is

$$\forall\mathbf{x}'\,\forall\mathbf{x}\,\exists\mathbf{y}'\,\exists\mathbf{y}(\mathbf{C}'[\mathbf{x}', \mathbf{y}', \mathbf{z}] \vee \mathbf{B}'[\mathbf{x}, \mathbf{y}, \mathbf{z}]).$$

By induction hypothesis, $\mathbf{B}^*$ is valid; so there is a sequence of terms $\mathbf{a}$ such that $\vdash \mathbf{B}'[\mathbf{x}, \mathbf{a}, \mathbf{z}]$. Hence for any sequence of terms $\mathbf{b}$,

$$\vdash \mathbf{C}'[\mathbf{x}', \mathbf{b}, \mathbf{z}] \vee \mathbf{B}'[\mathbf{x}, \mathbf{a}, \mathbf{z}].$$

It follows that $\mathbf{A}^*$ is valid.

Now suppose that $\mathbf{A}$ is inferred from $\mathbf{A} \vee \mathbf{A}$ by the contraction rule. If $\mathbf{A}$ is $\forall\mathbf{x}\,\exists\mathbf{y}\mathbf{A}'[\mathbf{x}, \mathbf{y}, \mathbf{z}]$, then $(\mathbf{A} \vee \mathbf{A})^*$ is

$$\forall\mathbf{x}\,\forall\mathbf{x}'\,\exists\mathbf{y}\,\exists\mathbf{y}'(\mathbf{A}'[\mathbf{x}, \mathbf{y}, \mathbf{z}] \vee \mathbf{A}'[\mathbf{x}', \mathbf{y}', \mathbf{z}]).$$

Since $(\mathbf{A} \vee \mathbf{A})^*$ is valid, there are sequences of terms $\mathbf{a}$ and $\mathbf{a}'$ such that

$$\vdash \mathbf{A}'[\mathbf{x}, \mathbf{a}, \mathbf{z}] \vee \mathbf{A}'[\mathbf{x}', \mathbf{a}', \mathbf{z}].$$

Substituting $\mathbf{x}$ for $\mathbf{x}'$, we have

$$\vdash \mathbf{A}'[\mathbf{x}, \mathbf{b}, \mathbf{z}] \vee \mathbf{A}'[\mathbf{x}, \mathbf{b}', \mathbf{z}] \tag{4}$$

for suitable **b** and **b**′. By means of (3), we can find a sequence of terms **c** such that

$$\mathbf{A}'[\mathbf{x}, \mathbf{b}, \mathbf{z}] \to \mathbf{c} = \mathbf{b},$$
$$\neg \mathbf{A}'[\mathbf{x}, \mathbf{b}, \mathbf{z}] \to \mathbf{c} = \mathbf{b}'.$$

From this and (4), $\vdash \mathbf{A}'[\mathbf{x}, \mathbf{c}, \mathbf{z}]$. Hence **A*** is valid.

Suppose that **A** is inferred from **B** by the associative rule. Then the matrix of **A*** can be inferred from the matrix of **B*** by the associative rule. It follows readily that the validity of **B*** implies the validity of **A***.

Suppose that **A** is inferred by the cut rule. Say that **A** is $\mathbf{C} \vee \mathbf{D}$, and is inferred from $\mathbf{B} \vee \mathbf{C}$ and $\neg\mathbf{B} \vee \mathbf{D}$. Let **B***, **C***, and **D*** be

$$\forall \mathbf{x}\, \exists \mathbf{y} \mathbf{B}'[\mathbf{x}, \mathbf{y}, \mathbf{z}], \quad \forall \mathbf{x}'\, \exists \mathbf{y}' \mathbf{C}'[\mathbf{x}', \mathbf{y}', \mathbf{z}], \quad \text{and} \quad \forall \mathbf{x}''\, \exists \mathbf{y}'' \mathbf{D}'[\mathbf{x}'', \mathbf{y}'', \mathbf{z}].$$

Then $(\mathbf{B} \vee \mathbf{C})^*$ is

$$\forall \mathbf{x}\, \forall \mathbf{x}'\, \exists \mathbf{y}\, \exists \mathbf{y}' \big(\mathbf{B}'[\mathbf{x}, \mathbf{y}, \mathbf{z}] \vee \mathbf{C}'[\mathbf{x}', \mathbf{y}', \mathbf{z}]\big)$$

and $(\neg\mathbf{B} \vee \mathbf{D})^*$ is

$$\forall \mathbf{y}_1\, \forall \mathbf{x}''\, \exists \mathbf{x}\, \exists \mathbf{y}'' \big(\neg \mathbf{B}'[\mathbf{x}, \mathbf{y}_1(\mathbf{x}), \mathbf{z}] \vee \mathbf{D}'[\mathbf{x}'', \mathbf{y}'', \mathbf{z}]\big).$$

Since both of these are valid by induction hypothesis, we have sequences of terms **a**, **a**′, **b** and **b**′ such that

$$\vdash \mathbf{B}'[\mathbf{x}, \mathbf{a}, \mathbf{z}] \vee \mathbf{C}'[\mathbf{x}', \mathbf{a}', \mathbf{z}], \tag{5}$$
$$\vdash \neg \mathbf{B}'[\mathbf{b}, \mathbf{y}_1(\mathbf{b}), \mathbf{z}] \vee \mathbf{D}'[\mathbf{x}'', \mathbf{b}', \mathbf{z}]. \tag{6}$$

We may suppose that **a** contains no variable not appearing in **x**, **z**, or **x**′; for such a variable could be replaced by a constant of the same type. We can then introduce a sequence **f** by the defining equations $\mathbf{f}(\mathbf{x}', \mathbf{z}, \mathbf{x}) = \mathbf{a}$. Putting $\mathbf{f}(\mathbf{x}', \mathbf{z}, \mathbf{x})$ for **a** in (5), we obtain

$$\vdash \mathbf{B}'[\mathbf{x}, \mathbf{f}(\mathbf{x}', \mathbf{z}, \mathbf{x}), \mathbf{z}] \vee \mathbf{C}'[\mathbf{x}', \mathbf{a}', \mathbf{z}]. \tag{7}$$

Substituting $\mathbf{f}(\mathbf{x}', \mathbf{z})$ for $\mathbf{y}_1$ in (6), we get

$$\vdash \neg \mathbf{B}'[\mathbf{b}_1, \mathbf{f}(\mathbf{x}', \mathbf{z}, \mathbf{b}_1), \mathbf{z}] \vee \mathbf{D}'[\mathbf{x}'', \mathbf{b}_1', \mathbf{z}]. \tag{8}$$

Substituting $\mathbf{b}_1$ for **x** in (7), we have

$$\vdash \mathbf{B}'[\mathbf{b}_1, \mathbf{f}(\mathbf{x}', \mathbf{z}, \mathbf{b}_1), \mathbf{z}] \vee \mathbf{C}'[\mathbf{x}', \mathbf{a}_1, \mathbf{z}]. \tag{9}$$

From (8) and (9),

$$\vdash \mathbf{C}'[\mathbf{x}', \mathbf{a}_1, \mathbf{z}] \vee \mathbf{D}'[\mathbf{x}'', \mathbf{b}_1', \mathbf{z}].$$

Since **A*** is

$$\forall \mathbf{x}'\, \forall \mathbf{x}''\, \exists \mathbf{y}'\, \exists \mathbf{y}'' (\mathbf{C}'[\mathbf{x}', \mathbf{y}', \mathbf{z}] \vee \mathbf{D}'[\mathbf{x}'', \mathbf{y}'', \mathbf{z}]),$$

it follows that **A*** is satisfiable.

Now suppose that **A** is inferred from $\mathbf{B} \vee \mathbf{C}$ by the $\forall$-introduction rule. Then **A** is $\forall \mathbf{w} \mathbf{B} \vee \mathbf{C}$ where **w** is not free in **C**. It is then easily checked that **A*** is $\forall \mathbf{w}\big((\mathbf{B} \vee \mathbf{C})^*\big)$. Hence the validity of $(\mathbf{B} \vee \mathbf{C})^*$ implies the validity of **A***.

Finally, suppose that $\mathbf{A}$ is inferred from $\mathbf{A_w}[0]$ and $\mathbf{A} \to \mathbf{A_w}[S\mathbf{w}]$ by the induction rule. Let $\mathbf{A}^*$ be $\forall \mathbf{x}\, \exists \mathbf{y} \mathbf{A}'[\mathbf{x}, \mathbf{y}, \mathbf{z}, \mathbf{w}]$. Then $(\mathbf{A_w}[0])^*$ is $\forall \mathbf{x}\, \exists \mathbf{y} \mathbf{A}'[\mathbf{x}, \mathbf{y}, \mathbf{z}, 0]$; so by induction hypothesis, there is a sequence of terms $\mathbf{a}$ such that $\vdash \mathbf{A}'[\mathbf{x}, \mathbf{a}, \mathbf{z}, 0]$. As above, we may define a sequence $\mathbf{g}$ by $\mathbf{g}(\mathbf{z}, \mathbf{x}) = \mathbf{a}$ and obtain

$$\vdash \mathbf{A}'[\mathbf{x}, \mathbf{g}(\mathbf{z}, \mathbf{x}), \mathbf{z}, 0]. \tag{10}$$

Now $(\mathbf{A} \to \mathbf{A_z}[S\mathbf{z}])^*$ is

$$\forall \mathbf{y}'\, \forall \mathbf{x}\, \exists \mathbf{x}'\, \exists \mathbf{y}\big(\mathbf{A}'[\mathbf{x}', \mathbf{y}'(\mathbf{x}'), \mathbf{z}, \mathbf{w}] \to \mathbf{A}'[\mathbf{x}, \mathbf{y}, \mathbf{z}, S(\mathbf{w})]\big);$$

so by induction hypothesis, there are sequences $\mathbf{b}$ and $\mathbf{c}$ such that

$$\vdash \mathbf{A}'[\mathbf{b}, \mathbf{y}'(\mathbf{b}), \mathbf{z}, \mathbf{w}] \to \mathbf{A}'[\mathbf{x}, \mathbf{c}, \mathbf{z}, S(\mathbf{w})].$$

Substituting $\mathbf{x}'(\mathbf{z})$ for $\mathbf{y}'$, we get

$$\vdash \mathbf{A}'[\mathbf{b}', \mathbf{x}'(\mathbf{z}, \mathbf{b}'), \mathbf{z}, \mathbf{w}] \to \mathbf{A}'[\mathbf{x}, \mathbf{c}', \mathbf{z}, S(\mathbf{w})].$$

As above, we may rewrite this as

$$\vdash \mathbf{A}'[\mathbf{b}', \mathbf{x}'(\mathbf{z}, \mathbf{b}'), \mathbf{z}, \mathbf{w}] \to \mathbf{A}'[\mathbf{x}, \mathbf{h}(\mathbf{x}', \mathbf{w}, \mathbf{z}, \mathbf{x}), \mathbf{z}, S(\mathbf{w})]. \tag{11}$$

Let $\mathbf{f}$ have the defining equations $\mathbf{f}(0) = \mathbf{g}$, $\mathbf{f}\big(S(\mathbf{w})\big) = \mathbf{h}\big(\mathbf{f}(\mathbf{w}), \mathbf{w}\big)$. Then (10) becomes

$$\vdash \mathbf{A}'[\mathbf{x}, \mathbf{f}(0, \mathbf{z}, \mathbf{x}), \mathbf{z}, 0], \tag{12}$$

while substituting $\mathbf{f}(\mathbf{w})$ for $\mathbf{x}'$ in (11) gives

$$\vdash \mathbf{A}'[\mathbf{b}_1, \mathbf{f}(\mathbf{w}, \mathbf{z}, \mathbf{b}_1), \mathbf{z}, \mathbf{w}] \to \mathbf{A}'[\mathbf{x}, \mathbf{f}\big(S(\mathbf{w}), \mathbf{z}, \mathbf{x}\big), \mathbf{z}, S(\mathbf{w})]. \tag{13}$$

We now show that $\vdash \mathbf{A}'[\mathbf{x}, \mathbf{f}(\mathbf{w}, \mathbf{z}, \mathbf{x}), \mathbf{z}, \mathbf{w}]$; this will imply that $\mathbf{A}^*$ is valid. We wish to show that for each n, if $\mathbf{w}$ has the value n, then $\mathbf{A}'[\mathbf{x}, \mathbf{f}(\mathbf{w}, \mathbf{z}, \mathbf{x}), \mathbf{z}, \mathbf{w}]$ is true for all values of $\mathbf{x}$ and $\mathbf{z}$. We prove this by induction on n. For $n = 0$ it holds by (12). Now suppose that it holds for some n. If $\mathbf{w}$ has the value n and $\mathbf{x}$ and $\mathbf{z}$ have any value, then

$$\mathbf{A}'[\mathbf{b}_1, \mathbf{f}(\mathbf{w}, \mathbf{z}, \mathbf{b}_1), \mathbf{z}, \mathbf{w}]$$

is true; so by (13),

$$\mathbf{A}'[\mathbf{x}, \mathbf{f}\big(S(\mathbf{w}), \mathbf{z}, \mathbf{x}\big), \mathbf{z}, S(\mathbf{w})]$$

is true. It follows that if $\mathbf{w}$ has the value $n + 1$ and $\mathbf{x}$ and $\mathbf{z}$ have any value, then $\mathbf{A}'[\mathbf{x}, \mathbf{f}(\mathbf{w}, \mathbf{z}, \mathbf{x}), \mathbf{z}, \mathbf{w}]$ is true. This completes the proof.

Our proof of the theorem has been constructive. In fact, we have shown how to construct the sequence $\mathbf{a}$ which shows that $\mathbf{A}^*$ is valid from a proof of $\mathbf{A}$ in P'.

The consistency proof is now easy. Let $\mathbf{A}$ be $0 \neq 0$. If $\vdash_P \mathbf{A}$, then $\vdash_{P'} \mathbf{A}$; so $\mathbf{A}^*$ is valid. But $\mathbf{A}^*$ is $0 \neq 0$, which is not valid. Hence $0 \neq 0$ is not a theorem of P; so P is consistent.

8.4 APPLICATIONS OF THE CONSISTENCY PROOF

We remarked earlier that a finitary proof often gives more information than a nonfinitary one because it requires us to prove more than is actually stated. The same is true of constructive consistency proofs; and our consistency proof for P is a good example. From the consistency proof by means of the standard model, we obtain a necessary condition for **A** to be a theorem of P, viz., that **A** be true. From our constructive proof we obtain another necessary condition, viz., that $\mathbf{A}^*$ be valid. Now if $\mathbf{A}^*$ is valid, it is true, and hence **A** is true; but the converse does not always hold (see Problem 8). Thus our constructive proof has led to additional information of a nonconstructive kind.

A sufficient condition that **A** be a theorem of P may be regarded as a partial solution of the characterization problem for P. The solution which we have obtained is not too satisfactory in one respect: if **A** is very complicated, the relation between **A** and the validity of $\mathbf{A}^*$ is not too clear. We shall see how this objection can be overcome.

As usual, it is sufficient to consider closed formulas in prenex form. Let **A** be such a formula, and let P_0 and P_0' be formed from P and P' by adding the new function symbols of $\mathbf{A}_H$. In the proof of Herbrand's theorem, we saw that $\mathbf{A} \to \mathbf{A}_H$ was provable without nonlogical axioms. Hence if $\vdash_P \mathbf{A}$, then $\mathbf{A}_H$ is a theorem of P_0 and hence of P_0'.

Now let us consider each of the new function symbols of $\mathbf{A}_H$ as a variable of Y of the appropriate type. We may then define $\mathbf{B}^*$ for **B** a formula of P_0' as before and prove the theorem of the last section for P_0'. We conclude that if $\vdash_P \mathbf{A}$, then $\mathbf{A}_H{}^*$ is valid.

Suppose that $\mathbf{f}_1, \ldots, \mathbf{f}_n$ are the new function symbols of $\mathbf{A}_H$ and that $\mathbf{A}_H$ is $\exists \mathbf{x}_1 \ldots \exists \mathbf{x}_m \mathbf{B}$ with **B** open. It is easy to see that $\mathbf{A}_H{}^*$ is $\exists \mathbf{x}_1 \ldots \exists \mathbf{x}_m \mathbf{B}'$, where $\mathbf{B}'$ results from **B** by adding $2m$ negation signs in front. Hence if $\vdash_P \mathbf{A}$, then there are terms $\mathbf{a}_1, \ldots, \mathbf{a}_m$ of Y such that $\vdash_Y \mathbf{B}'[\mathbf{a}_1, \ldots, \mathbf{a}_m]$ and hence $\vdash_Y \mathbf{B}[\mathbf{a}_1, \ldots, \mathbf{a}_m]$. We may suppose that $\mathbf{a}_i$ contains no variable other than $\mathbf{f}_1, \ldots, \mathbf{f}_n$; and we may then introduce a constant F_i by the defining equation $F_i(\mathbf{f}_1, \ldots, \mathbf{f}_n) = \mathbf{a}_i$. Then

$$\vdash_Y \mathbf{B}[F_1(\mathbf{f}_1, \ldots, \mathbf{f}_n), \ldots, F_m(\mathbf{f}_1, \ldots, \mathbf{f}_n)].$$

We call a functional *type recursive* if it is designated by some constant of Y. The result which we have proved may then be stated as follows.

No Counterexample Interpretation (*Kreisel*). Let **A** be a closed formula in prenex form which is a theorem of P. Let $\mathbf{A}_H$ be $\exists \mathbf{x}_1 \ldots \exists \mathbf{x}_m \mathbf{B}$ with **B** open, and let $\mathbf{f}_1, \ldots, \mathbf{f}_n$ be the new function symbols of $\mathbf{A}_H$. Then there are type recursive functionals $F_1, \ldots, F_m$ such that

$$\mathbf{B}_{\mathbf{x}_1, \ldots, \mathbf{x}_m}[F_1(\mathbf{f}_1, \ldots, \mathbf{f}_n), \ldots, F_m(\mathbf{f}_1, \ldots, \mathbf{f}_n)]$$

is true for every choice of $\mathbf{f}_1, \ldots, \mathbf{f}_n$.

To get a clearer picture of what this means, suppose that **A** is

$$\exists \mathbf{x}\, \forall \mathbf{y}\, \exists \mathbf{z}\, \forall \mathbf{w} \mathbf{B}[\mathbf{x}, \mathbf{y}, \mathbf{z}, \mathbf{w}].$$

Then $\mathbf{A}_H$ is $\exists \mathbf{x}\, \exists \mathbf{z} \mathbf{B}[\mathbf{x}, \mathbf{f}(\mathbf{x}), \mathbf{z}, \mathbf{g}(\mathbf{x}, \mathbf{z})]$. Hence if $\vdash_P \mathbf{A}$, then there are type recursive functionals F and G such that

$$\mathbf{B}[F(\mathbf{f}, \mathbf{g}), \mathbf{f}(F(\mathbf{f}, \mathbf{g})), G(\mathbf{f}, \mathbf{g}), \mathbf{g}(F(\mathbf{f}, \mathbf{g}), G(\mathbf{f}, \mathbf{g}))]. \tag{1}$$

Now $\neg\mathbf{A}$ is equivalent to $\forall \mathbf{x}\, \exists \mathbf{y}\, \forall \mathbf{z}\, \exists \mathbf{w} \neg \mathbf{B}[\mathbf{x}, \mathbf{y}, \mathbf{z}, \mathbf{w}]$; so **A** is false iff there are functions **f** and **g** such that

$$\neg \mathbf{B}[\mathbf{x}, \mathbf{f}(\mathbf{x}), \mathbf{z}, \mathbf{g}(\mathbf{x}, \mathbf{z})]$$

for all **x** and **z**. Let us call such an **f** and **g** a *counterexample* to **A**. Then **A** is true iff there is no counterexample to **A**, that is, iff for every **f** and **g**, we can find an $F(\mathbf{f}, \mathbf{g})$ and a $G(\mathbf{f}, \mathbf{g})$ such that (1) holds. The additional information which we get if **A** is provable in P is that F and G may be chosen type recursive.

Remark. If **A** is true, then there is a constant **e** and a function **f** such that $\mathbf{B}[\mathbf{e}, \mathbf{y}, \mathbf{f}(\mathbf{y}), \mathbf{z}]$ for all **y** and **z**. One might hope to conclude that if **A** is provable, then **f** may be chosen recursive. This is not the case; see Problem 9.

We shall extend the no counterexample interpretation to recursive extensions of P. For this, it obviously suffices to extend our consistency proof to such extensions. (Note that this extension of the consistency proof is pointless for proving consistency; the consistency of every recursive extension of P follows from the consistency of P by previous results.)

It will clearly suffice to suppose that we have extended the consistency proof to a recursive extension P' and show that it can be extended to a recursive extension P'' containing one more nonlogical symbol than P'. First suppose that the additional symbol is a predicate symbol **p** with a defining axiom $\mathbf{p}\mathbf{x}_1 \ldots \mathbf{x}_n \leftrightarrow \mathbf{A}$. Then **A** is open; so $\mathbf{A}^*$ is **A**. We introduce **p** as a defined symbol in Y by letting $\mathbf{p}\mathbf{a}_1 \ldots \mathbf{a}_n$ be an abbreviation of $\mathbf{A}[\mathbf{a}_1, \ldots, \mathbf{a}_n]$. Clearly **p** has the same meaning in Y as in P''. Moreover, $(\mathbf{p}\mathbf{x}_1 \ldots \mathbf{x}_n \leftrightarrow \mathbf{A})^*$ is $\mathbf{p}\mathbf{x}_1 \ldots \mathbf{x}_n \leftrightarrow \mathbf{A}$, which is certainly valid; so our consistency proof extends.

Now suppose that the new nonlogical symbol is a function symbol **f** with a defining axiom

$$\mathbf{f}\mathbf{x}_1 \ldots \mathbf{x}_n = \mu \mathbf{y} \mathbf{A}. \tag{2}$$

Then **A** is open and $\vdash_{P'} \exists \mathbf{y} \mathbf{A}$. Since our consistency proof has been extended to P', $(\exists \mathbf{y} \mathbf{A})^*$, which is $\exists \mathbf{y} \neg\neg \mathbf{A}$, is valid. Hence there is a term **a** such that $\vdash_Y \neg\neg \mathbf{A}_\mathbf{y}[\mathbf{a}]$ and hence $\vdash_Y \mathbf{A}_\mathbf{y}[\mathbf{a}]$. If **g** has the defining equation

$$\mathbf{g}(\mathbf{x}_1, \ldots, \mathbf{x}_n) = \mathbf{a},$$

we then have

$$\vdash_Y \mathbf{A}_\mathbf{y} [\mathbf{g}(\mathbf{x}_1, \ldots, \mathbf{x}_n)]. \tag{3}$$

We now introduce a constant **h** by

$$\mathbf{A} \to \mathbf{h}(\mathbf{y}, \mathbf{w}, \mathbf{x}_1, \ldots, \mathbf{x}_n) = \mathbf{y},$$
$$\neg\mathbf{A} \to \mathbf{h}(\mathbf{y}, \mathbf{w}, \mathbf{x}_1, \ldots, \mathbf{x}_n) = S(\mathbf{w}),$$

and a constant **f′** by

$$\mathbf{f}'(0, \mathbf{x}_1, \ldots, \mathbf{x}_n) = 0,$$
$$\mathbf{f}'\big(S(\mathbf{w}), \mathbf{x}_1, \ldots, \mathbf{x}_n\big) = \mathbf{h}\big(\mathbf{f}'(\mathbf{w}, \mathbf{x}_1, \ldots, \mathbf{x}_n), \mathbf{w}, \mathbf{x}_1, \ldots, \mathbf{x}_n\big).$$

Then

$$\vdash_Y \mathbf{A_y}[\mathbf{f}'(\mathbf{w}, \mathbf{x}_1, \ldots, \mathbf{x}_n)] \vee \mathbf{f}'(\mathbf{w}, \mathbf{x}_1, \ldots, \mathbf{x}_n) = \mathbf{w}, \tag{4}$$
$$\vdash_Y \mathbf{y} < \mathbf{f}'(\mathbf{w}, \mathbf{x}_1, \ldots, \mathbf{x}_n) \to \neg\mathbf{A}. \tag{5}$$

We let **f** be the constant of Y with the defining equation

$$\mathbf{f}(\mathbf{x}_1, \ldots, \mathbf{x}_n) = \mathbf{f}'\big(\mathbf{g}(\mathbf{x}_1, \ldots, \mathbf{x}_n), \mathbf{x}_1, \ldots, \mathbf{x}_n\big).$$

Then by (3), (4), and (5),

$$\vdash_Y \mathbf{A_y}[\mathbf{f}(\mathbf{x}_1, \ldots, \mathbf{x}_n)], \tag{6}$$
$$\vdash_Y \mathbf{y} < \mathbf{f}(\mathbf{x}_1, \ldots, \mathbf{x}_n) \to \neg\mathbf{A}. \tag{7}$$

From these it follows that **f** has the same meaning in Y as in P''.

We must still show that if **B** is the axiom (2), then **B*** is valid. Now **B** is

$$\mathbf{A_y}[\mathbf{f}\mathbf{x}_1 \ldots \mathbf{x}_n] \;\&\; \forall\mathbf{y}(\mathbf{y} < \mathbf{f}\mathbf{x}_1 \ldots \mathbf{x}_n \to \neg\mathbf{A});$$

so **B*** is

$$\forall\mathbf{y}\big(\mathbf{A_y}[\mathbf{f}(\mathbf{x}_1, \ldots, \mathbf{x}_n)] \;\&\; \big(\mathbf{y} < \mathbf{f}(\mathbf{x}_1, \ldots, \mathbf{x}_n) \to \neg\mathbf{A}\big)\big).$$

This is valid by (6) and (7).

In the course of extending the consistency proof, we have shown that every function which can be introduced in a recursive extension of P is type recursive. We shall now see that, conversely, every type recursive function can be introduced in a recursive extension of P.

We introduce a formal system F. The symbols of F are the constants of Y and the symbols Ap and $=$. We introduce the abbreviation $\mathbf{u}(\mathbf{v}_1, \ldots, \mathbf{v}_n)$ as in Y. The terms of F are defined by the generalized inductive definition:

a) a constant of type r is a term of type r;

b) if **u** is a term of type $(r \to s)$ and **v** is a term of type r, then **u**(**v**) is a term of type s.

We use the syntactical variables **f**, **g**, **h**, **a**, **b**, **c** as in Y. The formulas of F are the expressions **a** = **b** with **a** and **b** of the same type.

Every formula **a** = **a** is an axiom of F. If **f** has the defining equation

$$\mathbf{f}(\mathbf{x}_1, \ldots, \mathbf{x}_n) = \mathbf{a},$$

and $\mathbf{g}_1, \ldots, \mathbf{g}_n$ have the types of $\mathbf{x}_1, \ldots, \mathbf{x}_n$ respectively, then

$$\mathbf{f}(\mathbf{g}_1, \ldots, \mathbf{g}_n) = \mathbf{a}[\mathbf{g}_1, \ldots, \mathbf{g}_n]$$

is an axiom of F. If $\mathbf{f}$ has the defining equations $\mathbf{f}(0) = \mathbf{g}$ and $\mathbf{f}(S(\mathbf{x})) = \mathbf{h}(\mathbf{f}(\mathbf{x}), \mathbf{x})$, then $\mathbf{f}(0) = \mathbf{g}$ is an axiom of F, and all formulas $\mathbf{f}(\mathbf{k}_{n+1}) = \mathbf{h}(\mathbf{f}(\mathbf{k}_n), \mathbf{k}_n)$ are axioms of F. These are all the axioms of F. The only rule of F is: infer $\mathbf{B}$ from $\mathbf{A}$ and $\mathbf{a} = \mathbf{b}$ if $\mathbf{B}$ is obtained from $\mathbf{A}$ by replacing an occurrence of $\mathbf{a}$ by $\mathbf{b}$ or vice versa.

We define the notion of a *reducible* term of type r by induction on the length of r. A term $\mathbf{a}$ of type o is reducible if $\vdash_F \mathbf{a} = \mathbf{k}_n$ for some n. A term $\mathbf{a}$ of type $(r \to s)$ is reducible if $\mathbf{a}(\mathbf{b})$ is reducible for every reducible term $\mathbf{b}$ of type r.

We now prove some simple facts about reducibility.

i) If every constant occurring in $\mathbf{a}$ is reducible, then $\mathbf{a}$ is reducible We prove this by induction on the length of $\mathbf{a}$. If $\mathbf{a}$ is a constant it is immediate. Otherwise, $\mathbf{a}$ is $\mathbf{b}(\mathbf{c})$, where $\mathbf{b}$ and $\mathbf{c}$ are reducible by induction hypothesis. Then $\mathbf{a}$ is reducible by the definition of reducibility for $\mathbf{b}$.

ii) If $\vdash \mathbf{a} = \mathbf{b}$ and $\mathbf{b}$ is reducible, then $\mathbf{a}$ is reducible. We use induction on the type of $\mathbf{b}$. If $\mathbf{b}$ is of type o, then $\vdash \mathbf{b} = \mathbf{k}_n$ for some n; so $\vdash \mathbf{a} = \mathbf{k}_n$ by the rule of F. Thus $\mathbf{a}$ is reducible. Let $\mathbf{b}$ be of type $(r \to s)$, and let $\mathbf{c}$ be a reducible term of type r. Since $\vdash \mathbf{a}(\mathbf{c}) = \mathbf{a}(\mathbf{c})$ and $\vdash \mathbf{a} = \mathbf{b}$, we have $\vdash \mathbf{a}(\mathbf{c}) = \mathbf{b}(\mathbf{c})$. Since $\mathbf{b}$ is reducible, $\mathbf{b}(\mathbf{c})$ is reducible; so $\mathbf{a}(\mathbf{c})$ is reducible by the induction hypothesis. Thus $\mathbf{a}$ is reducible.

iii) If $\mathbf{a}(\mathbf{k}_n)$ is reducible for every n, then $\mathbf{a}$ is reducible. For let $\mathbf{b}$ be a reducible term of type o. Then $\vdash \mathbf{b} = \mathbf{k}_n$ for some n. Since $\vdash \mathbf{a}(\mathbf{b}) = \mathbf{a}(\mathbf{b})$, $\vdash \mathbf{a}(\mathbf{b}) = \mathbf{a}(\mathbf{k}_n)$. Hence $\mathbf{a}(\mathbf{b})$ is reducible by hypothesis and (ii).

We now show by induction on constants that every constant is reducible. Clearly 0 is reducible. Since $S(\mathbf{k}_n)$ is $\mathbf{k}_{n+1}$, it is reducible; so S is reducible by (iii). Let $\mathbf{f}$ have the defining equation $\mathbf{f}(\mathbf{x}_1, \ldots, \mathbf{x}_n) = \mathbf{a}$, where every constant in $\mathbf{a}$ is reducible. To prove that $\mathbf{f}$ is reducible, it will clearly suffice to show that $\mathbf{f}(\mathbf{g}_1, \ldots, \mathbf{g}_n)$ is reducible for all reducible terms $\mathbf{g}_1, \ldots, \mathbf{g}_n$. Now

$$\vdash \mathbf{f}(\mathbf{g}_1, \ldots, \mathbf{g}_n) = \mathbf{a}[\mathbf{g}_1, \ldots, \mathbf{g}_n];$$

and $\mathbf{a}[\mathbf{g}_1, \ldots, \mathbf{g}_n]$ is reducible by (i). Hence $\mathbf{f}(\mathbf{g}_1, \ldots, \mathbf{g}_n)$ is reducible by (ii).

Now let $\mathbf{f}$ have the defining equations $\mathbf{f}(0) = \mathbf{g}$, $\mathbf{f}(\mathbf{x}) = \mathbf{h}(\mathbf{f}(\mathbf{x}), \mathbf{x})$, where $\mathbf{g}$ and $\mathbf{h}$ are reducible. We shall show by induction on n that $\mathbf{f}(\mathbf{k}_n)$ is reducible; the reducibility of $\mathbf{f}$ will follow by (iii). Since $\vdash \mathbf{f}(0) = \mathbf{g}$, $\mathbf{f}(0)$ is reducible by (ii). Now suppose that $\mathbf{f}(\mathbf{k}_n)$ is reducible. Since $\mathbf{h}$ is reducible, $\mathbf{h}(\mathbf{f}(\mathbf{k}_n))$ is reducible. Since $\mathbf{k}_n$ is reducible, it follows that $\mathbf{h}(\mathbf{f}(\mathbf{k}_n), \mathbf{k}_n)$ is reducible. Since

$$\vdash \mathbf{f}(\mathbf{k}_{n+1}) = \mathbf{h}(\mathbf{f}(\mathbf{k}_n), \mathbf{k}_n),$$

it follows from (ii) that $\mathbf{f}(\mathbf{k}_{n+1})$ is reducible.

The next step is to assign expression numbers to the expressions of F and give definitions analogous to those of §6.6 for F. Since this involves nothing new, we shall omit it. If $\mathbf{f}$ is a constant designating an n-ary type recursive function, we can define a predicate R such that $R(a_1, \ldots, a_n, b)$ means that b is the number of a proof of a formula of the form $\mathbf{f}\mathbf{k}_{a_1} \ldots \mathbf{k}_{a_n} = \mathbf{k}_c$, and a function G such that if b

is the number of such a proof, then $G(b) = c$. Since **f** is reducible, we have $\exists bR(a_1, \ldots, a_n, b)$ for all $a_1, \ldots, a_n$; and it is clear that

$$G(\mu bR(a_1, \ldots, a_n, b)) \tag{8}$$

is just the function designated by **f**.

Using the results of §8.1, we can introduce R and G in a recursive extension P' of P. To introduce the function (8), we must be able to prove in P' the existence condition $\exists \mathbf{y}R(\mathbf{x}_1, \ldots, \mathbf{x}_n, \mathbf{y})$. Essentially this means that we must be able to prove the reducibility of **f** in P'.

For each type symbol r, we can introduce a predicate Red_r in a (nonrecursive) extension by definitions of P' such that $Red_r(x)$ means that x is the expression number of a reducible term of type r. We can then prove (i) for each type separately; i.e., for each r we can prove a formula which says that (i) holds for all terms **a** of type r. Similarly, (ii) and (iii) may be proved for each type separately. We can then show by induction on constants that if a is the expression number of a constant of type r, then $\vdash Red_r(\mathbf{k}_a)$. Taking a to be the expression number of **f**, we get the desired result.

8.5 SECOND-ORDER ARITHMETIC

We have seen that there is no axiom system in which we can prove all the true formulas of P and no false ones. Nevertheless, it is natural to look for ways of extending P which will enable us to prove more true results about natural numbers. One way to do this is to add variables which vary through functionals of higher types. We shall consider a somewhat simpler system, in which we add variables which vary through sets of numbers; this will illustrate the problems involved while avoiding some complications.

The most natural way to proceed would be to add a new kind of variable to P. However, we would then no longer have a theory, and hence would be unable to apply our previous results. Instead, we shall follow the procedure suggested in §2.1 for dealing with two kinds of individuals.

We shall describe a theory S, called *second-order arithmetic*. The nonlogical symbols of S are those of P, the binary predicate symbol $\in$, and the unary predicate symbols N and C. We intend that Nx shall mean that x is a number and that Cx shall mean that x is a set of numbers.

We shall allow ourselves to use all small Latin letters as if they were variables of S. Moreover, we shall generally use letters early in the alphabet when we expect the variable to designate a number rather than a set. This convention is only to make reading easier, and has no official status.

Our first two nonlogical axioms state that every individual is either a number or a set, but not both:

$$Nx \vee Cx, \tag{1}$$

$$\neg(Nx \;\&\; Cx). \tag{2}$$

The next three nonlogical axioms state that certain individuals are numbers or sets.

$$N0, \tag{3}$$

$$Na \rightarrow NSa, \tag{4}$$

$$a \in x \rightarrow Na \;\&\; Cx. \tag{5}$$

We define an interpretation I of $L(P)$ in $L(S)$ by letting U_I be N and letting $\mathbf{u}_I$ be $\mathbf{u}$ for $\mathbf{u}$ a nonlogical symbol of P. We then adopt as further nonlogical axioms the interpretations by I of N1 through N9(b).

The *induction axiom* of S is

$$Cx \;\&\; 0 \in x \;\&\; \forall a(Na \;\&\; a \in x \rightarrow Sa \in x) \rightarrow \forall a(Na \rightarrow a \in x).$$

The *extensionality axiom* is

$$Cx \;\&\; Cy \;\&\; \forall a(a \in x \leftrightarrow a \in y) \rightarrow x = y;$$

it asserts that two sets having the same members are identical.

We now need axioms which will enable us to obtain some sets. The principal way of obtaining a set is to take the set of all numbers having some property. We introduce some axioms, called *comprehension axioms*, which show that sets defined in this way exist. They are all formulas

$$\exists \mathbf{y}(C\mathbf{y} \;\&\; \forall \mathbf{x}(\mathbf{x} \in \mathbf{y} \leftrightarrow N\mathbf{x} \;\&\; \mathbf{A}))$$

where $\mathbf{y}$ is different from $\mathbf{x}$ and does not appear in $\mathbf{A}$.

Since a function designated by a function symbol must be defined for all individuals, we must have a meaning for Sx even when x is a set. We agree to set $Sx = 0$ in this case. We make similar arbitrary agreements about the behavior of $+$, $\cdot$, and $<$ for arguments which are not numbers. These are expressed in the axioms:

$$Cx \rightarrow Sx = 0, \tag{6}$$

$$Cx \vee Cy \rightarrow x + y = 0 \;\&\; x \cdot y = 0 \;\&\; \neg(x < y). \tag{7}$$

This completes our list of the nonlogical axioms of S.

We construct a model $\mathfrak{N}'$ of S as follows. We let $N_{\mathfrak{N}'}$ be the set of natural numbers, and let $C_{\mathfrak{N}'}$ be the set of all sets of natural numbers. We let $|\mathfrak{N}'|$ be the union of $N_{\mathfrak{N}'}$ and $C_{\mathfrak{N}'}$. We let $a \in_{\mathfrak{N}'} x$ hold if a is a number, x is a set, and $a \in x$. For $\mathbf{u}$ a nonlogical symbol of P, we let $\mathbf{u}_{\mathfrak{N}'}$ have the same meaning as $\mathbf{u}_{\mathfrak{N}}$ for arguments which are numbers, and the meaning given by (6) and (7) for other arguments. It is clear that $\mathfrak{N}'$ is indeed a model of S. We call it the *standard* model of S, and say that a formula of S is *true* if it is valid in $\mathfrak{N}'$.

We now investigate some syntactical properties of S. Suppose that $\mathbf{x}_1, \ldots, \mathbf{x}_n$, $\mathbf{y}$ are distinct variables including all those free in $\mathbf{A}$. We can then introduce a new function symbol $\mathbf{f}$ by the defining axiom

$$C\mathbf{f}\mathbf{x}_1 \ldots \mathbf{x}_n \;\&\; \forall \mathbf{y}(\mathbf{y} \in \mathbf{f}\mathbf{x}_1 \ldots \mathbf{x}_n \leftrightarrow N\mathbf{y} \;\&\; \mathbf{A}). \tag{8}$$

For the existence condition for $\mathbf{f}$ is a comprehension axiom; and the uniqueness condition is an easy consequence of the extensionality axiom and (5). From the induction axiom,

$$\vdash 0 \in \mathbf{f}\mathbf{x}_1 \ldots \mathbf{x}_n \ \&\ \forall \mathbf{y}(N\mathbf{y} \ \&\ \mathbf{y} \in \mathbf{f}\mathbf{x}_1 \ldots \mathbf{x}_n \to S\mathbf{y} \in \mathbf{f}\mathbf{x}_1 \ldots \mathbf{x}_n) \to \forall \mathbf{y}(N\mathbf{y} \to \mathbf{y} \in \mathbf{f}\mathbf{x}_1 \ldots \mathbf{x}_n);$$

so by (8) and the equivalence theorem,

$$\vdash N0 \ \&\ \mathbf{A}_\mathbf{y}[0] \ \&\ \forall \mathbf{y}(N\mathbf{y} \ \&\ \mathbf{A} \to NS\mathbf{y} \ \&\ \mathbf{A}_\mathbf{y}[S\mathbf{y}]) \to \forall \mathbf{y}(N\mathbf{y} \to N\mathbf{y} \ \&\ \mathbf{A}).$$

In view of (3) and (4), this can be simplified to

$$\vdash \mathbf{A}_\mathbf{y}[0] \ \&\ \forall \mathbf{y}(N\mathbf{y} \ \&\ \mathbf{A} \to \mathbf{A}_\mathbf{y}[S\mathbf{y}]) \to \forall \mathbf{y}(N\mathbf{y} \to \mathbf{A}). \tag{9}$$

From this we obtain the *induction rule*: if $\vdash \mathbf{A}_\mathbf{y}[0]$ and $\vdash N\mathbf{y} \ \&\ \mathbf{A} \to \mathbf{A}_\mathbf{y}[S\mathbf{y}]$, then $\vdash N\mathbf{y} \to \mathbf{A}$.

We shall now prove that I is an interpretation of P in S. From (3),

$$\vdash \exists a Na. \tag{10}$$

The interpretations of N3 and N4 are

$$Na \to a + 0 = a, \tag{11}$$

$$Na \to Nb \to a + Sb = S(a + b). \tag{12}$$

From (12) and (4),

$$\vdash Nb \to (Na \to N(a + b)) \to (Na \to N(a + Sb)). \tag{13}$$

By (11), (13), and the induction rule,

$$\vdash Nb \to Na \to N(a + b). \tag{14}$$

Similarly, using (14) and the interpretations of N5 and N6, we have

$$\vdash Nb \to Na \to N(a \cdot b). \tag{15}$$

From (10), (3), (4), (14) and (15), I is an interpretation of $L(P)$ in S.

It remains to show that the interpretation of an induction axiom of P is provable in S. Such an interpretation has the form

$$N\mathbf{y}_1 \to \cdots \to N\mathbf{y}_n \to (\mathbf{A}_\mathbf{x}[0] \ \&\ \forall \mathbf{x}(N\mathbf{x} \to \mathbf{A} \to \mathbf{A}_\mathbf{x}[S\mathbf{x}])) \to \forall \mathbf{x}(N\mathbf{x} \to \mathbf{A}),$$

and hence is provable by (9).

By combining this with the results of §4.7, we see that I can be extended to an interpretation of any recursive extension of P in an extension by definitions of S. Thus the functions and predicates which we have seen can be introduced in recursive extensions of P can also be introduced in extensions by definitions of S. We shall sometimes tacitly assume that some of these functions and predicates have been introduced.

These results, of course, only enable us to introduce functions with numbers as values. We shall indicate how one can introduce functions with sets as values which are defined inductively. For simplicity, we consider only unary functions.

Suppose that in an extension by definitions S' of S, we have a constant A and a binary function symbol G, and that we can prove $C(A)$ and $C(G(x, a))$. We shall show that we can introduce a function symbol F in an extension by definitions of S' and prove

$$F(0) = A,$$
$$Na \to F(Sa) = G(F(a), a).$$

Using (8), we define a function symbol *Cut* by the defining axiom

$$\forall b(b \in Cut(x, a) \leftrightarrow Nb \ \& \ \langle a, b\rangle \in x).$$

We can then write down a defining axiom for a function symbol H which says the following things. If x is a set, then $H(x) = 0$. If a is a number, then $H(a)$ is a set of numbers of the form $\langle b, c\rangle$ where $b \leqslant a$. Moreover, $Cut(H(a), 0) = A$, and, for each $b < a$, $Cut(H(a), Sb) = G(Cut(H(a), b), b)$. We can then prove the existence and uniqueness conditions for H by induction, using the comprehension axioms in the existence proof and the extensionality axiom in the uniqueness proof. We can also prove by induction on b that $Cut(H(Sa), b) = Cut(H(a), b)$ for $b \leqslant a$. We introduce F by the defining axiom $F(a) = Cut(H(a), a)$. Then

$$\begin{aligned} F(0) &= Cut(H(0), 0) = A, \\ F(Sa) &= Cut(H(Sa), Sa) = G(Cut(H(Sa), a), a) \\ &= G(Cut(H(a), a), a) = G(F(a), a). \end{aligned}$$

The proper way to "prove" a formula **A** of P in S is to prove $\mathbf{A}^{(I)}$. We have seen that every theorem of P may be proved in S in this sense. As the discussion at the beginning of the section indicates, not every true formula of P can be proved in S (see Problem 10). However, there are true formulas of P which are unprovable in P but provable in S. We demonstrate this by sketching a proof of $Con_P{}^{(I)}$ in S.

The idea is to formalize the proof of the consistency of P by means of the standard model. Put in another way, we prove by induction on theorems that every theorem of P is true, and conclude that $0 \neq 0$ is not a theorem of P. This is very straightforward once we have obtained a definition of the true formulas in P. For this it suffices to define the true closed formulas. This can be done by induction on the height of the formula, once we have defined the true variable-free atomic formulas. Now the set of expression numbers of true variable-free atomic formulas can be defined in a recursive extension of P, and hence in S. By the method described above, we can then introduce an F such that $F(a)$ is the set of expression numbers of true closed formulas of height a or less. It is then easy to define the set of expression numbers of true closed formulas.

Remark. This proof cannot, of course, be carried out in P. The reason is that the truth of a true closed instantiation depends upon the truth or falsity of an infinite number of formulas of smaller height. We have no tools in P for defining a function by induction where the value depends on infinitely many earlier values.

We now turn to models of S. If $\mathfrak{A}$ is a model of S, then $\mathfrak{A}_I$ (in the notation of §6.9) is a structure for $L(P)$. If $\mathbf{A}$ is an axiom of P, then $\vdash_S \mathbf{A}^{(I)}$; so $\mathbf{A}^{(I)}$ is valid in $\mathfrak{A}$; so $\mathbf{A}$ is valid in $\mathfrak{A}_I$. Thus $\mathfrak{A}_I$ is a model of P.

A model $\mathfrak{A}$ of S is *regular* if $C_\mathfrak{A}$ is a set of subsets of $N_\mathfrak{A}$ and $a \in_\mathfrak{A} x \leftrightarrow a \in x$ for a in $N_\mathfrak{A}$ and x in $C_\mathfrak{A}$. We shall show that every model $\mathfrak{A}$ of S is isomorphic to a regular model. Replacing $\mathfrak{A}$ by an isomorphic model, we may suppose that no element of $|\mathfrak{A}|$ is a subset of $|\mathfrak{A}|$. Define a mapping ϕ as follows: if a is in $N_\mathfrak{A}$, $\phi(a) = a$; if x is in $C_\mathfrak{A}$, $\phi(x)$ is the set of a in $N_\mathfrak{A}$ such that $a \in_\mathfrak{A} x$. Using (1), (2), and the extensionality axiom, we see that ϕ is a bijective mapping from $|\mathfrak{A}|$ to some set. Hence there is a model $\mathfrak{B}$ such that ϕ is an isomorphism of $\mathfrak{A}$ and $\mathfrak{B}$. Clearly $\mathfrak{B}$ is regular.

A regular model $\mathfrak{A}$ is completely determined by $\mathfrak{A}_I$ and $C_\mathfrak{A}$. For $\mathfrak{A}_I$ determines $N_\mathfrak{A} = |\mathfrak{A}_I|$ and the values of the functions and predicates other than $\in_\mathfrak{A}$ for arguments in $N_\mathfrak{A}$. By (1), $|\mathfrak{A}|$ must be the union of $N_\mathfrak{A}$ and $C_\mathfrak{A}$; and (5) and the regularity condition determine $\in_\mathfrak{A}$. The remaining functions and predicates are determined for arguments not in $N_\mathfrak{A}$ by (6) and (7).

Lemma. A model $\mathfrak{A}$ of P is isomorphic to $\mathfrak{N}$ iff every individual of $\mathfrak{A}$ is $\mathfrak{A}(\mathbf{k}_n)$ for some n.

Proof. The condition is clearly necessary. Suppose that it holds. We define a surjective mapping ϕ from $|\mathfrak{N}|$ to $|\mathfrak{A}|$ by $\phi(n) = \mathfrak{A}(\mathbf{k}_n)$. If $m \neq n$, then $\vdash_P \mathbf{k}_m \neq \mathbf{k}_n$; so $\mathfrak{A}(\mathbf{k}_m) \neq \mathfrak{A}(\mathbf{k}_n)$. This shows that ϕ is injective. The conditions for an isomorphism now follow from Lemma 3 of §8.2. For example, to prove that

$$\phi(m) +_\mathfrak{A} \phi(n) = \phi(m + n),$$

observe that $\vdash_P \mathbf{k}_m + \mathbf{k}_n = \mathbf{k}_{m+n}$, so that

$$\mathfrak{A}(\mathbf{k}_m) +_\mathfrak{A} \mathfrak{A}(\mathbf{k}_n) = \mathfrak{A}(\mathbf{k}_{m+n}).$$

If $\mathfrak{A}$ is a regular model of S, then $\mathfrak{A}(\mathbf{k}_n) = \mathfrak{A}_I(\mathbf{k}_n)$ for all n. Hence by Lemma 2, $\mathfrak{A}_I$ is isomorphic to $\mathfrak{N}$ iff every element of $|\mathfrak{A}_I| = N_\mathfrak{A}$ is $\mathfrak{A}(\mathbf{k}_n)$ for some n. When this holds, we say that $\mathfrak{A}$ is an *ω-model* of S.

A regular model $\mathfrak{A}$ of S is *total* if $C_\mathfrak{A}$ is the set of all subsets of $N_\mathfrak{A}$. We shall show that every total model of S is isomorphic to $\mathfrak{N}'$. (This is the well-known proof that the Peano axioms are categorical.) Let $\mathfrak{A}$ be a total model of S, and let x be the set of all $\mathfrak{A}(\mathbf{k}_n)$. Then $0_\mathfrak{A} \in x$, and if $a \in x$, then $S_\mathfrak{A} a \in x$. It follows by the induction axiom that $a \in x$ for every a in $N_\mathfrak{A}$; so $\mathfrak{A}$ is an ω-model. We may therefore suppose that $\mathfrak{A}_I = \mathfrak{N}$. But then $C_\mathfrak{A} = C_{\mathfrak{N}'}$; so $\mathfrak{A} = \mathfrak{N}'$.

We will now obtain a syntactical characterization of the sentences which are valid in every ω-model. The formal system S_ω is obtained from S by adding the ω-rule: infer $N\mathbf{x} \to \mathbf{A}$ from $\mathbf{A}[\mathbf{k}_0], \mathbf{A}[\mathbf{k}_1], \ldots$. Note that, unlike all rules which we have considered previously, the ω-rule is not finite; infinitely many hypotheses are required to obtain the conclusion.

Henkin-Orey Theorem. A formula of S is a theorem of S_ω iff it is valid in every ω-model of S.

Proof. To prove that every theorem of S_ω is valid in every ω-model $\mathcal{A}$ of S, it suffices to show that if all the hypotheses of the ω-rule are valid in $\mathcal{A}$, then the conclusion is valid in $\mathcal{A}$; and this is evident.

Now suppose that $\mathbf{A}$ is not a theorem of S_ω. Let S' be the theory with language $L(S)$ whose nonlogical axioms are the theorems of S_ω. Then every theorem of S' is a theorem of S_ω; so $\mathbf{A}$ is not a theorem of S'. Hence if $\mathbf{A}'$ is the closure of $\mathbf{A}$, then $T = S'[\neg\mathbf{A}']$ is consistent.

Let Γ be the subset of $S_1(T)$ consisting of the formula $N\mathbf{z}_1$ and all the formulas $\mathbf{z}_1 \neq \mathbf{k}_n$. We shall show that Γ is not principal. Suppose that $\mathbf{B}$ is a generator of Γ. For each n, $\vdash_T \mathbf{B} \rightarrow \mathbf{z}_1 \neq \mathbf{k}_n$. Substituting $\mathbf{k}_n$ for $\mathbf{z}_1$ and using the identity axioms, we get $\vdash_T \neg\mathbf{B}[\mathbf{k}_n]$. By the deduction theorem, $\neg\mathbf{A}' \rightarrow \neg\mathbf{B}[\mathbf{k}_n]$ is a theorem of S' and hence of S_ω. By the ω-rule, $N\mathbf{z}_1 \rightarrow \neg\mathbf{A}' \rightarrow \neg\mathbf{B}$ is a theorem of S_ω and hence of S'. Thus $\vdash_T N\mathbf{z}_1 \rightarrow \neg\mathbf{B}$. But $\vdash_T \mathbf{B} \rightarrow N\mathbf{z}_1$; so $\vdash_T \neg\mathbf{B}$ by the tautology theorem. This is impossible by the definition of a generator.

It follows by Ehrenfeucht's theorem that there is a model $\mathcal{A}$ of T such that no 1-type in $\mathcal{A}$ includes Γ. We may suppose that $\mathcal{A}$ is regular. Then $\mathcal{A}$ is an ω-model of S in which $\mathbf{A}$ is not valid.

If $\mathbf{A}$ is a true formula of P, then for every ω-model $\mathcal{A}$ we have

$$\mathcal{A}(\mathbf{A}^{(I)}) = \mathcal{A}_I(\mathbf{A}) = \mathfrak{N}(\mathbf{A}) = \mathbf{T};$$

so $\mathbf{A}^{(I)}$ is a theorem of S_ω. This does not contradict the conclusions which we drew from the incompleteness theorem because S_ω is not an axiom system in the sense mentioned there.

A formula $\mathbf{A}$ of S is a theorem of S_ω iff it belongs to every class Γ of formulas of S such that:

i) every axiom of S_ω is in Γ;

ii) if all of the hypotheses of a rule of S_ω are in Γ, then the conclusion of the rule is in Γ.

If we replace formulas by their expression numbers, we can formulate this definition in S. In other words, there is a formula $\mathbf{D}$ of S such that $\mathbf{D}[\mathbf{k}_n]$ is true iff n is the expression number of a theorem of S_ω. Using this, we can prove the following result just as in §8.2.

Rosser's Theorem. There is a true formula of S which is not a theorem of S_ω.

We conclude with a few remarks on consistency proofs for S. Excluding trivial proofs, the only such proof is due to Spector. This proof is essentially an extension of the consistency proof which we gave for P. In order to take care of the comprehension axioms, it is necessary to add further constants to Y. The defining equations for these new constants do not, like our previous defining equations, obviously define a functional. One can prove fairly easily that they do; but the proof is not constructive. Thus to make the proof constructive, one must give a constructive proof that these equations define functionals. Recent investigations have made it seem doubtful that this is possible. Hence at the present time, we do not know if there is a constructive consistency proof for S.

All of this raises a more general question: Exactly what is a constructive proof, and what are the properties of such proofs? This is the main question studied by intuitionism. Although no final answers have been given, a great deal of progress has been made. We shall not attempt to enter into the subject here.

PROBLEMS

All theories are assumed to have only finitely many nonlogical symbols.

1. A predicate symbol $\mathbf{p}$ in an extension by definitions P' of P is an *R-symbol* if $\mathbf{p}\mathbf{x}_1 \ldots \mathbf{x}_n$ is equivalent to an R-formula of P.

a) Show that if a predicate Q can be introduced as an R-symbol in P', then $Q\mathbf{x}_1 \ldots \mathbf{x}_n$ with $\mathbf{x}_1, \ldots, \mathbf{x}_n$ weakly represents Q in P'. [Use Lemma 3 of §8.2.] Conclude that Q is recursively enumerable.

b) Show that every primitive recursive function or predicate can be introduced in a recursive extension of P. In particular, the T_n can be so introduced.

c) Show that every recursively enumerable predicate may be introduced as an R-symbol in an extension by definitions of P. [Use (b) and the enumeration theorem.]

2. Let T be a theory such that $NLAx_T$ is recursively enumerable. By 1(c), $NLAx_T$ may be introduced as an R-symbol in an extension by definitions of P. We then construct Thm_T and Con_T like Thm_P and Con_P, using $NLAx_T$ in place of $NLAx_P$. (Note that Thm_T and Con_T depend not only on T, but also on the choice of the R-symbol $NLAx_T$.)

a) Show that Thm_T is an R-symbol.

b) Define $NLAx_N$ by a definition

$$NLAx_N(\mathbf{x}) \leftrightarrow \mathbf{x} = \mathbf{k}_{a_1} \vee \cdots \vee \mathbf{x} = \mathbf{k}_{a_n}.$$

Show that (8) of §8.2 holds when Thm_P is replaced by Thm_N.

c) Show that if T is an extension of N, then $\vdash Thm_N(x) \rightarrow Thm_T(x)$.

d) Show that if T is a consistent extension of P, then Con_T is not a theorem of T. [Follow the proof in §8.2, using (a) through (c).]

3. a) Let S and T be theories such that $NLAx_S$ and $NLAx_T$ are recursively enumerable. Suppose that S has an interpretation I in T. Show that for each choice of the R-symbol $NLAx_T$ there is a choice of the R-symbol $NLAx_S$ such that $\vdash Con_T \rightarrow Con_S$. [Introduce a function I' such that $I'(\ulcorner \mathbf{A} \urcorner) = \ulcorner \mathbf{A}^{(I)} \urcorner$ in a recursive extension of P. Introduce $NLAx_S$ as an R-symbol, and define a new R-symbol by

$$NLAx'_S(x) \leftrightarrow NLAx_S(x) \ \& \ Thm_T(I'(x)).$$

Follow the proof of the interpretation theorem to show that $\vdash Thm'_S(x) \rightarrow Thm_T(I'(x))$.]

b) Let T be a theory such that $NLAx_T$ is recursively enumerable. Show that if T is consistent, then $P[Con_T]$ is not interpretable in T. [Suppose not, and use (a) and 2 (d) to show that $P[Con_T]$ is inconsistent and hence that Con_T is false.]

c) Let T be a theory such that $NLAx_T$ is recursively enumerable. Show that if T is consistent and if I is an interpretation of P in T, then $Con_T^{(I)}$ is not a theorem of T. [Use (b).]

4. a) Show that if Q is a unary R-symbol in P', then there is a closed R-formula **A** such that $\vdash_{P'} \mathbf{A} \leftrightarrow Q(\mathbf{k}_{\ulcorner \mathbf{A} \urcorner})$. [Let **A** be $\mathbf{B_z}[\mathbf{k}_{\ulcorner \mathbf{B} \urcorner}]$, where **B** is an R-formula equivalent to $Q(Sub(\mathbf{z}, k_{\ulcorner \mathbf{z} \urcorner}, Num(\mathbf{z})))$.]

b) Show that if **B** is a closed formula of P such that $\vdash_P Thm(\mathbf{k}_{\ulcorner \mathbf{B} \urcorner}) \rightarrow \mathbf{B}$, then $\vdash_P \mathbf{B}$. [Define $Q(a) \leftrightarrow Thm_P(Imp(a, \ulcorner \mathbf{B} \urcorner))$ where $Imp(\ulcorner \mathbf{A} \urcorner, \ulcorner \mathbf{B} \urcorner) = \ulcorner \mathbf{A} \rightarrow \mathbf{B} \urcorner$. Introduce Q as an R-symbol and let **A** be as in (a). Show that

$$\vdash \mathbf{A} \rightarrow Thm(\mathbf{k}_{\ulcorner \mathbf{A} \urcorner}) \rightarrow Thm(\mathbf{k}_{\ulcorner \mathbf{B} \urcorner}) \qquad \text{and} \qquad \vdash \mathbf{A} \rightarrow Thm(\mathbf{k}_{\ulcorner \mathbf{A} \urcorner}),$$

and conclude successively that $\vdash \mathbf{A} \rightarrow \mathbf{B}$, $\vdash Q(\mathbf{k}_{\ulcorner \mathbf{A} \urcorner})$, $\vdash \mathbf{A}$, $\vdash \mathbf{B}$.]

c) Show that if Q is Thm_P and **A** is as in (a), then $\vdash_P \mathbf{A}$. [Use (b).]

5. Let **A** be a closed formula which is undecidable in P.

a) Let B be the set of expression numbers of theorems of $P[\mathbf{A}]$ which are not theorems of P. Show that B is not recursively enumerable. [Assume otherwise, and use the negation theorem to show that $P[\neg \mathbf{A}]$ is decidable.]

b) Let F be a recursive function. Show that there is a theorem **B** of P such that if m and n are the smallest numbers of proofs of **B** in P and $P[\mathbf{A}]$ respectively, then $m > F(n)$. [Assume otherwise and get a contradiction to (a).]

6. Let $\mathcal{A}$ be a model of P such that $|\mathcal{A}| = |\mathfrak{N}|$ but $\mathcal{A}$ is not isomorphic to $\mathfrak{N}$. By the lemma of §8.5, there are elements of $|\mathcal{A}|$ distinct from the $\mathcal{A}(\mathbf{k}_n)$; such elements are called *infinite* elements.

a) Show that if x is an infinite element of $|\mathcal{A}|$, then $\mathcal{A}(\mathbf{k}_n) <_{\mathbf{A}} x$ for all n. [Use N9 and Lemma 2 of §6.7.]

b) Show that if no variable except x is free in **A**, then

$$\vdash_P \exists z \, \exists y \, \forall x (x < w \rightarrow (\exists x'(y = x' \cdot (1 + x) \cdot z) \leftrightarrow \mathbf{A})).$$

[Similar to the lemma of §6.4]

c) Let Φ be the sequence of functions and predicates of $\mathcal{A}$. Show that if no variable except x is free in **A**, then the set A of n such that $\mathcal{A}(\mathbf{A}_x[\mathbf{k}_n]) = \mathbf{T}$ is recursive in Φ. [Show that $\mathcal{A}(\mathbf{k}_n)$, as a function of n, is recursive in Φ. Pick an infinite element w, and choose z, y, z', y' by (a) and (b) so that in $\mathcal{A}$

$$\exists x'(y = x' \cdot (1 + \mathcal{A}(\mathbf{k}_n)) \cdot z) \leftrightarrow \mathbf{A}[\mathbf{k}_n],$$
$$\exists x'(y' = x' \cdot (1 + \mathcal{A}(\mathbf{k}_n)) \cdot z') \leftrightarrow \neg \mathbf{A}[\mathbf{k}_n].$$

Use the negation theorem.]

d) Show that any two disjoint recursively enumerable sets can be separated by a set recursive in Φ. [Use (c) and Problem 8(b) of Chapter 6.] Conclude that some element of Φ is not recursive. [Use Problem 13(c) of Chapter 6.]

e) Show that if $\mathfrak{A}$ is elementarily equivalent to $\mathfrak{N}$, then some element of Φ is not arithmetical. [Show that every arithmetical set is representable in $Th(\mathfrak{N})$ and hence, by (c), is recursive in Φ. Then use the arithmetical hierarchy theorem.]

7. A *P-theory* is an extension of P in which every formula of the form

$$\mathbf{A_x}[0] \to \forall \mathbf{x}(\mathbf{A} \to \mathbf{A_x}[S\mathbf{x}]) \to \mathbf{A}$$

is a theorem. If T is a P-theory, we define recursive extensions as for P; we can then prove all the results of §8.1.

a) Show that an extension by definitions of a P-theory is a P-theory.

b) A unary function symbol $\mathbf{f}$ *dominates* an n-ary function symbol $\mathbf{g}$ in a P-theory T if

$$\vdash_T \mathbf{x}_1 < \mathbf{x} \to \cdots \to \mathbf{x}_n < \mathbf{x} \to \mathbf{g}\mathbf{x}_1 \ldots \mathbf{x}_n < \mathbf{f}\mathbf{x}.$$

Show that if T is a P-theory, then there is an extension by definitions T' of T containing a function symbol $\mathbf{f}$ which dominates every function symbol of T. [If $\mathbf{g}$ is an n-ary function symbol of T, define $\mathbf{h}_1, \ldots, \mathbf{h}_n$ in a recursive extension of T so that

$$\vdash \mathbf{x}_1 < \mathbf{y}_1 \to \cdots \to \mathbf{x}_i < \mathbf{y}_i \to \mathbf{g}\mathbf{x}_1 \ldots \mathbf{x}_n < \mathbf{h}\mathbf{y}_1 \ldots \mathbf{y}_i\mathbf{x}_{i+1} \ldots \mathbf{x}_n.]$$

c) Let T be a P-theory. Show that there is a binary function symbol $\mathbf{g}$ in an extension by definitions T' of T such that if $\mathbf{a}$ is a term of T containing at most m function symbols and $\mathbf{x}_1, \ldots, \mathbf{x}_n$ are the variables in $\mathbf{a}$, then

$$\vdash_{T'} \mathbf{x}_1 < \mathbf{x} \to \cdots \to \mathbf{x}_n < \mathbf{x} \to \mathbf{a} < \mathbf{g}(\mathbf{x}, \mathbf{k}_m),$$

and such that

$$\vdash_{T'} \mathbf{y} < \mathbf{z} \to \mathbf{g}(\mathbf{x}, \mathbf{y}) < \mathbf{g}(\mathbf{x}, \mathbf{z}).$$

[Define $\mathbf{g}(\mathbf{x}, \mathbf{y})$ by induction on $\mathbf{y}$, using the $\mathbf{f}$ of (b).]

d) Let $\exists \mathbf{y}\mathbf{A}$ be a theorem in a P-theory T, and let $\mathbf{x}_1, \ldots, \mathbf{x}_n$ be the variables free in $\exists \mathbf{y}\mathbf{A}$. Show that there is a function symbol $\mathbf{f}$ in an extension by definitions T' of T such that $\vdash_{T'} \mathbf{A_y}[\mathbf{f}\mathbf{x}_1 \ldots \mathbf{x}_n]$. [Use the least number principle.]

e) Show that if there is a finitely axiomatized consistent P-theory, then there is a finitely axiomatized consistent open P-theory. [Like the proof of Skolem's theorem, using (d).]

f) If $\mathfrak{A}$ is a model of a P-theory, an element x of $|\mathfrak{A}|$ such that $\mathfrak{A}(\mathbf{k}_n) \neq x$ for all n is called an *infinite* element. Show that if x is an infinite element, then $\mathfrak{A}(\mathbf{k}_n) <_{\mathfrak{A}} x$ for all n. [Like 6(a).] Show that if T is a consistent P-theory, then T has a model which contains an infinite element. [Use the cardinality theorem.]

g) Show that there is no finitely axiomatized consistent P-theory. [Suppose that T is such a theory. By (e), we may suppose that T is open. By (f), there is a model $\mathfrak{A}$ of T having an infinite element x. By the Łoś-Tarski theorem, we may suppose that every element of $|\mathfrak{A}|$ is $\mathfrak{A}(\mathbf{a})$, where $\mathbf{a}$ is a variable-free term of $L(\mathfrak{A})$ containing no name other than the name of x. Take $\mathbf{g}$ and T' as in (c), and expand $\mathfrak{A}$ to a model of T'. Show that $y <_{\mathfrak{A}} \mathbf{g}_{\mathfrak{A}}(S_{\mathfrak{A}}(x), x)$ for every individual y of $\mathfrak{A}$.]

8. a) Use the theorem of §8.3 to show that there is no closed theorem $\neg\forall \mathbf{w}\mathbf{A}$ of P such that $\vdash_P \mathbf{A}_\mathbf{w}[\mathbf{k}_n]$ for all n. [Assume otherwise. Let $\mathbf{A}^*$ be $\forall\mathbf{x}\,\exists\mathbf{y}\mathbf{A}'[\mathbf{x},\mathbf{y},\mathbf{z}]$. Then there are type-recursive $\mathbf{f}$, $\mathbf{g}$, and $\mathbf{h}_\mathbf{w}$ such that

$$\neg\mathbf{A}'[\mathbf{f}(\mathbf{y}'), \mathbf{y}'(\mathbf{f}(\mathbf{y}'), \mathbf{g}(\mathbf{y}')), \mathbf{g}(\mathbf{y}')]$$

and

$$\mathbf{A}'[\mathbf{x}, \mathbf{h}_\mathbf{w}(\mathbf{x}), \mathbf{w}]$$

are true for all $\mathbf{y}'$, $\mathbf{x}$, and $\mathbf{w}$. Obtain a contradiction by a proper choice of $\mathbf{y}'$, $\mathbf{x}$, and $\mathbf{w}$.]

b) Show that there is a closed theorem $\exists\mathbf{w}\mathbf{A}$ of P such that for each n, $\mathbf{A}_\mathbf{w}[\mathbf{k}_n]$ is not a theorem of P. [Let $\mathbf{B}$ be an R-sentence such that $\forall\mathbf{w}\mathbf{B}$ is true but not provable in P, and let $\mathbf{A}$ be $\neg\mathbf{B} \vee \forall\mathbf{w}\mathbf{B}$.]

9. a) Show that there is a recursive binary function F such that every unary type recursive function is $F_{(e)}$ for some e. [Use (8) of §8.4.] Conclude that there is a unary recursive function which is not type recursive. [Let $G(x) = F(x, x) + 1$.]

b) Show that there is a predicate symbol R in a recursive extension P' of P such that $\forall x\,\exists y R(x, y)$ is true, but such that for every unary type recursive function F there is an x such that $R(x, F(x))$ is false. [Let $R(x, y) \leftrightarrow T_1(e, x, y)$, where e is an index of a recursive function which is not type recursive.] Conclude that $\forall x\,\exists y R(x, y)$ is undecidable in P'.

c) Show that there is a predicate symbol R in a recursive extension P' of P such that $\vdash_{P'} \forall x\,\exists y\,\forall z R(x, y, z)$, but such that there is no recursive function F such that $R(x, F(x), z)$ is true for all x and z. [Let $R(x, y, z) \leftrightarrow T_1(x, x, y) \vee \neg T_1(x, x, z)$. If there is a recursive F such that $\forall x\,\forall z R(x, F(x), z)$, then $\exists y T_1(x, x, y)$ is a recursive predicate of x; and this is impossible by the diagonal lemma.]

10. a) Show that if T has a faithful interpretation in an axiomatized theory, then T is axiomatizable. [Use Problem 5(a) of Chapter 6.]

b) Show that if a complete theory T has an interpretation in a consistent axiomatized theory T', then T is decidable. [Show that the interpretation must be faithful, and use (a) and the lemma of §6.8.] Conclude that $Th(\mathfrak{N})$ does not have an interpretation in a consistent axiomatized theory.

c) Show that there is a true formula $\mathbf{A}$ of P such that $\mathbf{A}^{(I)}$ is not a theorem of S. [Use (b).]

11. a) Show that if Q is a $(1, n)$-ary arithmetical relation, then there is a sentence $\mathbf{A}$ of S such that for every ω-model $\mathfrak{A}$, $\mathfrak{A}(\mathbf{A}[\mathbf{i}, \mathbf{k}_{a_1}, \ldots, \mathbf{k}_{a_n}]) = \mathbf{T}$ iff $\mathbf{i}$ is the name of a set A such that $Q(K_A, a_1, \ldots, a_n)$. [Reduce to the case in which Q is recursively enumerable. Then note that

$$\begin{aligned} Q(\alpha, a) &\leftrightarrow \exists x R(\bar{\alpha}(x), a) \\ &\leftrightarrow \exists x\,\exists y(y = \bar{\alpha}(x)\ \&\ R(y, a)) \end{aligned}$$

with R recursive. Use the fact that R is representable in S.]

b) Show that every Π^1_1 predicate is weakly representable in S_ω. [Use (a), Problem 17(a) of Chapter 7, and the Henkin-Orey theorem.]

c) Show that the set of expression numbers of theorems of S_ω is Π^1_1. Conclude that every predicate weakly representable in S_ω is Π^1_1.

12. A set A of natural numbers is *in* an ω-model $\mathcal{A}$ (with $\mathcal{A}_I = \mathfrak{N}$) if A belongs to $C_{\mathcal{A}}$.

a) Show that if A is representable in S_ω, then A is in every ω-model. [Use the comprehension axiom and the Henkin-Orey theorem.]

b) Show that every hyperarithmetical set is representable in S_ω. [Use 11(a), Problem 27(b) of Chapter 7, and the Henkin-Orey theorem.]

c) Let T be a countable consistent theory, and let Γ and Δ be subsets of $S_1(T)$, neither of which is principal. Show that there is a countable model $\mathcal{A}$ of T such that no 1-type in $\mathcal{A}$ includes either Γ or Δ. [Like Ehrenfeucht's theorem.]

d) Show that a set A which is in every ω-model is hyperarithmetical. [Let S' and Γ be as in the proof of the Henkin-Orey theorem, and let Δ consist of $C\mathbf{z}_1$, the $\mathbf{k}_n \in \mathbf{z}_1$ for $n \in A$, and the $\mathbf{k}_n \notin \mathbf{z}_1$ for $n \notin A$. Use (c) to show that Δ has a generator. Use this and 11(c) to show that A and $\neg A$ are Π^1_1.]

CHAPTER 9

SET THEORY

9.1 AXIOMS FOR SETS

We now turn to an investigation of set theory. The great interest in sets is due partly to the important role which they have played in modern mathematics. But even without this, the notion of a set is so natural that it would call for investigation.

A *set* (or *class*) is a collection of objects. These objects may be numbers, functions, physical objects, or even sets. Since there are no restrictions on the objects which may be members of sets, it would seem that we can specify a set by specifying for each object in the universe whether or not that object is a member of the set. However, this leads immediately to the Russell paradox. For let us specify a set A by specifying that an object x is a member of A iff x is a set and x is not a member of x. Then A is a member of A iff A is not a member of A; and this is a contradiction.

A closer examination of the paradox shows that it does not really contradict the intuitive notion of a set. According to this notion, a set A is formed by gathering together certain objects to form a single object, which is the set A. Thus before the set A is formed, we must have available all of the objects which are to be members of A. It follows that the set A is not one of the possible members of A; so the Russell paradox disappears.

We are thus led to the following description of the construction of sets. We start with certain objects which are not sets and do not involve sets in their construction. We call these objects *urelements*. We then form sets in successive stages. At each stage we have available the urelements and the sets formed at earlier stages; and we form into sets all collections of these objects. A collection is to be a set only if it is formed at some stage in this construction.

We can carry out this construction with any collection of urelements. If we carry it out with no urelements, the sets which we obtain are called *pure* sets. It turns out that these are sufficient for mathematical purposes; and they are also sufficient to illustrate all the problems which arise in the general case. We shall therefore restrict ourselves to this case, and henceforth take *set* or *class* to mean *pure set*.

When can a collection of sets be formed into a set? For each set x in the collection, let S_x be the stage at which x is formed. Then we can form a set of this collection iff there is a stage S which follows all the S_x. However, such a stage may fail to exist. For example, *every* stage may be an S_x. Thus we want an answer to the following question: given a collection of stages, under what conditions is there a stage which follows every stage in the collection?

Since we wish to allow a set to be as arbitrary a collection as possible, we agree that there shall be such a stage whenever possible, i.e., whenever we can visualize a situation in which all the stages in the collection are completed. This is a rather vague principle; but we can conclude some precise results from it. For example, given a stage S, there is to be a stage following S. If a collection consists of an infinite sequence $S_1, S_2, \ldots$ of stages, then we can visualize a situation in which all of these stages are completed; so there is to be a stage after all the S_n.

Another important example is the following. Suppose that we have a set A, and that we have assigned a stage S_a to each element a of A. Since we can visualize the collection A as a single object (viz., the set A), we can also visualize the collection of stages S_a as a single object; so we can visualize a situation in which all these stages are completed. Hence there is to be a stage which follows all of the stages S_a. This result is called the *principle of cofinality.*

We are now going to use these principles to develop a theory. This theory is called *Zermelo-Fraenkel set theory*, and is designated by *ZF*.

The only nonlogical symbol of *ZF* is the binary predicate symbol $\in$. We intend that the individuals of *ZF* shall be the (pure) sets, and that $x \in y$ shall mean that x is a member of y.

The first nonlogical axiom of *ZF* states that if two sets have exactly the same members, then they are equal. This axiom, called the *extensionality axiom*, is

$$\forall z(z \in x \leftrightarrow z \in y) \rightarrow x = y.$$

The next nonlogical axiom, the *regularity axiom*, is

$$\exists y(y \in x) \rightarrow \exists y\big(y \in x \ \& \ \neg\exists z(z \in x \ \& \ z \in y)\big).$$

This says that if x has a member, then it has a member y such that x and y are disjoint (i.e., have no common member). Such a member of x will be called a *minimal element* of x. To see that the regularity axiom is true, let x be a nonempty set, and let y be a member of x formed at as early a stage as possible. Since the members of y must be formed at a still earlier stage, they are not members of x. Hence y is a minimal element of x.

The remaining axioms concern the existence of sets. First, we have the *subset axioms*. These are all formulas

$$\exists \mathbf{z}\, \forall \mathbf{x}(\mathbf{x} \in \mathbf{z} \leftrightarrow \mathbf{x} \in \mathbf{y} \ \& \ \mathbf{A}),$$

where $\mathbf{x}$, $\mathbf{y}$, and $\mathbf{z}$ are distinct and $\mathbf{y}$ and $\mathbf{z}$ do not occur in $\mathbf{A}$. To see what this means, let $\mathbf{x}$, $\mathbf{y}$, and $\mathbf{z}$ be x, y, and z, and let $\mathbf{A}$ be $..x...$ Then the axiom asserts that there is a set whose members are the members x of y such that $..x...$ To verify that this is true, let S be the stage at which y is constructed. Then every member x of y such that $..x..$ must be constructed before stage S; so z may be constructed at stage S.

We use $Set_{\mathbf{x}}\,\mathbf{A}$ as an abbreviation for $\exists\mathbf{y}\,\forall\mathbf{x}(\mathbf{A} \rightarrow \mathbf{x} \in \mathbf{y})$. (The variable **y** is to be one which is different from **x** and does not occur in **A**. Beyond this, the choice of **y** is immaterial by the variant theorem.) Clearly $Set_x \ldots x \ldots$ means that there is a set y which contains every x such that $\ldots x \ldots$

The *replacement axioms* are all formulas

$$\forall\mathbf{x}\,\exists\mathbf{z}\,\forall\mathbf{y}(\mathbf{A} \leftrightarrow \mathbf{y} \in \mathbf{z}) \rightarrow Set_{\mathbf{y}}\,\exists\mathbf{x}(\mathbf{x} \in \mathbf{w}\ \&\ \mathbf{A}),$$

where **x**, **y**, **z**, and **w** are distinct and **z** and **w** do not occur in **A**. To see what this means, let **x**, **y**, **z**, and **w** be x, y, z, and w, and let **A** be $\ldots x \ldots y \ldots$ The hypothesis states that for each x, there is a set z_x consisting of all y such that $\ldots x \ldots y \ldots$ The conclusion states that there is a set z (depending on w) which contains every y such that $\ldots x \ldots y \ldots$ for some x in w.

To verify this, let S_x be the stage at which z_x is constructed. By the cofinality principle, there is a stage S after all the stages S_x for $x \in w$. If $x \in w$ and $\ldots x \ldots y \ldots$, then $y \in z_x$; so y is constructed before stage S_x and hence before stage S. Then at stage S, we may construct the set of all y such that $\ldots x \ldots y \ldots$ for some $x \in w$.

The *power set axiom* is

$$Set_y\,\forall z(z \in y \rightarrow z \in x).$$

It says that given a set x, there is a set which contains every subset of x. Suppose that x is constructed at stage S. Then every member of x is constructed before stage S; so every subset of x may be constructed at stage S. Hence at any stage following S, we may construct a set which contains all subsets of x.

We need another axiom to guarantee the existence of an infinite set. The *axiom of infinity* is

$$\exists x(\exists y(y \in x\ \&\ \forall z(z \notin y)) \\ \&\ \forall y(y \in x \rightarrow \exists z(z \in x\ \&\ \forall w(w \in z \leftrightarrow w \in y \vee w = y))))$$

This says that there is a set x such that the empty set 0 is in x, and such that if $y \in x$, then the set $S(y)$ whose elements are y and the elements of y is also in x. Now it is clear that if y is constructed at some stage, then $S(y)$ can be constructed at the next stage. Thus there are stages $S_0, S_1, S_2, \ldots$ at which we can construct $0, S(0), S(S(0)), \ldots$. At a stage S following all of the stages $S_0, S_1, S_2, \ldots$, we can construct the set x whose elements are $0, S(0), S(S(0)), \ldots$. This set clearly has the properties required by the axiom of infinity.

This completes the description of *ZF*. We shall consider later some further axioms which may be added to *ZF*.

9.2 DEVELOPMENT OF SET THEORY

We assume that the reader is familiar with the results of elementary set theory. We shall be chiefly concerned with establishing two points:

a) that such basic concepts as *ordered pair*, *function*, and *natural number* can be defined in *ZF*;

b) that the sets used in elementary set theory can be proved to exist in *ZF*.

We shall often pass to extensions by definitions of *ZF*. In order not to consider this as extending the theory, we regard the new nonlogical symbols as defined symbols. Thus each formula of the extension will be a defined formula of *ZF*, abbreviating its translation into *ZF*. We may then consider the defining axiom for the new function or predicate symbol as being a definition of that symbol. Of course in proofs, we treat these defined function and predicate symbols just as we would ordinary function and predicate symbols; and we shall refer to them as function and predicate symbols of *ZF*.

Our definitions and proofs will usually be given in English; we assume that the reader knows how to translate these into the language of *ZF*. To increase readability, we will use all small Latin letters (sometimes with primes or subscripts) as variables of *ZF*. In the informal exposition, we shall restrict some of these variables to vary through special collections of sets.

We wish to avoid syntactical variables, since their use in an English context suggests that we are talking about *ZF* instead of translating *L*(*ZF*) into English. We shall therefore allow such statements as: *if Q is a predicate symbol, then . . . .* It is understood that whatever is proved about Q holds for all predicate symbols. Sometimes we use Q for a specific predicate symbol which is not important enough to be given a permanent name. In certain contexts, we use R, U, and M in the same manner as Q. We use F, G, and H in a similar manner, except that they are to be function symbols.

A suitable use of these letters enables us to dispense with syntactical variables for formulas. For example, consider a subset axiom

$$\exists \mathbf{z}\, \forall \mathbf{x}(\mathbf{x} \in \mathbf{z} \leftrightarrow \mathbf{x} \in \mathbf{y} \;\&\; \mathbf{A}).$$

Suppose that **x**, **y**, and **z** are x, y, and z. Define Q by

$$Q(x, v_1, \ldots, v_n) \leftrightarrow \mathbf{A}$$

where $v_1, \ldots, v_n$ are the remaining variables free in **A**. Then the axiom becomes

$$\exists z\, \forall x\big(x \in z \leftrightarrow x \in y \;\&\; Q(x, v_1, \ldots, v_n)\big).$$

Conversely, every formula of this form is a subset axiom. The corresponding form for replacement axioms is

$$\forall x\, \exists z\, \forall y\big(Q(x, y, v_1, \ldots, v_n) \leftrightarrow y \in z\big) \rightarrow Set_y\, \exists x\big(x \in w \;\&\; Q(x, y, v_1, \ldots, v_n)\big).$$

In both cases, the variables $v_1, \ldots, v_n$ are called *parameters*. Since they generally remain unchanged throughout a proof, we often omit writing them.

Suppose that Q has been defined and that we wish to define F so that $F(v_1, \ldots, v_n)$ is the set of all x such that $Q(x, v_1, \ldots, v_n)$. For this we need the defining axiom

$$\forall x\big(x \in F(v_1, \ldots, v_n) \leftrightarrow Q(x, v_1, \ldots, v_n)\big). \tag{1}$$

The uniqueness condition for this axiom is an easy consequence of the extensionality axiom. The existence condition is

$$\exists z\, \forall x\big(x \in z \leftrightarrow Q(x, v_1, \ldots, v_n)\big). \tag{2}$$

We show that (2) can be proved from

$$Set_x\, Q(x, v_1, \ldots, v_n). \tag{3}$$

We omit the parameters $v_1, \ldots, v_n$. Assume (3), and choose a set w such that $Q(x) \rightarrow x \in w$ for all x. By the subset axioms, there is a set z such that for all x,

$$x \in z \leftrightarrow x \in w \;\&\; Q(x).$$

Since $Q(x) \rightarrow x \in w$, this is equivalent to $x \in z \leftrightarrow Q(x)$. This proves (2).

We shall call (3) the *existence condition* for (1). When (3) is provable, we write the definition (1) in the form

$$F(v_1, \ldots, v_n) = [x \mid Q(x, v_1, \ldots, v_n)]. \tag{4}$$

We may also use $[x \mid Q(x, v_1, \ldots, v_n)]$ as a term in a definition or a proof. It is then understood to be an abbreviation for $F(v_1, \ldots, v_n)$, where F is defined by (4). Again, we may use $[x \mid \_x\_]$, where $\_x\_$ is a formula, as a term; this is understood as an abbreviation for $[x \mid Q(x, v_1, \ldots, v_n)]$, where Q is defined by

$$Q(x, v_1, \ldots, v_n) \leftrightarrow \_x\_.$$

Of course, $[x \mid \_x\_]$ can be used as an abbreviation only if the existence condition $Set_x \_x\_$ is provable. We note that this is always the case when $\_x\_$ is $x \in \ldots$ or $x \in \ldots \;\&\; \text{----}$, ... being a term not containing x.

We define

$$x \subset y \leftrightarrow \forall z(z \in x \rightarrow z \in y).$$

Then $x \subset y$ means that x is a subset of y. We define

$$P(x) = [y \mid y \subset x];$$

the existence condition for this definition is just the power set axiom. We call $P(x)$ the *power set* of x; it is the set of all subsets of x.

We use $[F(x, v_1, \ldots, v_n) \mid Q(x, v_1, \ldots, v_n)]_x$ as an abbreviation for

$$[y \mid \exists x(Q(x, v_1, \ldots, v_n) \;\&\; y = F(x, v_1, \ldots, v_n))].$$

Thus, omitting the parameters, $[F(x) \mid Q(x)]_x$ is the set of $F(x)$ for x such that $Q(x)$. We generally omit the subscript x. We often write $[..x.. \mid \_x\_]$; this is an abbreviation for $[F(x, v_1, \ldots, v_n) \mid Q(x, v_1, \ldots, v_n)]$, where F and Q are defined by

$$F(x, v_1, \ldots, v_n) = ..x..,$$
$$Q(x, v_1, \ldots, v_n) \leftrightarrow \_x\_.$$

We show that $Set_x\, Q(x, v_1, \ldots, v_n)$ implies the existence condition for $[F(x, v_1, \ldots, v_n) \mid Q(x, v_1, \ldots, v_n)]$. We omit the parameters. Now $F(x) \subset F(x)$; so $F(x) \in P(F(x))$; so

$$\forall y(y = F(x) \rightarrow y \in P(F(x))).$$

From this we get $\forall x\, Set_y(y = F(x))$, which, as seen above, implies

$$\forall x\, \exists z\, \forall y(y \in z \leftrightarrow y = F(x)).$$

By the replacement axioms, this implies

$$Set_y\, \exists x(x \in [x \mid Q(x)]\ \&\ y = F(x)),$$

which is equivalent to

$$Set_y\, \exists x(Q(x)\ \&\ y = F(x)).$$

This is the desired existence condition.

The empty set 0 is defined by

$$0 = [x \mid x \neq x].$$

The existence condition $\exists y\, \forall x(x \neq x \rightarrow x \in y)$ follows from the identity axioms.

We define the *unordered pair* $\{x, y\}$ of x and y by

$$\{x, y\} = [z \mid z = x \vee z = y].$$

To prove the existence condition, define F so that $F(0) = x$ and $F(z) = y$ for $z \neq 0$. (We are omitting the parameters x and y as arguments to F.) Let

$$w = [F(z) \mid z \in P(P(0))].$$

It will suffice to show that $\forall z(z = x \vee z = y \rightarrow z \in w)$, or, equivalently, $x \in w\ \&\ y \in w$. Now $0 \subset P(0)$; so $0 \in P(P(0))$; so $x = F(0) \in w$. Also $P(0) \subset P(0)$; so $P(0) \in P(P(0))$. Moreover, $0 \in P(0)$ and $0 \notin 0$; so $P(0) \neq 0$. Hence $y = F(P(0)) \in w$.

We define the *unit set* $\{x\}$ of x by

$$\{x\} = \{x, x\}.$$

We now define

$$Un(x) = [y \mid \exists z(z \in x\ \&\ y \in z)].$$

To prove the existence condition, note that $\forall z\, \exists v\, \forall y(y \in v \leftrightarrow y \in z)$; so by the replacement axioms, $Set_y\, \exists z(z \in x\ \&\ y \in z)$. We call $Un(x)$ the *union* of x; it is the union (in the usual sense) of the members of x.

We now define the operations of *union*, *intersection*, and *set difference* by

$$x \cup y = Un(\{x, y\}),$$
$$x \cap y = [z \mid z \in x\ \&\ z \in y],$$
$$x - y = [z \mid z \in x\ \&\ z \notin y].$$

The existence conditions for the last two are provable because they have the form $[z \mid z \in x\ \&\ \ldots]$.

We now define the *ordered pair* $\langle x, y\rangle$ of x and y by

$$\langle x, y\rangle = \{\{x\}, \{x, y\}\}.$$

We prove the basic property of ordered pairs:

$$\langle x, y\rangle = \langle x', y'\rangle \leftrightarrow x = x' \ \& \ y = y'. \tag{5}$$

The implication from right to left is trivial. Suppose that $\langle x, y\rangle = \langle x', y'\rangle$. Then $\{x\}$ is in $\langle x, y\rangle$ and hence in $\langle x', y'\rangle$; so $\{x\} = \{x'\}$ or $\{x\} = \{x', y'\}$. In either case, $x' \in \{x\}$; so $x = x'$. Again, $\{x, y\}$ is in $\langle x, y\rangle = \langle x', y'\rangle$; so $\{x, y\} = \{x'\}$ or $\{x, y\} = \{x', y'\}$. Thus y is in $\{x'\}$ or $\{x', y'\}$; so $y = x'$ or $y = y'$. By symmetry, $y' = x$ or $y' = y$. If we assume $y \neq y'$, we get $y = x' = x = y'$; so we must have $y = y'$.

If $x = \langle y, z\rangle$, we set $\pi_1(x) = y$ and $\pi_2(x) = z$; this is well-defined by (5). Since function symbols must be defined for all arguments, we must also decide on a value for $\pi_1(x)$ and $\pi_2(x)$ when x is not an ordered pair. We choose the value 0. In general, when we do not define the value of a function symbol for certain arguments, it is to be understood that that value is 0.

We define the *Cartesian product* $x \times y$ of x and y by

$$x \times y = [z \mid \exists a\, \exists b(a \in x \ \& \ b \in y \ \& \ z = \langle a, b\rangle)].$$

To prove the existence condition, note that

$$\begin{aligned} a \in x \ \& \ b \in y \rightarrow \{a\} \in P(x \cup y) \ \& \ \{a, b\} \in P(x \cup y) \\ \rightarrow \langle a, b\rangle \in P(P(x \cup y)). \end{aligned}$$

Hence

$$\exists a\, \exists b(a \in x \ \& \ b \in y \ \& \ z = \langle a, b\rangle) \rightarrow z \in P(P(x \cup y)).$$

We now define $\langle x_1, \ldots, x_n\rangle$ for each n. Proceeding by induction on n, we set for $n \geqslant 3$,

$$\langle x_1, \ldots, x_n\rangle = \langle x_1, \langle x_2, \ldots, x_n\rangle\rangle.$$

To include the case $n = 1$ we set

$$\langle x\rangle = x.$$

We then have

$$\langle x_1, \ldots, x_n, \langle y_1, \ldots, y_k\rangle\rangle = \langle x_1, \ldots, x_n, y_1, \ldots, y_k\rangle, \tag{6}$$

as is easily proved by induction on n. Using (5) and induction on n, we get

$$\langle x_1, \ldots, x_n\rangle = \langle y_1, \ldots, y_n\rangle \leftrightarrow x_1 = y_1 \ \& \cdots \& \ x_n = y_n. \tag{7}$$

We may thus define function symbols π_i^n $(1 \leqslant i \leqslant n)$ such that

$$\pi_i^n(\langle x_1, \ldots, x_n\rangle) = x_i.$$

We also extend the Cartesian product to more than two factors by setting

$$x_1 \times \cdots \times x_n = x_1 \times (x_2 \times \cdots \times x_n).$$

Then $x_1 \times \cdots \times x_n$ is the set of $\langle a_1, \ldots, a_n\rangle$ with $a_1 \in x_1, \ldots, a_n \in x_n$.

We let

$$[F(x_1, \ldots, x_n, v_1, \ldots v_m) \mid Q(x_1, \ldots, x_n, v_1, \ldots, v_m)]_{x_1,\ldots,x_n} \tag{8}$$

be an abbreviation for

$$[y \mid \exists x_1 \ldots \exists x_n(Q(x_1, \ldots, x_n, v_1, \ldots, v_m) \;\&\; y = F(x_1, \ldots, x_n, v_1, \ldots, v_m))].$$

The subscripts are generally omitted. Omitting parameters,

$$[F(x_1, \ldots, x_n) \mid Q(x_1, \ldots, x_n)]$$

is the set of $F(x_1, \ldots, x_n)$ for $x_1, \ldots, x_n$ such that $Q(x_1, \ldots, x_n)$.

We use

$$Set_{x_1,\ldots,x_n} Q(x_1, \ldots, x_n, v_1, \ldots, v_m) \tag{9}$$

as an abbreviation for

$$\exists w_1 \ldots \exists w_n \forall x_1 \ldots \forall x_n(Q(x_1, \ldots, x_n, v_1, \ldots, v_m) \\ \rightarrow x_1 \in w_1 \;\&\; \cdots \;\&\; x_n \in w_n).$$

We show that (9) implies the existence condition for (8). Define a function symbol $\langle F \rangle$ by

$$\langle F \rangle(z) = F(\pi_n^1(z), \ldots, \pi_n^n(z)).$$

Then

$$F(x_1, \ldots, x_n) = \langle F \rangle(\langle x_1, \ldots, x_n \rangle).$$

Assuming (9), it follows that for suitable $w_1, \ldots, w_n$,

$$\exists x_1 \ldots \exists x_n(Q(x_1, \ldots, x_n) \;\&\; y = F(x_1, \ldots, x_n)) \\ \rightarrow y \in [\langle F \rangle(z) \mid z \in w_1 \times \cdots \times w_n].$$

This gives the required existence condition. We shall therefore call (9) the *existence condition* for (8).

We shall identify a function with the set of ordered pairs $\langle a, b \rangle$, where a is the value of the function for the argument b. We therefore define the *domain* $Do(x)$ of x and the *range* $Ra(x)$ of x by

$$Do(x) = [\pi_2(y) \mid y \in x],$$
$$Ra(x) = [\pi_1(y) \mid y \in x].$$

We then say x is a *function*, and write $Func(x)$, if $x \subset Ra(x) \times Do(x)$, and if $\langle a, b \rangle$, $\langle a', b \rangle \in x$ implies $a = a'$. It is clear how to define an *injective* function; we write $IFunc(x)$ to mean that x is an injective function. We use $x`a$ to designate the value of the function $x$ at the argument $a$. We consider ‘ as a binary function symbol; in accordance with our convention, $x`a = 0$ if x is not a function or if a is not in the domain of x. The various notions concerned with mappings can now be defined in the obvious manner.

We shall obtain a definition of *natural number* in *ZF* in the next section. We will then have all the material necessary to proceed to any of the usual constructions of the real and complex numbers.

9.3 ORDINALS

For each stage S in the construction of §9.1, we choose a set x_S which is first constructed at stage S. We may assume that x_T has been chosen for each stage T preceding S. Then we let x_S be the set of all x_T for T a stage preceding S. This set can first be formed at stage S; for S is the first stage at which all of its members are available.

We shall define in *ZF* the property of being one of the sets x_S. We say that a set x is *transitive*, and write *Trans*(x), if every member of x is a subset of x; equivalently, if $\forall y \forall z(y \in x \mathbin{\&} z \in y \rightarrow z \in x)$. We say that x is an *ordinal* and write *Ord*(x) if x is transitive and every element of x is transitive. We let σ, τ, and ρ vary through ordinals. (These symbols are to be regarded as further variables of *ZF*.)

It is easy to see that each x_S is transitive; and from this it follows that each x_S is an ordinal. We shall prove the converse after obtaining some properties of ordinals.

We have

$$x \in \sigma \rightarrow Ord(x). \tag{1}$$

For let $x \in \sigma$. Then x is transitive. By the transitivity of σ, $x \subset \sigma$; so every element of x is an element of σ and hence transitive. Thus x is an ordinal.

We define

$$\sigma < \tau \leftrightarrow \sigma \in \tau.$$

(Note that $x_S < x_T$ iff stage S comes before stage T.) We have

$$\sigma < \tau \mathbin{\&} \tau < \rho \rightarrow \sigma < \rho \tag{2}$$

by the transitivity of ρ. Moreover,

$$\neg(\sigma < \sigma). \tag{3}$$

This is a special case of the result

$$x \notin x. \tag{4}$$

To prove this, note that $\{x\}$ has a minimal element, which must be x. Thus $x \cap \{x\} = 0$; so $x \notin x$. From (2) and (3) we obtain

$$\neg(\sigma < \tau \mathbin{\&} \tau < \sigma). \tag{5}$$

We have

$$\exists\sigma Q(\sigma, v_1, \ldots, v_n)$$
$$\rightarrow \exists\sigma\big(Q(\sigma, v_1, \ldots, v_n) \mathbin{\&} \forall\tau\big(\tau < \sigma \rightarrow \neg Q(\tau, v_1, \ldots, v_n)\big)\big).$$

We omit the parameters in the proof. Assume $\exists\sigma Q(\sigma)$, and choose σ so that $Q(\sigma)$. If $\forall\tau\big(\tau < \sigma \rightarrow \neg Q(\tau)\big)$, then σ is the desired ordinal. Otherwise, $[x \mid x < \sigma \mathbin{\&} Q(x)]$ is not empty, and hence has a minimal element ρ. Since $\rho \in \sigma$, ρ is an ordinal by (1). Clearly $Q(\rho)$; we show that $\tau < \rho \rightarrow \neg Q(\tau)$. Assume $\tau < \rho$. By (2), $\tau < \sigma$. By choice of ρ, $\tau \notin [x \mid x < \sigma \mathbin{\&} Q(x)]$; so $\neg Q(\tau)$.

An ordinal σ such that $Q(\sigma, v_1, \ldots, v_n)$ and $\neg Q(\tau, v_1, \ldots, v_n)$ for all $\tau < \sigma$ is called a *minimal* ordinal σ such that $Q(\sigma, v_1, \ldots, v_n)$. We may also speak of a minimal ordinal σ such that . . σ . . ; this is a minimal ordinal σ such that $Q(\sigma, v_1, \ldots, v_n)$, where Q is defined by

$$Q(\sigma, v_1, \ldots, v_n) \leftrightarrow \,.\,.\, \sigma \,.\,.\,.$$

Before assuming that a minimal ordinal σ such that . . σ . . exists, we must prove $\exists\sigma(\,.\,.\,\sigma\,.\,.)$.

We now prove

$$\sigma < \tau \vee \sigma = \tau \vee \tau < \sigma. \tag{6}$$

Abbreviate this to $C(\sigma, \tau)$. We shall assume $\exists\sigma\,\exists\tau\,\neg C(\sigma, \tau)$ and derive a contradiction. Let σ be a minimal ordinal such that $\exists\tau\,\neg C(\sigma, \tau)$, and let τ be a minimal ordinal such that $\neg C(\sigma, \tau)$. We shall first prove that $\tau \subset \sigma$. Let $\rho \in \tau$; ρ is an ordinal by (1). By choice of τ, $C(\sigma, \rho)$; so to prove $\rho \in \sigma$, it will suffice to show that $\sigma = \rho \vee \sigma < \rho$ leads to a contradiction. Since $\rho < \tau$, $\sigma = \rho \vee \sigma < \rho$ implies $\sigma < \tau$ by (2); and this contradicts $\neg C(\sigma, \tau)$.

Since $\tau \subset \sigma$ and $\neg C(\sigma, \tau)$, there is a ρ in $\sigma - \tau$. By (1), ρ is an ordinal. By choice of σ, $C(\rho, \tau)$. Since $\rho \notin \tau$, this implies that $\tau < \rho \vee \tau = \rho$. It follows by (2) that $\tau < \sigma$, contradicting $\neg C(\sigma, \tau)$.

We can now see that every ordinal is an x_S. Suppose that σ is first constructed at stage S. Since x_S is also first constructed at this stage, $\sigma \notin x_S$ and $x_S \notin \sigma$. Since σ and x_S are both ordinals, this implies $\sigma = x_S$ by (6).

Another consequence of (6) is that if $\exists\sigma Q(\sigma, v_1, \ldots, v_n)$, then the minimal ordinal σ such that $Q(\sigma, v_1, \ldots, v_n)$ is unique. We call it the *first* ordinal σ such that $Q(\sigma, v_1, \ldots, v_n)$, and designate it by $\mu\sigma Q(\sigma, v_1, \ldots, v_n)$.

We define

$$\sigma \leqslant \tau \leftrightarrow \sigma < \tau \vee \sigma = \tau.$$

The usual ordering properties of $\leqslant$ can be derived from the properties of $<$ proved above. In addition,

$$\sigma \leqslant \tau \leftrightarrow \sigma \subset \tau. \tag{7}$$

For if $\sigma < \tau$, then $\sigma \subset \tau$ by the transitivity of τ; so $\sigma \leqslant \tau \rightarrow \sigma \subset \tau$. Now suppose that $\neg(\sigma \leqslant \tau)$. By (6), $\tau < \sigma$; while by (3), $\neg(\tau < \tau)$. Thus $\tau \in \sigma - \tau$; so $\neg(\sigma \subset \tau)$.

Principle of Transfinite Induction. If

$$\forall\sigma(\forall\tau(\tau < \sigma \rightarrow Q(\tau, v_1, \ldots, v_n)) \rightarrow Q(\sigma, v_1, \ldots, v_n)),$$

then $\forall\sigma Q(\sigma, v_1, \ldots, v_n)$.

Proof. If the conclusion is false, then there is a first ordinal σ such that $\neg Q(\sigma, v_1, \ldots, v_n)$. But this contradicts the hypothesis.

The principle of transfinite induction tells us that if we wish to prove $Q(\sigma, v_1, \ldots, v_n)$, it suffices to prove

$$\forall\sigma(\forall\tau(\tau < \sigma \rightarrow Q(\tau, v_1, \ldots, v_n)) \rightarrow Q(\sigma, v_1, \ldots, v_n)).$$

In other words, in proving $Q(\sigma, v_1, \ldots, v_n)$, we may assume that $Q(\tau, v_1, \ldots, v_n)$ for all $\tau < \sigma$. A proof by this method is called a *proof by transfinite induction* on σ; the hypothesis that $Q(\tau, v_1, \ldots, v_n)$ for all $\tau < \sigma$ is called the *induction hypothesis*.

We can also prove a formula $..\sigma..$ by transfinite induction by defining a Q by $Q(\sigma, v_1, \ldots, v_n) \leftrightarrow ..\sigma..$; the induction hypothesis is then that $..\tau..$ for all $\tau < \sigma$.

There is also a form of induction appropriate to proving facts about sets (instead of only about ordinals). We say that H is an *ordinal function symbol* if* $\vdash Ord(H(x_1, \ldots, x_n))$. Suppose that this is the case. Then to prove

$$Q(x_1, \ldots, x_n, v_1, \ldots, v_m),$$

it suffices to prove it under the hypothesis that

$$\forall y_1 \ldots \forall y_n(H(y_1, \ldots, y_n) < H(x_1, \ldots, x_n) \rightarrow Q(y_1, \ldots, y_n, v_1, \ldots, v_m)).$$

For if we have proved it under this hypothesis, then we can prove

$$\forall x_1 \ldots \forall x_n(H(x_1, \ldots, x_n) = \sigma \rightarrow Q(x_1, \ldots, x_n, v_1, \ldots, v_m))$$

by transfinite induction on σ; and from this we can prove

$$Q(x_1, \ldots, x_n, v_1, \ldots, v_m).$$

If n is a natural number and S is the $(n + 1)$st stage of set construction, then x_S has exactly n members. We intend to identify n with the set x_S. Thus 0 is identified with the empty set. We must then define the successor operation and the property of being a natural number.

We define

$$S(\sigma) = \sigma \cup \{\sigma\}$$

and call $S(\sigma)$ the *successor* of σ. We have

$$S(\sigma) = S(\tau) \rightarrow \sigma = \tau.$$

For suppose that $S(\sigma) = S(\tau)$ and $\sigma \neq \tau$. Then $\sigma \in \sigma \cup \{\sigma\} = \tau \cup \{\tau\}$; so $\sigma \in \tau$. Similarly, $\tau \in \sigma$. This is impossible by (5).

We now show that $S(\sigma)$ is an ordinal. If $y \in S(\sigma)$, then $y \in \sigma$ or $y = \sigma$; so y is transitive and $y \subset \sigma \subset S(\sigma)$. Clearly $\sigma < S(\sigma)$. Moreover, $S(\sigma)$ is the first ordinal τ such that $\sigma < \tau$. For if $\tau < S(\sigma)$, then $\tau < \sigma \vee \tau = \sigma$, and hence $\neg(\sigma < \tau)$.

---

* Note that the statement that H is an ordinal function symbol is a statement about provability in *ZF*, not a statement in *ZF*. In general, whenever we attribute a property to a function or predicate symbol, we are making a statement about provability in *ZF*.

Let x be a set of ordinals. Then $Un(x)$ is an ordinal. For if $y \in Un(x)$, then $y \in \sigma$ for some $\sigma \in x$. Hence y is transitive and $y \subset \sigma \subset Un(x)$. Using (7), we see that $Un(x)$ is the first ordinal τ such that $\sigma \leqslant \tau$ for every $\sigma \in x$.

If x is a set, then there is an ordinal greater than every ordinal in x. For if y is the set of ordinals in x, then $S(Un(y))$ is such an ordinal. From this we can prove

$$Set_x\, \exists\sigma(x = F(\sigma)) \rightarrow \exists\sigma\, \exists\tau(\tau < \sigma\ \&\ F(\tau) = F(\sigma)). \tag{8}$$

For let $G(x) = \mu\sigma(F(\sigma) = x)$ if $\exists\sigma(F(\sigma) = x)$, and $G(x) = 0$ otherwise. Assume $Set_x\, \exists\sigma(x = F(\sigma))$, and choose w so that $F(\sigma) \in w$ for all σ. Choose σ not in the set $[G(x) \mid x \in w]$. Then $G(F(\sigma)) \neq \sigma$; so there is a τ such that $\tau < \sigma$ and $F(\tau) = F(\sigma)$.

An ordinal is a *limit ordinal* if it is not 0 and is not the successor of any ordinal. We shall prove that a limit ordinal exists. By the axiom of infinity, there is a set x such that $0 \in x$ and $\forall y(y \in x \rightarrow S(y) \in x)$. Let z be the set of ordinals in x and let $\sigma = Un(z)$. We have $S(0) \in x$ and hence $0 < S(0) \leqslant \sigma$; so $\sigma \neq 0$. Suppose that $\sigma = S(\tau)$. Then $\tau < \sigma = Un(z)$; so $\tau < \rho$ for some $\rho \in z$. It follows that $\sigma = S(\tau) \leqslant \rho < S(\rho)$. But $S(\rho) \in z$; so $S(\rho) \leqslant \sigma$. This contradiction shows that σ is a limit ordinal.

The first limit ordinal is designated by ω. The members of ω (that is, the ordinals less than ω) are called *natural numbers*. An ordinal is *finite* or *infinite* according as it is or is not a natural number. Thus ω is the first infinite ordinal.

It is now easy to prove the Peano postulates. Clearly 0 is the first ordinal and hence is a natural number. If σ is a natural number, then $\sigma < \omega$; so $S(\sigma) \leqslant \omega$. Since ω is not a successor, $S(\sigma) < \omega$; so $S(\sigma)$ is a natural number. Clearly $S(\sigma) \neq 0$; and $S(\sigma) = S(\tau) \rightarrow \sigma = \tau$ has already been proved. The fifth axiom says that if x is a set such that $0 \in x$ and $\forall\sigma(\sigma < \omega\ \&\ \sigma \in x \rightarrow S(\sigma) \in x)$, then x contains every natural number. We prove $\sigma < \omega \rightarrow \sigma \in x$ by transfinite induction on σ. If $\sigma = 0$, then $\sigma \in x$. Otherwise, since σ is less than the first limit ordinal, we must have $\sigma = S(\tau)$. Since $\tau < \sigma$, $\tau < \omega$ and $\tau \in x$ by the induction hypothesis. It follows that $\sigma = S(\tau) \in x$.

We can now define $1, 2, \ldots$ by $1 = S(0)$, $2 = S(1), \ldots$.

We now turn to definitions by transfinite induction. The idea is that we wish to define $F(\sigma)$ in terms of σ and the values of $F(\tau)$ for ordinals $\tau < \sigma$. Adding parameters, we arrive at the following situation: we have defined G, and wish to define F so that

$$F(\sigma, v_1, \ldots, v_n) = G(\sigma, [\langle F(\tau, v_1, \ldots, v_n), \tau\rangle \mid \tau < \sigma], v_1, \ldots, v_n). \tag{9}$$

We shall show that this can be done.

As usual, we omit the parameters. Let $G_{f,\sigma}$ abbreviate

$$G(\sigma, [\langle f`\tau, \tau\rangle \mid \tau < \sigma]).$$

Let $Q(f, \sigma)$ mean that f is a function whose domain includes σ and that for every $\tau < \sigma$, $f`\tau = G_{f,\tau}$. Clearly

$$Q(f, \sigma)\ \&\ \tau < \sigma \rightarrow Q(f, \tau). \tag{10}$$

We prove

$$Q(f, \sigma) \ \& \ Q(g, \sigma) \rightarrow G_{f,\sigma} = G_{g,\sigma} \tag{11}$$

by transfinite induction on σ. For $\tau < \sigma$, $Q(f, \tau)$ & $Q(g, \tau)$ by (10); so $G_{f,\tau} = G_{g,\tau}$ by induction hypothesis; so $f`\tau = G_{f,\tau} = G_{g,\tau} = g`\tau$. This implies that $G_{f,\sigma} = G_{g,\sigma}$.

We now define F as follows. If there is an f such that $Q(f, \sigma)$, then $F(\sigma) = G_{f,\sigma}$; otherwise, $F(\sigma) = 0$. This is well-defined by (11). Moreover,

$$\exists f Q(f, \sigma) \rightarrow F(\sigma) = G(\sigma, [\langle F(\tau), \tau\rangle \mid \tau < \sigma]). \tag{12}$$

For choose f so that $Q(f, \sigma)$. For each $\tau < \sigma$, $Q(f, \tau)$ by (10); so $F(\tau) = f`\tau$. Hence

$$F(\sigma) = G_{f,\sigma} = G(\sigma, [\langle F(\tau), \tau\rangle \mid \tau < \sigma]).$$

In view of (12), it only remains to prove $\exists f Q(f, \sigma)$. We prove this by transfinite induction on σ. Let $f = [\langle F(\tau), \tau\rangle \mid \tau < \sigma]$; we show that $Q(f, \sigma)$. Clearly f is a function with domain σ. If $\tau < \sigma$, then

$$F(\tau) = G(\tau, [\langle F(\rho), \rho\rangle \mid \rho < \tau])$$

by (12) and the induction hypothesis. Hence

$$f`\tau = F(\tau) = G(\tau, [\langle F(\rho), \rho\rangle \mid \rho < \tau]) = G_{f,\tau}.$$

We call (9) a definition of F by *transfinite induction* on σ. In practice, the right side of such a definition will not be in the form of the right side of (9); it will be necessary to define G properly to achieve this. Thus we might define

$$F(\sigma) = H([F(\tau) \mid \tau < \sigma]).$$

This becomes of the proper form if we define G by

$$G(\sigma, f) = H(Ra(f)).$$

Now suppose that H is an ordinal function symbol such that

$$Set_{x_1,\ldots,x_n}(H(x_1, \ldots, x_n) \leqslant \sigma).$$

We can then define a function symbol F such that

$$F(x_1, \ldots, x_n) = G(x_1, \ldots, x_n, [\langle F(y_1, \ldots, y_n), y_1, \ldots, y_n\rangle \mid H(y_1, \ldots, y_n) < H(x_1, \ldots, x_n)]). \tag{13}$$

To obtain such an F, we define

$$F(x_1, \ldots, x_n) = I(H(x_1, \ldots, x_n))`\langle x_1, \ldots, x_n\rangle \tag{14}$$

for a suitable I. We want $I(\sigma)$ to be a function with domain

$$[\langle x_1, \ldots, x_n\rangle \mid H(x_1, \ldots, x_n) = \sigma]$$

whose value at $\langle x_1, \ldots, x_n\rangle$ is $F(x_1, \ldots, x_n)$. We therefore define by transfinite induction

$$I(\sigma) = [\langle G(x_1, \ldots, x_n, Un([I(\tau) \mid \tau < \sigma])), x_1, \ldots, x_n\rangle \mid H(x_1, \ldots, x_n) = \sigma].$$

We can then prove (13) from this definition and (14). A definition of the form (13) (or a similar definition with parameters) will be called a definition by *induction on* $H(x_1, \ldots, x_n)$.

We can also define predicates by induction. For simplicity, we consider unary predicates. Let H be an ordinal function symbol such that $\vdash Set_x(H(x) \leqslant \sigma)$. Given Q, we wish to define R so that

$$R(x) \leftrightarrow Q(x, [y \mid H(y) < H(x) \;\&\; R(y)]). \tag{15}$$

We first define $K_Q(x, z)$ to be 0 if $Q(x, z)$ and 1 otherwise. Using induction on $H(x)$, we define

$$K_R(x) = K_Q(x, [y \mid H(y) < H(x) \;\&\; K_R(y) = 0]).$$

Finally we define

$$R(x) \leftrightarrow K_R(x) = 0.$$

We then easily derive (15). A definition of this form can also contain parameters.

We now set up a correspondence between ordinals and ordered pairs of ordinals. We first define $Max(\langle\sigma, \tau\rangle) = \sigma \cup \tau$. Then by (7), $Max(\langle\sigma, \tau\rangle) = \tau$ if $\sigma \leqslant \tau$ and $Max(\langle\sigma, \tau\rangle) = \sigma$ if $\tau \leqslant \sigma$.

We now define $MP(x)$ as follows. Let σ be the first ordinal such that $\neg(\sigma \times \sigma \subset x)$. This must exist, since otherwise $Ra(x)$ would contain every ordinal. Let τ be the first ordinal such that $\langle\tau, \rho\rangle \notin x$ for some $\rho \in \sigma$, and let ρ be the first ordinal such that $\langle\tau, \rho\rangle \notin x$. We set $MP(x) = \langle\tau, \rho\rangle$.

Clearly $MP(x)$ is an ordered pair of ordinals not in x. If σ is as above, then $Max(MP(x)) < \sigma$; so by choice of σ,

$$Max(MP(x)) \times Max(MP(x)) \subset x. \tag{16}$$

If σ and τ are as above and $\tau \neq 0$, then $\langle 0, \rho\rangle \in x$ for all $\rho < \sigma$, and hence for $\rho = Max(MP(x))$. Thus

$$\pi_1(MP(x)) \neq 0 \rightarrow \langle 0, Max(MP(x))\rangle \in x. \tag{17}$$

We now define K by transfinite induction as follows:

$$K(\sigma) = MP([K(\tau) \mid \tau < \sigma]).$$

Then $K(\sigma)$ is an ordered pair of ordinals not in $[K(\tau) \mid \tau < \sigma]$; so $K(\sigma) \neq K(\tau)$ for $\tau < \sigma$. From this,

$$\sigma \neq \tau \rightarrow K(\sigma) \neq K(\tau). \tag{18}$$

From (16),

$$Max(K(\sigma)) \times Max(K(\sigma)) \subset [K(\tau) \mid \tau < \sigma]. \tag{19}$$

We use this to show that every ordered pair x of ordinals is $K(\sigma)$ for some σ. By (18) and (8), $\neg Set_z \exists\sigma(z = K(\sigma))$. Hence there is a σ such that

$$K(\sigma) \notin S(Max(x)) \times S(Max(x)).$$

From this it readily follows that

$$x \in Max(K(\sigma)) \times Max(K(\sigma));$$

and the desired result then follows from (19).

We have

$$Max(K(\tau)) < Max(K(\sigma)) \to \tau < \sigma. \tag{20}$$

For the left-hand side and (19) imply that $K(\tau) \in [K(\rho) \mid \rho < \sigma]$; and this and (18) imply that $\tau < \sigma$.

Next we prove

$$Max(K(\sigma)) \leqslant \sigma \tag{21}$$

by transfinite induction on σ. Assume that $\sigma < Max(K(\sigma))$. Choose τ so that $K(\tau) = \langle 0, \sigma \rangle$. Then $Max(K(\tau)) = \sigma < Max(K(\sigma))$; so $\tau < \sigma$ by (20). Hence by the induction hypothesis, $Max(K(\tau)) \leqslant \tau$. Thus $\sigma \leqslant \tau < \sigma$, a contradiction.

Finally, we prove

$$\pi_1(K(\sigma)) \neq 0 \to Max(K(\sigma)) < \sigma. \tag{22}$$

Assume that $\pi_1(K(\sigma)) \neq 0$. By (17), there is a $\tau < \sigma$ such that

$$K(\tau) = \langle 0, Max(K(\sigma)) \rangle.$$

Then $Max(K(\sigma)) = Max(K(\tau)) \leqslant \tau < \sigma$ by (21).

9.4 CARDINALS

We say that x and y are *similar* if there is a bijective mapping from x to y; in symbols,

$$Sm(x, y) \leftrightarrow \exists f(IFunc(f) \;\&\; x = Do(f) \;\&\; y = Ra(f)).$$

It is easy to verify that similarity has the properties of an equivalence relation, that is,

$$Sm(x, x), \tag{1}$$

$$Sm(x, y) \to Sm(y, x), \tag{2}$$

$$Sm(x, y) \;\&\; Sm(y, z) \to Sm(x, z). \tag{3}$$

We intend the cardinal of a set to be a measure of its size; so we want two sets to have the same cardinal iff they are similar. This suggests that the cardinal of x should be the equivalence class of x under this relation. However, this equivalence class will not be a set. We therefore take the cardinal of x to be a particular member of this equivalence class, viz., the first ordinal in the equivalence class. We must, of course, first know that the equivalence class contains an ordinal; and this requires a new axiom.

We say that f is a *choice function on* x if f is a function with domain $P(x) - \{0\}$ and $f`y \in y$ for every y in the domain of f. In symbols,

$$CF(f, x) \leftrightarrow Func(f) \;\&\; Do(f) = P(x) - \{0\} \;\&\; \forall y(y \in Do(f) \rightarrow f`y \in y).$$

The *axiom of choice* is the formula

$$\forall x \, \exists f \, CF(f, x);$$

it says that for every set x, there is a choice function on x.

The axiom of choice is true for our meaning of set. For select for each nonempty subset y of x an element z_y of y, and let f be the collection of ordered pairs $\langle z_y, y\rangle$. Since every such ordered pair is in $x \times P(x)$, f is a set; and it is clearly a choice function on x.

The theory obtained from ZF by adding the axiom of choice is designated by ZFC. The remaining results of this section are proved in ZFC.

Well-Ordering Theorem (*Zermelo*). For every set x, there is a bijective mapping from an ordinal to x.

Proof. Let g be a choice function on x. Define by transfinite induction

$$F(\sigma) = g`\big(x - [F(\tau) \mid \tau < \sigma]\big)$$

(where we omit the parameters g and x as arguments of F). Since

$$F(\sigma) \in Ra(g) \cup \{0\},$$

we have

$$Set_z \, \exists \sigma\big(F(\sigma) = z\big).$$

Hence by (8) of §9.3, there are ordinals σ and τ such that $\tau < \sigma$ and $F(\tau) = F(\sigma)$. By the definition of $F(\sigma)$ and the choice of g, this implies that $x \subset [F(\tau) \mid \tau < \sigma]$.

Let σ be the smallest ordinal such that $x \subset [F(\tau) \mid \tau < \sigma]$, and let

$$f = [\langle F(\tau), \tau\rangle \mid \tau < \sigma].$$

Then f is a function with domain σ, and $x \subset Ra(f)$. If $\tau < \sigma$, it follows from the choice of σ that $x - [F(\rho) \mid \rho < \tau] \neq 0$; so

$$F(\tau) \in x - [F(\rho) \mid \rho < \tau].$$

This implies that $Ra(f) \subset x$; so $Ra(f) = x$. It also implies that

$$\rho < \tau < \sigma \rightarrow F(\rho) \neq F(\tau);$$

so f is injective.

Remark. We have also shown that if g is a choice function on x, then there is a bijective mapping f from an ordinal to x such that for every $\sigma \in Do(f)$,

$$x - [f`\tau \mid \tau < \sigma] \neq 0 \;\&\; f`\sigma = g`(x - [f`\tau \mid \tau < \sigma]).$$

We shall now prove

$$x \subset \sigma \rightarrow \exists\tau(\tau \leqslant \sigma \ \&\ Sm(\tau, x)). \tag{4}$$

Let $x \subset \sigma$. For y a nonempty subset of x, let $g`y = \mu\tau(\tau \in y)$. Then g is a choice function on x. Hence by the remark, there is a bijective mapping from an ordinal τ to x such that

$$f`\rho = \mu\sigma'(\sigma' \in x - [f`\rho' \mid \rho' < \rho])$$

for $\rho < \tau$. We complete the proof by showing that $\tau \leqslant \sigma$.

Suppose that $\sigma < \tau$. Then $f`\sigma \in x$; whence, since $x \subset \sigma$, $f`\sigma < \sigma$. Let ρ be the first ordinal such that $f`\rho < \rho$. Since f is injective,

$$f`\rho \in x - [f`\rho' \mid \rho' < f`\rho].$$

But $f`f`\rho$ is the first ordinal in this set; so $f`f`\rho \leqslant f`\rho$. Since $f`\rho < \rho$ and f is injective, $f`f`\rho < f`\rho$. But this with $f`\rho < \rho$ contradicts the choice of ρ.

The *cardinal of* a set x, designated by $Card(x)$, is the first ordinal similar to x. This is well-defined by the well-ordering theorem. We say that σ is a *cardinal*, and write $Cd(\sigma)$, if σ is the cardinal of some set.

Using (1) through (3),

$$Card(x) = Card(y) \leftrightarrow Sm(x, y). \tag{5}$$

From (1),

$$Card(\sigma) \leqslant \sigma. \tag{6}$$

Moreover,

$$Cd(\sigma) \leftrightarrow Card(\sigma) = \sigma. \tag{7}$$

The implication from right to left is immediate. If $Cd(\sigma)$, then $\sigma = Card(x)$ for some x. Then $Sm(\sigma, x)$; so $Card(\sigma) = Card(x) = \sigma$ by (5).

We have

$$x \subset y \rightarrow Card(x) \leqslant Card(y). \tag{8}$$

For let $\sigma = Card(y)$. Since y is similar to σ, x is similar to a subset of σ; so by (4), x is similar to an ordinal $\tau \leqslant \sigma$. Then $Card(x) \leqslant \tau \leqslant \sigma$.

We have

$$Func(f) \rightarrow Card(Ra(f)) \leqslant Card(Do(f)). \tag{9}$$

For let g be a choice function on $Do(f)$, and let h be the function with domain $Ra(f)$ defined by

$$h`z = g`[y \mid y \in Do(f) \ \&\ f`y = z].$$

Then h is a bijective mapping from $Ra(f)$ to a subset w of $Do(f)$; so, using (8),

$$Card(Ra(f)) = Card(w) \leqslant Card(Do(f)).$$

Next we show that

$$Sm(S(\sigma), S(\tau)) \rightarrow Sm(\sigma, \tau). \tag{10}$$

For let f be a bijective mapping from $S(\sigma) = \sigma \cup \{\sigma\}$ to $S(\tau) = \tau \cup \{\tau\}$. By interchanging two values of f we may suppose that $f`\sigma = \tau$. Then the restriction of f to σ is a bijective mapping from σ to τ; so $Sm(\sigma, \tau)$.

We shall now prove

$$\sigma \in \omega \rightarrow \neg Sm(\sigma, S(\sigma)) \tag{11}$$

by (ordinary) induction on σ. The case $\sigma = 0$ is easy; and the step from σ to $S(\sigma)$ follows from (10).

We can now prove

$$\sigma \in \omega \rightarrow Cd(\sigma) \tag{12}$$

by induction on σ. The case $\sigma = 0$ can be derived from (6) and (7). Now assume that $Cd(\sigma)$ with $\sigma \in \omega$. From (7) and (8), $\sigma = Card(\sigma) \leqslant Card(S(\sigma))$. But $Card(S(\sigma)) \neq Card(\sigma)$ by (11); so $\sigma < Card(S(\sigma))$ and hence $S(\sigma) \leqslant Card(S(\sigma))$. Using (6) and (7), we then get $Cd(S(\sigma))$.

We have thus shown that every natural number is a cardinal. Moreover, ω is a cardinal. If not, then $Card(\omega) < \omega$ by (6) and (7). Thus $Card(\omega)$ is a natural number σ. Since $S(\sigma) \leqslant \omega$, we have $S(\sigma) = Card(S(\sigma)) \leqslant Card(\omega) = \sigma$, a contradiction.

A set is *finite* or *infinite* according as its cardinal is finite or infinite. A set is *countable* if its cardinal is either a natural number or ω. The usual proofs of the elementary properties of finite and countable sets can now be carried out.

We define

$$\sigma + \tau = Card((\sigma \times \{0\}) \cup (\tau \times \{1\})),$$
$$\sigma \cdot \tau = Card(\sigma \times \tau).$$

It is easy to check that

$$x \cap y = 0 \rightarrow Card(x \cup y) = Card(x) + Card(y), \tag{13}$$
$$Card(x \times y) = Card(x) \cdot Card(y). \tag{14}$$

Writing $x \cup y = x \cup (y - x)$ and using (13) and (8), we get

$$Card(x \cup y) \leqslant Card(x) + Card(y). \tag{15}$$

Moreover,

$$Cd(\sigma) \ \& \ \forall y(y \in x \rightarrow Card(y) \leqslant \sigma) \rightarrow Card(Un(x)) \leqslant Card(x) \cdot \sigma. \tag{16}$$

We first note that if f is a bijective mapping from $Card(y)$ to y, where $y \in x$, then $f \in P(Un(x) \times \sigma)$. Hence using a choice function on $P(Un(x) \times \sigma)$, we obtain a function g with domain x such that for $y \in x$, $g`y$ is a bijective mapping from $Card(y)$ to $y$. Let $z$ be the set of $\langle y, \tau \rangle$ with $y \in x$ and $\tau < Card(y)$; and let $h$ be the function with domain $z$ defined by $h`\langle y, \tau \rangle = (g`y)`\tau$. Then h is a surjective mapping from z to $Un(x)$. Using (9) and (8), we conclude

$$Card(Un(x)) \leqslant Card(z) \leqslant Card(x \times \sigma) = Card(x) \cdot \sigma.$$

Applying (13) to $\sigma = \sigma \cup 0$ and $S(\sigma) = \sigma \cup \{\sigma\}$, we have

$$Card(\sigma) = Card(\sigma) + 0, \tag{17}$$

$$Card(S\sigma) = Card(\sigma) + 1. \tag{18}$$

From the definitions, we also have

$$\sigma + \sigma = \sigma \cdot 2. \tag{19}$$

If σ and τ are natural numbers, then

$$\sigma + 0 = \sigma,$$
$$\sigma + S(\tau) = S(\sigma + \tau),$$
$$\sigma \cdot 0 = 0,$$
$$\sigma \cdot S(\tau) = (\sigma \cdot \tau) + \sigma.$$

(The reader can easily provide proofs.) From this we obtain by induction on τ,

$$\sigma, \tau \in \omega \rightarrow \sigma + \tau, \sigma \cdot \tau \in \omega. \tag{20}$$

From (8),

$$\sigma \leqslant \tau \ \& \ \sigma' \leqslant \tau' \rightarrow \sigma + \sigma' \leqslant \tau + \tau' \ \& \ \sigma \cdot \sigma' \leqslant \tau \cdot \tau'. \tag{21}$$

We write $InfCd(\sigma)$ to mean that σ is an infinite cardinal. Using (21), (17), and (19), we obtain

$$InfCd(\sigma) \rightarrow \sigma = \sigma + 0 \leqslant \sigma + 1 \leqslant \sigma + \sigma = \sigma \cdot 2 \leqslant \sigma \cdot \sigma. \tag{22}$$

By the results of §9.3, $[\langle K(\tau), \tau \rangle \mid \tau < \sigma]$ is a bijective mapping from σ to $[K(\tau) \mid \tau < \sigma]$. Hence

$$Card(\sigma) = Card([K(\tau) \mid \tau < \sigma]).$$

We will use this to prove

$$InfCd(\sigma) \ \& \ Max(K(\sigma)) = \sigma \rightarrow \sigma \cdot \sigma = \sigma. \tag{23}$$

In view of (22), we need only show that $\sigma \cdot \sigma \leqslant \sigma$. By the hypothesis and (19) of §9.3, $\sigma \times \sigma \subset [K(\tau) \mid \tau < \sigma]$. Taking cardinals and using the previous equation, we get $\sigma \cdot \sigma \leqslant \sigma$.

We shall now prove

$$InfCd(\sigma) \rightarrow Max(K(\sigma)) = \sigma \tag{24}$$

by transfinite induction on σ. We first show that the induction hypothesis implies

$$\rho < \sigma \rightarrow S(\rho) \cdot S(\rho) < \sigma. \tag{25}$$

If $\rho < \omega$, then $S(\rho) \cdot S(\rho) < \omega \leqslant \sigma$ by (20). Now assume that $\omega \leqslant \rho$. Setting $\tau = Card(\rho)$, $\omega = Card(\omega) \leqslant \tau \leqslant \rho < \sigma$. The induction hypothesis then implies that $Max(K(\tau)) = \tau$; and this with (23) implies that $\tau \cdot \tau = \tau$. By (18) and (22), we have $Card(S(\rho)) = \tau + 1 = \tau$; so $S(\rho) \cdot S(\rho) = \tau \cdot \tau = \tau < \sigma$.

Returning to (24), assume that $Max(K(\sigma)) \neq \sigma$. Then

$$\rho = Max(K(\sigma)) < \sigma$$

by (21) of §9.3. If $\tau < \sigma$, then

$$Max(K(\tau)) \leqslant Max(K(\sigma)) < S(\rho)$$

by (20) of §9.3; so $K(\tau) \in S(\rho) \times S(\rho)$. From this and (25), we see that

$$\sigma = Card(\sigma) \leqslant Card(S(\rho) \times S(\rho)) < \sigma.$$

This is a contradiction.

Combining (23), (24), and (22), we get

$$InfCd(\sigma) \rightarrow \sigma \cdot \sigma = \sigma \;\&\; \sigma + \sigma = \sigma. \tag{26}$$

From this we conclude that

$$InfCd(\sigma) \;\&\; InfCd(\tau) \rightarrow \sigma \cdot \tau = \sigma + \tau = Max(\langle \sigma, \tau \rangle). \tag{27}$$

For let $\rho = Max(\langle \sigma, \tau \rangle)$. Then using (21), we find that

$$\rho = Max(\langle \sigma + 0, 0 + \tau \rangle) \leqslant \sigma + \tau \leqslant \rho + \rho = \rho,$$
$$\rho = Max(\langle \sigma \cdot 1, 1 \cdot \tau \rangle) \leqslant \sigma \cdot \tau \leqslant \rho \cdot \rho = \rho.$$

We also have

$$InfCd(\sigma) \;\&\; Card(w) \leqslant \sigma \rightarrow Card([F(x_1, \ldots, x_n) \mid x_1, \ldots, x_n \in w]) \leqslant \sigma. \tag{28}$$

For defining $\langle F \rangle$ as in §9.2, we have

$$[F(x_1, \ldots, x_n) \mid x_1, \ldots, x_n \in w] = [\langle F \rangle(a) \mid a \in w \times \cdots \times w].$$

Since $Card(w \times \cdots \times w) \leqslant \sigma \cdot \ldots \cdot \sigma = \sigma$, the desired result follows from (9).

We say that x is *F-closed* if for every $y_1, \ldots, y_n \in x$ we have $F(y_1, \ldots, y_n) \in x$.

Closure Theorem. Let $F_1, \ldots, F_k$ be function symbols. Given a set x and an infinite cardinal τ such that $Card(x) \leqslant \tau$, there is a set y such that $x \subset y$, $Card(y) \leqslant \tau$, and y is F_i-closed for $i = 1, \ldots, k$.

Proof. Define

$$G_i(z) = [F_i(w_1, \ldots, w_n) \mid w_1, \ldots, w_n \in z],$$
$$G(z) = G_1(z) \cup \cdots \cup G_k(z).$$

We define H by transfinite induction as follows: $H(0) = x$; $H(\sigma) = G(H(\rho))$ if $\sigma = S(\rho)$; $H(\sigma) = Un([H(\rho) \mid \rho < \sigma])$ if σ is a limit ordinal. We claim that $y = H(\omega)$ has the required properties. Obviously $x \subset y$. If $w_1, \ldots, w_n \in y$, then $w_1, \ldots, w_n \in H(\sigma)$ for some $\sigma < \omega$; so

$$F_i(w_1, \ldots, w_n) \in G(H(\sigma)) = H(S(\sigma)) \subset y.$$

Using (28) and (15), we get $Card(z) \leqslant \tau \to Card(G(z)) \leqslant \tau$. Using this, we easily prove by induction that $\sigma < \omega \to Card(H(\sigma)) \leqslant \tau$. Then by (16), we have $Card(y) \leqslant \tau \cdot \omega \leqslant \tau \cdot \tau = \tau$.

We shall now consider an operation which enables us to obtain cardinals larger than ω. We define

$$2^\sigma = Card(P(\sigma)).$$

(Here 2 is a function symbol, and has no connection with the constant 2. For the reason for this notation, see Problem 7.) Since $Sm(x, \sigma) \to Sm(P(x), P(\sigma))$, we have

$$Card(x) = \sigma \to Card(P(x)) = 2^\sigma. \tag{29}$$

The following result is known as *Cantor's theorem*:

$$Cd(\sigma) \to \sigma < 2^\sigma. \tag{30}$$

First of all, there is a bijective mapping f from σ to a subset of $P(\sigma)$ which maps τ into $\{\tau\}$. Thus $\sigma \leqslant 2^\sigma$; so it will suffice to show that $2^\sigma = \sigma$ leads to a contradiction. Suppose that g is a bijective mapping from σ to $P(\sigma)$. Let

$$x = [\tau \mid \tau \in \sigma \ \& \ \tau \notin g`\tau].$$

Then $x \in P(\sigma)$; so there is a τ such that $g`\tau = x$. By definition of x,

$$\tau \in x \leftrightarrow \tau \notin g`\tau \leftrightarrow \tau \notin x,$$

a contradiction.

More generally, we have

$$\sigma < 2^\sigma. \tag{31}$$

For if $2^\sigma \leqslant \sigma$, then, using (29), we get

$$2^{Card(\sigma)} = Card(P(\sigma)) = Card(2^\sigma) \leqslant Card(\sigma),$$

contradicting Cantor's theorem. As a consequence of (31), we see that for every ordinal there is a larger cardinal. It follows that for any set x, there is a cardinal greater than every element of x. Moreover, we may suppose that this cardinal is infinite; for if it is finite, we can replace it by the larger cardinal ω.

From the remark just made, we see that we may define a function symbol $\aleph$ by transfinite induction as follows:

$$\aleph(\sigma) = \mu\tau(InfCd(\tau) \ \& \ \tau \notin [\aleph(\rho) \mid \rho < \sigma]).$$

We generally write $\aleph_\sigma$ for $\aleph(\sigma)$. Then $\aleph_\sigma$ is an infinite cardinal. Moreover

$$\sigma < \tau \to \aleph_\sigma < \aleph_\tau. \tag{32}$$

For let $\sigma < \tau$. Then $\aleph_\tau$ is by definition distinct from $\aleph_\sigma$ and not in $[\aleph_\rho \mid \rho < \sigma]$. Since $\aleph_\sigma$ is the smallest infinite cardinal with the latter property, $\aleph_\sigma < \aleph_\tau$.

We now show that every infinite cardinal is $\aleph_\sigma$ for some σ (and hence, by (32), for a unique σ). By (32) and (8) of §9.3, $\neg Set_x \exists\sigma(x = \aleph_\sigma)$. Hence if τ is an infinite cardinal, then there is a σ such that $\aleph_\sigma \notin \tau$. Thus $\tau \leqslant \aleph_\sigma$. If equality holds, we are through. Suppose that $\tau < \aleph_\sigma$. Since $\aleph_\sigma$ is the first infinite cardinal not in $[\aleph_\rho \mid \rho < \sigma]$, we have $\tau \in [\aleph_\rho \mid \rho < \sigma]$, which gives the desired result.

Clearly $\aleph_0 = \omega$; and by (32), $\aleph_{S(\sigma)}$ is the first infinite cardinal larger than $\aleph_\sigma$. From this and Cantor's theorem, we get

$$\aleph_{S(\sigma)} \leqslant 2^{\aleph_\sigma}. \tag{33}$$

The formula

$$\forall\sigma(2^{\aleph_\sigma} = \aleph_{S(\sigma)}),$$

which says that equality always holds in (33), is called the *generalized continuum hypothesis*. The formula

$$2^{\aleph_0} = \aleph_1,$$

which expresses one case of this equality, is called the *continuum hypothesis*.

Although a great amount of effort has been expended on deciding whether or not these two formulas are true statements about sets, the problem is still unsolved. Logical investigations have at least shown why the problem is so difficult: neither of these formulas can be either proved or disproved in *ZFC*.

We shall devote the next few sections to a proof of this fact. First we note a difficulty. If we prove, say, that the continuum hypothesis is not a theorem of *ZFC*, it will follow that *ZFC* is consistent. Hence by the theorem on consistency proofs, such a proof could not be carried out in *ZFC*. It would therefore be very nonconstructive, and perhaps unacceptable to many mathematicians.

We avoid this difficulty by taking the consistency of *ZF* as a hypothesis. This is a reasonable approach; for even if one doubts the consistency of *ZF*, he must admit that this consistency is a very different problem from the independence of the continuum hypothesis. We will find that we can give a finitary proof of the statement: if *ZF* is consistent, then neither the (generalized) continuum hypothesis nor its negation can be proved in *ZFC*.

We shall also prove a related result: if *ZF* is consistent, then neither the axiom of choice nor its negation can be proved in *ZF*. One might ask why this is of special interest, since the axiom of choice is certainly true for sets. One answer is that the axiom of choice is of a special nature. The sets asserted to exist by the existence axioms of *ZF* (such as the power set of a set) can be explicitly described in *ZF*; in fact, they are of the form $[x \mid Q(x, v_1, \ldots, v_n)]$. On the other hand, there is no reason to suppose that for every set v, there is a choice function on v which can be described in this way. Thus it is conceivable that for some notion of set which involves using only collections which can be described, the axioms of *ZF* are true while the axiom of choice is false. Of course, this can only happen if the axiom of choice is not a theorem of *ZF*. Another reason for proving the result is that it reduces the difficult problem of giving some sort of useful consistency proof for *ZFC* to the possibly easier one of giving such a proof for *ZF*.

9.5 INTERPRETATIONS OF SET THEORY

The statement that **A** is not provable in *ZF* is, by the corollary to the reduction theorem for consistency, equivalent to the statement that a certain extension *ZF'* of *ZF* is consistent. Thus the results we want to prove have the form: if *ZF* is consistent, then *ZF'* is consistent. By the corollary to the interpretation theorem, this can be proved by giving an interpretation of *ZF'* in *ZF*. We therefore begin by studying interpretations of *ZF* and related theories.

It is convenient to generalize the notion of an interpretation slightly. An interpretation of $L(ZF)$ in T shall now consist of a unary predicate symbol U_I such that $\vdash \exists x U_I x$, and two binary predicate symbols $\in_I$ and $=_I$. We form $\mathbf{A}_I$ and $\mathbf{A}^{(I)}$ as before, except that we also replace $=$ by $=_I$. An interpretation of $L(ZF)$ will be called an *interpretation of ZF* if the interpretations of the identity axioms, equality axioms, and nonlogical axioms of *ZF* are provable. We can then prove the interpretation theorem and its corollary as before. If $=_I$ is $=$, we obtain the case previously considered. In this case, the interpretations of the identity and equality axioms are always provable.

We recall that if Q is defined by

$$Q(x_1, \ldots, x_n) \leftrightarrow \mathbf{D},$$

then Q_I is defined by

$$Q_I(x_1, \ldots, x_n) \leftrightarrow \mathbf{D}_I.$$

We can adopt this definition even when I is only an interpretation in the extended sense of $L(ZF)$. Then in forming $\mathbf{A}_I$ or $\mathbf{A}^{(I)}$, we simply replace Q by Q_I.

We shall take our interpretations to be in a theory T which is an extension of *ZF*. Moreover, we shall assume that the symbols of T which are not symbols of *ZF* are constants. This means that every subset or replacement axiom of T is an instance of a subset or replacement axiom of *ZF*, and hence can be proved in T. Therefore all the results which we have proved for *ZF* also hold for T. All proofs are to be given in T unless otherwise indicated.

Since an interpretation is a formal analogue of a structure, many notions concerning structures have analogues in the theory of interpretations. We shall consider an analogue of the notion of an isomorphism.

Let I and J be interpretations of $L(ZF)$ in T. An *isomorphism* of I and J is a unary function symbol F in T such that

$$\vdash U_J(y) \leftrightarrow \exists x\big(U_I(x) \ \&\ y = F(x)\big), \tag{1}$$

$$\vdash U_I(x) \ \&\ U_I(y) \rightarrow \big(x \in_I y \leftrightarrow F(x) \in_J F(y)\big), \tag{2}$$

$$\vdash U_I(x) \ \&\ U_I(y) \rightarrow \big(x =_I y \leftrightarrow F(x) =_J F(y)\big). \tag{3}$$

Lemma 1. Let I and J be interpretations of $L(ZF)$ in T; F an isomorphism of I and J; Q a predicate symbol of *ZF*. Then

$$\vdash U_I x_1 \ \&\ \cdots \ \&\ U_I x_n \rightarrow \big(Q_I(x_1, \ldots, x_n) \leftrightarrow Q_J(F(x_1), \ldots, F(x_n))\big).$$

Proof. We use induction on the length of the right side **A** of the definition of Q. If **A** is atomic, then Q is defined by

$$Q(x_1, \dots, x_n) \leftrightarrow x_i \in x_j$$

or

$$Q(x_1, \dots, x_n) \leftrightarrow x_i = x_j,$$

and the result follows from (2) and (3). If **A** is a negation or a disjunction, then Q is defined by

$$Q(x_1, \dots, x_n) \leftrightarrow \neg R(x_1, \dots, x_n)$$

where the result holds for R, or by

$$Q(x_1, \dots, x_n) \leftrightarrow R_1(x_1, \dots, x_n) \vee R_2(x_1, \dots, x_n)$$

where the result holds for R_1 and R_2. In either case, the result for Q follows easily from the induction hypothesis.

If **A** is an instantiation, then Q is defined by

$$Q(x_1, \dots, x_n) \leftrightarrow \exists y R(y, x_1, \dots, x_n)$$

where, by induction hypothesis,

$$U_I x_1 \,\&\, \cdots \,\&\, U_I x_n \,\&\, U_I y \rightarrow \big(R_I(y, x_1, \dots, x_n) \leftrightarrow R_J\big(F(y), F(x_1), \dots, F(x_n)\big)\big).$$

Hence, under the hypothesis that $U_I x_1 \,\&\, \cdots \,\&\, U_I x_n$,

$$\exists y\big(U_I(y) \,\&\, R_I(y, x_1, \dots, x_n)\big) \leftrightarrow \exists y\big(U_I(y) \,\&\, R_J\big(F(y), F(x_1), \dots, F(x_n)\big)\big).$$

The left-hand side of this equivalence is $Q_I(x_1, \dots, x_n)$; so we need only show that the right-hand side is equivalent to $Q_J\big(F(x_1), \dots, F(x_n)\big)$. Omitting the $F(x_1), \dots, F(x_n)$, we have

$$\begin{aligned}\exists y\big(U_I(y) \,\&\, R_J\big(F(y)\big)\big) &\leftrightarrow \exists y\big(U_I(y) \,\&\, \exists z(z = F(y) \,\&\, R_J(z))\big)\\ &\leftrightarrow \exists z\big(\exists y\big(U_I(y) \,\&\, z = F(y)\big) \,\&\, R_J(z)\big)\\ &\leftrightarrow \exists z\big(U_J(z) \,\&\, R_J(z)\big)\end{aligned}$$

by (1). The right-hand side is $Q_J\big(F(x_1), \dots, F(x_n)\big)$.

A proof that all predicate symbols have a certain property by an induction of the type used in the above proof will be called a proof by *induction on predicate symbols.*

If M is a unary predicate symbol of T such that $\vdash \exists x M(x)$, then we can form an interpretation I of $L(ZF)$ by taking U_I to be M, $\in_I$ to be $\in$, and $=_I$ to be $=$. We shall call this the *$\in$-interpretation M*, and write $\mathbf{A}_M$ and $\mathbf{A}^{(M)}$ for $\mathbf{A}_I$ and $\mathbf{A}^{(I)}$. The sets x such that $M(x)$ will be called *M-sets.*

Let A be a constant of T such that $\vdash A \neq 0$. Defining M by $M(x) \leftrightarrow x \in A$, we have $\vdash \exists x M(x)$. The $\in$-interpretation M will be called the *$\in$-interpretation A*; and we will write $\mathbf{A}_A$ and $\mathbf{A}^{(A)}$ for $\mathbf{A}_M$ and $\mathbf{A}^{(M)}$.

An $\in$-interpretation M is *transitive* if

$$\vdash M(x) \ \&\ y \in x \rightarrow M(y).$$

Then for A a constant, the $\in$-interpretation A is transitive iff $\vdash Trans(A)$.

We shall show that an interpretation satisfying suitable conditions is isomorphic to a transitive $\in$-interpretation. Assume that I is an interpretation of $L(ZF)$ such that $=_I$ is $=$ and such that the interpretation of the extensionality axiom holds. Suppose further that there is an ordinal function symbol H such that

$$\vdash Set_x(H(x) \leqslant \sigma)$$

and

$$\vdash y \in_I x \rightarrow H(y) < H(x).$$

We define

$$F(y) = [F(x) \mid U_I(x) \ \&\ x \in_I y]$$

by induction on $H(y)$, and then define

$$M(x) \leftrightarrow \exists y(U_I(y) \ \&\ x = F(y)). \tag{4}$$

Clearly $\vdash \exists x M(x)$; so M is an $\in$-interpretation. Moreover, it is transitive. For suppose that $M(x) \ \&\ z \in x$. Then $x = F(y)$ for some y; so $z \in F(y)$; so $z = F(w)$ for some w such that $U_I(w)$; so $M(z)$.

We show that F is an isomorphism of I and the $\in$-interpretation M. We must prove (1) through (3) when J is M. Now (1) is just (4). For (3), we must prove

$$x = y \leftrightarrow F(x) = F(y)$$

under the hypotheses $U_I(x)$ and $U_I(y)$. We use induction on $H(x)$. The implication from left to right is immediate. By the interpretation under I of the extensionality axiom,

$$\forall z(U_I(z) \rightarrow (z \in_I x \leftrightarrow z \in_I y)) \rightarrow x = y;$$

so it will suffice to show that if $F(x) = F(y)$ and $U_I(z)$, then $z \in_I x \leftrightarrow z \in_I y$. Suppose that $z \in_I x$. Then $F(z) \in F(x) = F(y)$; so $F(z) = F(w)$ for some w such that $U_I(w)$ and $w \in_I y$. By induction hypothesis, $z = w$; so $z \in_I y$. The implication $z \in_I y \rightarrow z \in_I x$ is proved similarly.

For (2), we must prove

$$x \in_I y \leftrightarrow F(x) \in F(y)$$

under the hypotheses $U_I(x)$ and $U_I(y)$. The implication from left to right is clear. If $F(x) \in F(y)$, then $F(x) = F(z)$ for some z such that $U_I(z)$ and $z \in_I y$. By the above, $x = z$; so $x \in_I y$.

We are now going to obtain some sufficient conditions for the interpretations of the axioms of ZF by a transitive $\in$-interpretation to be provable.

Lemma 2. If M is a transitive $\in$-interpretation of $L(ZF)$, then the interpretations of the extensionality axiom and the regularity axiom hold.

Proof. The interpretation of the extensionality axiom is

$$M(x) \rightarrow M(y) \rightarrow \forall z\big(M(z) \rightarrow (z \in x \leftrightarrow z \in y)\big) \rightarrow x = y.$$

Assume the hypotheses. If $z \in x$, then $M(z)$ by the transitivity of x; so $z \in y$. Similarly, $z \in y$ implies $z \in x$; so $x = y$.

The interpretation of the regularity axiom is

$$M(x) \rightarrow \exists y\big(M(y) \;\&\; y \in x\big) \rightarrow \exists y\big(M(y) \;\&\; y \in x \;\&\; \neg\exists z\big(M(z) \;\&\; z \in x \;\&\; z \in y\big)\big).$$

Assume the two hypotheses. Then x has a minimal element y, which is an M-set by the transitivity of M. Thus y satisfies the conclusion.

A function symbol F of T is *M-invariant* if

$$\vdash M(x_1) \;\&\; \cdots \;\&\; M(x_n) \rightarrow M\big(F(x_1, \ldots, x_n)\big).$$

Lemma 3. Let M be a transitive $\in$-interpretation of $L(ZF)$ such that for each predicate symbol Q of $L(ZF)$, the F defined by

$$F(y, v_1, \ldots, v_n) = [x \mid x \in y \;\&\; Q_M(x, v_1, \ldots, v_n)]$$

is M-invariant. Then the interpretation of each subset axiom of ZF holds.

Proof. A subset axiom has the form

$$\exists z\, \forall x\big(x \in z \leftrightarrow x \in y \;\&\; Q(x)\big)$$

(where we have omitted the parameters). The M-interpretation of this is

$$M(y) \rightarrow \exists z\big(M(z) \;\&\; \forall x\big(M(x) \rightarrow \big(x \in z \leftrightarrow x \in y \;\&\; Q_M(x)\big)\big)\big). \tag{5}$$

Given an M-set y, set $z = F(y)$ with F as in the lemma. Then z is an M-set; and for all M-sets x, $x \in z \leftrightarrow x \in y \;\&\; Q_M(x)$. Thus (5) holds.

Lemma 4. Let M be a transitive $\in$-interpretation of $L(ZF)$ such that for each M-invariant function symbol F,

$$\vdash M(w) \;\&\; M(v_1) \;\&\; \cdots \;\&\; M(v_n) \\ \rightarrow \exists z\big(M(z) \;\&\; \forall x\big(x \in w \rightarrow F(x, v_1, \ldots, v_n) \subset z\big)\big).$$

Then the interpretation of each replacement axiom of ZF holds.

Proof. A replacement axiom has the form

$$\forall x\, \exists z\, \forall y\big(y \in z \leftrightarrow Q(x, y)\big) \rightarrow Set_y\, \exists x\big(x \in w \;\&\; Q(x, y)\big)$$

(where we have omitted the parameters). The M-interpretation of this has the hypothesis

$$M(w) \;\&\; \forall x\big(M(x) \rightarrow \exists z\big(M(z) \;\&\; \forall y\big(M(y) \rightarrow \big(y \in z \leftrightarrow Q_M(x, y)\big)\big)\big)\big) \tag{6}$$

and the conclusion

$$\exists z\big(M(z) \;\&\; \forall y\big(M(y) \rightarrow \exists x\big(M(x) \;\&\; x \in w \;\&\; Q_M(x, y)\big) \rightarrow y \in z\big)\big). \tag{7}$$

Define F by

$$F(x) = [y \mid M(y) \,\&\, Q_M(x, y)]$$

if the set on the right exists and is an M-set, and $F(x) = x$ otherwise. Then F is M-invariant. Assume (6). Then for each M-set x there is an M-set z such that

$$\forall y(M(y) \rightarrow (y \in z \leftrightarrow Q_M(x, y))).$$

Now $y \in z \rightarrow M(y)$ by the transitivity of M; so $z = [y \mid M(y) \,\&\, Q_M(x, y)]$. It follows that

$$M(x) \rightarrow F(x) = [y \mid M(y) \,\&\, Q_M(x, y)]. \tag{8}$$

By the hypothesis of the theorem, we can choose an M-set z such that $\forall x(x \in w \rightarrow F(x) \subset z)$. To prove (7), we must show that

$$M(y) \,\&\, M(x) \,\&\, x \in w \,\&\, Q_M(x, y) \rightarrow y \in z.$$

By (8), the hypotheses imply $y \in F(x)$; so they imply $y \in z$.

Lemma 5. Let M be a transitive $\in$-interpretation of $L(ZF)$ such that

$$\vdash M(y) \rightarrow \exists w(M(w) \,\&\, [x \mid M(x) \,\&\, x \subset y] \subset w).$$

Then the interpretation of the power set axiom holds.

Proof. The interpretation of the power set axiom is

$$M(y) \rightarrow \exists w(M(w) \,\&\, \forall x(M(x) \rightarrow \forall z(M(z) \rightarrow z \in x \rightarrow z \in y) \rightarrow x \in w)). \tag{9}$$

If $M(x)$, then $z \in x \rightarrow M(z)$ by transitivity of M. Hence in (9), we may drop the $M(z) \rightarrow$. Then (9) holds by the hypothesis of the lemma.

Lemma 6. If M is a transitive interpretation of $L(ZF)$ such that $M(\omega)$, then the interpretation of the axiom of infinity holds.

Proof. Since $M(\omega)$, we have $M(\sigma)$ for every natural number σ by transitivity of M. It is then easy to prove that ω is an M-set satisfying the conditions required by the interpretation of the axiom of infinity.

We now suppose that M is a fixed transitive $\in$-interpretation of ZF. We can then form F_M for each function symbol F of ZF. (This requires choosing a constant in T; we can choose the constant 0.) Now let $\mathbf{A}$ be a formula containing defined nonlogical symbols, and let $\mathbf{A}^*$ be its translation into the language of ZF. Then $\vdash \mathbf{A} \leftrightarrow \mathbf{A}^*$ in ZF; so

$$\vdash M(\mathbf{x}_1) \,\&\, \cdots \,\&\, M(\mathbf{x}_n) \rightarrow (\mathbf{A}_M \leftrightarrow \mathbf{A}^*_M)$$

(where $\mathbf{x}_1, \ldots, \mathbf{x}_n$ are the variables free in $\mathbf{A}$). It follows that $\vdash \mathbf{A}^{(M)}$ iff $\vdash \mathbf{A}^{*(M)}$. This shows that in forming $\mathbf{A}^{(M)}$, we do not need to eliminate the defined symbols, but can replace Q and F by Q_M and F_M.

Now $\exists y(y = F(x_1, \ldots, x_n))$ is provable in *ZF*. The *M*-interpretation of this is

$$M(x_1) \,\&\, \cdots \,\&\, M(x_n) \rightarrow \exists y(M(y) \,\&\, y = F_M(x_1, \ldots, x_n)).$$

This shows that F_M is *M*-invariant.

If *F* is defined by

$$F(v_1, \ldots, v_n) = [x \mid Q(x, v_1, \ldots, v_n)],$$

then

$$F_M(v_1, \ldots, v_n) = [x \mid M(x) \,\&\, Q_M(x, v_1, \ldots, v_n)] \tag{10}$$

for any *M*-sets $v_1, \ldots, v_n$. For the interpretation of the defining axiom of *F* states that for such $v_1, \ldots, v_n$,

$$\forall x(M(x) \rightarrow (x \in F_M(v_1, \ldots, v_n) \leftrightarrow Q_M(x, v_1, \ldots, v_n))).$$

Since F_M is *M*-invariant, $F_M(v_1, \ldots, v_n)$ is an *M*-set; so

$$x \in F_M(v_1, \ldots, v_n) \rightarrow M(x)$$

by the transitivity of *M*. Combining these two results, we get (10).

If *F* is defined by

$$F(v_1, \ldots, v_n) = [G(x, v_1, \ldots, v_n) \mid Q(x, v_1, \ldots, v_n)],$$

then

$$F_M(v_1, \ldots, v_n) = [G_M(x, v_1, \ldots, v_n) \mid M(x) \,\&\, Q_M(x, v_1, \ldots, v_n)] \tag{11}$$

for any *M*-sets $v_1, \ldots, v_n$. For by (10),

$$F_M(v_1, \ldots, v_n)$$
$$= [y \mid M(y) \,\&\, \exists x(M(x) \,\&\, y = G_M(x, v_1, \ldots, v_n) \,\&\, Q_M(x, v_1, \ldots, v_n))].$$

Using the *M*-invariance of G_M, we get (11).

We say that a predicate symbol *Q* of *ZF* is *absolute for M* if

$$\vdash M(x_1) \,\&\, \cdots \,\&\, M(x_n) \rightarrow (Q(x_1, \ldots, x_n) \leftrightarrow Q_M(x_1, \ldots, x_n)). \tag{12}$$

We say that a function symbol *F* of *ZF* is *absolute for M* if

$$\vdash M(x_1) \,\&\, \cdots \,\&\, M(x_n) \rightarrow F(x_1, \ldots, x_n) = F_M(x_1, \ldots, x_n).$$

When *M* is fixed, we say simply *absolute* for *absolute for M*.

If *Q* is absolute, we may replace Q_M by *Q* whenever the hypotheses ensure that the arguments to Q_M are *M*-sets; and similarly for function symbols. This is of value in studying the interpretations of axioms.

If *F* is absolute, then *F* is *M*-invariant; this follows easily from the fact that F_M is *M*-invariant.

We now consider methods for proving that nonlogical symbols are absolute. Clearly $\in$ and $=$ are absolute. If *Q* is defined by

$$Q(x_1, \ldots, x_n) \leftrightarrow \underline{\qquad} \tag{13}$$

where ____ consists of $x_1, \ldots, x_n$ and absolute nonlogical symbols, then Q is absolute. For the interpretation of (13) says that

$$Q_M(x_1, \ldots, x_n) \leftrightarrow (\_\_\_\_)_M$$

for $x_1, \ldots, x_n$ M-sets. Now the terms in $(\_\_\_\_)_M$ are built from the x_i and function symbols G_M. Since the latter are M-invariant, the hypothesis that the x_i are M-sets implies that these terms represent M-sets. Hence by the above remarks, we may replace each G_M or R_M in $(\_\_\_\_)_M$ by G or R. We thus get

$$Q_M(x_1, \ldots, x_n) \leftrightarrow \_\_\_\_,$$

which with (13) gives (12).

Next we show that if F is defined by

$$F(x_1, \ldots, x_n) = y \leftrightarrow Q(y, x_1, \ldots, x_n) \tag{14}$$

where Q is absolute, then F is absolute. Assume that $x_1, \ldots, x_n$ are M-sets. Putting $F_M(x_1, \ldots, x_n)$ for y in (14),

$$F(x_1, \ldots, x_n) = F_M(x_1, \ldots, x_n) \leftrightarrow Q\big(F_M(x_1, \ldots, x_n), x_1, \ldots, x_n\big).$$

Thus we need only prove the right-hand side of this equivalence. By the absoluteness of Q and the M-invariance of F_M, this is equivalent to

$$Q_M\big(F_M(x_1, \ldots, x_n), x_1, \ldots, x_n\big). \tag{15}$$

Now by (14), $Q\big(F(x_1, \ldots, x_n), x_1, \ldots, x_n\big)$ is provable in ZF. Taking its interpretation, we get (15).

A particular case is when F is defined by

$$F(x_1, \ldots, x_n) = \_\_\_\_$$

where ____ consists of $x_1, \ldots, x_n$ and absolute function symbols. For this definition is equivalent to (14) where Q is defined by

$$Q(y, x_1, \ldots, x_n) \leftrightarrow y = \_\_\_\_;$$

and we have seen above that such a Q is absolute.

If Q is defined by

$$Q(x_1, \ldots, x_n) \leftrightarrow \neg R(x_1, \ldots, x_n)$$

where R is absolute, or by

$$Q(x_1, \ldots, x_n) \leftrightarrow R_1(x_1, \ldots, x_n) \vee R_2(x_1, \ldots, x_n)$$

where R_1 and R_2 are absolute, then Q is absolute; this is easily proved. From this we obtain a similar result with $\rightarrow$, &, or $\leftrightarrow$ in place of $\vee$.

We say that Q is *complete for M* (or simply *complete*) if

$$\vdash M(y_1) \;\&\; \cdots \;\&\; M(y_n) \;\&\; Q(x, y_1, \ldots, y_n) \rightarrow M(x).$$

For example, $=$ is clearly complete, and $\in$ is complete by the transitivity of M. We sometimes say that $\_\_x\_\_$ is *complete in* x; this means that the Q defined by

$$Q(x, v_1, \dots, v_n) \leftrightarrow \_\_x\_\_$$

is complete. Thus by the completeness of $\in$, we see that

$$x \in y \;\&\; \_\_x\_\_$$

and

$$\neg(x \in y \rightarrow \_\_x\_\_)$$

are complete in x.

If Q is defined by

$$Q(x_1, \dots, x_n) \leftrightarrow \exists y R(y, x_1, \dots, x_n) \tag{16}$$

where R is absolute and complete, then Q is absolute. For let $x_1, \dots, x_n$ be M-sets. The interpretation of (16) then gives

$$Q_M(x_1, \dots, x_n) \leftrightarrow \exists y\big(M(y) \;\&\; R_M(y, x_1, \dots, x_n)\big).$$

By the absoluteness of R, we may replace R_M by R; and then by the completeness of R, we may drop the $M(y)\&$. Combining the result with (16), we get (12). It follows that if Q is defined by

$$Q(x_1, \dots, x_n) \leftrightarrow \forall y R(y, x_1, \dots, x_n) \tag{17}$$

where R is absolute and $\neg R$ is complete, then Q is absolute.

If F is defined by

$$F(v_1, \dots, v_n) = [x \mid Q(x, v_1, \dots, v_n)]$$

where Q is absolute and complete, then F is absolute. For if $v_1, \dots, v_n$ are M-sets, then (10) holds. As above, we can first replace Q_M by Q and then drop the $M(x)\&$. This gives

$$F_M(v_1, \dots, v_n) = [x \mid Q(x, v_1, \dots, v_n)]$$

which implies that F is absolute. A similar proof, using (11), shows that if F is defined by

$$F(v_1, \dots, v_n) = [G(x, v_1, \dots, v_n) \mid Q(x, v_1, \dots, v_n)]$$

where G and Q are absolute and Q is complete, then F is absolute.

We shall now use these results to show that many of the nonlogical symbols which we have defined are absolute. For each nonlogical symbol we shall give a defining axiom. We leave most of the task of verifying that the above results apply to the reader, only pointing out some less obvious steps. Sometimes the defining axiom which we give is slightly different from (but clearly equivalent to) the defining axiom which we originally used.

A. $x \subset y \leftrightarrow \forall z(z \in x \rightarrow z \in y)$.

The application of the rules in this case makes use of the fact that $\neg(z \in x \rightarrow z \in y)$ is complete in z.

B. $0 = [x \mid x \neq x]$.

It is clear that $x \neq x$ is complete in x.

C. $\{x, y\} = [z \mid z = x \lor z = y]$.

D. $\{x\} = \{x, x\}$.

E. $Un(x) = [y \mid \exists z(z \in x \ \&\ y \in z)]$.

Clearly $z \in x \ \&\ y \in z$ is complete in z. We must also show that $\exists z(z \in x \ \&\ y \in z)$ is complete in y; that is,

$$M(x) \ \&\ \exists z(z \in x \ \&\ y \in z) \rightarrow M(y).$$

This follows from the transitivity of M.

F. $x \cup y = Un(\{x, y\})$.

G. $x \cap y = [z \mid z \in x \ \&\ z \in y]$.

H. $x - y = [z \mid z \in x \ \&\ z \notin y]$.

I. $\langle x, y \rangle = \{\{x\}, \{x, y\}\}$.

J. $\exists y(x = \langle \pi_1(x), y \rangle) \lor (\neg \exists y \, \exists z(x = \langle y, z \rangle) \ \&\ \pi_1(x) = 0)$.

The necessary completeness results for this definition follow from

$$M(\langle a, b \rangle) \rightarrow M(a) \ \&\ M(b). \tag{18}$$

This follows from the transitivity of M, since $a, b \in \{a, b\}$ and $\{a, b\} \in \langle a, b \rangle$. We treat π_2 similarly.

K. $x \times y = [z \mid \exists a \, \exists b(a \in x \ \&\ b \in y \ \&\ z = \langle a, b \rangle)]$.

For this, we need to prove

$$\exists a \, \exists b(a \in x \ \&\ b \in y \ \&\ z = \langle a, b \rangle) \tag{19}$$

is complete in z. Assume that (19) holds with x and y M-sets. Then a and b are M-sets. Since $\langle \ \rangle$ is absolute, it is M-invariant; so $z = \langle a, b \rangle$ is an M-set.

L. $\langle x_1, \ldots, x_n \rangle = \langle x_1, \langle x_2, \ldots, x_n \rangle \rangle$.

We can treat π_i^n like π_1.

M. $Do(x) = [\pi_2(y) \mid y \in x]$.

N. $Ra(x) = [\pi_1(y) \mid y \in x]$.

O. $Func(x) \leftrightarrow x \subset Ra(x) \times Do(x)$
$\qquad \ \&\ \forall a \, \forall b(a \in x \ \&\ b \in x \ \&\ \pi_2(a) = \pi_2(b) \rightarrow a = b)$.

P. $IFunc(x) \leftrightarrow Func(x) \ \&\ \forall a \, \forall b(a \in x \ \&\ b \in x \ \&\ \pi_1(a) = \pi_1(b) \rightarrow a = b)$.

Q. $(Func(x) \mathbin{\&} y \in Do(x) \mathbin{\&} \langle x`y, y\rangle \in x)$
$\vee (\neg(Func(x) \mathbin{\&} y \in Do(x)) \mathbin{\&} x`y = 0)$.

R. $Trans(x) \leftrightarrow \forall y(y \in x \rightarrow y \subset x)$.

S. $Ord(x) \leftrightarrow Trans(x) \mathbin{\&} \forall y(y \in x \rightarrow Trans(v))$.

T. $\sigma < \tau \leftrightarrow \sigma \in \tau$.

U. $\sigma \leqslant \tau \leftrightarrow \sigma < \tau \vee \sigma = \tau$.

If F is defined by

$$F(x_1, \dots, x_n) = \mu\sigma Q(\sigma, x_1, \dots, x_n),$$

where Q is absolute, then F is absolute. For this defining axiom is

$$F(x_1, \dots, x_n) = y \leftrightarrow R(y, x_1, \dots, x_n),$$

where R is defined by

$$R(y, x_1, \dots, y_n) \leftrightarrow Ord(y) \mathbin{\&} Q(y, x_1, \dots, x_n) \mathbin{\&} \forall z(z \in y \rightarrow \neg Q(z, x_1, \dots, x_n))$$

and hence is absolute.

V. $S(\sigma) = \sigma \cup \{\sigma\}$.

W. $\omega = \mu\sigma(\sigma \neq 0 \mathbin{\&} \neg\exists x(Ord(x) \mathbin{\&} \sigma = S(x)))$.

The completeness of $Ord(x) \mathbin{\&} \sigma = S(x)$ in x follows from the implication $\sigma = S(x) \rightarrow x \in \sigma$.

X. $1 = S(0)$, $2 = S(1)$, ...

We shall now prove that if F is defined by transfinite induction in terms of an absolute function symbol, then F is absolute. As usual, we shall omit the parameters. Thus F is defined by

$$F(\sigma) = G(\sigma, [\langle F(\tau), \tau\rangle \mid \tau < \sigma]),$$

where G is absolute. Set

$$H(\sigma) = [\langle F(\tau), \tau\rangle \mid \tau < \sigma].$$

Then F can be defined by $F(\sigma) = H(S(\sigma))`\sigma$; so it will suffice to show that H is absolute. We can define H by

$$H(x) = y \leftrightarrow Q(y, x),$$

where Q is defined by

$$\begin{aligned} Q(y, x) \leftrightarrow (&Ord(x) \mathbin{\&} Func(y) \mathbin{\&} Do(y) = x \\ &\mathbin{\&} \forall z(z \in x \rightarrow y`z = G(z, [\langle y`w, w\rangle \mid w \in z]))) \\ &\vee (\neg Ord(x) \mathbin{\&} y = 0). \end{aligned}$$

Then Q is absolute; so H is absolute.

Y. $Max(x) = \pi_1(x) \cup \pi_2(x)$.

Z. $MP(x) = \langle \pi_1(MP(x)), \pi_2(MP(x)) \rangle$
$\& \; \pi_1(MP(x)) = \mu\tau(\exists\rho(\rho \in \mu\sigma(\neg(\sigma \times \sigma \subset x)) \; \& \; \langle \tau, \rho \rangle \notin x))$
$\& \; \pi_2(MP(x)) = \mu\rho(\langle \pi_1(MP(x)), \rho \rangle \notin x)$.

A′. $K(\sigma) = MP(Ra([\langle K(\tau), \tau \rangle \mid \tau < \sigma]))$.

We say that M is *supertransitive* if it is transitive and $\subset$ is complete for M, that is,

$$\vdash M(y) \;\&\; x \subset y \rightarrow M(x).$$

We shall now assume that M is supertransitive and prove that certain further nonlogical symbols are absolute.

B′. $P(x) = [y \mid y \subset x]$.

C′. $Sm(x, y) \leftrightarrow \exists f(IFunc(f) \;\&\; x = Do(f) \;\&\; y = Ra(f))$.

For completeness, we must show that

$$M(x) \;\&\; M(y) \;\&\; IFunc(f) \;\&\; x = Do(f) \;\&\; y = Ra(f) \rightarrow M(f).$$

Since $\times$ is absolute and hence M-invariant, the hypotheses imply that $M(y \times x)$. Since they also imply that $f \subset y \times x$, we have $M(f)$ by the supertransitivity of M.

D′. $CF(f, x) \leftrightarrow Func(f) \;\&\; Do(f) = P(x) - \{0\}$
$\& \; \forall y(y \in Do(f) \rightarrow f`y \in y)$.

We also note that CF is complete. For this, we observe that

$$CF(f, x) \rightarrow f \subset x \times P(x)$$

and then proceed as above.

The remaining nonlogical symbols are concerned with cardinals. We shall therefore suppose that M is a supertransitive interpretation of ZFC.

E′. $Card(x) = \mu\sigma(Sm(\sigma, x))$.

F′. $Cd(x) \leftrightarrow x = Card(x)$.

G′. $InfCd(x) \leftrightarrow Cd(x) \;\&\; x \notin \omega$.

H′. $2^\sigma = Card(P(\sigma))$.

I′. $\aleph(\sigma) = \mu\tau(InfCd(\tau) \;\&\; \tau \notin Ra([\langle \aleph(\rho), \rho \rangle \mid \rho < \sigma]))$.

9.6 CONSTRUCTIBLE SETS

We are going to construct a transitive $\in$-interpretation L of ZF in ZF. We shall do this by assigning a set to each ordinal; the sets so assigned will be the L-sets. The members of each L-set will be earlier L-sets (i.e., L-sets assigned to smaller ordinals); this will ensure that L is transitive. As a result of this, the interpretations of the extensionality axiom and the regularity axiom will hold.

Since the remaining axioms are existence axioms, the remaining problem is to ensure that there are sufficiently many L-sets. In particular, we want to insure that there are sufficiently large L-sets. To achieve this, we shall take the L-set assigned to certain ordinals σ to be the set of all earlier L-sets. Moreover, there will be arbitrarily large σ for which this is done. This will imply that every set of L-sets is included in an L-set.

Next we must make sure that there are sufficiently many small L-sets; i.e., that if a is an L-set, then sufficiently many subsets of a are L-sets. The subset axioms require that for each Q, the set of x in a such that $Q_L(x)$ be an L-set (and similarly with parameters). To achieve this, we repeatedly apply certain operations to the L-sets already obtained and make the resulting sets into L-sets. The reasons for the exact choice of these operations will appear as the proof proceeds; for the moment, we only give some idea of where these operations come from.

We intend to prove that the required sets are L-sets by induction on predicate symbols. Now if Q is defined in terms of R, then R may have more arguments than Q. Thus it is inconvenient to try to deal only with unary predicate symbols. We shall therefore prove that if a is an L-set, then the set of $\langle x_1, \ldots, x_n \rangle$ such that $x_1, \ldots, x_n \in a$ and $Q_L(x_1, \ldots, x_n)$ is an L-set. This requires us to be able to form ordered n-tuples. Since these are formed by repeatedly taking unordered pairs, one of our operations will be taking unordered pairs.

Let us consider the most difficult step in the induction, viz., when Q is defined by $Q(x) \leftrightarrow \exists y R(y, x)$. (We take Q unary here for simplicity.) Now

$$Q_L(x) \leftrightarrow \exists y(L(y) \,\&\, R_L(y, x))$$

for x an L-set. Thus each x such that $Q_L(x)$ is the second element in an ordered pair $\langle y, x \rangle$ such that $R_L(y, x)$. This suggests that $[x \mid x \in a \,\&\, Q_L(x)]$ should be obtained as the domain of a set $[\langle y, x \rangle \mid y, x \in b \,\&\, R_L(y, x)]$. There is no problem in choosing the operations so that the domain of an L-set is an L-set; the only problem is to find the set b. We want to know that b includes a, and that for each x in a, if there is an L-set y such that $R_L(y, x)$, then there is such an L-set y in b. Using the replacement axioms, we can easily produce a set b of L-sets with this property; and we can then enlarge b to an L-set.

We define the binary function symbols $\mathfrak{F}_i$, $i = 1, \ldots, 9$, by

$$\begin{aligned}
\mathfrak{F}_1(x, y) &= [\langle a, b \rangle \mid \langle a, b \rangle \in x \,\&\, a \in b],\\
\mathfrak{F}_2(x, y) &= [\langle a, a \rangle \mid \langle a, a \rangle \in x],\\
\mathfrak{F}_3(x, y) &= [\langle a, b \rangle \mid \langle a, b \rangle \in x \,\&\, a \in y],\\
\mathfrak{F}_4(x, y) &= [\langle a, b \rangle \mid \langle a, b \rangle \in x \,\&\, b \in y],\\
\mathfrak{F}_5(x, y) &= [\langle a, b \rangle \mid \langle a, b \rangle \in x \,\&\, \langle b, a \rangle \in y],\\
\mathfrak{F}_6(x, y) &= [\langle a, b, c \rangle \mid \langle a, b, c \rangle \in x \,\&\, \langle b, a, c \rangle \in y],\\
\mathfrak{F}_7(x, y) &= [\langle a, b, c \rangle \mid \langle a, b, c \rangle \in x \,\&\, \langle c, a, b \rangle \in y],\\
\mathfrak{F}_8(x, y) &= x - y,\\
\mathfrak{F}_9(x, y) &= x \cap Do(y).
\end{aligned}$$

(The argument y is inserted in $\mathfrak{F}_1$ and $\mathfrak{F}_2$ so that all the operations will be binary.) Note that $\mathfrak{F}_i(x, y) \subset x$ for $i = 1, \ldots, 9$.

We also define

$$J_0(\sigma) = \pi_1(K(\sigma)),$$
$$J_1(\sigma) = \pi_1(K(\pi_2(K(\sigma)))),$$
$$J_2(\sigma) = \pi_2(K(\pi_2(K(\sigma)))).$$

Then for any σ_0, σ_1, and σ_2, there is a σ such that $J_i(\sigma) = \sigma_i$ for $i = 0, 1, 2$. By (21) and (22) of §9.3,

$$J_0(\sigma) \neq 0 \rightarrow J_1(\sigma) < \sigma \ \&\ J_2(\sigma) < \sigma. \tag{1}$$

We now define a function symbol C by transfinite induction as follows:

$$\begin{aligned} C(\sigma) &= [C(\tau) \mid \tau < \sigma] && \text{if } J_0(\sigma) = 0, \\ &= \mathfrak{F}_i(C(J_1(\sigma)), C(J_2(\sigma))) && \text{if } J_0(\sigma) = i,\ i = 1, \ldots, 9, \\ &= \{C(J_1(\sigma)), C(J_2(\sigma))\} && \text{if } 9 < J_0(\sigma). \end{aligned}$$

We need (1) to see that this is a valid definition by transfinite induction. We also set

$$C^*(\sigma) = [C(\tau) \mid \tau < \sigma].$$

We define

$$L(x) \leftrightarrow \exists\sigma(x = C(\sigma)).$$

A set is *constructible* if it is an L-set. If x is constructible, then the first σ such that $x = C(\sigma)$ is called the *order* of x, and is designated by $Od(x)$.

We shall now prove that the constructible sets have the required properties.

Lemma 1. If x is constructible and $y \in x$, then y is constructible and $Od(y) < Od(x)$.

Proof. We use transfinite induction on $\sigma = Od(x)$. If $J_0(\sigma) = 0$, then

$$x = C(\sigma) = C^*(\sigma).$$

Hence $y = C(\tau)$ for some $\tau < \sigma$; so $Od(y) \leqslant \tau < \sigma$. If $J_0(\sigma) = i, i = 1, \ldots, 9$, then $x = \mathfrak{F}_i(C(J_1(\sigma)), C(J_2(\sigma))) \subset C(J_1(\sigma))$. Since $J_1(\sigma) < \sigma$ by (1), the induction hypothesis shows that y is constructible and that $Od(y) < J_1(\sigma) < \sigma$. If $9 < J_0(\sigma)$, then $y = C(J_i(\sigma))$ with $i = 1$ or $i = 2$. In either case, y is constructible and $Od(y) \leqslant J_i(\sigma) < \sigma$.

Lemma 2. If every member of x is constructible, then x is included in a constructible set.

Proof. Let σ be an ordinal larger than every ordinal in $[Od(y) \mid y \in x]$, and choose τ so that $K(\tau) = \langle 0, \sigma \rangle$. By (21) of §9.3, $\sigma \leqslant \tau$. It follows that

$$x \subset C^*(\sigma) \subset C^*(\tau) = C(\tau).$$

We have

$$L(x) \; \& \; L(y) \rightarrow L(\mathfrak{F}_i(x, y)), \qquad i = 1, \ldots, 9. \tag{2}$$

For let $x = C(\sigma)$, $y = C(\tau)$, and choose ρ so that $J_0(\rho) = i$, $J_1(\rho) = \sigma$, $J_2(\rho) = \tau$. Then $C(\rho) = \mathfrak{F}_i(x, y)$. Similarly,

$$L(x) \; \& \; L(y) \rightarrow L(\{x, y\}), \tag{3}$$

from which it follows that

$$L(x_1) \; \& \cdots \& \; L(x_n) \rightarrow L(\langle x_1, \ldots, x_n \rangle). \tag{4}$$

Next,

$$L(x) \; \& \; L(y) \rightarrow L(x \times y). \tag{5}$$

For by (4), Lemma 1, and Lemma 2, there is a constructible set z such that $x \times y \subset z$. Then

$$x \times y = \mathfrak{F}_4(\mathfrak{F}_3(z, x), y);$$

so $x \times y$ is constructible by (2).

We define

$$x \times_1 y = [\langle a, b, c \rangle \mid b \in x \; \& \; \langle a, c \rangle \in y],$$
$$x \times_2 y = [\langle a, b, c \rangle \mid c \in x \; \& \; \langle a, b \rangle \in y].$$

Then

$$L(x) \; \& \; L(y) \rightarrow L(x \times_1 y) \; \& \; L(x \times_2 y). \tag{6}$$

For by (4), Lemma 1, and Lemma 2, there is a constructible z such that $x \times_1 y \subset z$; and then $x \times_1 y = \mathfrak{F}_6(z, x \times y)$. The proof for $x \times_2 y$ is similar, but uses $\mathfrak{F}_7$ instead of $\mathfrak{F}_6$.

We set

$$Cv(x) = [\langle a, b \rangle \mid \langle b, a \rangle \in x].$$

Then

$$L(x) \rightarrow L(Cv(x)). \tag{7}$$

For if we choose a constructible z such that $Cv(x) \subset z$, then $Cv(x) = \mathfrak{F}_5(z, x)$.

From (2),

$$L(x) \; \& \; L(y) \rightarrow L(x - y). \tag{8}$$

From this we get

$$L(x) \; \& \; L(y) \rightarrow L(x \cup y) \; \& \; L(x \cap y). \tag{9}$$

For by Lemma 2, we can choose a constructible z such that $x \cup y \subset z$; and then

$$x \cup y = z - ((z - x) - y) \quad \text{and} \quad x \cap y = x - (x - y).$$

We also have

$$L(x) \rightarrow L(Do(x)). \tag{10}$$

We first note that if $a \in Do(x)$, then there is a b such that $\langle a, b \rangle \in x$. Since $a \in \{a\} \; \& \; \{a\} \in \langle a, b \rangle$, it follows by Lemma 1 that $L(a)$. Hence by Lemma 2, there is a constructible z such that $Do(x) \subset z$; and $Do(x) = \mathfrak{F}_9(z, x)$.

Lemma 3. Let $1 \leqslant i \leqslant n$, $1 \leqslant j \leqslant n$. If a and b are constructible, then there is a constructible set c such that for all $x_1, \ldots, x_n$ in a,

$$\langle x_i, x_j\rangle \in b \leftrightarrow \langle x_1, \ldots, x_n\rangle \in c.$$

Proof. We first assume that $i < j$ and give the proof by induction on n. First let $i > 1$. By the induction hypothesis, there is a constructible d such that for $x_2, \ldots, x_n$ in a,

$$\langle x_i, x_j\rangle \in b \leftrightarrow \langle x_2, \ldots, x_n\rangle \in d.$$

Then for $x_1, \ldots, x_n$ in a,

$$\langle x_i, x_j\rangle \in b \leftrightarrow \langle x_1, \ldots, x_n\rangle \in a \times d.$$

We therefore take $c = a \times d$, using (5) to show that c is constructible.

Now let $i = 1$ and $j > 2$. By the induction hypothesis, there is a constructible d such that for $x_1, x_3, \ldots, x_n$ in a,

$$\langle x_i, x_j\rangle \in b \leftrightarrow \langle x_1, x_3, \ldots, x_n\rangle \in d.$$

Then for $x_1, \ldots, x_n$ in a,

$$\langle x_i, x_j\rangle \in b \leftrightarrow \langle x_1, \ldots, x_n\rangle \in a \times_1 d.$$

We take $c = a \times_1 d$, using (6).

Finally, let $i = 1, j = 2$. Then for $x_1, \ldots, x_n$ in a,

$$\langle x_i, x_j\rangle \in b \leftrightarrow \langle x_1, \ldots, x_n\rangle \in a^{n-2} \times_2 b$$

(where $a^{n-2} = a \times \cdots \times a$ with $n - 2$ factors). We take $c = a^{n-2} \times_2 b$, using (5) and (6). (Of course, if $n = 2$, we take $c = b$.)

Now suppose that $j < i$. By (7) and the above, there is a constructible set c such that for $x_1, \ldots, x_n$ in a,

$$\langle x_j, x_i\rangle \in Cv(b) \leftrightarrow \langle x_1, \ldots, x_n\rangle \in c.$$

Then

$$\langle x_i, x_j\rangle \in b \leftrightarrow \langle x_1, \ldots, x_n\rangle \in c.$$

Finally, let $i = j$. Then for $x_1, \ldots, x_n$ in a,

$$\begin{aligned}\langle x_i, x_j\rangle \in b &\leftrightarrow x_i \in Do(\mathfrak{F}_2(b, b))\\ &\leftrightarrow \langle x_1, \ldots, x_n\rangle \in a^{i-1} \times Do(\mathfrak{F}_2(b, b)) \times a^{n-i}.\end{aligned}$$

Lemma 4. Let Q be a predicate symbol of ZF. For every constructible set a, there is a constructible set b such that for $x_1, \ldots, x_n \in a$,

$$\langle x_1, \ldots, x_n\rangle \in b \leftrightarrow Q_L(x_1, \ldots, x_n).$$

Proof. We use induction on predicate symbols. Suppose that

$$Q(x_1, \ldots, x_n) \leftrightarrow x_i \in x_j$$

or

$$Q(x_1, \ldots, x_n) \leftrightarrow x_i = x_j.$$

Then $Q_L(x_1, \ldots, x_n) \leftrightarrow Q(x_1, \ldots, x_n)$. In view of Lemma 3, it will suffice to find constructible sets c and d such that

$$\langle x, y\rangle \in c \leftrightarrow x \in y$$

and

$$\langle x, y\rangle \in d \leftrightarrow x = y$$

for x and y in a. We take $c = \mathfrak{F}_1(a \times a, a)$, $d = \mathfrak{F}_2(a \times a, a)$.

Suppose that Q is defined by

$$Q(x_1, \ldots, x_n) \leftrightarrow \neg R(x_1, \ldots, x_n)$$

so that

$$Q_L(x_1, \ldots, x_n) \leftrightarrow \neg R_L(x_1, \ldots, x_n).$$

By induction hypothesis, there is a constructible set c such that for $x_1, \ldots, x_n$ in a,

$$\langle x_1, \ldots, x_n\rangle \in c \leftrightarrow R_L(x_1, \ldots, x_n).$$

We take $b = a^n - c$.

If Q is defined by

$$Q(x_1, \ldots, x_n) \leftrightarrow R_1(x_1, \ldots, x_n) \vee R_2(x_1, \ldots, x_n),$$

then we take sets c_1 and c_2 corresponding to R_1 and R_2 and set $b = c_1 \cup c_2$.

Now suppose that Q is defined by

$$Q(x_1, \ldots, x_n) \leftrightarrow \exists y R(y, x_1, \ldots, x_n)$$

so that

$$Q_L(x_1, \ldots, x_n) \leftrightarrow \exists y\big(L(y) \,\&\, R_L(y, x_1, \ldots, x_n)\big).$$

Define a function symbol F as follows. If $\exists y\big(L(y) \,\&\, R_L(y, x_1, \ldots, x_n)\big)$, then $F(x_1, \ldots, x_n)$ is the constructible set y of smallest order such that

$$R_L(y, x_1, \ldots, x_n);$$

otherwise, $F(x_1, \ldots, x_n) = C(0)$. Let

$$c = [F(x_1, \ldots, x_n) \mid x_1 \in a \,\&\, \cdots \,\&\, x_n \in a].$$

By Lemma 2, there is a constructible set d such that $a \cup c \subset d$. By the induction hypothesis, there is a constructible set e such that for $y, x_1, \ldots, x_n$ in d,

$$\langle y, x_1, \ldots, x_n\rangle \in e \leftrightarrow R_L(y, x_1, \ldots, x_n).$$

Replacing e by $e \cap d^{n+1}$, we may suppose that $e \subset d^{n+1}$. Then for $x_1, \ldots, x_n$ in a,

$$\begin{aligned}\langle x_1, \ldots, x_n\rangle \in Do(e) &\leftrightarrow \exists y(y \in d \,\&\, \langle y, x_1, \ldots, x_n\rangle \in e)\\ &\leftrightarrow \exists y\big(y \in d \,\&\, R_L(y, x_1, \ldots, x_n)\big)\\ &\leftrightarrow \exists y\big(L(y) \,\&\, R_L(y, x_1, \ldots, x_n)\big)\\ &\leftrightarrow Q_L(x_1, \ldots, x_n).\end{aligned}$$

We may therefore take $b = Do(e)$.

We can now prove that L is an interpretation of ZF in ZF. Clearly $\exists x L(x)$. By Lemma 1, L is transitive; so by Lemma 2 of §9.5, the interpretations of the extensionality axiom and the regularity axiom hold.

To prove the interpretations of the subset axioms, it suffices, by Lemma 3 of §9.5, to show that if $y, v_1, \ldots, v_n$ are constructible, then

$$[x \mid x \in y \ \& \ Q_L(x, v_1, \ldots, v_n)] \tag{11}$$

is constructible. By Lemma 2, there is a constructible set which includes y and contains $v_1, \ldots, v_n$. Hence by Lemma 4, there is a constructible set z such that for $x \in y$,

$$\langle x, v_1, \ldots, v_n \rangle \in z \leftrightarrow Q_L(x, v_1, \ldots, v_n).$$

Replacing z by $z \cap (y \times \{v_1\} \times \cdots \times \{v_n\})$, we have for all x,

$$x \in Ra(z) \leftrightarrow x \in y \ \& \ Q_L(x, v_1, \ldots, v_n).$$

Thus the set (11) is $Ra(z) = Do(Cv(z))$ and hence is constructible.

To prove the interpretations of the replacement axioms, it suffices, by Lemma 4 of §9.5, to show that if F is L-invariant and $w, v_1, \ldots, v_n$ are constructible, then there is a constructible set z such that $F(x, v_1, \ldots, v_n) \subset z$ for all $x \in w$. Let $u = Un([F(x, v_1, \ldots, v_n) \mid x \in w])$. By Lemma 1, every set in u is constructible. Hence by Lemma 2, there is a constructible set z such that $u \subset z$. Then for $x \in w$,

$$F(x, v_1, \ldots, v_n) \subset u \subset z.$$

To prove the interpretation of the power set axiom, it suffices, by Lemma 5 of §9.5, to show that if y is constructible, then there is a constructible set which includes $[x \mid L(x) \ \& \ x \subset y]$. This follows from Lemma 2.

To prove the interpretation of the axiom of infinity, it suffices, by Lemma 6 of §9.5, to show that ω is constructible. We shall actually prove that every ordinal is constructible.

We first recall that we proved

$$\exists x(Ord(x) \ \& \ x \notin y)$$

in ZF. The proof did not require the axiom of infinity; so its interpretation

$$L(y) \rightarrow \exists x(L(x) \ \& \ Ord_L(x) \ \& \ x \notin y)$$

is provable. Now Ord is absolute for L (since our proof of the absoluteness did not require the interpretation of the axiom of infinity). Hence

$$L(y) \rightarrow \exists x(L(x) \ \& \ Ord(x) \ \& \ x \notin y). \tag{12}$$

We can now prove $L(\sigma)$ by transfinite induction on σ. The induction hypothesis shows that every member of σ is constructible; so by Lemma 2, there is a constructible set y such that $\sigma \subset y$. By (12), there is a constructible τ such that $\tau \notin y$. The $\sigma \leqslant \tau$; so $\sigma = \tau$ or $\sigma \in \tau$. Since L is transitive, it follows that σ is constructible.

9.7 THE AXIOM OF CONSTRUCTIBILITY

The formula $\forall x L(x)$, which asserts that every set is constructible, is called the *axiom of constructibility*. The theory obtained from *ZF* by adding this axiom is designated by *ZFL*.

We do not propose adopting the axiom of constructibility as an axiom of set theory, since there is no reason to believe that it is true. However, it is useful for investigations of consistency, as the following two theorems of Gödel show.

Theorem 1. If *ZF* is consistent, then *ZFL* is consistent.

Theorem 2. The axiom of choice and the generalized continuum hypothesis are theorems of *ZFL*.

It follows that if *ZF* is consistent, then neither the negation of the axiom of choice nor the negation of the generalized continuum hypothesis is provable in *ZF* (or in *ZFC*).

To prove Theorem 1, we show that L is an $\in$-interpretation of *ZFL* in *ZF*. Since we have already shown that it is an interpretation of *ZF*, it is only necessary to prove the interpretation of the axiom of constructibility.

We shall first show that C is absolute (for all transitive $\in$-interpretations of *ZF*). The absoluteness of the $\mathfrak{F}_i$ follows from the results of §9.5. To see this, it is best to rewrite the definitions, using the π_i^n. Thus

$$\mathfrak{F}_2(x, y) = [z \mid z \in x \;\&\; z = \langle \pi_1(z), \pi_1(z) \rangle],$$
$$\mathfrak{F}_6(x, y) = [z \mid z \in x \;\&\; z = \langle \pi_1^3(z), \pi_2^3(z), \pi_3^3(z) \rangle \;\&\; \langle \pi_2^3(z), \pi_1^3(z), \pi_3^3(z) \rangle \in y].$$

The absoluteness of the J_i is immediate. Now C is defined by

$$C(\sigma) = G(\sigma, [\langle C(\tau), \tau \rangle \mid \tau < \sigma])$$

for a certain G; we must show that this G is absolute. We can define G by

$$G(\sigma, f) = y \leftrightarrow R(y, \sigma, f),$$

where R is defined by

$$\begin{aligned} R(y, \sigma, f) \leftrightarrow & \big(J_0(\sigma) = 0 \;\&\; y = Ra(f)\big) \\ & \vee \big(J_0(\sigma) = 1 \;\&\; y = \mathfrak{F}_1(f`J_1(\sigma), f`J_2(\sigma))\big) \vee \cdots \\ & \vee \big(9 < J_0(\sigma) \;\&\; y = \{f`J_1(\sigma), f`J_2(\sigma)\}\big). \end{aligned}$$

Then R is absolute; so G is absolute.

We now prove that L is absolute for L (but not for all transitive $\in$-interpretations). We have

$$L(x) \leftrightarrow \exists y\big(Ord(y) \;\&\; x = C(y)\big).$$

Thus it suffices to show that $Ord(y) \;\&\; x = C(y)$ is complete in y for L. This follows from the fact that every ordinal is constructible.

The interpretation under L of the axiom of constructibility is

$$\forall x(L(x) \to L_L(x)). \tag{1}$$

The absoluteness of L for L means that

$$L(x) \to (L(x) \leftrightarrow L_L(x));$$

and this clearly implies (1).

Now we turn to the proof of Theorem 2. The proof of the axiom of choice in ZFL is quite easy. We define

$$Ch(x) = C(\mu\sigma(x \neq 0 \to C(\sigma) \in x)).$$

The axiom of constructibility shows that Ch is well-defined and that

$$x \neq 0 \to Ch(x) \in x.$$

From this it follows that a choice function for x is given by

$$[\langle Ch(y), y\rangle \mid y \in P(x) - \{0\}].$$

To prove the generalized continuum hypothesis, we first connect cardinals with constructible sets by showing that

$$Card(C^*(\aleph_\sigma)) = \aleph_\sigma. \tag{2}$$

Since $[\langle C(\tau), \tau\rangle \mid \tau < \aleph_\sigma]$ is a surjective mapping from $\aleph_\sigma$ to $C^*(\aleph_\sigma)$, we have $Card(C^*(\aleph_\sigma)) \leqslant Card(\aleph_\sigma) = \aleph_\sigma$. To prove the reverse inequality, it will suffice to define an injective mapping from $\aleph_\sigma$ to $C^*(\aleph_\sigma)$. Let $F(\tau)$ be the (unique) ordinal such that $K(F(\tau)) = \langle 0, \tau\rangle$; and let $f`\tau = C(F(\tau))$ for $\tau \in \aleph_\sigma$. If $\tau < \aleph_\sigma$, then $Max(K(F(\tau)) = \tau < \aleph_\sigma = Max(K(\aleph_\sigma))$ by (24) of §9.4; so $F(\tau) < \aleph_\sigma$ by (20) of §9.3. Hence f is a mapping from $\aleph_\sigma$ to $C^*(\aleph_\sigma)$. Suppose $\tau, \tau' \in \aleph_\sigma$ and $\tau \neq \tau'$. Then $F(\tau) \neq F(\tau')$. If $F(\tau) < F(\tau')$, then

$$f`\tau = C(F(\tau)) \in C^*(F(\tau')) = C(F(\tau')) = f`\tau'.$$

Similarly, if $F(\tau') < F(\tau)$, then $f`\tau' \in f`\tau$. In either case, $f`\tau \neq f`\tau'$ by (4) of §9.3. Thus f is injective.

By (29) of §9.4 and (2),

$$Card(P(C^*(\aleph_\sigma))) = 2^{\aleph_\sigma}. \tag{3}$$

We shall prove

$$P(C^*(\aleph_\sigma)) \subset C^*(\aleph_{S(\sigma)}). \tag{4}$$

From (4), (3), and (2),

$$2^{\aleph_\sigma} \leqslant \aleph_{S(\sigma)}.$$

From this and (33) of §9.4, we get the generalized continuum hypothesis.

The proof of (4) is based on the following lemma, which is a formal analog of the cardinality theorem.

Lemma. Let T be the theory obtained from ZFL by adding a constant B and an axiom $Trans(B)$. Then in a suitable conservative extension T' of T, we can define a constant A and prove

$$B \subset A,$$
$$Trans(A),$$
$$Card(B) \leqslant \aleph_\sigma \rightarrow Card(A) \leqslant \aleph_\sigma,$$

and $\mathbf{A} \leftrightarrow \mathbf{A}_A$ for every closed formula $\mathbf{A}$ of ZFL.

Proof. To form T' from T, we add a constant D and axioms

$$B \subset D, \tag{5}$$
$$Card(B) \leqslant \aleph_\sigma \rightarrow Card(D) \leqslant \aleph_\sigma, \tag{6}$$

and

$$x_1, \ldots, x_n \in D \rightarrow F(x_1, \ldots, x_n) \in D \tag{7}$$

for each function symbol F of ZFL. We first prove that T' is a conservative extension of T. Suppose that $\mathbf{A}$ is a formula of T provable in T'. Let T_1 be obtained from T by adding the constant D. By the reduction theorem, $\vdash_{T_1} \mathbf{B} \rightarrow \mathbf{A}$, where $\mathbf{B}$ is a conjunction of closures of axioms (5), (6), and (7). By the theorem on constants and the $\exists$-introduction rule, $\vdash_T \exists \mathbf{x}\mathbf{B}' \rightarrow \mathbf{A}$, where $\mathbf{B}'$ results from $\mathbf{B}$ by replacing D by a new variable $\mathbf{x}$. But $\vdash_T \exists \mathbf{x}\mathbf{B}'$ by the closure theorem; so $\vdash_T \mathbf{A}$.

We now show that

$$\vdash_{T'} x_1, \ldots, x_n \in D \rightarrow \big(Q(x_1, \ldots, x_n) \leftrightarrow Q_D(x_1, \ldots, x_n)\big)$$

for every predicate symbol Q of ZFL. We use induction on predicate symbols. If Q is defined by $Q(x_1, \ldots, x_n) \leftrightarrow x_i \in x_j$ or $Q(x_1, \ldots, x_n) \leftrightarrow x_i = x_j$, then Q_D is Q and the result is evident. If Q is defined as a negation or a disjunction, the result follows from the induction hypothesis. Now suppose that Q is defined by

$$Q(x_1, \ldots, x_n) \leftrightarrow \exists y R(y, x_1, \ldots, x_n).$$

By induction hypothesis,

$$x_1, \ldots, x_n, y \in D \rightarrow \big(R(y, x_1, \ldots, x_n) \leftrightarrow R_D(y, x_1, \ldots, x_n)\big);$$

so, under the hypothesis $x_1, \ldots, x_n \in D$,

$$\exists y\big(y \in D \ \& \ R(y, x_1, \ldots, x_n)\big) \leftrightarrow \exists y\big(y \in D \ \& \ R_D(y, x_1, \ldots, x_n)\big).$$

The right-hand side is $Q_D(x_1, \ldots, x_n)$; so we need only prove that the left-hand side is equivalent to $Q(x_1, \ldots, x_n)$. For this, it suffices to prove (under the hypothesis $x_1, \ldots, x_n \in D$)

$$\exists y R(y, x_1, \ldots, x_n) \rightarrow \exists y\big(y \in D \ \& \ R(y, x_1, \ldots, x_n)\big).$$

Define $F(x_1, \ldots, x_n)$ to be the set y of smallest order such that $R(y, x_1, \ldots, x_n)$, if such a y exists, and to be 0 otherwise. If

$$\exists y R(y, x_1, \ldots, x_n),$$

then

$$R(F(x_1, \ldots, x_n), x_1, \ldots, x_n).$$

But $F(x_1, \ldots, x_n) \in D$ by (7); so $\exists y(y \in D \ \& \ R(y, x_1, \ldots, x_n))$.

Now let **A** be a closed formula of *ZFL*, and define a 0-ary predicate symbol Q by $Q \leftrightarrow \mathbf{A}$. By the above, $\vdash Q \leftrightarrow Q_D$; so $\vdash \mathbf{A} \leftrightarrow \mathbf{A}_D$. From this we conclude that D is an $\in$-interpretation of *ZFL* in T'. For let **B** be an axiom of *ZFL*, and let **A** be the closure of **B**. Then $\mathbf{A}_D$ is the closure of $\mathbf{B}^{(D)}$. We conclude successively that **B**, **A**, $\mathbf{A}_D$, and $\mathbf{B}^{(D)}$ are provable in T'.

In particular, the interpretation of the extensionality axiom under D is provable. Now $y \in x \to Od(y) < Od(x)$ by Lemma 1 of §9.6; and Od is an ordinal function symbol such that $Set_x(Od(x) \leqslant \sigma)$, since $Od(x) \leqslant \sigma \to x \in C^*(S(\sigma))$. Thus we may apply the results of §9.5. We define

$$F(y) = [F(x) \mid x \in D \ \& \ x \in y]$$

and

$$A = [F(x) \mid x \in D].$$

We conclude that A is transitive and that F is an isomorphism of the $\in$-interpretations D and A. It follows that $Card(A) = Card(D)$; so from (6) we get

$$Card(B) \leqslant \aleph_\sigma \to Card(A) \leqslant \aleph_\sigma.$$

Using Lemma 1 of §9.5 (applied to 0-ary predicate symbols), we show as above that $\vdash \mathbf{A}_D \leftrightarrow \mathbf{A}_A$ for **A** a closed formula of *ZFL*. Hence $\vdash \mathbf{A} \leftrightarrow \mathbf{A}_A$ for such **A**.

It remains to prove $B \subset A$. For this it suffices to show

$$x \in B \to F(x) = x.$$

We prove this by induction on $Od(x)$. Let $x \in B$. If $y \in x$, then $Od(y) < Od(x)$, and $y \in B$ by the transitivity of B. Thus by the induction hypothesis and the inclusion $B \subset D$,

$$\begin{aligned} F(x) &= [F(y) \mid y \in D \ \& \ y \in x] \\ &= [y \mid y \in x] = x. \end{aligned}$$

In the notation of the lemma, A is an $\in$-interpretation of *ZFL* in T'; this is proved like the corresponding fact for D. The definition

$$Od(x) = \mu\sigma(x = C(\sigma))$$

together with the absoluteness of C shows that Od is absolute for A. (Of course the axiom of constructibility is needed to justify this defining axiom for Od.) It follows that Od is A-invariant. From this and the transitivity of A,

$$\begin{aligned} x \in A &\to Od(x) \subset A \\ &\to Card(Od(x)) \leqslant Card(A). \end{aligned}$$

Using this with $B \subset A$ and $Card(B) \leqslant \aleph_\sigma \to Card(A) \leqslant \aleph_\sigma$, we get

$$Card(B) \leqslant \aleph_\sigma \ \& \ x \in B \to Card(Od(x)) \leqslant \aleph_\sigma.$$

The right-hand side implies that $x \in C^*(\aleph_{S(\sigma)})$. For if $x \notin C^*(\aleph_{S(\sigma)})$, then $\aleph_{S(\sigma)} \leqslant Od(x)$; so

$$\aleph_\sigma < \aleph_{S(\sigma)} = Card(\aleph_{S(\sigma)}) \leqslant Card(Od(x)).$$

Thus

$$Card(B) \leqslant \aleph_\sigma \to B \subset C^*(\aleph_{S(\sigma)})$$

is provable in T'. By the lemma, it is also provable in T. Hence by the deduction theorem and the theorem on constants,

$$Trans(b) \ \& \ Card(b) \leqslant \aleph_\sigma \to b \subset C^*(\aleph_{S(\sigma)}) \tag{8}$$

is provable in *ZFL*.

We can now prove (4). Let $a \in P(C^*(\aleph_\sigma))$, that is, $a \subset C^*(\aleph_\sigma)$. Let $b = C^*(\aleph_\sigma) \cup \{a\}$. By Lemma 1 of §9.6, $C^*(\aleph_\sigma)$ is transitive. From this and $a \subset C^*(\aleph_\sigma)$, we see that b is transitive. Also

$$Card(b) \leqslant Card(C^*(\aleph_\sigma)) + Card(\{a\}) = \aleph_\sigma + 1 = \aleph_\sigma$$

by (2). Hence $b \subset C^*(\aleph_{S(\sigma)})$ by (8); so

$$a \in C^*(\aleph_{S(\sigma)}).$$

This completes the proof of theorem 2.

We conclude this section by using the lemma to derive a formal version of the Löwenheim-Skolem theorem. Let ZFL_A be the theory obtained from *ZFL* by adding a constant A and the axioms

$$Trans(A),$$
$$Card(A) \leqslant \aleph_0,$$

and $\mathbf{A} \leftrightarrow \mathbf{A}_A$ for every closed formula $\mathbf{A}$ of *ZFL*. We shall prove that ZFL_A is a conservative extension of *ZFL*.

To prove this, let T and T' be as in the lemma, and let T'' be obtained from T' by adding the axiom $B = 0$. Then all the axioms of ZFL_A are provable in T''; so it will suffice to show that T'' is a conservative extension of *ZFL*. Suppose that $\mathbf{A}$ is a formula of *ZFL* provable in T''. By the deduction theorem, $B = 0 \to \mathbf{A}$ is provable in T' and hence in T. By the deduction theorem and the theorem on constants, $Trans(b) \ \& \ b = 0 \to \mathbf{A}$ is provable in *ZFL*. Substituting 0 for b, we find that $\mathbf{A}$ is provable in *ZFL*.

We can show as above that A is an $\in$-interpretation of *ZFL* in ZFL_A. Thus we have a countable transitive $\in$-interpretation of *ZFL* in its conservative extension ZFL_A.

9.8 FORCING

We now describe a method of constructing interpretations of *ZFC* which will be used to prove the independence of the axiom of choice and the continuum hypothesis.*

We will construct an extension of *ZFL* in which the interpretation will be given. We assume that a constant *CD* of *ZFL* such that $\vdash_{ZFL} 0 \in CD$ is fixed. (The actual choice of *CD* will depend on the application.) The elements of *CD* are called *conditions*. We let p, q, and r vary through conditions. If $p \subset q$, we say that q is an *extension* of p.

To each condition will correspond a (partial) description of the interpretation *I*. Moreover, the description corresponding to an extension of p will include the description corresponding to p. However, not all these descriptions will be correct descriptions of *I*. We shall pick out certain conditions, which we call the *correct* conditions, and build *I* to fit the description given by the correct conditions.

To ensure that there is at least one correct condition, we require that 0 be correct. To ensure that the descriptions corresponding to two correct conditions do not contradict one another, we require that $p \cup q$ be correct whenever both p and q are correct.

There is a third requirement which is designed to ensure that, as far as possible, every set x contains a correct condition. Now a situation in which this is not possible is the following: we have chosen p as a correct condition, and there is no condition q in x such that $p \cup q$ is a condition. This suggests that we take the requirement to be: for every set x, there is a correct condition p such that either p is in x or no extension of p is in x.

We construct ZFL_{Cor} from *ZFL* by adding a unary predicate symbol *Cor* and four new nonlogical axioms:

Cor₁. $Cor(p) \rightarrow p \in CD$,

Cor₂. $Cor(0)$,

Cor₃. $Cor(p) \;\&\; Cor(q) \rightarrow Cor(p \cup q)$,

Cor₄. $\exists p(Cor(p) \;\&\; (p \in x \vee \forall q(p \subset q \;\&\; q \in CD \rightarrow q \notin x)))$.

Note that we have not added the subset and replacement axioms containing the new symbol *Cor*; so we cannot apply our results on set formation to formulas containing this symbol.

Our first task is to prove that if *ZF* is consistent, then ZFL_{Cor} is consistent. The consistency of *ZF* implies the consistency of *ZFL* and hence of its conservative extension ZFL_A. Moreover, *A* is an interpretation of *ZFL* in ZFL_A. We will extend this to an interpretation of $L(ZFL_{Cor})$ by defining Cor_A in ZFL_A. We shall then prove the interpretations of the new axioms, thus showing that we have an interpretation of ZFL_{Cor} in ZFL_A. This will give the desired result.

---

* The proof given here is basically the original proof of Cohen; but use has been made of some simplifications discovered by Feferman, Scott and Solovay.

In ZFL_A, we let p, q, and r vary through elements of CD_A. A subset c of CD_A will be called *generic* if it satisfies the three requirements:

$$0 \in c,$$
$$p \in c \;\&\; q \in c \rightarrow p \cup q \in c,$$
$$\forall x(x \in A \rightarrow \exists p(p \in c \;\&\; (p \in x \vee \forall q(p \subset q \;\&\; q \in CD_A \rightarrow q \notin x)))).$$

We first prove that a generic set exists. Let the elements of A be arranged in a sequence $x_0, x_1, \ldots$. We define a sequence $p_0, p_1, \ldots$ of elements of CD_A inductively. Let $p_0 = 0$. Now suppose that p_n is chosen. If there is a q such that $p_n \subset q$ and $q \in x_n$, we let p_{n+1} be a q with this property. (To be specific, we can let it be the q with this property having the smallest order.) If there is no such q, we let $p_{n+1} = p_n$. It is then clear that $[p_n \mid n \in \omega]$ is generic.

We define the constant G by an axiom saying that G is the generic set having the smallest order. We then define

$$Cor_A(p) \leftrightarrow p \in G.$$

Since $CD_A \in A$ and A is transitive, every element of CD_A is in A. Using this with the absoluteness of 0, $\cup$, and $\subset$, we find that the interpretations of Cor_1 through Cor_4 are equivalent to

$$Cor_A(p) \rightarrow p \in CD_A,$$
$$Cor_A(0),$$
$$Cor_A(p) \;\&\; Cor_A(q) \rightarrow Cor_A(p \cup q),$$
$$x \in A \rightarrow \exists p(Cor_A(p) \;\&\; (p \in x \vee \forall q(p \subset q \;\&\; q \in CD_A \rightarrow q \notin x))).$$

These all follow from the fact that G is generic.

We now turn to the construction of our interpretation of ZFC in ZFL_{Cor}. We must first specify the description of I which corresponds to p. We shall define *p forces* $Q(x_1, \ldots, x_n)$ for each predicate symbol Q of ZFC. The description corresponding to p then says that $Q_I(x_1, \ldots, x_n)$ is true whenever p forces $Q(x_1, \ldots, x_n)$.

First consider the case in which Q is $\in$. We would like to be able to construct a set y such that for each p, the set of x such that p forces $x \in y$ is a predetermined set z_p. We can do this by taking y to be the set of $\langle x, p\rangle$ with $x \in z_p$, and then defining p forces $x \in y$ to mean that $\langle x, p\rangle \in y$.

This definition must be modified so that the description corresponding to an extension of p includes the description corresponding to p. We therefore define

$$x \in_p y \leftrightarrow \forall q(p \subset q \rightarrow \langle x, q\rangle \in y).$$

Clearly

$$x \in_p y \;\&\; p \subset q \rightarrow x \in_q y \tag{1}$$

and

$$x \in_p y \rightarrow x \in Ra(y). \tag{2}$$

We want p to force $x \in y$ whenever $x \in_p y$. It is then natural to also make p *forces* $x \in y$ true when p forces $x = z$ for a z such that $z \in_p y$.

Let us picture the selection of the correct conditions as taking place successively, with each correct condition being an extension of the previous ones. If q is selected and $z \in_q x$, then z will have to be in x. If no extension of q forces $z \in y$, then we can never force z to be in y. Hence x and y will be unequal. The same holds if $z \in_q y$ and no extension of q forces $z \in x$. In either of these cases, let us say that q *prevents* $x = y$. We then want p to force $x = y$ if no extension of p prevents $x = y$.

We are thus led to the following definitions.

A. A condition p forces $x \in y$ if for some z, $z \in_p y$ and p forces $x = z$.

B. A condition p forces $x = y$ if for every z and every extension q of p, the following hold:

i) if $z \in_q x$, then some extension of q forces $z \in y$;

ii) if $z \in_q y$, then some extension of q forces $z \in x$.

We must show how the circularity in these definitions can be eliminated. We consider A as a definition of the set $|x \in y|$ of conditions forcing $x \in y$, and B as a definition of the set $|x = y|$ of conditions forcing $x = y$. Then B defines $|x = y|$ in terms of $|z \in x|$ and $|z \in y|$, where

$$\exists q(z \in_q x \vee z \in_q y). \tag{3}$$

If we replace $|z \in x|$ and $|z \in y|$ by their definitions according to A, we obtain a definition B′ of $|x = y|$ in terms of $|z = w|$, where

$$\exists q(w \in_q x \vee w \in_q y). \tag{4}$$

Now

$$z \in Ra(x) \rightarrow Od(z) < Od(x). \tag{5}$$

For the hypothesis implies that $\langle z, w\rangle \in x$ for some w; since

$$z \in \{z\} \qquad \text{and} \qquad \{z\} \in \langle z, w\rangle,$$

we have $Od(z) < Od(x)$ by Lemma 1 of §9.6. From (2), (3), (4), and (5),

$$Max(\langle Od(z), Od(w)\rangle) < Max(\langle Od(x), Od(y)\rangle).$$

Thus we may consider B′ as a definition by induction on $Max(\langle Od(x), Od(y)\rangle)$. For this, we must prove

$$Set_{x,y}(Max(\langle Od(x), Od(y)\rangle) \leqslant \sigma);$$

but this follows from

$$Max(\langle Od(x), Od(y)\rangle) \leqslant \sigma \rightarrow x, y \in C^*(S(\sigma)).$$

Thus we may adopt B′ as a definition of $|x = y|$. If we then adopt A as a definition of $|x \in y|$, we can prove B from B′ and A.

We now define *p forces* $Q(x_1, \ldots, x_n)$ by induction on function symbols as follows.

i) If $Q(x_1, \ldots, x_n) \leftrightarrow x_i \in x_j$, then p forces $Q(x_1, \ldots, x_n)$ if p forces $x_i \in x_j$.

ii) If $Q(x_1, \ldots, x_n) \leftrightarrow x_i = x_j$, then p forces $Q(x_1, \ldots, x_n)$ if p forces $x_i = x_j$.

iii) If $Q(x_1, \ldots, x_n) \leftrightarrow \neg R(x_1, \ldots, x_n)$, then p forces $Q(x_1, \ldots, x_n)$ if no extension of p forces $R(x_1, \ldots, x_n)$.

iv) If
$$Q(x_1, \ldots, x_n) \leftrightarrow R(x_1, \ldots, x_n) \vee R'(x_1, \ldots, x_n),$$
then p forces $Q(x_1, \ldots, x_n)$ if either p forces $R(x_1, \ldots, x_n)$ or p forces $R'(x_1, \ldots, x_n)$.

v) If $Q(x_1, \ldots, x_n) \leftrightarrow \exists y R(y, x_1, \ldots, x_n)$, then p forces $Q(x_1, \ldots, x_n)$ if for some y, p forces $R(y, x_1, \ldots, x_n)$.

If these definitions were given formally, *p forces* $Q(x_1, \ldots, x_n)$ would be an $(n + 1)$-ary predicate symbol applied to the arguments $p, x_1, \ldots, x_n$. Note that the definitions do not make use of the new symbol *Cor*; so we may apply our results on set existence to sentences about forcing.

We say that $Q(x_1, \ldots, x_n)$ *is forced* if some correct condition forces $Q(x_1, \ldots, x_n)$. We now define our interpretation by:

$$x \in_I y \leftrightarrow x \in y \text{ is forced},$$
$$x =_I y \leftrightarrow x = y \text{ is forced},$$
$$U_I(x) \leftrightarrow x = x.$$

We shall sometimes say that *p forces* ____; this means that p forces $Q(x_1, \ldots, x_n)$, where Q is defined by
$$Q(x_1, \ldots, x_n) \leftrightarrow \text{\_\_\_\_}.$$
We interpret ____ *is forced* similarly.

Lemma 1. If p forces $Q(x_1, \ldots, x_n)$, then every extension of p forces $Q(x_1, \ldots, x_n)$.

Proof. The proof is by induction on predicate symbols, using (1) when Q is $\in$. Details are left to the reader.

Lemma 2. If $Q(x_1, \ldots, x_n)$ and $R(x_1, \ldots, x_n)$ are forced, then there is a correct p which forces $Q(x_1, \ldots, x_n)$ and $R(x_1, \ldots, x_n)$.

Proof. Choose a correct q which forces $Q(x_1, \ldots, x_n)$ and a correct r which forces $R(x_1, \ldots, x_n)$. Then $p = q \cup r$ is correct by Cor_3; and p forces
$$Q(x_1, \ldots, x_n) \quad \text{and} \quad R(x_1, \ldots, x_n)$$
by Lemma 1.

Truth Lemma. For every predicate symbol Q of ZFC,

$$Q_I(x_1, \ldots, x_n) \quad \text{iff} \quad Q(x_1, \ldots, x_n) \textit{ is forced.}$$

Proof. We use induction on predicate symbols. If $Q(x_1, \ldots, x_n) \leftrightarrow x_i \in x_j$ or $Q(x_1, \ldots, x_n) \leftrightarrow x_i = x_j$, then the result follows from the definitions of $\in_I$ and $=_I$ and (i) and (ii).

Suppose that $Q(x_1, \ldots, x_n) \leftrightarrow \neg R(x_1, \ldots, x_n)$. Then by the induction hypothesis

$$\begin{aligned} Q_I(x_1, \ldots, x_n) &\leftrightarrow \neg R_I(x_1, \ldots, x_n) \\ &\leftrightarrow R(x_1, \ldots, x_n) \textit{ is not forced.} \end{aligned}$$

Hence we need only show that exactly one of $R(x_1, \ldots, x_n)$ and $\neg R(x_1, \ldots, x_n)$ is forced. By Cor_4, there is a correct p such that either p forces $R(x_1, \ldots, x_n)$ or no extension of p forces $R(x_1, \ldots, x_n)$. It follows that at least one of $R(x_1, \ldots, x_n)$ and $\neg R(x_1, \ldots, x_n)$ is forced. Now suppose that both are forced. By Lemma 2, some p forces $R(x_1, \ldots, x_n)$ and $\neg R(x_1, \ldots, x_n)$. This is impossible by (iii).

If $Q(x_1, \ldots, x_n) \leftrightarrow R(x_1, \ldots, x_n) \vee R'(x_1, \ldots, x_n)$, then by the induction hypothesis and (iv),

$$\begin{aligned} Q_I(x_1, \ldots, x_n) &\leftrightarrow R_I(x_1, \ldots, x_n) \vee R'_I(x_1, \ldots, x_n) \\ &\leftrightarrow R(x_1, \ldots, x_n) \textit{ is forced or } R'(x_1, \ldots, x_n) \textit{ is forced} \\ &\leftrightarrow Q(x_1, \ldots, x_n) \textit{ is forced.} \end{aligned}$$

If $Q(x_1, \ldots, x_n) \leftrightarrow \exists y R(y, x_1, \ldots, x_n)$, then by the induction hypothesis and (v),

$$\begin{aligned} Q_I(x_1, \ldots, x_n) &\leftrightarrow \exists y R_I(y, x_1, \ldots, x_n) \\ &\leftrightarrow \exists y \big(R(y, x_1, \ldots, x_n) \textit{ is forced}\big) \\ &\leftrightarrow Q(x_1, \ldots, x_n) \textit{ is forced.} \end{aligned}$$

It is clear that

$$p \textit{ forces } x = y \rightarrow p \textit{ forces } y = x. \tag{6}$$

We prove

$$z \in_p x \rightarrow p \textit{ forces } z \in x \tag{7}$$

and

$$p \textit{ forces } x = x \tag{8}$$

by induction on $Od(x)$. If $z \in_p x$, then $Od(z) < Od(x)$ by (2) and (5); so p forces $z = z$ by the induction hypothesis; so p forces $z \in x$. Now if $p \subset q$ and $z \in_q x$, then, as just proved, q forces $z \in x$; so p forces $x = x$.

Next we prove

$$p \textit{ forces } x = y \textit{ and } p \textit{ forces } y = z \rightarrow p \textit{ forces } x = z \tag{9}$$

by induction on $Od(z)$. By symmetry, it will suffice to show that if $p \subset q$ and $a \in_q x$, then some extension of q forces $a \in z$. Since p forces $x = y$, some extension

r of q forces $a \in y$. Then r forces $a = b$ where $b \in_r y$. Since p forces $y = z$, some extension r' of r forces $b \in z$. Then r' forces $b = c$ where $c \in_{r'} z$. By (2) and (5), $Od(c) < Od(z)$; and by Lemma 1, r' forces $a = b$. Hence by the induction hypothesis, r' forces $a = c$. It follows that r' forces $a \in z$.

We now turn to the proofs of the interpretations of the axioms of *ZFC*. The interpretation of the identity axiom is $x =_I x$. Since 0 is correct by Cor_2 and forces $x = x$ by (8), $x =_I x$.

The interpretation of the equality axiom for $=$ is

$$x =_I y \rightarrow z =_I w \rightarrow x =_I z \rightarrow y =_I w.$$

This is easily proved from

$$x =_I y \rightarrow y =_I x \tag{10}$$

and

$$x =_I y \;\&\; y =_I z \rightarrow x =_I z. \tag{11}$$

Now (10) follows from (6). To prove (11), let $x =_I y$ and $y =_I z$. By Lemma 2, there is a correct p which forces $x = y$ and $y = z$. By (9), p forces $x = z$; so $x =_I z$.

The interpretation of the equality axiom for $\in$ is

$$x =_I y \rightarrow z =_I w \rightarrow x \in_I z \rightarrow y \in_I w.$$

This follows from

$$x =_I y \;\&\; x \in_I z \rightarrow y \in_I z \tag{12}$$

and

$$x =_I y \;\&\; z \in_I x \rightarrow z \in_I y. \tag{13}$$

To prove (12), assume that $x =_I y$ and $x \in_I z$. By Lemma 2, there is a correct p which forces $x = y$ and $x \in z$. Hence for some w, p forces $x = w$ and $w \in_p z$. By (6) and (9), p forces $y = w$; so p forces $y \in z$; so $y \in_I z$.

To prove (13), let $x =_I y$ and $z \in_I x$, and choose a correct p which forces $x = y$ and $z \in x$. Then for some w, p forces $z = w$ and $w \in_p x$. For each extension q of p, $w \in_q x$, and hence some extension of q forces $w \in y$. This shows that p forces $\neg\neg(w \in y)$. Hence by the truth lemma $\neg\neg(w \in_I y)$. Thus $w \in_I y$. Since also $z =_I w$, $z \in_I y$ by (12).

Since we have now proved the interpretations of the identity axioms and the equality axioms, we can conclude that the interpretation of the equality theorem holds. Thus we have

$$x =_I y \rightarrow \big(Q_I(x, v_1, \ldots, v_n) \leftrightarrow Q_I(y, v_1, \ldots, v_n)\big). \tag{14}$$

The interpretation of the extensionality axiom is

$$\forall z(z \in_I x \leftrightarrow z \in_I y) \rightarrow x =_I y. \tag{15}$$

Let a be the set of p such that

$$\exists z\big((z \in_p x \;\&\; p \textit{ forces } z \notin y) \vee (z \in_p y \;\&\; p \textit{ forces } z \notin x)\big).$$

First suppose that a contains a correct condition p. Then there is a z such that, say, $z \in_p x$ and p forces $z \notin y$. By (7) and the truth lemma, $z \in_I x \ \& \ \neg(z \in_I y)$; so the left-hand side of (15) is false. Now suppose that a contains no correct condition. By Cor_4, there is a correct condition p having no extension in a. Then p forces $x = y$; so $x =_I y$.

We have

$$x \in_I y \rightarrow \exists z\big(x =_I z \ \& \ z \in Ra(y)\big). \tag{16}$$

For choose a correct p forcing $x \in y$. Then for some z, p forces $x = z$ and $z \in_p y$. Hence $x =_I z \ \& \ z \in Ra(y)$.

Lemma 3. Let p be a correct condition forcing $\exists y(y \in x)$. Then there is a y and a correct extension q of p such that q forces $y \in x$ and q forces $z \notin x$ for every z such that $Od(z) < Od(y)$.

Proof. Let a be the set of extensions q of p such that

$$\exists y\big(q \textit{ forces } y \in x \ \& \ \forall z\big(Od(z) < Od(y) \rightarrow q \textit{ forces } z \notin x\big)\big).$$

We want to show that a contains a correct condition. Suppose not. By Cor_4 there is a correct condition q having no extension in a. By Cor_3, $p \cup q$ is correct; and by Lemma 1, $p \cup q$ forces $y \in x$ for some y. Thus the set of y such that

$$\exists r(p \cup q \subset r \ \& \ r \textit{ forces } y \in x)$$

is not empty. Let y be the element of this set having the smallest order, and let r be an extension of $p \cup q$ forcing $y \in x$. Since $r \notin a$, there is a z such that $Od(z) < Od(y)$ and r does not force $z \notin x$. Thus some extension of r, and hence of $p \cup q$, forces $z \in x$. This is impossible by the choice of y.

The interpretation of the regularity axiom is

$$\exists y(y \in_I x) \rightarrow \exists y\big(y \in_I x \ \& \ \neg\exists z(z \in_I x \ \& \ z \in_I y)\big).$$

Assume that $\exists y(y \in_I x)$. By the truth lemma, there is a correct p forcing $\exists y(y \in x)$. Choose y and q as in Lemma 3. Then $y \in_I x$. Suppose that $z \in_I y$; we must show that $\neg(z \in_I x)$. By (16) and (14) we may suppose that $z \in Ra(y)$; so $Od(z) < Od(y)$ by (5). Hence q forces $z \notin x$; so $\neg(z \in_I x)$ by the truth lemma.

Lemma 4. If

$$z = [\langle u, p\rangle \mid u \in Ra(y) \ \& \ p \textit{ forces } u \in y \ \& \ p \textit{ forces } Q(u, v_1, \ldots, v_n)],$$

then for all x,

$$x \in_I z \leftrightarrow x \in_I y \ \& \ Q_I(x, v_1, \ldots, v_n).$$

Proof. We omit the parameters. By Lemma 1, $u \in_p z \leftrightarrow \langle u, p\rangle \in z$. Hence $x \in_I z$ is equivalent to

$$\exists p \, \exists u\big(Cor(p) \ \& \ p \textit{ forces } x = u \ \& \ \langle u, p\rangle \in z\big)$$

and hence to

$$\exists p\, \exists u(Cor(p) \,\&\, p \textit{ forces } x = u \,\&\, u \in Ra(y) \,\&\, p \textit{ forces } u \in y \,\&\, p \textit{ forces } Q(u)).$$

By the truth lemma and Lemma 2, this is equivalent to

$$\exists u(x =_I u \,\&\, u \in Ra(y) \,\&\, u \in_I y \,\&\, Q_I(u)).$$

By (14), this is equivalent to

$$\exists u(x =_I u \,\&\, u \in Ra(y)) \,\&\, x \in_I y \,\&\, Q_I(x).$$

This is equivalent to $x \in_I y \,\&\, Q_I(x)$ by (16).

The interpretation of a subset axiom has the form

$$\exists z\, \forall x(x \in_I z \leftrightarrow x \in_I y \,\&\, Q_I(x, v_1, \ldots, v_n)).$$

This is a consequence of Lemma 4.

We have

$$x \in_I (y \times CD) \leftrightarrow \exists z(z \in y \,\&\, x =_I z). \tag{17}$$

The implication from left to right follows from (16). Now suppose that

$$z \in y \,\&\, x =_I z,$$

and choose a correct p which forces $x = z$. Then $z \in_p (y \times CD)$; so p forces $x \in (y \times CD)$; so $x \in_I (y \times CD)$.

The interpretation of a replacement axiom has the hypothesis

$$\forall x\, \exists u\, \forall y(y \in_I u \leftrightarrow Q_I(x, y)) \tag{18}$$

and the conclusion

$$\exists z\, \forall y(\exists x(x \in_I w \,\&\, Q_I(x, y)) \rightarrow y \in_I z) \tag{19}$$

(where we have omitted the parameters). Define $F(x, p)$ to be the u of smallest order such that

$$p \textit{ forces } \forall y(y \in u \leftrightarrow Q(x, y)), \tag{20}$$

provided that such a u exists. Assume (18). Then for each x, there is a u such that $\forall y(y \in_I u \leftrightarrow Q_I(x, y))$; so by the truth lemma, there is a correct p such that (20) holds. This implies that (20) holds for $u = F(x, p)$. Hence, using the truth lemma again,

$$\forall x\, \exists p\, \forall y(y \in_I F(x, p) \leftrightarrow Q_I(x, y)). \tag{21}$$

Let w be given, and set

$$z = Un([Ra(F(x, p)) \mid x \in Ra(w)]_{x,p}) \times CD.$$

To prove (19), we must show that

$$x \in_I w \,\&\, Q_I(x, y) \rightarrow y \in_I z.$$

By (16) and (14) we may suppose that $x \in Ra(w)$. Choosing p by (21), we have $y \in_I F(x, p)$. By (16), $y =_I y'$ for some y' in $Ra(F(x, p))$; so $y \in_I z$ by (17).

The interpretation of the power set axiom is

$$\exists w \, \forall x(\forall z(z \in_I x \to z \in_I y) \to x \in_I w).$$

Let $w = P(Ra(y) \times CD) \times CD$. Assume that $\forall z(z \in_I x \to z \in_I y)$, and set

$$v = [\langle u, p\rangle \mid u \in Ra(y) \;\&\; p \text{ forces } u \in y \;\&\; p \text{ forces } u \in x].$$

By Lemma 4 we have for all z,

$$\begin{aligned} z \in_I v &\leftrightarrow z \in_I y \;\&\; z \in_I x \\ &\leftrightarrow z \in_I x. \end{aligned}$$

Hence by the interpreted extensionality axiom, $x =_I v$. Since $v \subset Ra(y) \times CD$, $x \in_I w$ by (17).

We let ZF_0 be the theory obtained from ZF by omitting the axiom of infinity.

Lemma 5. Let J be an interpretation of ZF_0 in ZFL_{Cor}, and let O be a function symbol such that in ZFL_{Cor},

$$\vdash U_J(O(\sigma)), \tag{22}$$

$$\vdash x \in_J O(\sigma) \leftrightarrow \exists \tau(\tau < \sigma \;\&\; x =_J O(\tau)). \tag{23}$$

Then J is an interpretation of ZF. Moreover, the interpretation of $x = \omega$ under J is $x =_J O(\omega)$.

Proof. Define (in ZF_0)

$$\begin{aligned} &Zer(x) \leftrightarrow \forall y(y \notin x), \\ &Suc(x, y) \leftrightarrow \forall z(z \in x \leftrightarrow z \in y \vee z = y). \end{aligned}$$

From (23),

$$Zer_J(O(0)), \tag{24}$$

$$Suc_J(O(S(\sigma)), O(\sigma)). \tag{25}$$

Using these and (22) and (23), we get

$$\exists y(U_J(y) \;\&\; y \in_J O(\omega) \;\&\; Zer_J(y)), \tag{26}$$

$$\forall y(U_J(y) \;\&\; y \in_J O(\omega) \to \exists z(U_J(z) \;\&\; z \in_J O(\omega) \;\&\; Suc_J(z, y))). \tag{27}$$

The axiom of infinity is

$$\exists x(\exists y(y \in x \;\&\; Zer(y)) \;\&\; \forall y(y \in x \to \exists z(z \in x \;\&\; Suc(z, y)))).$$

Its interpretation follows easily from (26), (27), and (22).

Now define

$$Nn(x) \leftrightarrow x \in \omega.$$

In ZF we can prove

$$\exists y(y \in w \;\&\; Zer(y)) \;\&\; \forall y(y \in w \to \exists z(z \in w \;\&\; Suc(z, y))) \;\&\; Nn(x) \to x \in w.$$

If we take the interpretation of this, substitute $O(\omega)$ for w, and use (26) and (27), we get

$$U_J(x) \;\&\; Nn_J(x) \rightarrow x \in_J O(\omega). \tag{28}$$

We can also prove in ZF

$$Zer(x) \rightarrow Nn(x),$$
$$Nn(x) \;\&\; Suc(y, x) \rightarrow Nn(y).$$

Taking the interpretations of these and using (24) and (25), we have

$$Nn_J(O(0)), \tag{29}$$
$$Nn_J(O(\sigma)) \rightarrow Nn_J(O(S(\sigma))). \tag{30}$$

Now assume that

$$\exists x(U_J(x) \;\&\; x \in_J O(\omega) \;\&\; \neg Nn_J(x)). \tag{31}$$

In ZF we can prove from the regularity axiom

$$\exists x(x \in w \;\&\; \neg Nn(x)) \rightarrow \exists x(x \in w \;\&\; \neg Nn(x) \;\&\; \forall z(z \in x \;\&\; z \in w \rightarrow Nn(z))).$$

Taking the interpretation, substituting $O(\omega)$ for w, and using (31), we find that there is an x such that

$$U_J(x) \;\&\; x \in_J O(\omega) \;\&\; \neg Nn_J(x)$$

and such that for all z,

$$U_J(z) \;\&\; z \in_J x \;\&\; z \in_J O(\omega) \rightarrow Nn_J(z). \tag{32}$$

By (23), $x =_J O(\sigma)$ for some natural number σ; so by the interpreted equality theorem, we may suppose x is $O(\sigma)$. From $\neg Nn_J(O(\sigma))$ and (29) and (30), we see that $\sigma = S(\tau)$ where $\neg Nn_J(O(\tau))$. Putting $O(\tau)$ for z in (32) and using (22) and (23), we get $Nn_J(O(\tau))$. Thus we have derived a contradiction from (31). From this and (28),

$$U_J(x) \rightarrow (Nn_J(x) \leftrightarrow x \in_J O(\omega)). \tag{33}$$

In ZF, $x = \omega$ is equivalent to $\forall y(y \in x \leftrightarrow Nn(y))$. It follows that the interpretation of $x = \omega$ is equivalent to

$$\forall y(U_J(y) \rightarrow (y \in_J x \leftrightarrow Nn_J(y))).$$

By (33) this is equivalent to

$$\forall y(U_J(y) \rightarrow (y \in_J x \leftrightarrow y \in_J O(\omega))).$$

This is equivalent to $x =_J O(\omega)$ by the interpreted equality and extensionality axioms.

To apply the lemma to our case, we define by transfinite induction on σ

$$O(\sigma) = [O(\tau) \mid \tau < \sigma] \times CD.$$

Then (22) is obvious and (23) follows from (17).

The *multiplicative axiom* is the following statement: if $0 \notin z$ and if every two distinct members of z are disjoint, then there is a set y having exactly one element in common with each member of z. In *ZF*, this axiom implies the axiom of choice. For let x be given, and for $a \subset x$ let

$$z_a = [\langle b, a\rangle \mid b \in a].$$

Let $z = [z_a \mid a \in P(x) - \{0\}]$, and let y be as in the multiplicative axiom. Then $y \cap z$ is a choice function on x.

It follows that we need only prove the interpretation of the multiplicative axiom. We are thus given a z such that

$$w \in_I z \to \exists a(a \in_I w) \tag{34}$$

and

$$w \in_I z \;\&\; a \in_I w \;\&\; w' \in_I z \;\&\; a \in_I w' \to w =_I w'. \tag{35}$$

We are to find a y such that

$$w \in_I z \to \exists a(a \in_I w \;\&\; a \in_I y) \tag{36}$$

and

$$w \in_I z \;\&\; a \in_I w \;\&\; a \in_I y \;\&\; b \in_I w \;\&\; b \in_I y \to a =_I b. \tag{37}$$

We let y be the set of $\langle a, p\rangle$ such that for some w, p forces $w \in z$ and $a \in w$, and p forces $b \notin w$ for all b such that $Od(b) < Od(a)$. Then $a \in_p y \leftrightarrow \langle a, p\rangle \in y$ by Lemma 1.

We first prove (36). Let $w \in_I z$. By (34) and the truth lemma, there is a correct p forcing $w \in z$ and $\exists a(a \in w)$. Hence by Lemma 3, there is an a and a correct extension q of p such that q forces $a \in w$ and q forces $b \notin w$ for all b such that $Od(b) < Od(a)$. Then $\langle a, q\rangle \in y$; so $a \in_q y$; so q forces $a \in y$. Hence $a \in_I w \;\&\; a \in_I y$.

Now we prove (37). Assume the hypothesis and choose a correct p forcing $a \in y$. Then for some c, p forces $a = c$ and $\langle c, p\rangle \in y$. Thus for some w', p forces $w' \in z$ and $c \in w'$ and $d \notin w'$ for all d such that $Od(d) < Od(c)$; whence

$$a =_I c \;\&\; w' \in_I z \;\&\; c \in_I w' \;\&\; \forall d\big(Od(d) < Od(c) \to \neg(d \in_I w')\big).$$

By (35), $w =_I w'$; so

$$a =_I c \;\&\; c \in_I w \;\&\; \forall d\big(Od(d) < Od(c) \to \neg(d \in_I w)\big).$$

In a similar way, we find a d such that

$$b =_I d \;\&\; d \in_I w \;\&\; \forall c\big(Od(c) < Od(d) \to \neg(c \in_I w)\big).$$

From these we get $Od(c) = Od(d)$; so $c = d$. Since $a =_I c$ and $b =_I d$, it follows that $a =_I b$.

9.9 THE INDEPENDENCE PROOFS

We are now going to prove the two following theorems of Cohen.

Theorem 1. If *ZF* is consistent, then the continuum hypothesis is not a theorem of *ZFC*.

Theorem 2. If *ZF* is consistent, then the axiom of choice is not a theorem of *ZF*.

To prove Theorem 1, we will make a choice of *CD*, and then show that the interpretation of the negation of the continuum hypothesis holds. Actually, we shall prove the interpretation of a sentence implying the negation of the continuum hypothesis.

We define (in *ZFC*)

$$Im(f, x, y) \leftrightarrow \langle y, x\rangle \in f,$$
$$Sur(a, b) \leftrightarrow \exists f\, \forall y\big(y \in b \rightarrow \exists x(x \in a \;\&\; Im(f, x, y) \;\&\; \forall z(Im(f, x, z) \rightarrow z = y))\big).$$

Then *Sur*(*a*, *b*) means that there is a surjective mapping from *a* to *b*. We have

$$b \neq 0 \;\&\; Card(b) \leqslant Card(a) \rightarrow Sur(a, b). \tag{1}$$

For the hypothesis implies that there is a bijective mapping from a subset of *a* to *b*; and this can be extended to a surjective mapping from *a* to *b*.

Now assume the continuum hypothesis, and let

$$0 \neq a \subset b \subset P(\omega).$$

If $Card(a) \leqslant \aleph_0$, then $Sur(\omega, a)$ by (1). If $\aleph_0 < Card(a)$, then

$$\aleph_1 \leqslant Card(a) \leqslant Card(b) \leqslant Card\big(P(\omega)\big) = 2^{\aleph_0} = \aleph_1.$$

Hence $Card(a) = Card(b)$; so $Sur(a, b)$ by (1). Thus the continuum hypothesis implies

$$\forall a\, \forall b\big(a \neq 0 \;\&\; a \subset b \;\&\; b \subset P(\omega) \rightarrow Sur(\omega, a) \vee Sur(a, b)\big).$$

Hence the negation of the continuum hypothesis is implied by

$$\exists a\, \exists b\big(a \neq 0 \;\&\; a \subset b \;\&\; b \subset P(\omega) \;\&\; \neg Sur(\omega, a) \;\&\; \neg Sur(a, b)\big).$$

We rewrite this as

$$\exists a\, \exists b\, \exists c\big(c = \omega \;\&\; \exists x(x \in a) \;\&\; a \subset b \;\&\; \forall x(x \in b \rightarrow x \subset c) \;\&\; \neg Sur(c, a) \;\&\; \neg Sur(a, b)\big).$$

Taking the interpretation and using Lemma 5 of §9.8, we get

$$\exists a\, \exists b\, \exists c\big(c =_I O(\omega) \;\&\; \exists x(x \in_I a) \;\&\; a \subset_I b \;\&\; \forall x(x \in_I b \rightarrow x \subset_I c) \;\&\; \neg Sur_I(c, a) \;\&\; \neg Sur_I(a, b)\big).$$

Making use of the interpreted equality theorem, we see that it will suffice to define constants A and B and prove

A. $\exists x(x \in_I A)$,

B. $A \subset_I B$,

C. $x \in_I B \rightarrow x \subset_I O(\omega)$,

D. $\neg Sur_I(O(\omega), A)$,

E. $\neg Sur_I(A, B)$.

We let CD be the set of mappings from finite subsets of $\omega \times \aleph_2$ to $\{0, 1\}$. We let i and j vary through natural numbers. We define

$$\begin{aligned} N(\sigma) &= [\langle O(i), p\rangle \mid p`\langle i, \sigma\rangle = 1], \\ A &= [N(\sigma) \mid \sigma < \aleph_1] \times CD, \\ B &= [N(\sigma) \mid \sigma < \aleph_2] \times CD. \end{aligned}$$

Then A and B follow from (17) of §9.8.

To prove C, we must show

$$x \in_I B \;\&\; y \in_I x \rightarrow y \in_I O(\omega).$$

By (17) of §9.8 and the interpreted equality theorem, we may suppose that $x = N(\sigma)$. By (16) of §9.8, $y =_I O(i)$ for some i; since $O(i) \in_I O(\omega)$ by (17) of §9.8, $y \in_I O(\omega)$.

To prove D and E we shall need some lemmas.

Lemma 1. Let x be a set of conditions such that for all p and q in x, $p \neq q$ implies that $p \cup q$ is not a condition. Then x is countable.

Proof. We define inductively a sequence $x_0, x_1, \ldots$ of finite subsets of x. Let a_n be the set of $\langle i, z\rangle$ in $\{0, 1\} \times (\omega \times \aleph_2)$ such that $\langle 1 - i, z\rangle$ is in some condition in $x_0 \cup \cdots \cup x_{n-1}$. Since each condition is a finite set and $x_0 \cup \cdots \cup x_{n-1}$ is a finite set of conditions, the set a_n is finite. We may therefore choose a finite subset x_n of x such that for every $p \in x$, there is a $q \in x_n$ such that $p \cap a_n = q \cap a_n$.

It will now suffice to show that every p in x is in some x_n. Since p is finite and $a_n \subset a_{n+1}$, we may choose n so that $p \cap a_n = p \cap a_{n+1}$. Choose $q \in x_n$ so that $p \cap a_n = q \cap a_n$. If $\langle i, z\rangle$ is in q, then $\langle 1 - i, z\rangle$ is in $a_{n+1} - q$. Since

$$p \cap a_{n+1} = q \cap a_n,$$

it follows that $\langle 1 - i, z\rangle \notin p$. This implies that $p \cup q$ is a condition; so by hypothesis, $p = q$.

Lemma 2. If p forces $Q(x, v_1, \ldots, v_n)$ and $\forall z(Q(z, v_1, \ldots, v_n) \rightarrow z = y)$, then p forces $x = y$.

Proof. We omit the parameters. Let q be an extension of p. Then q does not force $\exists z \neg(Q(z) \rightarrow z = y)$ and hence does not force $\neg(Q(x) \rightarrow x = y)$; so some extension r of q forces $Q(x) \to x = y$. This means that r forces either $\neg Q(x)$ or $x = y$. The former is impossible, since, as an extension of p, r forces $Q(x)$. Thus r forces $x = y$.

To prove that p forces $x = y$, suppose that $p \subset q$ and, say, $z \in_q x$. Choose r as above. Since r forces $x = y$ and $z \in_r x$, there is an extension of r which forces $z \in y$. Thus some extension of q forces $z \in y$.

A set a is *separated* if

$$\forall p\, \forall x\, \forall y(x, y \in a \;\&\; p \textit{ forces } x = y \to x = y).$$

Lemma 3. If $Sur_I(a \times CD, b \times CD)$ where $Card(a)$ is infinite and b is separated, then $Card(b) \leqslant Card(a)$.

Proof. By the hypothesis and (17) of §9.8, there is an f such that for each $y \in b$, there is an x such that

$$x \in_I a \times CD \;\&\; Im_I(f, x, y) \;\&\; \forall w\big(Im_I(f, x, w) \to w =_I y\big).$$

It follows that there is an $x \in a$ such that

$$Im_I(f, x, y) \;\&\; \forall w\big(Im_I(f, x, w) \to w =_I y\big).$$

Then by the truth lemma there is a correct p such that

$$p \textit{ forces } Im(f, x, y) \textit{ and } \forall w\big(Im(f, x, w) \to w = y\big). \tag{2}$$

For $x \in a$, let b_x be the set of $y \in b$ such that (2) holds for some p (not necessarily correct). We have shown that $b = Un([b_x \mid x \in a])$. We show that b_x is countable; this will imply that $Card(b) \leqslant Card(a) \cdot \omega = Card(a)$. For each $y \in b_x$, let p_y be a p such that (2) holds. Suppose that $y, z \in b_x$ and that

$$p = p_y \cup p_z$$

is a condition. Then p forces

$$Im(f, x, y) \qquad \text{and} \qquad \forall w\big(Im(f, x, w) \to w = z\big).$$

By Lemma 2, p forces $y = z$; so, since b is separated, $y = z$. This shows that $p_y = p_z \to y = z$; so to show that b_x is countable, it suffices to show that $[p_y \mid y \in b_x]$ is countable. But this follows from Lemma 1 and the result just proved.

We shall now prove by transfinite induction on τ that

$$\sigma < \tau \to p \textit{ does not force } O(\sigma) = O(\tau). \tag{3}$$

Since $\sigma < \tau$, $O(\sigma) \in_p O(\tau)$. If p forces $O(\sigma) = O(\tau)$, it follows that some extension q of p forces $O(\sigma) \in O(\sigma)$. Then q forces $O(\sigma) = O(\rho)$ for some $\rho < \sigma$. This contradicts the induction hypothesis.

From (3) we have

$$p \textit{ forces } O(\tau) = O(\sigma) \to \tau = \sigma. \tag{4}$$

We use this to show that

$$p \text{ forces } N(\tau) = N(\sigma) \rightarrow \tau = \sigma. \tag{5}$$

For suppose $\tau \neq \sigma$. Choose i so large that no $\langle i, \rho\rangle$ is in the domain of p. Let q be an extension of p such that $\langle i, \tau\rangle$ and $\langle i, \sigma\rangle$ are in the domain of q and

$$q`\langle i, \tau\rangle = 1, \qquad q`\langle i, \sigma\rangle = 0.$$

From the former, $O(i) \in_q N(\tau)$. Hence some extension r of q forces $O(i) \in N(\sigma)$. This implies that for some j, r forces $O(i) = O(j)$ and $O(j) \in_r N(\sigma)$. By (4), $i = j$; so $r`\langle i, \sigma\rangle = 1$. This is impossible, since r is an extension of q.

From (4) and (5),

$$\tau \neq \sigma \rightarrow O(\tau) \neq O(\sigma) \ \& \ N(\tau) \neq N(\sigma).$$

Hence

$$Card([O(i) \mid i \in \omega]) = \aleph_0,$$
$$Card([N(\sigma) \mid \sigma < \aleph_1]) = \aleph_1,$$
$$Card([N(\sigma) \mid \sigma < \aleph_2]) = \aleph_2.$$

Moreover $[N(\sigma) \mid \sigma < \aleph_1]$ and $[N(\sigma) \mid \sigma < \aleph_2]$ are separated by (5). Recalling that

$$O(\omega) = [O(i) \mid i \in \omega] \times CD,$$

we see that D and E follow from Lemma 3.

We remark that many extensions of Theorem 1 can be proved by the same method. For example, we can prove that *ZFC* remains consistent upon adding the axioms $2^{\aleph_0} = \aleph_1$ and $2^{\aleph_1} = \aleph_3$.

We now turn to the proof of Theorem 2. We let *CD* be the set of mappings from finite subsets of $\omega \times \omega$ to $\{0, 1\}$. We then construct *I* as before. We shall construct a new interpretation *J* of *ZF* in ZFL_{Cor} so that the interpretation of the negation of the axiom of choice holds.

By a *permutation*, we mean a bijective mapping from ω to ω. We let f and g vary through permutations. We use $f \circ g$ for the composition of f and g (so that $(f \circ g)`i = f`(g`i)$); $f^*$ for the inverse of $f$ (so that $f^*`i = j$ iff $f`j = i$); and $I$ for the identity permutation (so that $I`i = i$). We say that f is a *k-permutation* if $f`i = i$ for $i \leqslant k$.

We set

$$\pi_f(p) = [\langle i, j, f`k\rangle \mid \langle i, j, k\rangle \in p].$$

(Thus π is a binary function symbol, one of whose arguments is written as a subscript.) Clearly $\pi_f(p)$ is a condition; and we have $\pi_{f\circ g}(p) = \pi_f(\pi_g(p))$ and $\pi_I(p) = p$.

We define

$$\Pi_f(x) = [\langle \Pi_f(y), \pi_f(p)\rangle \mid \langle y, p\rangle \in x].$$

This is a definition by induction on $Od(x)$. We have $\Pi_{f\circ g}(x) = \Pi_f(\Pi_g(x))$.

We say x is *invariant* and write $Iv(x)$ if there is a k such that $\Pi_f(x) = x$ for every k-permutation f. This obviously implies that $\Pi_I(x) = x$.

If x is invariant, then $\Pi_f(x)$ is invariant. For choose k so that $\Pi_g(x) = x$ for every k-permutation g, and choose k' so that $i \leqslant k \rightarrow f`i \leqslant k'$. Suppose that g is a k'-permutation. Then $f^* \circ g \circ f$ is a k-permutation; so

$$\Pi_{f^*}(\Pi_g(\Pi_f(x))) = x.$$

Applying Π_f to both sides, $\Pi_g(\Pi_f(x)) = \Pi_f(x)$.

We define U_J by induction on $Od(x)$ as follows:

$$U_J(x) \leftrightarrow Iv(x) \;\&\; \forall y(y \in Ra(x) \rightarrow U_J(y)). \tag{6}$$

We now complete the definition of the interpretation J by taking $\in_J$ to be $\in_I$ and $=_J$ to be $=_I$.

For every open formula $\mathbf{A}$, $\mathbf{A}_I$ is the same as $\mathbf{A}_J$; so if $\vdash \mathbf{A}^{(I)}$, then $\vdash \mathbf{A}^{(J)}$. From this we obtain the interpretations under J of the identity and equality axioms. Then we easily prove

$$x =_J y \rightarrow (Q_J(x, v_1, \ldots, v_n) \leftrightarrow Q_J(y, v_1, \ldots, v_n)) \tag{7}$$

by induction on predicate symbols. From (16) of §9.8 and (6), we get

$$U_J(y) \;\&\; x \in_J y \rightarrow \exists z(U_J(z) \;\&\; x =_J z). \tag{8}$$

The interpretation of the extensionality axiom under J is

$$U_J(x) \;\&\; U_J(y) \;\&\; \forall z(U_J(z) \rightarrow (z \in_J x \leftrightarrow z \in_J y)) \rightarrow x =_J y.$$

Suppose that the hypotheses hold and the conclusion is false. Using the interpretation under I of the extensionality axiom, we find that there is a z such that $\neg(z \in_J x \leftrightarrow z \in_J y)$. From (8) and (7), this z may be chosen so that $U_J(z)$; and this contradicts the hypotheses.

The interpretation under J of the regularity axiom is

$$U_J(x) \;\&\; \exists y(U_J(y) \;\&\; y \in_J x)$$
$$\rightarrow \exists y(U_J(y) \;\&\; y \in_J x \;\&\; \neg\exists z(U_J(z) \;\&\; z \in_J x \;\&\; z \in_J y)).$$

Assume the hypothesis. By the interpretation under I of the regularity axiom, there is a y such that $y \in_J x$ and $\neg\exists z(z \in_J x \;\&\; z \in_J y)$. But by (8) and (7) we may also suppose that $U_J(y)$.

We now introduce the notion of *J-forcing*. We define *p J-forces* $Q(x_1, \ldots, x_n)$ just like *p forces* $Q(x_1, \ldots, x_n)$, except in the case in which

$$Q(x_1, \ldots, x_n) \leftrightarrow \exists y R(y, x_1, \ldots, x_n).$$

In this case, p J-forces $Q(x_1, \ldots, x_n)$ if for some y, $U_J(y)$ and p J-forces $R(y, x_1, \ldots, x_n)$. We say that $Q(x_1, \ldots, x_n)$ *is J-forced* if some correct condition J-forces $Q(x_1, \ldots, x_n)$.

Lemmas 1 and 2 of the last section hold for J-forcing. The truth lemma also holds:

$$Q_J(x_1, \ldots, x_n) \leftrightarrow Q(x_1, \ldots, x_n) \textit{ is J-forced.}$$

We prove

$$U_J(x) \to U_J(\Pi_f(x)) \tag{9}$$

by induction on $Od(x)$. If $y \in Ra(\Pi_f(x))$, then $y = \Pi_f(z)$ for some $z \in Ra(x)$. Then $U_J(z)$; so $U_J(y)$ by induction hypothesis. As previously proved, $\Pi_f(x)$ is invariant; so $U_J(\Pi_f(x))$.

We now prove

$$p \textit{ J-forces } Q(x_1, \ldots, x_n) \leftrightarrow \pi_f(p) \textit{ J-forces } Q(\Pi_f(x_1), \ldots, \Pi_f(x_n)) \tag{10}$$

by induction on predicate symbols. For the atomic case, we must prove

$$p \textit{ J-forces } x \in y \leftrightarrow \pi_f(p) \textit{ J-forces } \Pi_f(x) \in \Pi_f(y), \tag{11}$$

$$p \textit{ J-forces } x = y \leftrightarrow \pi_f(p) \textit{ J-forces } \Pi_f(x) = \Pi_f(y). \tag{12}$$

Assume that (12) holds when $Od(x) < \sigma$ and $Od(y) < \sigma$. We show that (11) holds when $Od(x) < \sigma$, $Od(y) \leqslant \sigma$. By definition, p *J-forces* $x \in y$ is equivalent to

$$\exists z(p \textit{ J-forces } x = z \;\&\; z \in_p y). \tag{13}$$

Now $z \in_p y \to Od(z) < Od(y) \leqslant \sigma$. Hence by hypothesis and the definition of $\in_p$, (13) is equivalent to

$$\exists z(\pi_f(p) \textit{ J-forces } \Pi_f(x) = \Pi_f(z) \;\&\; \Pi_f(z) \in_{\pi_f(p)} \Pi_f(x)). \tag{14}$$

Now if $z' \in_q \Pi_f(y)$, then $z' \in Ra(\Pi_f(y))$ and hence $z' = \Pi_f(z)$ for some z. Thus (14) is equivalent to

$$\exists z'(\pi_f(p) \textit{ J-forces } \Pi_f(x) = z' \;\&\; z' \in_{\pi_f(p)} \Pi_f(y))$$

and hence to $\pi_f(p)$ *J-forces* $\Pi_f(x) \in \Pi_f(y)$.

We now show that, under the same assumption, (12) holds for $Od(x) \leqslant \sigma$ and $Od(y) \leqslant \sigma$. This will prove (12) by induction on $Max(\langle Od(x), Od(y)\rangle)$; and (11) will follow from the above.

By definition, p *J-forces* $x = y$ iff

$$\forall q \, \forall z(p \subset q \;\&\; z \in_q x \to \exists r(q \subset r \;\&\; r \textit{ J-forces } z \in y)) \tag{15}$$

and a similar statement with x and y interchanged holds. From the result just proved, (15) is equivalent to

$$\begin{aligned} &\forall q \, \forall z(\pi_f(p) \subset \pi_f(q) \;\&\; \Pi_f(z) \in_{\pi_f(q)} \Pi_f(x) \\ &\qquad \to \exists r(\pi_f(q) \subset \pi_f(r) \;\&\; \pi_f(r) \textit{ J-forces } \Pi_f(z) \in \Pi_f(y))). \end{aligned} \tag{16}$$

Now as q varies through all conditions, $\pi_f(q)$ varies through all conditions (since $q = \pi_f(\pi_{f^*}(q))$). Hence, as above, (16) is equivalent to

$$\forall q \, \forall z'(\pi_f(p) \subset q \;\&\; z' \in_q \Pi_f(x) \to \exists r(q \subset r \;\&\; r \textit{ J-forces } z' \in \Pi_f(y))). \tag{17}$$

From the equivalence of (15) and (17) and the corresponding result with x and y interchanged, we get (12).

We now return to (10). The case in which Q is a negation or a disjunction is quite easy. Now suppose that $Q(x_1, \ldots, x_n) \leftrightarrow \exists y R(y, x_1, \ldots, x_n)$. Then p *J-forces* $Q(x_1, \ldots, x_n)$ is equivalent to

$$\exists y\big(U_J(y) \mathbin{\&} p \textit{ J-forces } R(y, x_1, \ldots, x_n)\big)$$

which, by induction hypothesis, is equivalent to

$$\exists y\big(U_J(y) \mathbin{\&} \pi_f(p) \textit{ J-forces } R\big(\Pi_f(y), \Pi_f(x_1), \ldots, \Pi_f(x_n)\big)\big). \tag{18}$$

We want to show that this is equivalent to

$$\exists y'\big(U_J(y') \mathbin{\&} \pi_f(p) \textit{ J-forces } R\big(y', \Pi_f(x_1), \ldots, \Pi_f(x_n)\big)\big). \tag{19}$$

Now (18) implies (19) by (9). If (19) holds for some y', set $y = \Pi_{f^*}(y')$. Then $U_J(y)$ by (9); and $\Pi_f(y) = \Pi_I(y') = y'$ because y' is invariant. Thus (18) holds.

Lemma 4. Let $y, v_1, \ldots, v_n$ be U_J-sets, and set

$$z = [\langle u, p\rangle \mid u \in Ra(y) \mathbin{\&} p \textit{ J-forces } u \in y \mathbin{\&} p \textit{ J-forces } Q(u, v_1, \ldots, v_n)].$$

Then z is a U_J-set, and for all x,

$$x \in_J z \leftrightarrow x \in_J y \mathbin{\&} Q_J(x, v_1, \ldots, v_n).$$

Proof. The proof of the equivalence is as in Lemma 4 of §9.8, using (7). Since $Ra(z) \subset Ra(y)$, every element of $Ra(z)$ is a U_J-set. Thus we need only show that z is invariant.

Choose k so that $\Pi_f(y) = y$, $\Pi_f(v_1) = v_1, \ldots, \Pi_f(v_n) = v_n$ for every k-permutation f. Then for f a k-permutation,

$$\Pi_f(z) = [\langle \Pi_f(u), \pi_f(p)\rangle \mid u \in Ra(y) \mathbin{\&} p \textit{ J-forces } u \in y \\ \mathbin{\&} p \textit{ J-forces } Q(u, v_1, \ldots, v_n)].$$

Using (10), we get

$$\Pi_f(z) = [\langle \Pi_f(u), \pi_f(p)\rangle \mid u \in Ra(y) \mathbin{\&} \pi_f(p) \textit{ J-forces } \Pi_f(u) \in \Pi_f(y) \\ \mathbin{\&} \pi_f(p) \textit{ J-forces } Q\big(\Pi_f(u), \Pi_f(v_1), \ldots, \Pi_f(v_n)\big)]. \tag{20}$$

As p varies through all conditions, $\pi_f(p)$ varies through all conditions. As u varies through all elements of $Ra(y)$, $\Pi_f(u)$ varies through all elements of

$$[\Pi_f(w) \mid w \in Ra(y)] = Ra\big(\Pi_f(y)\big) = Ra(y).$$

Hence (20) shows that $\Pi_f(z) = z$.

The interpretation under J of a subset axiom is

$$U_J(y) \mathbin{\&} U_J(v_1) \mathbin{\&} \cdots \mathbin{\&} U_J(v_n) \\ \rightarrow \exists z\big(U_J(z) \mathbin{\&} \forall x\big(U_J(x) \rightarrow \big(x \in_J z \leftrightarrow x \in_J y \mathbin{\&} Q_J(u, v_1, \ldots, v_m)\big)\big)\big).$$

This follows from Lemma 4.

We define

$$\mathfrak{I}(z) = Un[\Pi_f(z) \mid f \text{ a permutation}].$$

Then

$$U_J(x) \ \&\ x \in_J z \rightarrow x \in_J \mathfrak{I}(z). \tag{21}$$

For by the hypothesis, there is a correct p such that p J-forces $x \in z$. By (10),

$$\pi_I(p) \ J\text{-forces } \Pi_I(x) \in \Pi_I(z).$$

Since x is invariant, this means that p J-forces $x \in \Pi_I(z)$. Hence for some y, p J-forces $x = y$ and $y \in_p \Pi_I(z)$. Then $y \in_p \mathfrak{I}(z)$; so p J-forces $x \in \mathfrak{I}(z)$; so $x \in_J \mathfrak{I}(z)$.

Now

$$\begin{aligned}\Pi_g(\mathfrak{I}(z)) &= [\langle \Pi_g(x), \pi_g(p)\rangle \mid \langle x, p\rangle \in \mathfrak{I}(z)]_{x,p} \\ &= [\langle \Pi_g(x), \pi_g(p)\rangle \mid \langle x, p\rangle \in \Pi_f(z)]_{x,p,f} \\ &= [\langle \Pi_{g\circ f}(x), \pi_{g\circ f}(p)\rangle \mid \langle x, p\rangle \in z]_{x,p,f} \\ &= [\langle \Pi_f(x), \pi_f(p)\rangle \mid \langle x, p\rangle \in z]_{x,p,f} = \mathfrak{I}(z)\end{aligned}$$

(where we have used the fact that as f varies through all permutations, $g \circ f$ varies through all permutations). Thus $\mathfrak{I}(z)$ is invariant. From this,

$$\forall y(y \in Ra(z) \rightarrow U_J(y)) \rightarrow U_J(\mathfrak{I}(z)). \tag{22}$$

For if $x \in Ra(\mathfrak{I}(z))$, then $x \in Ra(\Pi_f(z))$ for some f; so $x = \Pi_f(y)$ for some $y \in Ra(z)$; so $U_J(x)$ by (9).

The interpretation under J of a replacement axiom has the hypotheses $U_J(w)$ and

$$\forall x(U_J(x) \rightarrow \exists u(U_J(u) \ \&\ \forall y(U_J(y) \rightarrow (y \in_J u \leftrightarrow Q_J(x, y))))), \tag{23}$$

and the conclusion

$$\exists z(U_J(z) \ \&\ \forall y(U_J(y) \rightarrow \exists x(U_J(x) \ \&\ x \in_J y \ \&\ Q_J(x, y)) \rightarrow y \in_J z)) \tag{24}$$

(where we have omitted the parameters). Let $F(x, p)$ be the u of smallest order such that $U_J(u)$ and p J-forces $\forall y(y \in u \leftrightarrow Q(x, y))$ if such a u exists. As in the last section

$$\forall x(U_J(x) \rightarrow \exists p \, \forall y(U_J(y) \rightarrow (y \in_J F(x, p) \leftrightarrow Q_J(x, y)))). \tag{25}$$

We set

$$z = \mathfrak{I}(Un([Ra(F(x, p)) \mid x \in Ra(w)]_{x,p}) \times CD).$$

Since $U_J(F(x, p))$, every element of $Ra(F(x, p))$ is a U_J-set. From this and (22), $U_J(z)$. It remains to show that

$$U_J(y) \ \&\ U_J(x) \ \&\ x \in_J w \ \&\ Q_J(x, y) \rightarrow y \in_J z.$$

Assuming the hypotheses, choose p as in (25). Then $y \in_J F(x, p)$. By (16) of §9.8, $y =_J y'$ with $y' \in Ra(F(x, p))$. By (17) of §9.8 and (21), $y' \in_J z$; so $y \in_J z$.

The interpretation under J of the power set axiom is

$$U_J(y) \to \exists w\big(U_J(w) \;\&\; \forall x\big(U_J(x) \to \forall z\big(U_J(z) \to z \in_J x \to z \in_J y\big) \to x \in_J w\big)\big).$$

Let

$$w = \mathfrak{J}\big([v \mid U_J(v) \;\&\; v \subset Ra(y) \times CD] \times CD\big).$$

Then $U_J(w)$ by (22). Suppose that $U_J(x)$ and that $z \in_J x \to z \in_J y$ for every U_J-set z. Let

$$v = [\langle u, p\rangle \mid u \in Ra(y) \;\&\; p \textit{ J-forces } u \in y \;\&\; p \textit{ J-forces } u \in x].$$

By Lemma 4, $U_J(v)$; and, for every U_J-set z,

$$\begin{aligned} z \in_J v &\leftrightarrow z \in_J y \;\&\; z \in_J x \\ &\leftrightarrow z \in_J x. \end{aligned}$$

Thus $v =_J x$ by the interpreted extensionality axiom. Since $v \subset Ra(y) \times CD$, $v \in_J w$ by (17) of §9.8 and (21); so $x \in_J w$.

We have

$$\Pi_f\big(O(\sigma)\big) = [\Pi_f\big(O(\tau)\big) \mid \tau < \sigma] \times CD.$$

From this we readily obtain

$$\Pi_f\big(O(\sigma)\big) = O(\sigma) \tag{26}$$

by transfinite induction on σ. Thus $O(\sigma)$ is invariant. Using this, we prove

$$U_J\big(O(\sigma)\big) \tag{27}$$

by transfinite induction on σ. Every element of $Ra\big(O(\sigma)\big)$ is $O(\tau)$ for some $\tau < \sigma$, and hence is a U_J-set by the induction hypothesis. Hence $O(\sigma)$ is a U_J-set. The interpreted axiom of infinity follows from (27) and Lemma 5 of §9.8.

It remains to prove the interpretation of the negation of the axiom of choice. We actually prove the interpretation of

$$\neg \exists z CF\big(z, P(\omega)\big). \tag{28}$$

(Note that, by contrast, $\exists z CF(z, \omega)$ is a theorem of ZF. For to obtain a choice function z on ω, we let z‘x be the smallest natural number in x for $x \in P(\omega) - \{0\}$.)

Lemma 5. Let p be a correct condition, and let $k < i$. Then there is a k-permutation f such that f‘$i \neq i$ and $\pi_f(p)$ has a correct extension.

Proof. Let a be the set of extensions of conditions $\pi_f(p)$, where f is a k-permutation such that f‘$i \neq i$. We want to show that a contains a correct condition. Suppose not. Then by Cor_4, there is a correct condition q having no extension in a. By Cor_3, $p \cup q$ is a condition. Choose j so large that $i < j$ and no $\langle x, j\rangle$ is in $Do(p) \cup Do(q)$; and let f be the permutation which interchanges i and j and leaves all other natural numbers fixed. Clearly $\pi_f(p) \cup q$ is a condition. This contradicts the choice of q, since f is a k-permutation and f‘$i \neq i$.

If $CF(z, P(\omega))$, then

$$a \subset P(\omega) \;\&\; a \neq 0 \rightarrow \exists x(x \in a \;\&\; Im(z, a, x) \;\&\; \forall y(Im(z, a, y) \rightarrow y = x)).$$

Thus (28) is implied by

$$\forall z\, \exists a(a \subset P(\omega) \;\&\; a \neq 0 \;\&\; \neg\exists x(x \in a \;\&\; Im(z, a, x) \\ \;\&\; \forall y(Im(z, a, y) \rightarrow y = x))).$$

We thus need only prove the interpretation of this sentence. If we form this interpretation by the methods used earlier in this section, we see that we must produce for each U_J-set z a U_J-set a such that

$$U_J(x) \;\&\; x \in_J a \rightarrow x \subset_J O(\omega), \tag{29}$$

$$\exists x(U_J(x) \;\&\; x \in_J a), \tag{30}$$

and such that there is no U_J-set x for which

$$x \in_J a \;\&\; Im_J(z, a, x) \;\&\; \forall y(U_J(y) \rightarrow Im_J(z, a, y) \rightarrow y =_J x). \tag{31}$$

We define

$$N(i) = [\langle O(j), p\rangle \mid p`\langle j, i\rangle = 1].$$

As before, p J-forces $N(i) = N(j)$ implies $i = j$; so

$$N(i) =_J N(j) \rightarrow i = j. \tag{32}$$

Also, using (26) and the definition of $\pi_{f^*}(q)$, we get

$$\begin{aligned} \Pi_f(N(i)) &= [\langle O(j), \pi_f(p)\rangle \mid p`\langle j, i\rangle = 1] \\ &= [\langle O(j), q\rangle \mid \pi_{f^*}(q)`\langle j, i\rangle = 1] \\ &= [\langle O(j), q\rangle \mid q`\langle j, f`i\rangle = 1]; \end{aligned}$$

so

$$\Pi_f(N(i)) = N(f`i). \tag{33}$$

It follows that $N(i)$ is invariant. From this and (27), $U_J(N(i))$.

Given the U_J-set z, choose k so that $\Pi_f(z) = z$ for every k-permutation f, and set

$$a = [N(i) \mid k < i] \times CD.$$

Every element of $Ra(a)$ is an $N(i)$ and hence a U_J-set. Using (33), we see that $\Pi_f(a) = a$ for f a k-permutation; so a is invariant and hence a U_J-set. The proof of (29) is similar to the proof of B. Since $N(S(i)) \in_J a$, we have (30).

Now assume that there is a U_J-set x satisfying (31). From $x \in_J a$ we have $x =_J N(i)$ for some $i > k$. We may as well suppose that x is $N(i)$. Choose p by the truth lemma so that p J-forces $\forall y(Im(z, a, y) \rightarrow y = N(i))$. Choose f as in Lemma 5, and let q be a correct extension of $\pi_f(p)$. Since $\Pi_f(z) = z$, $\Pi_f(a) = a$, and $\Pi_f(N(i)) = N(f`i)$, we have by (10)

$$q \textit{ forces } \forall y(Im(z, a, y) \rightarrow y = N(f`i)).$$

Hence by the truth lemma

$$\forall y(U_J(y) \to Im_J(z, a, y) \to y =_J N(f`i)).$$

Substituting $N(i)$ for y and using (31), we get $N(i) =_J N(f`i)$. Since $i \neq f`i$, this contradicts (32).

9.10 LARGE CARDINALS

Since the axioms of *ZFC* do not settle the continuum hypothesis and the axiom of constructibility, it is natural to look for axioms which do. Now both the continuum hypothesis and the axiom of constructibility restrict the number of sets. Thus the unprovable part of the continuum hypothesis says that the number of subsets of ω does not exceed $\aleph_1$. Hence if we hope to prove these results, we must find new axioms which restrict the number of sets.

The extensionality axiom shows that an individual is determined by its members and hence can be identified with the collection whose elements are these members. The only further restriction we wish to make on the individuals is that they shall be sets, i.e., shall occur at some stage of our construction. But we will show that this can actually be proved in *ZF*.

We recall that at each stage, exactly one new ordinal is constructed. Let S_σ be the stage at which σ is constructed and let $Stg(\sigma)$ be the set of all sets which may be constructed at stage S_σ. We can then define Stg in *ZFC* by transfinite induction as follows:

$$Stg(\sigma) = P(Un([Stg(\tau) \mid \tau < \sigma])). \tag{1}$$

The result which we want to prove is then

$$\exists\sigma(x \in Stg(\sigma)). \tag{2}$$

We shall first prove that there is a transitive set containing x. Define by transfinite induction

$$F(\sigma) = Un[Un(F(\tau)) \cup \{x\} \mid \tau < \sigma].$$

Clearly $x \in F(\omega)$. If $z \in F(\omega)$, then $z \in Un(F(i)) \cup \{x\}$ for some i. Then $z \in F(S(i))$; so $z \subset Un(F(S(i)))$; so $z \subset F(\omega)$. This shows that $F(\omega)$ is transitive.

Now we prove (2). Let a be a transitive set containing x, and let b be the set of elements in a which belong to no $Stg(\sigma)$. If b is empty, we are through. Otherwise, let y be a minimal element of b. If $w \in y$, then $w \in a$ by the transitivity of a, but $w \notin b$ by the choice of y. Hence $w \in Stg(\sigma_w)$ for some σ_w. Let σ be larger than any ordinal in $[\sigma_w \mid w \in y]$. It follows from (1) that $y \in Stg(\sigma)$, contradicting $y \in b$.

In view of this, there seems to be little hope of proving the continuum hypothesis or the axiom of constructibility without changing our notion of a set. If we try to disprove the continuum hypothesis and the axiom of constructibility, the situation appears much more hopeful. For this, we need axioms which guarantee the existence of more sets. Now the existence axioms of *ZFC* are very deficient in some respects. For example, the subset axioms do not really guarantee the existence of all subsets of x, but only of those subsets which can be described in *ZFC* (using parameters). If we introduced symbols for new operations which cannot be defined in *ZFC*, we would increase our ability to describe sets and hence increase the power of the subset (and replacement) axioms. This appears to be a natural approach; but so far no one has been able to propose any suitable operations.

An approach which is more promising at the moment is to make fuller use of our principle of existence of stages: if we can imagine a situation in which all of the stages in a collection are completed, then there must be a stage after all the stages in the collection. In *ZFC* we have used only a few special cases of this. Now the existence of further stages is equivalent to the existence of larger ordinals; or, since there is a cardinal larger than each ordinal, to the existence of larger cardinals. Thus we must look for axioms guaranteeing the existence of cardinals larger than those which may be proved to exist in *ZFC*. Such axioms will be called *generalized axioms of infinity*.

Let us consider how we may obtain cardinals in *ZFC*. We obtain $\aleph_0$ from the axiom of infinity. Having obtained an infinite cardinal σ, we can obtain a larger one by passing to 2^σ. Another method is to take $Un(x)$, where x is a set of cardinals. Of course, this should only be viewed as a method for obtaining a new cardinal when $Un(x)$ is larger than every element of x and also larger than $Card(x)$. This can happen; for example, x may be $[\aleph_\sigma \mid \sigma \in \omega]$, so that $Un(x) = \aleph_\omega$.

We say that a cardinal is inaccessible if it cannot be obtained by these methods. More precisely, a cardinal σ is *inaccessible* if:

i) $\omega < \sigma$;

ii) for every cardinal τ, $\tau < \sigma \rightarrow 2^\tau < \sigma$;

iii) for every subset x of σ, $Card(x) < \sigma \rightarrow Un(x) < \sigma$.

Our first generalized axiom of infinity, the *axiom of inaccessibility*, says that there exists an inaccessible cardinal. It is not immediately apparent that this is really a new axiom; for we might be able to prove it in *ZFC* by using constructions other than those discussed above. However, we shall prove that this is not the case; if *ZFC* is consistent, then the axiom of inaccessibility is not a theorem of *ZFC*. To prove this, we construct an interpretation of *ZFC* in *ZFC* for which the interpretation of the negation of the axiom of inaccessibility holds.

The *rank* of a set x is the first σ such that $x \in Stg(\sigma)$:

$$rk(x) = \mu\sigma\big(x \in Stg(\sigma)\big).$$

Then

$$x \in y \rightarrow rk(x) < rk(y). \tag{3}$$

For $y \in Stg\big(rk(y)\big)$; so if $x \in y$, then $x \in Stg(\sigma)$ for some $\sigma < rk(y)$ by (1). Hence $rk(x) \leqslant \sigma < rk(y)$. Moreover, $Set_x\big(rk(x) \leqslant \sigma\big)$, since

$$rk(x) \leqslant \sigma \rightarrow x \in Un\big([Stg(\tau) \mid \tau \leqslant \sigma]\big).$$

Thus we may give definitions by induction on $rk(x)$.

We write $Inac(\sigma)$ to mean that σ is an inaccessible cardinal. We say that x is *accessible* if $Card(x)$ is less than every inaccessible cardinal:

$$Ac(x) \leftrightarrow \forall\sigma\big(Inac(\sigma) \rightarrow Card(x) < \sigma\big).$$

We then define M by induction on $rk(x)$ as follows:

$$M(x) \leftrightarrow Ac(x) \ \& \ \forall y\big(y \in x \rightarrow M(y)\big). \tag{4}$$

Clearly $M(0)$; so $\exists x M(x)$ and hence M is an $\in$-interpretation. It is transitive by (4). Since a subset of an accessible set is accessible, it follows from (4) that M is supertransitive.

Lemma 1. If x is accessible and every member of x is accessible, then $Un(x)$ is accessible.

Proof. Let $Inac(\sigma)$. Then $Card(x) < \sigma$ and $Card(y) < \sigma$ for every member y of x. It follows that

$$Card\big(Un(x)\big) \leqslant \sigma \cdot \sigma = \sigma.$$

Hence we may as well suppose that $Un(x)$ is a subset of σ. If $y \in x$, then $y \subset \sigma$ and $Card(y) < \sigma$; so $Un(y) < \sigma$. Since $Card\big([Un(y) \mid y \in x]\big) \leqslant Card(x) < \sigma$,

$$\tau = Un\big([Un(y) \mid y \in x]\big) < \sigma.$$

If $\rho \in Un(x)$, then $\rho \in y$ for some $y \in x$; so $\rho \leqslant Un(y) \leqslant \tau$. It follows that $Un(x) \subset S(\tau)$; so

$$Card\big(Un(x)\big) \leqslant Card\big(S(\tau)\big) = Card(\tau) + 1 \leqslant Max\big(\langle Card(\tau), \omega\rangle\big) < \sigma.$$

We now prove the interpretations of the axioms of ZF by means of the lemmas of §9.5. The interpretation of the extensionality axiom and the regularity axiom hold because M is transitive. For the subset axioms, we must show that if $y, v_1, \ldots, v_n$ are M-sets, then

$$z = [x \mid x \in y \ \& \ Q_M(x, v_1, \ldots, v_n)]$$

is an M-set. Since $z \subset y$, this follows from the supertransitivity of M.

For the replacement axioms, it suffices to show that if F is M-invariant and $w, v_1, \ldots, v_n$ are M-sets, then

$$z = Un\big([F(x, v_1, \ldots, v_n) \mid x \in w]\big)$$

is an M-set. From the transitivity of M and the M-invariance of F, we find that every element of z is an M-set. By Lemma 1, z is accessible. Hence z is an M-set.

For the power set axiom, we must show that if y is an M-set, then

$$z = [x \mid M(x) \ \& \ x \subset y]$$

is an M-set. Obviously every element of z is an M-set; so we need only show that z is accessible. Let σ be an inaccessible cardinal. Then $Card(y) < \sigma$; so

$$Card(z) \leqslant Card\big(P(y)\big) = 2^{Card(y)} < \sigma.$$

If σ is less than every inaccessible cardinal, then σ is an *M*-set; this is easily proved by transfinite induction. Hence $M(\omega)$. This gives the interpretation of the axiom of infinity.

Since *CF* is absolute, the interpretation of the axiom of choice is

$$\forall x(M(x) \rightarrow \exists y(M(y) \;\&\; CF(y, x))).$$

Since *CF* is complete (for supertransitive interpretations), this is equivalent to

$$\forall x(M(x) \rightarrow \exists y CF(y, x)).$$

But this is a consequence of the axiom of choice.

We must still prove the interpretation of the negation of the axiom of inaccessibility. Now the defining axiom of *Inac* is

$$Inac(\sigma) \leftrightarrow Cd(\sigma) \;\&\; \omega < \sigma \;\&\; \forall\tau(\tau < \sigma \;\&\; Cd(\tau) \rightarrow 2^{\tau} < \sigma) \\ \;\&\; \forall x(x \subset \sigma \;\&\; Card(x) < \sigma \rightarrow Un(x) < \sigma).$$

Using the results of §9.5, we conclude that *Inac* is absolute for supertransitive interpretations. Since the negation of the axiom of constructibility is

$$\neg\exists x\, Inac(x),$$

its interpretation is equivalent to

$$\neg\exists x(M(x) \;\&\; Inac(x)).$$

This is evident; for if $M(x)$ & $Inac(x)$, then $x = Card(x) < x$.

Although not provable in *ZFC*, the axiom of inaccessibility is justified by our general principles; for we may certainly imagine a situation in which the operations described above for obtaining new cardinals have been repeated until no more new cardinals can be obtained in this way. We might even hope to prove the consistency of this axiom. More exactly, we might hope to prove that if *ZFC* is consistent, then so is the theory *ZFI* obtained from *ZFC* by adding the axiom of inaccessibility. However, no such proof can be carried out in *ZFC*, for reasons which we now discuss briefly.

The main point is that the consistency of *ZFC* can be proved in *ZFI*. In *ZFI*, let σ be the first inaccessible cardinal. Then $rk(x) < \sigma$ for every *M*-set *x*, as is easily proved by induction on $rk(x)$. It follows that $Set_x\, M(x)$. This means that in *ZFI* we can construct the structure for *ZFC* whose universe is $[x \mid M(x)]$ and whose $\in$-relation is the usual $\in$-relation. We can use essentially the proof just given to show that this is a model of *ZFC*. We can then prove the validity theorem in *ZFI*, and conclude that *ZFC* is consistent.

Suppose that the statement *if ZFC is consistent, then ZFI is consistent* were provable in *ZFC*, or even in *ZFI*. Then the consistency of *ZFI* would be provable in *ZFI*; so it would follow from the theorem on consistency proofs that *ZFI* is inconsistent.

The next and most important question is whether the continuum hypothesis and the axiom of constructibility can be proved or disproved in *ZFI*. Unfortunately, the answer is no. The proof is essentially the same as the proof for *ZFC*. The only new point is to prove the interpretations of the axiom of inaccessibility; and this involves nothing essentially new.

There are many more generalized axioms of infinity of a similar nature. For example, we could assume the existence of two inaccessible cardinals. It can be proved as above that this axiom is not provable in *ZFI*. We could assume much stronger axioms; e.g., that the cardinal of the set of inaccessible cardinals is itself inaccessible. However, none of these axioms serve to settle either the continuum hypothesis or the axiom of constructibility.

We shall now introduce an axiom which does settle the axiom of constructibility. Before giving the axiom, we need some definitions.

By a *measure ideal* on a set x, we mean a subset y of $P(x)$ such that:

i) for each subset z of y such that $Card(z) < Card(x)$, $Un(z) \in y$;

ii) for every subset z of x, either $z \in y$ or $x - z \in y$;

iii) $Un(y) = x$;

iv) $x \notin y$.

One may think of a measure ideal on x as providing a division of the subsets of x into small and large subsets; the subsets in the measure ideal are small and the remaining subsets are large.

A set x is *measurable* if it is uncountable and there is a measure ideal on x. Clearly every set similar to a measurable set is measurable; so we shall only consider measurable cardinals. The *axiom of measurability* states that there exists a measurable cardinal.

It is not immediately clear that the axiom of measurability is a generalized axiom of infinity. However, we shall see that measurable cardinals are extremely large, much larger than any of the cardinals which we have considered so far. On the other hand, investigations have not revealed any contradictions, or even any implausible results, which follow from this axiom. This at least suggests that the axiom of measurability is simply a very strong generalized axiom of infinity. If so, it would be a reasonable new axiom to adopt.

We let *ZFM* be the theory obtained from *ZFC* by adding the axiom of measurability. We let *ZFM** be the theory obtained from *ZFM* by adding two constants χ and V and axioms stating that χ is an uncountable cardinal and that V is a measure ideal on χ. By the theorem on functional extensions, *ZFM** is a conservative extension of *ZFM*. We shall now prove some results in *ZFM**.

$$x \in V \ \& \ y \subset x \rightarrow y \in V. \tag{5}$$

For if $y \notin V$, then $\chi - y \in V$ by (ii); so $\chi = (\chi - y) \cup x \in V$ by (i), contradicting (iv).

$$x \subset \chi \ \& \ Card(x) < \chi \rightarrow x \in V. \tag{6}$$

If $\sigma < \chi$, then $\{\sigma\} \in V$ by (iii) and (5). We then get (6) from (i).

$$x \subset \chi \rightarrow (x \in V \leftrightarrow \chi - x \notin V). \tag{7}$$

In view of (ii), we need only derive a contradiction from $x \in V$ and $\chi - x \in V$. But these imply $\chi \in V$ by (i), contradicting (iv).

$$x \subset \chi \;\&\; y \subset \chi \rightarrow (x \cap y \in V \leftrightarrow x \in V \vee y \in V). \tag{8}$$

If $x \in V$ or $y \in V$, then $x \cap y \in V$ by (5). Now suppose that $x \cap y \in V$, $x \notin V$, and $y \notin V$. By (7) and (i), $\chi = (x \cap y) \cup (\chi - x) \cup (\chi - y) \in V$, contradicting (iv).

Banach-Ulam Theorem. The cardinal χ is inaccessible.

Proof. Let σ be a cardinal such that $\sigma < \chi$; we must show that $2^\sigma < \chi$. Suppose that $\chi \leqslant 2^\sigma$. Then there is an injective mapping f from χ to $P(\sigma)$. For $\tau < \sigma$, let

$$z_\tau = [\rho \mid \rho < \chi \;\&\; \tau \in f`\rho].$$

Let $y_\tau = z_\tau$ if $z_\tau \in V$, and let $y_\tau = \chi - z_\tau$ otherwise. Then $y_\tau \in V$ by (7). Since $Card(\sigma) \leqslant \sigma < \chi$, we have $Un([y_\tau \mid \tau < \sigma]) \in V$ by (i). We shall show that at most one element of χ is not in $Un([y_\tau \mid \tau < \sigma])$. With (6) and (i) this will imply that $\chi \in V$, contradicting (iv).

Suppose that ρ is in no y_τ. Then for each τ,

$$\begin{aligned} \tau \in f`\rho &\leftrightarrow \rho \in z_\tau \\ &\leftrightarrow y_\tau = \chi - z_\tau \\ &\leftrightarrow z_\tau \notin V. \end{aligned}$$

Hence $f`\rho = [\tau \mid \tau < \sigma \;\&\; z_\tau \notin V]$. This determines $f`\rho$ and hence ρ.

Now let $x \subset \chi$ and $Card(x) < \chi$; we must show that $Un(x) < \chi$. Every element of x is an ordinal less than χ, and hence a subset of χ whose cardinal is less than χ. By (6), every member of x is in V; so by (i), $Un(x) \in V$. Then $Un(x) \neq \chi$ by (iv). Since clearly $Un(x) \leqslant \chi$, we have $Un(x) < \chi$.

We are now going to define an interpretation of *ZFM* in *ZFM**. The interpretation I is defined by

$$\begin{aligned} x \in_I y &\leftrightarrow [\sigma \mid \sigma < \chi \;\&\; x`\sigma \notin y`\sigma] \in V, \\ x =_I y &\leftrightarrow [\sigma \mid \sigma < \chi \;\&\; x`\sigma \neq y`\sigma] \in V, \\ U_I(x) &\leftrightarrow x = x. \end{aligned}$$

We shall first prove that for every predicate symbol Q of *ZFM*,

$$Q_I(x_1, \ldots, x_n) \leftrightarrow [\sigma \mid \sigma < \chi \;\&\; \neg Q(x_1`\sigma, \ldots, x_n`\sigma)] \in V. \tag{9}$$

We use induction on function symbols. If

$$Q(x_1, \ldots, x_n) \leftrightarrow x_i \in x_j$$

or

$$Q(x_1, \ldots, x_n) \leftrightarrow x_i = x_j,$$

the result follows from the definitions of $\in_I$ and $=_I$. Suppose that

$$Q(x_1, \ldots, x_n) \leftrightarrow \neg R(x_1, \ldots, x_n).$$

Then by the induction hypothesis and (7),

$$\begin{aligned} Q_I(x_1, \ldots, x_n) &\leftrightarrow \neg R_I(x_1, \ldots, x_n) \\ &\leftrightarrow [\sigma \mid \sigma < \chi \;\&\; \neg R(x_1`\sigma, \ldots, x_n`\sigma)] \notin V \\ &\leftrightarrow [\sigma \mid \sigma < \chi \;\&\; \neg Q(x_1`\sigma, \ldots, x_n`\sigma)] \in V. \end{aligned}$$

If Q is a disjunction, then the proof is similar, using (8) in place of (7). Now suppose that

$$Q(x_1, \ldots, x_n) \leftrightarrow \exists y R(y, x_1, \ldots, x_n).$$

Then

$$\begin{aligned} Q_I(x_1, \ldots, x_n) &\leftrightarrow \exists y R_I(y, x_1, \ldots, x_n) \\ &\leftrightarrow \exists y\big([\sigma \mid \sigma < \chi \;\&\; \neg R(y`\sigma, x_1`\sigma, \ldots, x_n`\sigma)] \in V\big). \end{aligned}$$

Set

$$\begin{aligned} a_y &= [\sigma \mid \sigma < \chi \;\&\; \neg R(y`\sigma, x_1`\sigma, \ldots, x_n`\sigma)], \\ a &= [\sigma \mid \sigma < \chi \;\&\; \neg \exists z R(z, x_1`\sigma, \ldots, x_n`\sigma)]. \end{aligned}$$

We must show that $\exists y(a_y \in V)$ iff $a \in V$. Since $a \subset a_y$ for all y,

$$\exists y(a_y \in V) \rightarrow a \in V$$

by (5). Now suppose that $a \in V$. If $\sigma < \chi$ and $\sigma \notin a$, there is a z_σ such that $R(z_\sigma, x_1`\sigma, \ldots, x_n`\sigma)$. Let y be such that $y`\sigma = z_\sigma$ for $\sigma < \chi$ and $\sigma \notin a$. Then $a_y \subset a$; so $a_y \in V$ by (5).

It follows from (9) that

$$Q_I(\{x_1\} \times \chi, \ldots, \{x_n\} \times \chi) \leftrightarrow [\sigma \mid \sigma < \chi \;\&\; \neg Q(x_1, \ldots, x_n)] \in V.$$

Now $[\sigma \mid \sigma < \chi \;\&\; \neg Q(x_1, \ldots, x_n)]$ is 0 or χ according as $Q(x_1, \ldots, x_n)$ or $\neg Q(x_1, \ldots, x_n)$. Since $0 \in V$ and $\chi \notin V$ by (iv) and (7),

$$Q_I(\{x_1\} \times \chi, \ldots, \{x_n\} \times \chi) \leftrightarrow Q(x_1, \ldots, x_n). \tag{10}$$

In particular, it follows that $Q_I \leftrightarrow Q$ for a 0-ary Q. As seen in §9.7, this implies that I is an interpretation of *ZFM*.

We are now going to show that I is isomorphic to a transitive $\in$-interpretation. In view of the results of §9.5, it will suffice to show that I is isomorphic to an interpretation J, where $=_J$ is $=$ and there is an ordinal function symbol H such that $Set_x\big(H(x) \leqslant \sigma\big)$ and $x \in_J y \rightarrow H(x) < H(y)$.

We have

$$x =_I x, \tag{11}$$

$$x =_I y \rightarrow y =_I x, \tag{12}$$

$$x =_I y \;\&\; y =_I z \rightarrow x =_I z; \tag{13}$$

for these are interpretations of theorems of *ZF*. Although $=_I$ has the properties of an equivalence relation, we cannot take equivalence classes in the usual sense, since they are not sets. We therefore define the equivalence class $EC(x)$ to be the set of all $y \in Stg(\sigma)$ such that $x =_I y$, where σ is the first ordinal such that $\exists y(y \in Stg(\sigma) \ \& \ x =_I y)$. (Such an ordinal exists by (11).) Using (11) through (13), it is then easy to prove the fundamental property of equivalence classes:

$$EC(x) = EC(y) \leftrightarrow x =_I y. \tag{14}$$

We now define the interpretation J by

$$x \in_J y \leftrightarrow \exists x' \exists y'(x = EC(x') \ \& \ y = EC(y') \ \& \ x' \in_I y'),$$
$$x =_J y \leftrightarrow x = y,$$
$$U_J(x) \leftrightarrow \exists x'(x = EC(x')).$$

We claim that EC is an isomorphism of I and J. In view of (14) and the definition of U_J, it is only necessary to prove

$$x \in_I y \leftrightarrow EC(x) \in_J EC(y). \tag{15}$$

The implication from left to right is obvious. Suppose that $EC(x) \in_J EC(y)$. Then $EC(x) = EC(x')$, $EC(y) = EC(y')$, and $x' \in_I y'$. By (14), $x =_I x'$ and $y =_I y'$. Thus we need only prove

$$x =_I x' \ \& \ y =_I y' \ \& \ x' \in_I y' \rightarrow x \in_I y.$$

But this is the interpretation of a theorem of ZF.

We must still define H. We first define by transfinite induction

$$Z(\sigma) = [x \mid x \in Stg(\sigma) \ \& \ \forall y(y \in_J x \rightarrow \exists \tau(\tau < \sigma \ \& \ y \in Z(\tau)))].$$

We will show that every set belongs to some $Z(\sigma)$. We then define

$$H(x) = \mu\sigma(x \in Z(\sigma)). \tag{16}$$

Then $Set_x(H(x) \leqslant \sigma)$ because

$$H(x) \leqslant \sigma \rightarrow x \in Un([Z(\tau) \mid \tau < S(\sigma)]);$$

and $x \in_J y \rightarrow H(x) < H(y)$ follows from the definition of Z.

We shall assume that x is in no $Z(\sigma)$ and derive a contradiction. If $\neg U_J(x)$, then $\forall y(y \notin_J x)$ and hence $x \in Z(rk(x))$. Thus $U_J(x)$; so $x = EC(y)$ for some y. Choose σ greater than every ordinal in $[rk(z) \mid z \in Ra(y)]$. Then if $z \in Ra(y)$, $z \in Stg(rk(z)) \subset Stg(\sigma)$. Hence if $y' = y \cap (Stg(\sigma) \times \chi)$, we have $y' =_J y$ and hence $x = EC(y')$. This shows that the set

$$a = [EC(u) \mid u \subset Stg(\sigma) \times \chi]$$

contains an element in no $Z(\tau)$.

Next we note that if $y \in a$ and $z \in_I y$, then $z \in a$. For we have $y = EC(u)$ with $u \subset Stg(\sigma) \times \chi$. By (15), $z = EC(v)$ with $v \in_I u$. Let v' be the set of $\langle v`\tau, \tau\rangle$ with $\tau < \chi$ and $v`\tau \in u`\tau$. Clearly $v' =_I v$; so $z = EC(v')$. By (3), $v' \subset Stg(\sigma) \times \chi$; so $z \in a$.

If $y \in a$ and y is in no $Z(\tau)$, then there is a $z \in a$ such that $z \in_I y$ and z is in no $Z(\tau)$. For suppose that this is not the case. Then for each $z \in_I y$, we have $z \in a$, and hence we have $z \in Z(\tau_z)$ for some τ_z. Choose τ larger than every ordinal in $[\tau_z \mid z \in_I y \ \& \ z \in a]$ and larger than $rk(y)$. Then $y \in Z(\tau)$, contrary to the hypothesis.

Using the results proved and the axiom of choice, we can find a sequence $y_0, y_1, \ldots$ of elements of a such that y_i is in no $Z(\tau)$ and $y_{S(i)} \in_I y_i$. Then

$$[\rho \mid \rho < \chi \ \& \ y_{S(i)}`\rho \notin y_i`\rho] \in V$$

for all i. Since $\omega < \chi$, the union of these sets is in V and hence is not equal to χ. Thus there is a $\rho < \chi$ such that $y_{S(i)}`\rho \in y_i`\rho$ for all i. This means that the set $[y_i`\rho \mid i \in \omega]$ has no minimal element. We have thus reached the desired contradiction.

We conclude that there is a transitive $\in$-interpretation M and an isomorphism G of I and M. We define

$$F(x) = G(\{x\} \times \chi).$$

Then from (10) and the fact that G is an isomorphism,

$$Q_M(F(x_1), \ldots, F(x_n)) \leftrightarrow Q(x_1, \ldots, x_n). \tag{17}$$

Now $F(x_1), \ldots, F(x_n)$ are M-sets; so when Q is absolute for M, we may drop the subscript M. In particular,

$$F(x) \in F(y) \leftrightarrow x \in y, \tag{18}$$

$$Ord(F(x)) \leftrightarrow Ord(x). \tag{19}$$

We have

$$G(y) \in F(x) \ \& \ Card(x) < \chi \rightarrow \exists z(z \in x \ \& \ G(y) = F(z)). \tag{20}$$

For assume the hypothesis. Since G is an isomorphism,

$$y \in_I \{x\} \times \chi,$$

so

$$[\sigma \mid \sigma < \chi \ \& \ y`\sigma \in x] \notin V$$

by (7). But

$$[\sigma \mid \sigma < \chi \ \& \ y`\sigma \in x] = Un[a_z \mid z \in x]$$

where

$$a_z = [\sigma \mid \sigma < \chi \ \& \ y`\sigma = z].$$

Since $Card(x) < \chi$, it follows from (i) that for some $z \in x$, $a_z \notin V$. From this and (7), $y =_I \{z\} \times \chi$. Applying the isomorphism G, we have $G(y) = F(z)$.

We shall use this to prove

$$Card(x) < \chi \ \& \ \forall z(z \in x \rightarrow F(z) = z) \rightarrow F(x) = x. \tag{21}$$

Assume the hypotheses. If $z \in x$, then $F(z) = z$ by hypothesis and $F(z) \in F(x)$ by (18); so $z \in F(x)$. This proves that $x \subset F(x)$. Now let $u \in F(x)$. Since $M(F(x))$ and M is transitive, u is an M-set; so $u = G(y)$ for some y. By (20), $u = F(z)$ for some $z \in x$. But $F(z) = z$ by hypothesis; so $u \in x$. Thus $F(x) \subset x$.

From (21) we easily prove by transfinite induction

$$\sigma < \chi \rightarrow F(\sigma) = \sigma. \tag{22}$$

On the other hand, we shall prove that

$$\chi < F(\chi). \tag{23}$$

For this, let $i = [\langle\sigma, \sigma\rangle \mid \sigma < \chi]$. Then clearly $i \in_I \{\chi\} \times \chi$. Applying G, $G(i) \in F(\chi)$. Since $F(\chi)$ is an ordinal by (19), $G(i)$ is an ordinal and

$$G(i) < F(\chi). \tag{24}$$

Now suppose that $\sigma < \chi$. Then

$$[\tau \mid \tau < \chi \;\&\; \sigma \notin i`\tau] = \sigma \cup \{\sigma\},$$

and $\sigma \cup \{\sigma\} \in V$ by (6) and (i). Hence $\{\sigma\} \times \chi \in_I i$. Applying G, $F(\sigma) \in G(i)$. Using (22), this gives $\sigma < G(i)$. Thus $\sigma < \chi \rightarrow \sigma < G(i)$. Substituting $G(i)$ for σ, we see that $\chi \leqslant G(i)$. From this and (24) we get (23).

Lemma 2. Let M be a transitive interpretation of ZFC. Then

$$M(\sigma) \;\&\; Inac(\sigma) \rightarrow Inac_M(\sigma).$$

Proof. Taking the interpretation of the defining axiom of Sm and using our results on absoluteness, we obtain

$$Sm_M(x, y) \leftrightarrow \exists z\big(M(z) \;\&\; IFunc(z) \;\&\; Do(z) = x \;\&\; Ra(z) = y\big)$$

for x and y M-sets. Hence

$$M(x) \;\&\; M(y) \;\&\; Sm_M(x, y) \rightarrow Sm(x, y). \tag{25}$$

Taking the interpretation of the theorem

$$Ord\big(Card(x)\big) \;\&\; Sm\big(Card(x), x\big)$$

and using (25), we get

$$M(x) \rightarrow Ord\big(Card_M(x)\big) \;\&\; Sm\big(Card_M(x), x\big);$$

so

$$M(x) \rightarrow Ord\big(Card_M(x)\big) \;\&\; Card(x) \leqslant Card_M(x). \tag{26}$$

The interpretation of the theorem $\sigma \leqslant Card(\sigma) \rightarrow Cd(\sigma)$ gives

$$\sigma \leqslant Card_M(\sigma) \rightarrow Cd_M(\sigma)$$

for σ an M-set. If $Cd(\sigma)$, then

$$\sigma = Card(\sigma) \leqslant Card_M(\sigma)$$

by (26); so

$$M(\sigma) \;\&\; Cd(\sigma) \rightarrow Cd_M(\sigma). \tag{27}$$

By (10) of §9.5 and the absoluteness of $\subset$,

$$P_M(x) = [y \mid M(y) \;\&\; y \subset x]$$

for x an M-set; so

$$M(x) \rightarrow P_M(x) \subset P(x). \tag{28}$$

The interpretation of the defining axiom of *Inac* says that for σ an M-set

$$\begin{aligned} Inac_M(\sigma) \leftrightarrow Cd_M(\sigma) \;\&\; \omega < \sigma \\ \&\; \forall\tau\big(Cd_M(\tau) \;\&\; \tau < \sigma \rightarrow Card_M(P_M(\tau)) < \sigma\big) \\ \&\; \forall x\big(M(x) \;\&\; x \subset \sigma \;\&\; Card_M(x) < \sigma \rightarrow Un(x) < \sigma\big). \end{aligned}$$

We must show that if σ is inaccessible, then the right-hand side holds. By (27), $Cd_M(\sigma)$; and clearly $\omega < \sigma$. Suppose that $Cd_M(\tau)$ and $\tau < \sigma$. Then $Card(\tau) < \sigma$; so

$$Card\big(P(\tau)\big)' = 2^{Card\,(\tau)} < \sigma.$$

Hence by (28), $Card\big(P_M(\tau)\big) < \sigma$. This and (25) imply

$$\neg\exists y\big(M(y) \;\&\; y \subset P_M(\tau) \;\&\; Sm_M(\sigma, y)\big).$$

If we take the interpretation of the theorem

$$\neg\exists y\big(y \subset P(\tau) \;\&\; Sm(\sigma, y)\big) \;\&\; Cd(\sigma) \rightarrow Card\big(P(\tau)\big) < \sigma,$$

we find that $Card_M\big(P_M(\tau)\big) < \sigma$.

Finally, suppose that $M(x)$ and $x \subset \sigma$ and $Card_M(x) < \sigma$. By (26), $Card(x) < \sigma$; so $Un(x) < \sigma$.

Hanf-Tarski Theorem. The cardinal of the set of inaccessible numbers less than χ is χ.

Proof. Define Q by

$$Q(\tau, a) \leftrightarrow \forall\sigma\big(Inac(\sigma) \;\&\; \sigma < \tau \rightarrow \sigma \in a\big).$$

From (17)

$$Q_M\big(F(\chi), F(a)\big) \leftrightarrow Q(\chi, a);$$

that is,

$$\begin{aligned} \forall\sigma\big(M(\sigma) \;\&\; Inac_M(\sigma) \;\&\; \sigma < F(\chi) \rightarrow \sigma \in F(a)\big) \\ \leftrightarrow \forall\sigma\big(Inac(\sigma) \;\&\; \sigma < \chi \rightarrow \sigma \in a\big). \end{aligned}$$

Taking a to be the set of inaccessible cardinals less than χ, we find that the right-hand side certainly holds; so the left-hand side holds also. Substituting χ for σ, we get

$$M(\chi) \;\&\; Inac_M(\chi) \;\&\; \chi < F(\chi) \rightarrow \chi \in F(a).$$

Now $\chi < F(\chi)$ by (23). Since $M\big(F(\chi)\big)$ and M is transitive, this implies $M(\chi)$. Also $Inac_M(\chi)$ by Lemma 2 and the Banach-Ulam theorem. Thus we have $\chi \in F(a)$. But $\chi \notin a$; so $F(a) \neq a$. If $x \in a$, x is a cardinal less than χ; so $F(x) = x$ by (22). From these facts and (21), we conclude that $\chi \leqslant Card(a)$. Since $a \subset \chi$, we have $Card(a) = \chi$.

The above theorem is the promised result showing how large a measurable cardinal must be. Actually, much stronger results in the same direction have been proved by the same method.

Scott's Theorem. The negation of the axiom of constructibility is a theorem of *ZFM*.

Proof. We add to *ZFM** an axiom saying that χ is the first measurable cardinal. It is still a conservative extension of *ZFM*; and the results stated above can still be proved.

By (19), $Ord(F(\sigma))$ for all σ. Moreover, $\sigma \leqslant F(\sigma)$ for all σ. For otherwise, there would be a first ordinal σ such that $F(\sigma) < \sigma$. By (18), $F(F(\sigma)) < F(\sigma)$. This and $F(\sigma) < \sigma$ contradict the choice of σ. From $\sigma \leqslant F(\sigma)$ and the transitivity of M, we conclude that $M(\sigma)$ for all σ. Now C is absolute and hence M-invariant; so $M(C(\sigma))$ for all o. This shows that every constructible set is an M-set.

It will thus suffice to get a contradiction from the assumption $\forall x M(x)$. From this assumption it follows that $Q_M(x) \leftrightarrow Q(x)$ for all Q and x. Hence by (17), $Q(F(x)) \leftrightarrow Q(x)$ for all Q and x. If we let $Q(x)$ mean that x is the first measurable cardinal, then $Q(F(x)) \leftrightarrow Q(x)$ for all x; so $Q(F(\chi)) \leftrightarrow Q(\chi)$. But $Q(\chi)$ is true, while $Q(F(\chi))$ is false by (23). This is the desired contradiction.

Since measurable cardinals are very large, it might be thought that the nonconstructible sets produced by the axiom of measurability would be very far removed from the sets usually used in mathematics. However, Rowbottom has proved in *ZFM* that there are nonconstructible sets of natural numbers; in fact, that there are only countably many constructible sets of natural numbers. Several further extensions of Scott's theorem along these lines are known.

At this point, one might hope to settle the continuum hypothesis by means of the axiom of measurability. However, Lévy and Solovay have shown that the continuum hypothesis and its negation are unprovable in *ZFM* (provided that *ZFM* is consistent).

There are several other known generalized axioms of infinity; and for some of them results similar to the above have been proven. However, none of them settles the continuum hypothesis.

The situation at present can thus be summarized as follows. We have some axioms which imply the negation of the axiom of constructibility. There is some evidence that these axioms are true; but this is still an open question. We have no reasonable axioms which settle the continuum hypothesis.

Although these results may seem rather meager, the situation is really quite promising, especially when compared with the situation a few years ago. We know that the continuum hypothesis cannot be settled on the basis of presently available axioms, and we have some idea of what sort of new axioms to look for. An optimist would expect that the continuum hypothesis, which has baffled mathematicians for almost a century, will be settled in the next few years.

PROBLEMS

1. a) Let ZF_1 be obtained from ZF by omitting the subset and replacement axioms and adding the axiom

$$Set_x\, \exists z(x \in z \;\&\; z \in y)$$

and the axioms

$$\forall \mathbf{x}\, \exists \mathbf{z}\, \forall \mathbf{y}(\mathbf{A} \rightarrow \mathbf{y} = \mathbf{z}) \rightarrow \exists \mathbf{v}\, \forall \mathbf{y}(\mathbf{y} \in \mathbf{v} \leftrightarrow \exists \mathbf{x}(\mathbf{x} \in \mathbf{w} \;\&\; \mathbf{A}))$$

where $\mathbf{x}$, $\mathbf{y}$, $\mathbf{z}$, $\mathbf{v}$ and $\mathbf{w}$ are distinct, and $\mathbf{z}$, $\mathbf{v}$, and $\mathbf{w}$ do not appear in $\mathbf{A}$. Show that ZF_1 is equivalent to ZF.

b) Let ZF_2 be obtained from ZF by omitting the axiom of infinity and adding the axiom

$$\exists x(\exists y(y \in x) \;\&\; \forall y(y \in x \rightarrow \exists z(z \in x \;\&\; y \subset z \;\&\; y \neq z))).$$

Show that ZF_2 is equivalent to ZF. [In ZF_2, define a natural number to be an ordinal which is less than every limit number. Prove by transfinite induction that no natural number is similar to a smaller natural number. Let x be as in the above axiom. Prove that the set of natural numbers similar to a subset of an element of x exists and that this set contains all natural numbers.]

2. Let ZF' be ZF with the regularity axiom omitted. A set a is *regular* if every nonempty subset of a has a minimal element.

a) Define x to be an ordinal if x is transitive and regular, and every element of x is transitive. Show that the results of §9.3 can then be proved in ZF'.

b) Show that if ZF' is consistent, then ZF is consistent. [Let $M(x) \leftrightarrow \exists \sigma(x \in Stg(\sigma))$. Show that M is an interpretation of ZF in ZF'.]

3. Let E be a binary predicate of ZF such that

$$\vdash x \neq 0 \rightarrow \exists y(y \in x \;\&\; \neg\exists z(z \in x \;\&\; z\, E\, y))$$

and

$$\vdash Set_x(x\, E\, y).$$

Show that there is an ordinal function symbol H such that $\vdash Set_x(H(x) \leqslant \sigma)$ and $\vdash x\, E\, y \rightarrow H(x) < H(y)$. [Similar to the method used in §9.10 for $\in_I$.]

4. A set a is a *chain* if $\forall x\, \forall y(x, y \in a \rightarrow x \subset y \vee y \subset x)$. A set a is *inductive* if for every nonempty chain b included in a, $Un(b) \in a$. A *maximal element* of a is an element of a not included in any other element of a. *Zorn's lemma* says that if a is a nonempty inductive set, then a has a maximal element.

a) Prove Zorn's lemma in ZFC. [Suppose that a is a nonempty inductive set having no maximal element. Show that a is not a chain. Show that there is a choice function g on a such that if $b \subset a$ and b is a chain, then $Un(b) \subset g‘(a - b)$. Let f be as in the remark after the well-ordering theorem. Show that $\sigma < \tau \rightarrow f‘\sigma \subset f‘\tau$ for $\sigma, \tau \in Do(f)$, and derive a contradiction.]

b) Prove the Teichmüller-Tukey lemma in ZFC. [Use (a).]

c) Show that the axiom of choice follows from the Teichmüller-Tukey lemma in ZF. [Given x, let z be the set of functions f with $Do(f) \subset P(x)$ and $f‘y \in y$ for $y \in Do(f)$. Pick a maximal element of z.]

5. Let y be a subset of $x \times x$, and write $a <_y b$ for $\langle a, b\rangle \in y$. If $z \subset x$, a *y-first* element of z is an element a of z such that $a <_y b$ for all $b \in z - \{a\}$. We say that y is a *well-ordering* of x if $\neg(a <_y a)$ for all a, and every nonempty subset of x has a y-first element.

a) Let y be a well-ordering of x. Show that if $a, b \in x$, then exactly one of $a <_y b$, $a = b$, and $b <_y a$ holds. Show that if $a <_y b$ and $b <_y c$, then $a <_y c$. Show that every nonempty subset of x has a unique y-first element.

b) Let f be a mapping from σ to σ' such that $\tau < \rho < \sigma \rightarrow f`\tau < f`\rho$. Show that $\tau \leqslant f`\tau$ for $\tau < \sigma$. [Use transfinite induction.] Conclude that $\sigma \leqslant \sigma'$.

c) Let y be a well-ordering of x. Show that there is a unique bijective mapping f from an ordinal σ to x such that $\tau < \rho < \sigma \rightarrow f`\tau <_y f`\rho$. [Use the remark after the well-ordering theorem and (b).]

d) A *descending y-sequence* in x is a sequence $a_0, a_1, \ldots$ of elements of x such that $a_{S(i)} <_y a_i$ for all i. Prove in *ZFC* that a subset y of $x \times x$ is a well-ordering of x iff it is a linear ordering of x and there are no descending y-sequences in x.

6. a) Show that *Card'* may be defined in *ZF* so that $Card'(x) = Card'(y) \leftrightarrow Sm(x, y)$. [Like the definition of *EC* in §9.10.]

b) Define $x \leqslant' y \leftrightarrow \exists a\, \exists b(x = Card'(a)\ \&\ y = Card'(b)\ \&\ a \subset b)$. Prove (in *ZF*)

$$\exists a(x = Card'(a)) \rightarrow x \leqslant' x,$$
$$x \leqslant' y\ \&\ y \leqslant' z \rightarrow x \leqslant' z.$$

c) Prove in *ZF* that if there is an injective mapping f from a to b and an injective mapping g from b to a, then a is similar to b. [Let

$$d = [c \mid c \subset a\ \&\ \forall x(x \in b\ \&\ g`x \in c \rightarrow \exists y(y \in c\ \&\ x = f`y))].$$

Let $a_1 = Un(d)$, $a_2 = a - a_1$, $b_1 = [f`x \mid x \in a_1]$, $b_2 = b - b_1$. Show that $a_1 \in d$, and conclude that $g$ maps $b_2$ into $a_2$. Let $e$ be the set of element of $a$ not of the form $g`x$ with $x \in b_2$. Show that $e \in d$, and conclude that g maps b_2 onto a_2.] Conclude that $x \leqslant' y\ \&\ y \leqslant' x \rightarrow x = y$.

d) Show that for each x, the set of ordinals similar to a subset of x exists. [Note that the set of well-orderings of subsets of x exists, and use 5(c).]

e) Show that the formula

$$\forall x\, \forall y(\exists a(x = Card'(a))\ \&\ \exists b(y = Card'(b)) \rightarrow x \leqslant' y \vee y \leqslant' x)$$

is equivalent to the axiom of choice. [Assuming this sentence, use (d) to prove the well-ordering theorem and derive the axiom of choice.]

7. In *ZFC*, define σ^τ to be the cardinal of the set of mappings from τ to σ.

a) Prove that 2^σ has the same value under this definition as under the old definition.

b) For σ, τ, and ρ cardinals, show that

$$\sigma^{\tau+\rho} = \sigma^\tau \cdot \sigma^\rho,$$
$$\sigma^{\tau\cdot\rho} = (\sigma^\tau)^\rho,$$
$$(\sigma \cdot \tau)^\rho = \sigma^\rho \cdot \tau^\rho.$$

c) Show that

$$\sigma \leqslant \sigma' \ \&\ \tau \leqslant \tau' \rightarrow \sigma^{\tau} \leqslant \sigma'^{\tau'}.$$

d) Show that

$$\sigma \leqslant \tau \rightarrow \aleph_\sigma^{\aleph_\tau} = 2^{\aleph_\tau}.$$

[Use (b) and (c).]

8. For σ a cardinal, let $cf(\sigma)$ be the first cardinal τ such that

$$\exists x(x \subset \sigma \ \&\ Un(x) = \sigma \ \&\ Card(x) = \tau).$$

A cardinal σ is *regular* if $cf(\sigma) = \sigma$ and *singular* if $cf(\sigma) < \sigma$. A cardinal $\aleph_\sigma$ is *weakly inaccessible* if it is regular and σ is a limit number.

a) Show that $cf(\sigma) \leqslant \sigma$.

b) Show that $\aleph_{S(\sigma)}$ is regular.

c) Show that every inaccessible cardinal is weakly inaccessible. Show that the generalized continuum hypothesis implies that every weakly inaccessible cardinal is inaccessible.

d) Show that if *ZF* is consistent, then the statement that a weakly inaccessible cardinal exists cannot be proved in *ZFC*. [In *ZFL*, define *M* as in (4) of §9.10, except that inaccessible is replaced by weakly inaccessible. Use (4) of §9.7 to prove the interpretation of the power set axiom.]

e) Show that

$$cf(\aleph_\sigma) \leqslant \tau \rightarrow \aleph_\sigma < \aleph_\sigma^{\tau}.$$

[Assume that $\tau = cf(\sigma)$ and $\aleph_\sigma = \aleph_\sigma^{\tau}$. Let

$$x \subset \aleph_\sigma, \qquad Card(x) = \tau, \qquad Un(x) = \aleph_\sigma;$$

and let $\rho \rightarrow f_\rho$ be a bijective mapping from $\aleph_\sigma$ to the set of mappings from x to $\aleph_\sigma$. For each $\nu \in x$, choose $f`\nu$ in $\aleph_\sigma - [f_\rho`\nu \mid \rho < \nu]$. Show that f is not an f_ρ.]

f) Show that

$$cf(\aleph_\sigma) \leqslant \aleph_\tau \rightarrow 2^{\aleph_\tau} \neq \aleph_\sigma.$$

[Use (e).] Conclude that $2^{\aleph_0} \neq \aleph_\omega$.

9. a) Prove in *ZFL*

$$\omega \leqslant \rho \ \&\ a \subset C^*(\rho) \ \&\ \rho < Card(\tau) \rightarrow a \in C^*(\tau).$$

[Let $Card(\rho) = \aleph_\sigma$, and apply (8) of §9.7 to $b = C^*(\rho) \cup \{a\}$.]

b) Prove in *ZFC*

$$L(a) \ \&\ a \subset C^*(\aleph_{S(\sigma)}) \ \&\ Card(a) \leqslant \aleph_\sigma \rightarrow a \in C^*(\aleph_{S(\sigma)}).$$

[Take the interpretation under L of the theorem of (a), and use (26) of §9.10 and 8(b).]

c) Prove in *ZFC* that every constructible subset of ω and every constructible mapping from ω to ω has countable order. [Use (b).] Conclude that the formula $\forall x(x \subset \omega \rightarrow L(x))$ implies the continuum hypothesis.

10. a) Show that there is an arithmetical predicate P such that for A a tree,

$$P(K_A, i, j) \leftrightarrow A(i) \;\&\; \|A_{[i]}\| = j.$$

b) Show that there is an arithmetical predicate P such that if A is a tree, then $P(K_A, \alpha)$ defines $\langle K_{R_A}\rangle$ implicitly, where

$$R_A(i, j, k) \leftrightarrow A(i) \;\&\; A(j) \;\&\; A(k) \;\&\; K(\|A_{[i]}\|) = \langle \|A_{[j]}\|, \|A_{[k]}\|\rangle.$$

Conclude that there is a hyperarithmetical relation Q such that for A a tree,

$$Q(K_A, i, j, k) \leftrightarrow R_A(i, j, k).$$

c) Show that there are arithmetical relations $P_1 - P_4$ such that if A is a tree and C_A is defined by

$$C_A(i, j) \leftrightarrow A(i) \;\&\; A(j) \;\&\; C(\|A_{[i]}\|) \in C(\|A_{[j]}\|),$$

then

$$P_1(K_A, \langle K_{C_A}\rangle, i, j) \leftrightarrow A(i) \;\&\; A(j) \;\&\; C(\|A_{[i]}\|) = C(\|A_{[j]}\|),$$
$$P_2(K_A, \langle K_{C_A}\rangle, i,\ j,\ k) \leftrightarrow A(i) \;\&\; A(j) \;\&\; A(k) \;\&\; C(\|A_{[i]}\|) = \{C(\|A_{[j]}\|), C(\|A_{[k]}\|)\},$$
$$P_3(K_A, \langle K_{C_A}\rangle, i, j, k) \leftrightarrow A(i) \;\&\; A(j) \;\&\; A(k) \;\&\; C(\|A_{[i]}\|) = \langle C(\|A_{[j]}\|), C(\|A_{[k]}\|)\rangle,$$
$$P_4(K_A, \langle K_{C_A}\rangle, i, j, k, m) \leftrightarrow A(i) \;\&\; A(j) \;\&\; A(k) \;\&\; A(m) \;\&\; C(\|A_{[i]}\|) = \langle C(\|A_{[j]}\|), C(\|A_{[k]}\|), C(\|A_{[m]}\|)\rangle.$$

d) Show that there is a hyperarithmetical relation P such that for A a tree and C_A as in (c), $P(K_A, \alpha)$ defines $\langle K_{C_A}\rangle$ implicitly. [Use (a), (b), and (c).]

e) Show that there is an arithmetical relation P such that if A is a tree, then for C_A as in (c),

$$P(K_A, \langle K_{C_A}\rangle, i, j) \leftrightarrow A(i) \;\&\; C(\|A_{[i]}\|) = j.$$

f) Show that there is a hyperarithmetical relation P such that for A a tree,

$$P(K_A, \alpha, i) \leftrightarrow A(i) \;\&\; C(\|A_{[i]}\|) = \alpha.$$

[Use (d) and (e).]

g) Show that the class of constructible sets of natural numbers is Σ^1_2. [Use (f) and 9(c).]

h) Show that if α and β are constructible, then either α is Δ^1_2 in β or β is Δ^1_2 in α. [Suppose that $Od(\alpha) < Od(\beta)$. Let P be as in (f). Use 9(c) and the uniformization theorem to find a γ such that $Tr(\gamma)$ & $\exists i P(\gamma, \beta, i)$ and γ is Δ^1_2 in β. Show that α is hyperarithmetical in γ.]

11. Assume the axiom of constructibility. Let $\alpha <_L \beta$ mean that $Od(\alpha) < Od(\beta)$.

a) Prove that $<_L$ is a Δ^1_2 (2, 0)-ary relation. [Use 10(f) and 9(c).]

b) Prove that for $n \geqslant 2$, the class of Δ^1_{n+1} functions is a basis for the collection of Π^1_n classes of functions. [If P is Π^1_n and nonempty, let α be the element of P of smallest order, and use (a).]

c) Prove that for $n \geqslant 3$, the collection of Σ^1_n subsets of $N_{m,k}$ satisfies the reduction principle. [Similar to Problem 25(b) of Chapter 7, using (a).]

12. a) Let M be a transitive $\in$-interpretation of ZF; Q and Q' predicate symbols absolute for M; R a predicate symbol such that

$$\vdash_{ZF} R(x_1, \ldots, x_n) \leftrightarrow \exists y Q(y, x_1, \ldots, x_n),$$
$$\vdash_{ZF} R(x_1, \ldots, x_n) \leftrightarrow \forall y Q'(y, x_1, \ldots, x_n).$$

Show that R is absolute for M. [Take the interpretation of these two theorems.]

b) Let $We(y, x)$ mean that y is a well-ordering of x. Show that We is absolute for all transitive $\in$-interpretations of ZF. [Use (a) and 5(c).]

c) Show that there is an interpretation I of P in ZF such that for each defined predicate symbol Q of P, Q_I is absolute for all transitive $\in$-interpretations of ZF. Conclude that each recursive predicate can be represented in ZF by a predicate symbol absolute for all transitive $\in$-interpretations of ZF. Extend this to arithmetical relations. [Use the method of Problem 11(a) of Chapter 8.]

d) Show that a (2, 1)-ary Π^1_1 relation can be represented in ZF by a predicate symbol Q defined by

$$Q(\alpha, \beta, i) \leftrightarrow We(F(\alpha, \beta, i), G(\alpha, \beta, i))$$

where F and G are absolute. [Use Problem 19(f) of Chapter 7.] Conclude that Q is absolute. [Use (b).]

e) Show that if Q is as in (d) and R is defined by

$$R(\alpha, i) \leftrightarrow \exists \beta Q(\alpha, \beta, i),$$

then R is absolute for L. [Using 5(c),

$$R(\alpha, i) \leftrightarrow \exists \beta\, \exists f\, \exists \sigma (IFunc(f)\ \&\ Do(f) = \omega\ \&\ Ra(f) \subset \sigma$$
$$\&\ \forall j\, \forall k (j, k \in G(\alpha, \beta, i) \rightarrow (\langle j, k \rangle \in F(\alpha, \beta, i) \leftrightarrow f`j < f`k))).$$

Letting $f \restriction j$ be the restriction of f to j, rewrite this as

$$R(\alpha, i) \leftrightarrow \exists f\, \exists \sigma (Func(f)\ \&\ Do(f) = \omega\ \&\ Ra(f) \subset \sigma \times \omega\ \&\ \forall j R'(f \restriction j, \sigma, \alpha, i))$$

where R' is absolute. Using the idea of the proof of (d) and a well-ordering of $\sigma \times \omega$ provided by K, rewrite this as

$$R(\alpha, i) \leftrightarrow \exists \sigma\ \neg We(H_1(\sigma, \alpha, i), H_2(\sigma, \alpha, i))$$

with H_1 and H_2 absolute, and use (b), recalling that $\forall \sigma L(\sigma)$.]

f) Show that every Σ^1_2 or Π^1_2 predicate is constructible. [Define R as in (e) with α missing, and use the absoluteness of R.]

g) An ordinal σ is a Δ^1_2 ordinal if it is finite or there is a bijective mapping f from ω to σ such that $f`i < f`j$ is a Δ^1_2 predicate. Show that the order of a Δ^1_2 predicate is a Δ^1_2 ordinal. [Use the corollary to the uniformization theorem.]

13. Let ZF^Y be obtained from ZF by adding a constant Y and axioms $Y \subset \omega$ and $\neg L(Y)$. Define C^Y like C, except that a new operation $\mathfrak{F}_{10}(x, y) = x \cap Y$ is added; and define C^{*Y} and L^Y similarly. Let ZFL^Y be obtained from ZF^Y by adding the axiom $\forall x L^Y(x)$.

a) Show that if ZF is consistent, then ZF^Y is consistent. [Use Theorem 1 of §9.9 and 9(c).]

b) Show that L^Y is an interpretation of ZF in ZF^Y and that $\forall\sigma L^Y(\sigma)$.

c) Show that L is absolute for L^Y and that $L(x) \to L^Y(x)$. [Note that C is absolute for L^Y and hence L^Y-invariant, and use (b).] Conclude that $L^Y(Y)$. [Use an analogue of Lemma 2 of §9.6.]

d) If M is an $\in$-interpretation of ZF in ZF^Y such that $M(Y)$, we extend M to an interpretation of $L(ZF^Y)$ by taking Y_M to be Y. Show that M is then an interpretation of ZF^Y. We define absoluteness for M of nonlogical symbols of ZF^Y as before. Show that Y is absolute, and that if M is transitive, then C^Y is absolute.

e) Show that L^Y is absolute for L^Y. Conclude that L^Y is an interpretation of ZFL^Y in ZF^Y.

f) Prove the axiom of choice in ZFL^Y.

g) Prove in ZFL^Y

$$Y \in b \ \&\ Trans(b) \ \&\ Card(b) \leqslant \aleph_\sigma \to b \subset C^{*Y}(\aleph_{S(\sigma)}).$$

[Like (8) of §9.7. In proving the lemma, note that $F(Y) = Y$.]

h) Prove the generalized continuum hypothesis in ZFL^Y. [Like Theorem 2 of §9.7. In proving the analogue of (4) of §9.7, take $b = C^*(\aleph_\sigma) \cup \{a\} \cup \{Y\} \cup \omega$.]

i) Let ZFC' be obtained from ZFC by adding the generalized continuum hypothesis. Show that if ZF is consistent, then $\forall x(x \subset \omega \to L(x))$ is not a theorem of ZFC'. [Use (a), (e), (f), and (h).]

14. A *δ-ideal* on a set x is a subset y of $P(x)$ satisfying (ii), (iii), and (iv) of the definition of a measure ideal and, in addition: (i′) for every countable subset z of y, $Un(z) \in y$. Suppose that χ is a cardinal, that y is a δ-ideal on χ, and that there is no δ-ideal on any cardinal less than χ. Show that y is a measure ideal. [Assume not. Let σ be the first cardinal such that for some subset z of y, $Card(z) = \sigma$ and $Un(z) \notin y$. Show that this z may be chosen so that $Un(z) = \chi$. Let f be a bijective mapping from σ to z, and let w be the set of all subsets v of σ such that $Un([f`\tau \mid \tau \in v] \notin y)$. Show that w is a δ-ideal on σ.]

APPENDIX

THE WORD PROBLEM

We are going to consider a decision problem which arises in group theory. We shall assume that the reader is familiar with groups defined by generators and relations and with free products. We review the facts which we need.

The unit element of any group is designated by e. We call $\{e\}$ the *zero* subgroup, and say that an element is *nonzero* if it is different from e. We use $\{A\}$ to designate the subgroup generated by A. If ϕ is a mapping, then $\phi \mid A$ represents the restriction of ϕ to A. If G is a group, G_1 will designate an isomorphic copy of G under an isomorphism which takes g into g_1; then A_1 will designate the image of A under this isomorphism.

The *free group* $[A]$ on a set A is a group including A which has this property: every element of $[A]$ can be uniquely written in the form

$$a_1^{\pm 1} a_2^{\pm 1} \ldots a_n^{\pm 1}$$

where the a_i are in A and no a appears adjacent to an a^{-1}. (We allow $n = 0$; this gives the expression for e.) These expressions are called *words* on A. We may think of them as being the elements of $[A]$. We multiply two words by juxtaposing them and then crossing out expressions aa^{-1} and $a^{-1}a$ until we obtain a word; this crossing out process is called *reduction*. We have $[A] = \{A\}$. More generally, if $B \subset A$, then $\{B\}$ consists of the words on B, and can be identified with $[B]$.

If ϕ is a mapping from A to a group G, then there is a unique extension ϕ' of ϕ to a homomorphism from $[A]$ to G; and ϕ' is surjective iff $\phi(A)$ generates G. In particular, if A is a generating set in a group G, then there is a unique homomorphism from $[A]$ to G which is the identity on A; and this homomorphism is surjective. Thus G is naturally isomorphic to a factor group of $[A]$. It follows that every finitely generated group is isomorphic to a factor group of a free group on a finite set.

A subset A of a group G is *free* if the homomorphism from $[A]$ to G which is the identity on A is injective. In this case, we may identify the subgroup $\{A\}$ of G with $[A]$.

Example. If $z \notin A$, then the set of all XzX^{-1} in $[A, z]$ with X a word on A is free.

A *relation* on A is an expression $X = Y$, where X and Y are words on A. This relation *holds* in a factor group $[A]/K$ if X and Y lie in the same coset of K; equivalently, if $XY^{-1} \in K$.

Let R be a set of relations on A. A relation on A is a *consequence* of R if it holds in every factor group of $[A]$ in which all the relations in R hold. The set of consequences of R is designated by $C(R)$. There is a unique factor group $[A]/K_R$ of $[A]$ in which the relations which hold are just the consequences of R; K_R is the normal subgroup generated by the XY^{-1} for $X = Y$ in R. We call $[A]/K_R$ the group with the set of *generators* A and the set of *defining relations* R, and designate it by $[A; R]$.

We extend the notation $[A; R]$ to allow several generators or sets of generators before the semicolon and several relations or sets of relations after the semicolon. Thus $[A, t; R, X = Y]$ has the set of generators $A \cup \{t\}$ and the set of defining relations $R \cup \{X = Y\}$. It is understood that no generator is repeated; thus in the above example we must have $t \notin A$. If A appears before the semicolon and a after the semicolon, it is understood that a varies through A; and similarly for other letters. Thus in $[A, t; at = ta]$, the defining relations are all the $at = ta$ for a in A.

Since $[A]$ is a subgroup of $[A, B]$ and K_R is a subgroup of $K_{R \cup S}$, there is a natural homomorphism from $[A; R]$ to $[A, B; R, S]$ which maps the coset in $[A; R]$ of a word on A into the coset of that word in $[A, B; R, S]$. If this homomorphism is bijective, we identify these two groups. This will certainly happen in the following case: S consists of one relation $b = X$ with X a word on A for each b in B.

We recall that a group G is naturally identified with a factor group of $[G]$, and hence with the group $[G; R_G]$, where R_G is the set of all relations on G which hold in this factor group. If G appears before the semicolon, it is understood that the relations in R_G are among the defining relations, even if they do not appear explicitly after the semicolon.

Example. The group $[G, H; gh = hg]$ is the direct product of G and H. (We assume that G and H are disjoint; otherwise they must first be replaced by isomorphic groups.)

Let G and G' be groups, and let ϕ be an isomorphism of a subgroup H of G and a subgroup H' of G'. The *free product* of the groups G and G' with the *amalgamation* ϕ is the group

$$G *_\phi G' = [G, G'; h = \phi(h)].$$

The natural mappings of G and G' into $G *_\phi G'$ are injective; so we identify G and G' with their images under these mappings. Then H and H' are identified via the isomorphism ϕ. We have $G *_\phi G' = \{G, G'\}$ and $G \cap G' = H = H'$. This last group is called the *amalgam*.

Let T consist of one element in each right coset of H in G other than H itself; and let T' be formed similarly from H' and G'. A word on $G \cup G'$ is in *normal form* if it is $ht_1t_2 \ldots t_n$ where $h \in H$; $t_1, t_2, \ldots, t_n \in T \cup T'$; and $t_i \in T$ iff $t_{i+1} \in T'$ for $1 \leqslant i < n$. Then Schreier's theorem states that every coset in $G *_\phi G'$ contains exactly one word in normal form.

Let K and K' be subgroups of G and G' respectively such that

$$\phi(H \cap K) = \phi(H) \cap K'.$$

(This means that in $G *_\phi G'$, K and K' have the same intersection with the amalgam.) The restriction $\psi = \phi \mid (H \cap K)$ is then an isomorphism of $H \cap K$ and $\phi(H) \cap K$. We may thus form $K *_\psi K'$; and there is a natural mapping χ from $K *_\psi K'$ to $G *_\phi G'$ whose image is $\{K, K'\}$. We show that χ is injective. Since elements in different cosets of $H \cap K$ in K lie in different cosets of H in G, we may suppose that T contains one element in each coset of $H \cap K$ in K other than $H \cap K$ itself. We may suppose that T' is chosen similarly. Then the words on $K \cup K'$ in normal form are among the words on $G \cup G'$ in normal form. Since χ maps the coset of a word into the coset of the same word, it follows from Schreier's theorem that χ is injective.

We may thus identify $K *_\psi K'$ with $\{K, K'\} \subset G *_\phi G'$. Then

$$G \cap \{K, K'\} = K. \tag{1}$$

It will suffice to show that the left side is included in the right side. The normal form of an element g of G is h or ht with $t \in T$. If this is a normal form of an element in $\{K, K'\}$, then $h \in H \cap K$ and $t \in K$; so $g \in K$.

If ϕ is the isomorphism of the zero subgroups, we write $G * G'$ for $G *_\phi G'$, and call $G * G'$ the *free product* of G and G'. Thus $G * G' = [G, G']$. The normal forms become the products of nonzero elements which are alternately in G and G' (if we omit the initial e). The K and K' described above can then be any subgroups of G and G'.

Let G and G' be subgroups of L. Let $H = G \cap G'$, and let ϕ be the identity mapping from H to H. Then there is a unique homomorphism from $G *_\phi G'$ to L which is the identity on G and G'; and its image is $\{G, G'\}$. If this homomorphism is injective, we identify $G *_\phi G'$ and $\{G, G'\}$, and say that $\{G, G'\}$ is the free product of G and G' with the amalgam H (omitting mention of H if H is the zero subgroup).

Example. If h is a nonzero element of H, then in $G * H$ the subgroup $\{G, hGh^{-1}\}$ is the free product of G and hGh^{-1}; this follows easily from Schreier's theorem.

A group is *finitely presented* if it is isomorphic to a group $[A; R]$ with A and R finite. The direct product of two finitely presented groups is finitely presented; for

$$[A; R] \times [B; S] = [A, B; R, S, ab = ba].$$

A free product of two finitely presented groups with a finitely generated amalgam is finitely presented. For let the product be $[A; R] *_\phi [B; S]$, where the domain of ϕ has a finite number of generators $h_1, \ldots, h_n$. Let X_i be a word on A in the coset of h_i and let Y_i be a word on B in the coset of $\phi(h_i)$. Then

$$[A; R] *_\phi [B; S] = [A, B; R, S, X_1 = Y_1, \ldots, X_n = Y_n].$$

Let R be a set of relations on a finite set A. The *word problem* for R is the decision problem for $C(R)$. Since $C(R)$ is the set of relations holding in $[A; R]$, we also call this problem the *word problem for* $[A; R]$.

To translate this problem into recursion theory, we identify the symbols a and a^{-1} ($a \in A$) and the symbol $=$ with natural numbers. A word or relation on A then becomes a finite sequence of natural numbers, and hence has a sequence number. A set P of words or relations is *recursive* (or *recursively enumerable*) if the set of sequence numbers of elements of P is recursive (or recursively enumerable). It is easy to check that this is independent of the numbers with which the symbols are identified. The word problem for R is then solvable iff $C(R)$ is recursive.

The relation $X = Y$ is in $C(R)$ iff XY^{-1} is in K_R; and X is in K_R iff $X = e$ is in $C(R)$. It follows that K_R is recursive (recursively enumerable) iff $C(R)$ is recursive (recursively enumerable). Moreover, if R is recursively enumerable, then $C(R)$ is recursively enumerable. For the set J of XY^{-1} with $X = Y$ in R is recursively enumerable. We obtain the words in K_R by reducing expressions $X_1Y_1^{\pm 1}X_1^{-1} \cdot \ldots \cdot X_nY_n^{\pm 1}X_n^{-1}$, where the Y_i are in J. It follows that K_R is recursively enumerable; so $C(R)$ is recursively enumerable.

A group is *recursively presented* if it is isomorphic to a group $[A; R]$ where A is finite and R is recursively enumerable. Obviously every finitely presented group is recursively presented. (The converse is known to be false.)

Suppose that $[A; R]$ is embedded in $[B; S]$ (where A and B are finite). For each a in A we pick a word Y_a on B such that the cosets of a and Y_a correspond under the embedding isomorphism. For X a word on A, let X' be the word on B obtained from X by replacing a by Y_a and a^{-1} by Y_a^{-1} and then reducing. Then the cosets of X and X' correspond under the embedding isomorphism; so $X = Y$ holds in $[A; R]$ iff $X' = Y'$ holds in $[B; S]$. A first consequence is that if the word problem for $[B; S]$ is solvable, then the word problem for $[A; R]$ is solvable. A second consequence is that if $C(S)$ is recursively enumerable, then $C(R)$ is recursively enumerable. Since $[A; R] = [A; C(R)]$, this implies that a subgroup of a recursively presented group is recursively presented.

If G is a finitely generated group, then G is isomorphic to a group $[A; R]$ with A finite. We say that the word problem for G is *solvable* or *unsolvable* according as the word problem for $[A; R]$ is solvable or unsolvable. By the result just proved, this is independent of the choice of $[A; R]$.

Our principal object is to prove the following result.

Novikoff's Theorem. There is a finitely presented group which has an unsolvable word problem.

We obtain a recursively presented group with an unsolvable word problem as follows. Let A be a recursively enumerable set, and let H be the subgroup of $G = [a, b]$ generated by the a^nba^{-n} for $n \in A$. Let ϕ be the identity mapping from H to H. Then $G *_\phi G$ is

$$[a, b, a_1, b_1; a^nba^{-n} = a_1^nba_1^{-n} \quad \textit{for} \quad n \in A]$$

and hence is recursively presented. Now $a^nba^{-n} = a_1^nba_1^{-n}$ holds in $G *_\phi G$ iff $n \in A$. For if it holds, then a^nba^{-n} belongs to the amalgam H; since the a^mba^{-m} are free, this implies that $n \in A$. Thus the decision problem for A is reducible to the word problem for $G *_\phi G$. If we choose A nonrecursive, it follows that $G *_\phi G$ has an unsolvable word problem.

In order to prove Novikoff's theorem, it will suffice to embed $G *_\phi G$ in a finitely presented group. We do this by means of the following theorem.

Higman's Theorem. A finitely generated group is embeddable in a finitely presented group iff it is recursively presented.

There is one remarkable aspect to Higman's theorem. Let us call a group a *Higman group* if it is finitely generated and embeddable in a finitely presented group. This is a purely algebraic notion. By Higman's theorem, it coincides with the notion of a recursively presented group, which is defined by means of recursion theory.

The "only if" part of Higman's theorem is trivial. A Higman group is isomorphic to a subgroup of a finitely presented and hence recursively presented group; so it is recursively presented. The rest of the appendix is devoted to the proof of the converse.

Lemma 1. If G and H are Higman groups, then $G \times H$ is Higman.

Proof. Clearly $G \times H = \{G, H\}$ is finitely generated. If G and H are embedded in the finitely presented groups L and M, then $G \times H$ is embedded in $L \times M$, which is finitely presented.

Lemma 2. If G and G' are Higman groups, and ϕ is an isomorphism from a finitely generated subgroup of G into G', then $G *_\phi G'$ is Higman.

Proof. Clearly $G *_\phi G' = \{G, G'\}$ is finitely generated. If G and G' are embedded in the finitely presented groups K and K', then $G *_\phi G'$ is embedded in $K *_\phi K'$, which is finitely presented.

An isomorphism *in* a group G is a isomorphism from a subgroup H of G into G. For such an isomorphism ϕ, we set

$$G_\phi = [G, t; tht^{-1} = \phi(h)].$$

To study G_ϕ, we embed it in a larger group. In $[G, r] = G * [r]$, $\{G, rHr^{-1}\}$ is the free product of G and rHr^{-1}. Similarly, in $[G_1, s]$, $\{G_1, s\phi(H)_1s^{-1}\}$ is the free product of G_1 and $s\phi(H)_1s^{-1}$. Hence there is a isomorphism ψ of $\{G, rHr^{-1}\}$ and $\{G_1, s\phi(H)_1s^{-1}\}$ defined by $\psi(g) = g_1$, $\psi(rhr^{-1}) = s\phi(h)_1s^{-1}$. Then

$$\begin{aligned}
[G, r] *_\psi [G_1, s] &= [G, G_1, r, s; g = g_1, rhr^{-1} = s\phi(h)_1s^{-1}] \\
&= [G, r, s; rhr^{-1} = s\phi(h)s^{-1}] \\
&= [G, r, s, t; rhr^{-1} = s\phi(h)s^{-1}, t = s^{-1}r] \\
&= [G, t, s, r; tht^{-1} = \phi(h), r = st] \\
&= [G, t, s; tht^{-1} = \phi(h)] = G_\phi * [s].
\end{aligned}$$

Thus there is an isomorphism from G_ϕ into $[G, r] *_\psi [G_1, s]$ which maps the coset of g into g and maps t into $s^{-1}r$.

It follows that different elements of G have different cosets in G_ϕ; so the natural mapping from G to G_ϕ is injective. We therefore identify G with a subgroup of G_ϕ. We also identify t with its coset in G_ϕ, and call t the *ϕ-element*. Then G_ϕ is generated by G and t, and $tht^{-1} = \phi(t)$ for $h \in H$.

Since $\{G, rGr^{-1}\}$ is the free product of G and rGr^{-1}, we have

$$\{G, rHr^{-1}\} \cap rGr^{-1} = rHr^{-1}$$

by (1). Similarly,

$$\{G_1, s\phi(H)_1 s^{-1}\} \cap sG_1s^{-1} = s\phi(H)_1 s^{-1}.$$

It follows that $rGr^{-1} *_\psi sG_1s^{-1}$ is embedded in $[G, r] *_\psi [G, s]$ as

$$\{rGr^{-1}, sG_1s^{-1}\} = \{rGr^{-1}, sGs^{-1}\}.$$

Thus this group is the free product of rGr^{-1} and sGs^{-1} with the amalgam rHr^{-1}. Applying the inner automorphism through r^{-1} and recalling that $r^{-1}s = t$, we see that $\{G, t^{-1}Gt\}$ is the free product of G and $t^{-1}Gt$ with the amalgam H. Hence

$$G \cap t^{-1}Gt = H. \tag{2}$$

Let G, ϕ, and H be as above. A subgroup K of G is *invariant* under ϕ if $\phi(H \cap K) = \phi(H) \cap K$. (This means that for $h \in H$, $h \in K \leftrightarrow \phi(h) \in K$.) If K is invariant under ϕ, then $\phi' = \phi \mid (H \cap K)$ is an isomorphism in K. We show that the natural mapping from $K_{\phi'}$ to G_ϕ is an embedding. It will suffice to show that the natural mapping from $[K, r] *_{\psi'} [K_1, s]$ to $[G, r] *_\psi [G_1, s]$ is an embedding (where ψ' is defined like ψ). This mapping certainly embeds $[K, r]$ in $[G, r]$ as $\{K, r\}$ and $[K_1, s]$ in $[G_1, s]$ as $\{K_1, s\}$. Hence we must check that $\{K, r\}$ and $\{K_1, s\}$ have the same intersection with the amalgam. Now

$$\{K, r\} \cap \{G, rHr^{-1}\} = \{K, r(H \cap K)r^{-1}\}. \tag{3}$$

To see this, we note that the normal form of an element of the free product $\{G, rHr^{-1}\}$ is of the form $\dots g_1rh_1r^{-1}g_2rh_2r^{-1}\dots$. This is a normal form in the free product $[G, r]$, and lies in $\{K, r\} = [K, r]$ iff the g_i and h_i are in K. This proves (3). Similarly,

$$\{K_1, s\} \cap \{G_1, s\phi(H)_1s^{-1}\} = \{K_1, s(\phi(H)_1 \cap K_1)s^{-1}\}. \tag{4}$$

Since $\phi(H \cap K) = \phi(H) \cap K$, the right-hand sides of (3) and (4) correspond under ψ, as required.

We may thus identify $K_{\phi'}$ with the subgroup $\{K, t\}$ of G_ϕ. It also follows from (1) that

$$\{K, r, s\} \cap \{G, r\} = \{K, r\}. \tag{5}$$

By applying (1) to $\{K, r\} = K * [r]$ as a subgroup of $G * [r]$, we get

$$\{K, r\} \cap G = K.$$

From this and (5), $\{K, r, s\} \cap G \subset K$. It follows that

$$\{K, t\} \cap G = K \tag{6}$$

in G_ϕ.

Suppose that we are given a set $\phi, \psi, \ldots$ of isomorphisms in G. Then

$$G_{\phi,\psi,\ldots} = [G, t_\phi, t_\psi, \ldots ; t_\phi h t_\phi^{-1} = \phi(h), \ldots].$$

Then G is naturally embedded in $G_{\phi,\psi,\ldots}$. For if not, then some relation $g = g'$ holds in $G_{\phi,\psi,\ldots}$ which does not hold in G; and it must be a consequence of a finite number of defining relations. Thus we need only consider the case in which there are only finitely many isomorphisms. This case is easily proved by induction, since

$$G_{\phi_1,\ldots,\phi_n,\phi_{n+1}} = (G_{\phi_1,\ldots,\phi_n})_{\phi_{n+1}}.$$

If K is a subgroup of G invariant under all of $\phi, \psi, \ldots$, then

$$\{K, t_\phi, t_\psi, \ldots\} \cap G = K. \tag{7}$$

The right-hand side is included in the left. Since an element of $\{K, t_\phi, t_\psi, \ldots\}$ is already generated by K and a finite number of the t's, we need only prove the reverse inclusion for a finite number of isomorphisms. We do this by induction on the number n of isomorphisms. The case $n = 1$ is (6). If $n > 1$,

$$\{K, t_1, \ldots, t_{n-1}\} \cap G = K$$

by the induction hypothesis. It follows that $\{K, t_1, \ldots, t_{n-1}\}$ is invariant under ϕ_n; so by (6),

$$\{K, t_1, \ldots, t_n\} \cap G_{\phi_1,\ldots,\phi_{n-1}} = \{K, t_1, \ldots, t_{n-1}\}.$$

Intersecting both sides with G and using the induction hypothesis again, we get

$$\{K, t_1, \ldots, t_n\} \cap G \subset K.$$

If ϕ is the identity mapping from H to H, we write G_H for G_ϕ. Then any subgroup K of G is invariant under ϕ, and $K_{H \cap K}$ is embedded in G_H.

Let G be a Higman group. An isomorphism ϕ in G is *benign* if G_ϕ is Higman. A subgroup H of G is *benign* if G_H is Higman, i.e., if the identity mapping from H to H is benign. Note that G_ϕ and G_H are certainly finitely generated (since G is).

Lemma 3. If G is a Higman group, and ϕ is an isomorphism of a finitely generated subgroup H of G into G, then ϕ is benign in G.

Proof. The groups $[G, r]$ and $[G_1, s]$ are Higman by Lemma 2. Since $\{G, rHr^{-1}\}$ is finitely generated, $[G, r] *_\psi [G_1, s]$ is Higman by Lemma 2. Hence its finitely generated subgroup G_ϕ is Higman.

Corollary. A finitely generated subgroup of a Higman group G is benign in G.

Lemma 4. Let K be a Higman group and let G be a Higman subgroup of K. Then an isomorphism ϕ in G is benign in G iff it is benign in K. Hence a subgroup of G is benign in G iff it is benign in K.

Proof. Suppose that K_ϕ is Higman. Then G_ϕ is embedded in K_ϕ and hence is Higman. Now suppose that G_ϕ is Higman. If $\psi(g) = g_1$, then $G_\phi *_\psi K_1$ is Higman by Lemma 2. Now if H is the domain of ϕ,

$$\begin{aligned} G_\phi *_\psi K_1 &= [G, t, K_1; tht^{-1} = \phi(h), g = g_1] \\ &= [G, t, K_1; th_1t^{-1} = \phi(h)_1, g = g_1] \\ &= [K_1, t; th_1t^{-1} = \phi(h)_1] \simeq K_\phi. \end{aligned}$$

We often make tacit use of Lemma 4 by not specifying in what Higman group an isomorphism or subgroup is benign.

Lemma 5. If H and K are benign subgroups of the Higman group G, then $H \cap K$ and $\{H, K\}$ are benign in G.

Proof. First suppose that one of the subgroups, say K, is finitely generated and hence Higman. Then $K_{H\cap K}$ is embedded in G_H and hence is Higman; so $H \cap K$ is benign. In G_H, $\{G, t^{-1}Gt\}$ is the free product of G and $t^{-1}Gt$ with the amalgam $H = t^{-1}Ht$. Since $\{H, K\}$ and $t^{-1}Gt$ include the amalgam, we have by (1)

$$\{H, K, t^{-1}Gt\} \cap G = \{H, K\},$$

whence

$$\{K, t^{-1}Gt\} \cap G = \{H, K\}. \tag{8}$$

Now the two groups on the left are finitely generated; so their intersection $\{H, K\}$ is benign.

In the general case, we have $G \cap t^{-1}Gt = H$ in G_H; so

$$H \cap K = G \cap (t^{-1}Gt \cap K).$$

Since G and $t^{-1}Gt$ are finitely generated and hence benign (by the corollary to Lemma 3), two applications of the special case show that $H \cap K$ is benign. We still have the equality (8). By the special case, $\{K, t^{-1}Gt\}$ is benign. Thus $\{H, K\}$ is the intersection of two benign subgroups and is therefore benign.

Lemma 6. Let ϕ be a homomorphism from the Higman group G to the Higman group H. If L is a benign subgroup of G, then $\phi(L)$ is a benign subgroup of H. If M is a benign subgroup of H, then $\phi^{-1}(M)$ is a benign subgroup of G.

Proof. By Lemma 1, $G \times H$ is Higman. Let Q be the subgroup of $G \times H$ consisting of all $(g, \phi(g))$. Then Q is isomorphic to G (by the mapping $(g, \phi(g)) \to g$). Thus G, H, and Q are finitely generated and hence benign in $G \times H$. Since

$$\begin{aligned} \phi(L) &= \{\{L, H\} \cap Q, G\} \cap H, \\ \phi^{-1}(M) &= \{\{M, G\} \cap Q, H\} \cap G, \end{aligned}$$

the lemma follows from Lemma 5.

Lemma 7. If ϕ is an isomorphism of the Higman group G into G and H is a benign subgroup of G, then $\phi \mid H$ is benign in G.

Proof. By Lemma 3, ϕ is benign in G; so G_ϕ is Higman. Thus H is benign in G_ϕ; so $(G_\phi)_H$ is Higman. But

$$\begin{aligned}(G_\phi)_H &= [G, s, t; sgs^{-1} = \phi(g), tht^{-1} = h]\\ &= [G, s, t, r; sgs^{-1} = \phi(g), stht^{-1}s^{-1} = \phi(h), r = st]\\ &= [G, r, s, t; rhr^{-1} = \phi(h), sgs^{-1} = \phi(g), t = s^{-1}r]\\ &= [G, r, s; rhr^{-1} = \phi(h), sgs^{-1} = \phi(g)] = (G_{\phi|H})_\phi.\end{aligned}$$

Thus $G_{\phi|H}$ is embedded in a Higman group and consequently is Higman.

A set $\phi, \psi, \ldots$ of isomorphisms in a Higman group G is *benign* if $G_{\phi,\psi,\ldots}$ may be embedded in a Higman group H so that $\{t_\phi, t_\psi, \ldots\}$ is a benign subgroup of H. A finite set of benign isomorphisms in G is benign; we may take $H = G_{\phi,\psi,\ldots}$ and use the corollary to Lemma 3.

Lemma 8. Let G be a Higman group; H a benign subgroup of G; $\phi, \psi, \ldots$ a benign set of isomorphisms in G; K the smallest subgroup of G which includes H and is invariant under $\phi, \psi, \ldots$. Then K is benign.

Proof. Embed $G_{\phi,\psi,\ldots}$ in a Higman group L so that $\{t_\phi, t_\psi, \ldots\}$ is benign. We show that

$$K = G \cap \{H, t_\phi, t_\psi, \ldots\};$$

the lemma will follow by Lemma 5. The right-hand side is clearly invariant under $\phi, \psi, \ldots$ and hence includes K. By (7),

$$G \cap \{H, t_\phi, t_\psi, \ldots\} \subset G \cap \{K, t_\phi, t_\psi, \ldots\} = K.$$

Let A be a finite set, and choose an element z not in A. For P a set of words on A, E_P is the subgroup of $[A, z]$ generated by the words XzX^{-1} for X in P. Since the XzX^{-1} for X a word on A form a free set, $XzX^{-1} \in E_P$ iff $X \in P$.

A subset P of $[A]$ is *benign* in $[A]$ if E_P is benign in $[A, z]$. This gives two definitions of *benign* if P is a subgroup of $[A]$; we must show that they agree. Suppose that P is benign as a subgroup. Then $\{P, z\}$ is benign in $[A, z]$ by Lemma 5. Now $E_{[A]}$ is the smallest subgroup of $[A, z]$ which contains z and is invariant under the inner automorphisms through elements of A. Hence $E_{[A]}$ is benign by Lemmas 3 and 8. We show that $E_P = \{P, z\} \cap E_{[A]}$; it will follow by Lemma 5 that E_P is benign. Clearly $E_P \subset \{P, z\} \cap E_{[A]}$. Now E_P contains z and is invariant under the inner automorphisms through elements of P; so it is a normal subgroup of $\{P, z\}$. Since $z \in E_P$, the natural mapping from P to $\{P, z\}/E_P$ is surjective. Hence if $x \in \{P, z\}$, then $x = py$ with $p \in P$, $y \in E_P$. The homomorphism ϕ from $[A, z]$ to $[A]$ defined by $\phi(a) = a$, $\phi(z) = e$ maps $E_{[A]}$ into the zero subgroup. Hence if $x = py$ is in $E_{[A]}$, then $e = \phi(x) = \phi(p)\phi(y) = \phi(p) = p$. Hence $x = y$; so $x \in E_P$.

Now suppose that E_P is benign; we must show that P is a benign subgroup. Let c and d be new elements. For X a word on A, let ϕ_X be the isomorphism of $\{c\}$ and $\{dX\}$ defined by $\phi_X(c) = dX$. We will show that the set of ϕ_X for X in P

is benign. Assuming this, the smallest subgroup which contains c and d and is invariant under the ϕ_X for X in P is benign. This subgroup is $\{P, c, d\}$. Since

$$P = \{P, c, d\} \cap G,$$

P is benign by Lemma 5.

Let $H = [A, c, d]_{\phi_X, \phi_Y, \ldots}$ (using all words on A). Then

$$H = [A, c, d, t_X, t_Y, \ldots ; t_X c t_X^{-1} = dX, \ldots].$$

It will suffice to embed H in a Higman group in which the subgroup generated by the t_X for X in P is benign. Suppose $A = \{a_1, \ldots, a_n\}$. Define an automorphism ψ_i of $[A, c, d, t_X, t_Y, \ldots]$ by $\psi_i(a) = a$, $\psi_i(c) = c$, $\psi_i(d) = da_i$, $\psi_i(t_X) = t_{a_i X}$. Applying ψ_i to the relation $t_X c t_X^{-1} = dX$ gives $t_{a_i X} c t_{a_i X}^{-1} = da_i X$. Thus ψ_i permutes the defining relations of H and hence induces an automorphism of H. Hence we can embed H in a group K with generators A, c, d, the t_X, and $a'_1, \ldots, a'_n$ and with the defining relations

$$t_X c t_X^{-1} = dX, \tag{9}$$
$$a'_i a a'^{-1}_i = a, \tag{10}$$
$$a'_i c a'^{-1}_i = c, \tag{11}$$
$$a'_i d a'^{-1}_i = da_i, \tag{12}$$
$$a'_i t_X a'^{-1}_i = t_{a_i X}. \tag{13}$$

Let $t = t_e$; and for X a word on A, let X' be the word obtained from X by replacing each a_i by a'_i. By (13), $X' t X'^{-1} = t_X$. From the hypothesis that E_P is benign and Lemma 6, we see that the subgroup of K generated by the $X' t X'^{-1} = t_X$ is benign. The equality $X' t X'^{-1} = t_X$ also shows that we may drop the generators t_X other than t, replacing (9) by

$$(X' t X'^{-1}) c (X' t X'^{-1})^{-1} = dX \tag{9'}$$

and omitting (13). We will show that (9′) is a consequence of (10), (11), (12) and $tct^{-1} = d$; this will imply that K is finitely presented and hence Higman. From (10), (11), and (12) we get $X' c X'^{-1} = c$, $X' d X'^{-1} = dX$. Hence applying the inner automorphism through X' to $tct^{-1} = d$, we get (9′).

Principal Lemma. If A is finite and P is a recursively enumerable set of words on A, then P is benign.

Let us first see how the principal lemma enables us to complete the proof of Higman's theorem. A recursively presented group can be written as G/K, where G is a free group on a finite set and K is a recursively enumerable normal subgroup of G. By the principal lemma, K is benign. Hence we may embed G_K in a finitely presented group H. Now in G_K, $\{G, t^{-1}Gt\}$ is the free product of G and $t^{-1}Gt$ with the amalgam $K = t^{-1}Kt$. The natural mapping from G to G/K and the mapping from $t^{-1}Gt$ to the zero subgroup of G/K agree on the amalgam. Hence we have a homomorphism ϕ from $\{G, t^{-1}Gt\}$ to G/K such that $\phi(g) = gK$,

$\phi(t^{-1}gt) = eK$. Define a homomorphism ψ from $\{G, t^{-1}Gt\}$ to $H \times G/K$ by $\psi(x) = (x, \phi(x))$. Then ψ is an isomorphism in $H \times G/K$. Since G/K is embedded in $(H \times G/K)_\psi$, it will suffice to show that the latter is finitely presented.

The group H has a finite number of generators and defining relations. By adding more if necessary, we may suppose that these generators include t and a set of generators of G. To obtain generators for $(H \times G/K)_\psi$, we add a finite number of generators of G_1 and a ψ-element s. Besides the defining relations of H, we need as defining relations of $(H \times G/K)_\psi$ relations which say that the elements of K_1 are equal to e; relations which say that the generators of H commute with the generators of G_1; and relations which give the value of shs^{-1} for h a generator of $\{G, t^{-1}Gt\}$. We show that the relations which say that the elements of K are equal to e are superfluous; we will then be left with a finite number of defining relations.

Let X_1 be a word on the generators of G_1 which represents an element of K_1. The equations for the values of shs^{-1} give $sXs^{-1} = X \cdot X_1$ and $stXt^{-1}s^{-1} = X$. Since X is in K, the relations in H imply that $tXt^{-1} = X$. From these three equations we get $X_1 = e$.

We now prove some lemmas on benign sets of words. Throughout A and B are finite.

Lemma 9. Every finite subset of $[A]$ is benign in $[A]$.

Proof. By the corollary to Lemma 3.

Lemma 10. If $A \subset B$ and P is a set of words on A, then P is benign in $[A]$ iff P is benign in $[B]$.

Proof. By Lemma 4.

Lemma 11. If P and Q are benign subsets of $[A]$, then $P \cap Q$ and $P \cup Q$ are benign.

Proof. Since the XzX^{-1} are free, $E_{P\cap Q} = E_P \cap E_Q$ and $E_{P\cup Q} = \{E_P, E_Q\}$. Now use Lemma 5.

An *associate* of a mapping from ϕ from $[A]$ to $[B]$ is a homomorphism ψ from $[A, z]$ to $[B, z]$ such that $\psi(XzX^{-1}) = \phi(X)z\phi(X)^{-1}$ for X a word on A. If ϕ is a homomorphism, then it has an associate. For we can extend ϕ to a homomorphism ψ from $[A, z]$ to $[B, z]$ by setting $\psi(z) = z$; and ψ is then an associate of ϕ.

A bijective mapping from $[A]$ to $[A]$ is *nice* if it has an associate which is an automorphism of $[A, z]$. Clearly the composition of two nice mappings is nice. An automorphism ϕ of $[A]$ is nice; for ϕ can be extended to an automorphism ψ of $[A, z]$ by setting $\psi(z) = z$, and ψ is an associate of ϕ.

For Y a word on A, we define mappings L_Y and R_Y from $[A]$ to $[A]$ by $L_Y(X) = Y \cdot X$, $R_Y(X) = X \cdot Y$. These mappings are nice. For the inner automorphism through Y is an associate of L_Y, and an associate ϕ of R_Y is defined by $\phi(a) = a$, $\phi(z) = YzY^{-1}$.

Lemma 12. Let ϕ be a mapping from $[A]$ to $[B]$ which has an associate. If P is a benign subset of $[A]$, then $\phi(P)$ is a benign subset of $[B]$.

Proof. If ψ is an associate of ϕ, then $E_{\phi(P)} = \psi(E_P)$. Now use Lemma 6.

Let P and Q be subsets of $[A]$, and let ϕ be a mapping from $[A]$ to $[A]$. We say that P is (ϕ, Q)-*invariant* if for each X in Q, $X \in P$ iff $\phi(X) \in P$. If $Q = [A]$, we say *invariant under* ϕ for (ϕ, Q)-*invariant*.

Lemma 13. Let $P, Q_1, \ldots, Q_n$ be benign subsets of $[A]$; $\phi_1, \ldots, \phi_n$ nice mappings from $[A]$ to $[A]$; R the smallest subset of $[A]$ which includes P and is (ϕ_i, Q_i)-invariant for $i = 1, \ldots, n$. Then R is benign.

Proof. Let ψ_i be an automorphism of $[A, z]$ which is an associate of ϕ_i. By Lemma 7, $\psi_i \mid E_{Q_i}$ is benign. Hence by Lemma 8, the smallest subgroup which includes E_P and is invariant under the $\psi_i \mid E_{Q_i}$ is benign. We show that this subgroup is E_R. An element g of E_{Q_i} is a product of words $Xz^{\pm 1}X^{-1}$ with X in Q_i. To obtain $\psi_i(g)$, we replace each X by $\phi_i(X)$. Then $g \in E_R$ iff $\psi_i(g) \in E_R$. Thus E_R is invariant under $\psi_i \mid E_{Q_i}$. Now an element of R is obtained from an element of P by repeatedly applying the ϕ_i and ϕ_i^{-1}, subject to the condition that ϕ_i is applied only to a word in Q_i and ϕ_i^{-1} is applied only to a word in $\phi_i(Q_i)$. It follows that if $X \in R$, then XzX^{-1} is in every subgroup which includes E_P and is invariant under the $\psi_i \mid E_{Q_i}$. The desired result follows.

Since $[A]$ is benign as a subgroup and hence as a subset, we may replace (ϕ_i, Q_i)-*invariant* by *invariant under* ϕ_i in Lemma 13.

Lemma 14. Let $b_i = b^i a b^{-i}$, and let P be the set of all words $b_{i_1} b_{i_2} \ldots b_{i_n}$ with $0 \leqslant i_1 < i_2 < \cdots < i_n$. Then P is a benign subset of $[a, b]$.

Proof. Let H, H^+, and H' be the subgroups generated by the b_i, the b_i for $i > 0$, and the b_i for $i \geqslant 0$. We show that these subgroups are benign. Since H is the smallest subgroup which contains a and is invariant under the inner automorphism through b, it is benign by Lemma 8. Now $H' = \{H^+, a\}$; so, in view of Lemma 5, it will suffice to consider H^+.

Define homomorphisms ϕ and χ from $[a, b]$ to $[a, b]$ by

$$\phi(a) = a, \qquad \chi(a) = bab^{-1}, \qquad \phi(b) = \chi(b) = b^2.$$

It is easy to see that ϕ and χ are injective; so by Lemma 7, $\phi \mid H$ and $\chi \mid H$ are benign. Hence by Lemma 8, it will suffice to show that H^+ is the smallest subgroup which contains b_1 and is invariant under $\phi \mid H$ and $\chi \mid H$. Now $\phi(b_i) = b_{2i}$, $\chi(b_i) = b_{2i+1}$. It is then easy to prove by induction on i that any subgroup which contains b_1 and is invariant under $\phi \mid H$ and $\chi \mid H$ must contain b_i for $i > 0$. An element x of H is a product of the b_i and their inverses; and $x \in H^+$ iff all the b_i used have $i > 0$. From this we find that $x \in H^+ \leftrightarrow \phi(x) \in H^+ \leftrightarrow \chi(x) \in H^+$. This shows that H^+ is invariant under $\phi \mid H$ and $\chi \mid H$.

Now let ψ be the automorphism of $[a, b]$ defined by

$$\psi(a) = bab^{-1}, \qquad \psi(b) = b.$$

Then $\psi(b_i) = b_{i+1}$. In view, of Lemma 13, it will suffice to show that P is the smallest subset which contains e and is (L_a, H^+)-invariant and (ψ, H')-invariant. We show, for example, that P is (L_a, H^+)-invariant. An element h of H^+ is $b_{i_1}^{\pm 1} \ldots b_{i_n}^{\pm 1}$ with the i_j positive. Then $L_a(h)$ is $b_0 b_{i_1}^{\pm 1} \ldots b_{i_n}^{\pm 1}$. Recalling that the b_i are free, we see that these are both in P if the exponents are all $+1$ and $i_1 < \cdots < i_n$, and both not in P otherwise.

A word on A is *positive* if it does not contain any a^{-1} with a in A.

Lemma 15. The set of all positive words on A is benign.

Proof. Let $A = \{a_1, \ldots, a_n\}$. Define an automorphism ϕ of $[A, z]$ by

$$\phi(a_i) = a_{i+1} \quad \text{for} \quad i < n, \qquad \phi(a_n) = a_1, \qquad \phi(z) = z.$$

Then $[A, z]_\phi$ is Higman by Lemma 3. Let t be the ϕ-element, and let ψ be the homomorphism from $[a, b, z]$ to $[A, z]_\phi$ defined by

$$\psi(a) = a_1, \qquad \psi(b) = t, \qquad \psi(z) = z.$$

If P is as in Lemma 14, it is easy to see that $\psi(P)$ is the set Q of positive words on A. Then $\psi(E_P) = E_Q$; so E_Q is benign by Lemma 14 and Lemma 6.

We identify each k-tuple of natural numbers with a positive word on $\{a, b\}$ by identifying $x_1, x_2, \ldots, x_k$ with $a^{x_1}ba^{x_2}b \ldots ba^{x_k}$. It then makes sense to say that a k-ary predicate is benign. We collect some results on benign predicates. We use W for the set of positive words and W_b for the set of words in W beginning with b. These are benign by Lemma 15 and Lemma 12 (since $W_b = L_b(W)$). Throughout i is an integer and m and n are natural numbers.

A. The predicates $=$, $\mathfrak{G}_+$, and $\mathfrak{G}_.$ are benign.

Proof. The smallest set which contains b and is invariant under L_aR_a is the set of a^iba^i. Its intersection with W is the set of a^nba^n; and this set is the predicate $=$. The smallest set which includes $L_b(=)$ and is invariant under L_aR_a is the set of $a^iba^nba^{i+n}$; the intersection of this set with W is $\mathfrak{G}_+$.

The smallest set of words on $\{a, b, c\}$ which includes b and is invariant under L_aR_c is the set of a^ibc^i. If we intersect this set with W and take the image under L_{cb}, we get the set Q of cba^nbc^n. Let ϕ be the automorphism of $[a, b, c]$ defined by

$$\phi(a) = a, \qquad \phi(b) = b, \qquad \phi(c) = ca.$$

If we take the smallest set which includes Q and is invariant under ϕ, and intersect it with W, we get the set P of $ca^mba^nb(ca^m)^n$. If ψ is the homomorphism from $[a, b, c]$ to $[a, b]$ defined by $\psi(a) = a$, $\psi(b) = b$, $\psi(c) = e$, then $\psi(P) = \mathfrak{G}_.$.

B. If P is benign and Q is defined by $Q(\mathfrak{a}, x) \leftrightarrow P(x, \mathfrak{a})$, then Q is benign.

Proof. We may suppose that $\mathfrak{a}$ is not empty, since otherwise Q is P. The smallest set including $R_b(P)$ which is invariant under $L_{a^{-1}}R_a$ is the set of $a^{n-i}bXba^i$ with a^nbX in P. If we intersect this set with W_b and take the image of the result under $L_{b^{-1}}$, we get Q.

C. If P is benign and Q is defined by $Q(x, \mathfrak{a}) \leftrightarrow P(\mathfrak{a})$, then Q is benign.

Proof. Take the smallest set which includes $L_b(P)$ and is invariant under L_a, and intersect it with W.

D. If P is benign and Q is defined by $Q(\mathfrak{a}) \leftrightarrow \exists x P(x, \mathfrak{a})$, then Q is benign.

Proof. Take the smallest set which includes P and is invariant under L_a, intersect it with W_b, and take the image under $L_{b^{-1}}$.

E. If P is benign and Q is defined by

$$Q(x, \mathfrak{a}) \leftrightarrow \forall y_{y<x} P(y, \mathfrak{a}),$$

then Q is benign.

Proof. The set of all $\mathfrak{a}$ is benign; this is proved by induction on the number of variables in $\mathfrak{a}$, using C. Applying L_b, we find that the predicate R defined by $R(x, \mathfrak{a}) \leftrightarrow x = 0$ is benign. Then Q is the smallest set which includes R and is (L_a, P)-invariant.

We now show that certain explicit definitions of predicates lead to benign predicates. First suppose that the definition uses only variables and symbols for benign predicates. It then has the form

$$P(x_1, \ldots, x_k) \leftrightarrow Q(x_{j_1}, \ldots, x_{j_p})$$

with Q benign. We may rewrite this as

$$P(x_1, \ldots, x_k) \leftrightarrow \exists y_1 \ldots \exists y_p(y_1 = x_{j_1} \mathbin{\&} \cdots \mathbin{\&} y_p = x_{j_p} \mathbin{\&} Q(y_1, \ldots, y_p)).$$

To show that P is benign it will suffice, in view of Lemma 11 and D, to show that $y_r = x_{j_r}$ and $Q(y_1, \ldots, y_p)$ are benign predicates of $y_1, \ldots, y_p, x_1, \ldots, x_k$ Since $=$ and Q are benign, this follows easily from B and C.

By Lemma 11, D, and E, we may also use $\vee$, &, existential quantifiers and bounded universal quantifiers in explicit definitions of benign predicates. We may also use constants. For example, we may replace . . 0 . . by $\exists x(x = 0 \mathbin{\&} \ldots x \ldots)$, and then observe that $x = 0$ is a benign predicate by Lemma 9.

Lemma 16. If F is a recursive function, then $\mathfrak{G}_F$ is benign.

Proof. We use induction on recursive functions. If F is I_i^n, the result follows from A, B, and C. If F is $+$ or $\cdot$, it follows from A. Now the predicate $x \neq 0$ is benign;

for it is the image under L_a of the predicate $x = x$. From this and the explicit definitions

$$x \leqslant y \leftrightarrow \exists z\, \mathfrak{G}_+(x, z, y),$$
$$x < y \leftrightarrow \exists z(z \neq 0 \;\&\; \mathfrak{G}_+(x, y, z)),$$

we see that $\leqslant$ and $<$ are benign. Hence if F is $K_<$, we have the explicit definition

$$\mathfrak{G}_F(x, y, z) \leftrightarrow (x < y \;\&\; z = 0) \vee (y \leqslant x \;\&\; z = 1).$$

Suppose that F is defined by

$$F(\mathfrak{a}) = G(H_1(\mathfrak{a}), \ldots, H_k(\mathfrak{a})),$$

where $G, H_1, \ldots, H_k$ are benign. Then G_F has the explicit definition

$$\mathfrak{G}_F(\mathfrak{a}, x) \leftrightarrow \exists y_1 \ldots \exists y_k(G_{H_1}(\mathfrak{a}, y_1) \;\&\; \cdots \;\&\; G_{H_k}(\mathfrak{a}, y_k) \;\&\; G_G(y_1, \ldots, y_k, x)).$$

Suppose that F is defined by

$$F(\mathfrak{a}) = \mu x(G(\mathfrak{a}, x) = 0)$$

where G is benign. Then $\mathfrak{G}_F$ has the explicit definition

$$\mathfrak{G}_F(\mathfrak{a}, x) \leftrightarrow \mathfrak{G}_G(\mathfrak{a}, x, 0) \;\&\; \forall y_{y<x} \exists z(z \neq 0 \;\&\; \mathfrak{G}_G(\mathfrak{a}, x, z)).$$

Lemma 17. Every recursively enumerable predicate is benign.

Proof. In view of D, it will suffice to consider a recursive predicate P. Since $P(\mathfrak{a}) \leftrightarrow \mathfrak{G}_{K_P}(\mathfrak{a}, 0)$, P is benign by Lemma 16.

Lemma 18. If Q is a recursively enumerable set of positive words on A, then Q is benign.

Proof. Let P be as in Lemma 14, and let ψ be as in the proof of Lemma 15. Since $\psi(P)$ is the set of positive words on A,

$$\psi(\psi^{-1}(Q) \cap P) = Q$$

and hence

$$\psi(E_{\psi^{-1}(Q) \cap P}) = E_Q.$$

It will therefore suffice to show that $\psi^{-1}(Q) \cap P$ is benign. Since $\psi^{-1}(Q) \cap P$ is clearly recursively enumerable, this will follow if we show that every recursively enumerable subset R of P is benign.

Let ϕ be the homomorphism from $[a, z]$ to $[a, z]$ defined by $\phi(a) = a^2$, $\phi(z) = z$. Then ϕ is injective. By Lemma 3, $[a, z]_\phi$ is Higman. Let b be the ϕ-element, and let χ be the natural mapping from $[a, b, z]$ to $[a, z]_\phi$. In the notation of Lemma 14, $\chi(b_i) = a^{2^i}$ for $i \geqslant 0$. Hence if $X = b_{i_1} \ldots b_{i_n}$ is a word in P, then $\chi(X) = a^x$ with $x = 2^{i_1} + \cdots + 2^{i_n}$; so $\chi(XzX^{-1}) = a^x z a^{-x}$. Now a number x can be written in the form $2^{i_1} + \cdots + 2^{i_n}$ with $0 \leqslant i_1 < \cdots < i_n$ in only one way. It follows that χ is injective on E_P; so $E_R = E_P \cap \chi^{-1}(\chi(E_R))$. Thus it will suffice to show that $\chi(E_R)$ is benign. Now $\chi(E_R) = E_{\chi(R)}$ where $\chi(R)$ is a set of positive words on $\{a\}$. Since R is recursively enumerable, $\chi(R)$ is recursively enumerable. This means that $\chi(R)$ is a recursively enumerable unary predicate. Hence $\chi(R)$ is benign by Lemma 17.

We can now prove the principal lemma. Let P be a recursively enumerable set of words on A. Let A' consist of an element a' for each element a of A, and let ϕ be the homomorphism from $[A, A']$ to $[A]$ defined by $\phi(a) = a$, $\phi(a') = a^{-1}$. Let P' be the set of positive words X on $A \cup A'$ such that $\phi(X) \in P$. Then P' is recursively enumerable and hence benign by Lemma 18. Since $P = \phi(P')$, P is benign by Lemma 12.

INDEX